AF247631

THE ADRENAL MEDULLA 1986 - 1988

THE ADRENAL MEDULLA 1986 - 1988

Stephen W. Carmichael
Mayo Medical School
Mayo Clinic and Foundation
Rochester, Minnesota

and

Susan L. Stoddard
Indiana University School of Medicine
Fort Wayne, Indiana

THE TELFORD PRESS, INC.
Post Office Box 287, Caldwell, New Jersey 07006

ISBN 0-936-923-288

LC Card Number 86-640508

TABLE OF CONTENTS

This book is dedicated to Andi, Cathy, Forrest, and Sally, my siblings,

SWC

and to my parents, Rachel and Leland, for their love, understanding, and unfailing good humor,

SLS

ACKNOWLEDGMENTS

The authors gratefully acknowledge the efforts of many people who contributed to the book, including:

Julia R. Kettelkamp for her dedication and precision in clarifying the citations in this volume;

Monica M. Zeien for processing the manuscript;

Dr. Bernard K. Forscher for his editorial assistance and the Section of Publications, Mayo Clinic;

Alan D. Scott for his assistance and advice with word processing, computer processing, and indexing;

Gail E. Bowdish, Dean R. Fisher, and Michael J. Link for organizing the references, and Marsha M. Kuhns for indexing;

The staffs of the Mayo Medical and Indiana University School of Medicine Libraries, including especially Patricia J. Erwin and Anita C. Thomas at Mayo and Marjorie Collins, Beth Grunewald, and Marilyn Walker at Indiana, for their assistance and cooperation;

Our colleagues who reviewed portions of the text, including Drs. Jack C. Brooks, Stewart L. Chritton, David E. Clapham, Miloslava K. Dousa, Torgier Flatmark, Patrick J. Fleming, Patrick J. Kelly, Robert L. Perlman, Gertrude M. Tyce, and Joel A. Vilensky;

Our colleagues who sent us reprints and discussed their work freely; Rev. Charles Swartz for translating from Russian;

Dr. Patrick J. Kelly from Telford Press for his cooperation;

and E. Stuart Eickelberg for his administrative help.

The terminology used in this review was chosen carefully. The term *adrenal medulla* is not technically correct for the organ in man, (**Nomina Anatomica** *medulla glandulae suprarenalis*), but it is in common usage. Here, it refers to the catecholamine- secreting portion of the suprarenal, interrenal, or adrenal gland.

The principal catecholamines of the adrenal medulla are referred to by the **United States Pharmacopeia** terms epinephrine and norepinephrine. The **British Pharmacopeia** terms are adrenaline and noradrenaline, respectively.

The parenchymal cells of the adrenal medulla are referred to as the adrenal chromaffin cells. This term has been in use since Kohn described the chromium salt-reducing property of these cells (*die chromaffine Zelle*) in 1902. Other chromaffin cells exist in the body, but this review primarily covers the chromaffin cells of the adrenal medulla.

The catecholamine-storing organelles of the chromaffin cell are referred to as chromaffin vesicles. These vesicles are often called chromaffin granules, since by light microscopy they impart a granular appearance to the cytoplasm of the chromaffin cells. However, the appearance of these organelles as visualized by electron microscopy is clearly that of a membrane-bound cavity, properly referred to as a vesicle.

The subject index was compiled by computer, using *Wordstar 5*. Thus, to facilitate referencing, similar forms of the same term have been combined, *i.e.* cosecreted and cosecretion. Additionally, most animal names are included, although *Genus* and *species* are omitted.

This review covers all 1056 listings under *Adrenal Medulla* in Index Medicus during 1986, 1987, and 1988. In an attempt to be as complete as possible, 626 other references are also reviewed. A total of 3611 different scientists have authored these publications. Very few earlier works are included; rather, appropriate reviews of earlier works are referred to as needed.

Just as the studies addressed in this review do not stand isolated in the history of our understanding of the adrenal medulla, this review is not intended to stand alone. Earlier publications and reviews, including previous volumes in this series (*The Adrenal Medulla, Volumes 1-4*), should be consulted for the most complete understanding of the adrenal medulla.

Generally, the chapters have been organized by disciplines as well as by the sequence of events in the secretory cycle. While this approach was somewhat arbitrary and some articles overlapped chapters, it appeared to offer the best means of compiling information from diverse sources. However, due to the overlap of some disciplines, such as biochemistry and pharmacology, assignments of articles to particular chapters involved some subjective choices.

CHAPTER 1

OVERVIEW

Widespread interest in the adrenal medulla was sparked recently by the report of success in treating Parkinson's disease by autologous transplantation of the adrenal medulla to the brain (Madrazo, Drucker-Colín, Díaz et al., 1987). While Madrazo et al. reported success with two patients, they added that their results are preliminary and that further work is necessary to determine whether transplantation of the adrenal medulla to the caudate nucleus will be successful in the long term. Subsequently, Madrazo, León, Torres et al. (1988) transplanted human fetal adrenal tissue in another patient and reported sustained clinical improvement. Backlund, Olson, Seiger et al. (1987) reported on two patients in whom the adrenal medulla had been transplanted to the putamen. Neither of these patients were significantly improved 6 months after the operation. Zhang, Ding and Jiao (1988) reported successful transplantation in six patients in China. Penn, Goetz, Tanner et al. (1988) reported on five patients who underwent this surgical procedure in the United States; the long term efficacy of the procedure was not assessed, but important side effects were noted. It was discovered that catecholamine levels in the adrenal medulla of parkinsonian patients are lower than levels in controls, which may indicate that autologous grafts do not provide ideal donor tissue (Carmichael, Wilson, Brimijoin et al., 1988; Cervera, Rascol, Ploska et al., 1988). Concurrently, several investigators are having misgivings about these procedures, since basic research with animal models is far from complete. For example, Joynt and Gash (1987) and Sladek and Shoulson (1988) stressed that fundamental questions remain concerning the appropriateness of this extensive surgical procedure.

Several laboratories are working on animal models pertaining to transplantation of the adrenal medulla for Parkinson's disease. Earlier studies in this area were reviewed by Olson, Backlund, Freed et al. (1985) and Bjorklund, Lindvall, Isacson et al. (1987). Morihisa, Nakamura, Freed et al. (1987) demonstrated that adrenal medullary grafts to the brains of monkeys show at least some catecholamine histofluorescence up to 8 months after transplantation. Similar results in a rat model were reported by Jiao, Wang, Cai et al. (1987). Kamo, Kim, McGeer et al. (1987) demonstrated that grafts of human fetal adrenal chromaffin cell cultures into the brain of rats remain viable and appear to have behavioral consequences. In contrast, extensive studies in monkeys and rats by Hansen, Kordower, Fiandaca et al. (1988) and Bing, Notter, Hansen et al. (1988) indicated that few chromaffin cells survive for a prolonged period after transplantation. It has been suggested that behavioral recovery in the parkinsonian model is due to the release of a substance other than dopamine from the graft. Other experiments suggest that trophic factors play an important role in behavioral recovery (i.e., Pezzoli, Fahn, Dwork et al., 1988).

The blood-brain barrier in adrenal medullary grafts was studied by Rosenstein (1987a,b) who showed that grafts of adrenal medullary tissue into rat brain disturbed the blood-brain barrier. Becker and Freed (1988) suggested that adrenal medullary grafts may enhance functional activity in rat models of Parkinson's disease by producing dopamine which then leaks into the host striatum through permeable blood vessels.

Transplantation of the adrenal medulla and other neural tissues has been suggested for neurological diseases other than parkinsonism. For example, Stevens, Phillips, Freed et al. (1988) transplanted adrenal medulla and brain into the ventricles of rat models of epilepsy; the induced seizures were not appreciably reduced. Tulipan (1988) reviewed the literature on neural transplantation as therapy for Parkinson's disease, Huntington's disease, Alzheimer's disease, aging, trauma, hormonal disorders, and pain.

The neurotoxin 1-methyl-4-phenyl-1,2,3,6-tetrahydropyridine (MPTP) has proven to be very useful in the development of a model for Parkinson's disease since it is toxic to the same cells in the brain that degenerate during the course of this disease. The effect of MPTP and its metabolite, referred to as MPP^+, on adrenal chromaffin cells, has been examined in several laboratories. Reinhard, Diliberto, Viveros et al. (1987), Daniels, Reinhard and Painter (1988), and Wilson and Beeler (1988) showed that MPP^+ is accumulated within chromaffin vesicles, thereby protecting adrenal chromaffin cells from the toxic effects of this agent. This observation was supported by the findings of Scherman, Darchen, Desnos et al. (1988) who demonstrated that MPP^+ is a substrate for the transporter that takes catecholamines up into chromaffin vesicles. Kiuchi, Hagiwara, Hidaka et al. (1988) showed that MPP^+ acutely inhibits the phosphorylation of tyrosine hydroxylase, resulting in decreased enzyme activity.

In transplant studies unrelated to Parkinson's disease, Sagen, Pappas and Perlow (1986), showed that transplants of adrenal medullary tissue to the rat spinal cord reduced pain sensitivity. Such an effect appeared to be due to the release of enkephalins from adrenal chromaffin cells. Additional studies from this laboratory demonstrated that adrenal medullary grafts may be useful in the control of pain. A clinical trial involving a terminally ill patient was performed by Vaquero, Martínez, Oya et al. (1988) with disappointing results.

The adrenal chromaffin cell continues to be useful as a model for studying stimulus-secretion coupling, as discussed in a collection of papers edited by Rosenheck and Lelkes (1987). A difference in stimulus-secretion coupling between epinephrine-containing cells and norepinephrine-containing cells was identified by Marley and Livett (1987a). Under standard conditions, they found a significant difference in the secretion of epinephrine and norepinephrine, providing the first evidence that postsynaptic mechanisms may contribute to the differential secretion of these two catecholamines from the adrenal medulla. Critchley, Ellis, Henderson et al. (1986) showed that muscarinic and nicotinic agonists selectively released epinephrine and norepinephrine, respectively, from the perfused adrenal gland. These and other studies strongly suggest that at least some of the selective secretion of epinephrine and norepinephrine is effected within the adrenal medulla itself.

The phenomenon of desensitization of the nicotinic secretory response of adrenal chromaffin cells was reviewed by Marley (1988). Several groups, including Malhotra, Wakade and Wakade (1988a) are investigating this phenomenon. Higgins and Berg (1988a,b) characterized the desensitized nicotinic receptor on chromaffin cells and found that it is similar to desensitized receptors in muscle and the electric organ of fish.

Several laboratories, including those of Kumakura, Ohara and Satô (1986) and Boarder and McArdle (1986), developed new techniques to aid in studying the time course of stimulus-secretion coupling. Livett, Marley, Mitchelhill et al. (1987) reviewed various applications of these new techniques.

The importance of muscarinic receptors in stimulus-secretion coupling has been shown by several investigators. For example, Knight and Baker (1986) demonstrated that catecholamine secretion from the adrenal chromaffin cells of the chicken is activated almost entirely by acetylcholine acting on muscarinic receptors. This is in contrast to results reported in mammals, in which acetylcholine acts primarily at nicotinic receptors. Harish, Kao, Raffaniello et al. (1987) presented evidence to suggest that muscarine-evoked catecholamine secretion from perfused rat adrenal glands is mediated by intracellular calcium. Barron, Murrin and Hexum (1986) demonstrated that more than one type of muscarinic binding site is present in the bovine adrenal medulla. Nakaki, Sasakawa, Yamamoto et al. (1988) reported a functional shift from muscarinic to nicotinic receptors over time in cultured chromaffin cells.

The role of adrenergic and dopaminergic receptors in stimulus-secretion coupling is also of current interest. While some laboratories have demonstrated that adrenergic receptors do not play a regulatory role in secretion from chromaffin cells (Powis and Baker, 1986; Sharma, Wakade, Malhotra et al., 1986; Nguyen and De Léan, 1987), others argue for a role of these receptors (Foucart, Nadeau and de Champlain, 1987). González, Artalejo, Montiel et al. (1986), Lyon, Titeler, Bigornia et al. (1987), and Quik, Bergeron, Mount et al. (1987) described the dopaminergic receptors on adrenal chromaffin cells.

γ-Aminobutyric acid (GABA) has been shown to play an important role in the modulation of catecholamine secretion from the adrenal medulla. This phenomenon has been reviewed by Alho, Fujimoto, Guidotti et al. (1986) and Kitayama and Tsujimoto (1986). Several laboratories presented evidence that the $GABA_A$ receptor is involved in the inhibition of catecholamine secretion caused by endogenous GABA (Kataoka, Fujimoto, Alho et al., 1986; Kitayama, Koyama, Morita et al., 1987; and Callachan, Cottrell, Hather et al., 1987). Kataoka, Ohara-Imaizumi, Ueki et al. (1988) used real-time monitoring to characterize GABA-evoked secretion.

Peptides also have a modulatory role on secretion from chromaffin cells. For example, Wilson (1988) and Malhotra, Wakade and Wakade (1988b) demonstrated that vasoactive intestinal polypeptide influences the secretory apparatus of adrenal chromaffin cells.

The role of ions in stimulus-secretion coupling continues to be an area of active investigation. These events have been reviewed by Baker and Knight (1987a, b) and others. Ahnert-Hilger and Gratzl (1988) specifically reviewed the use of pore-forming proteins as a tool to dissect ionic events in stimulus-secretion coupling. Cobbold, Cheek, Cuthbertson et al. (1987) suggested that the intracellular rise in calcium may not be the only trigger in stimulus-secretion coupling. They hypothesized that activation of the nicotinic receptor generates an additional uncharacterized signal, other than an increase in calcium, that leads to exocytosis.

Increased intracellular calcium apparently interacts with protein(s) within the cell to mediate exocytosis. Creutz, Zaks, Hamman et al. (1987a,b) examined several of these proteins and labeled them "chromobindins." The roles of these calcium-binding proteins and related proteins are the subject of intense investigations.

Cytoskeletal proteins may also play a role in stimulus-secretion coupling. Burgoyne and Cheek (1987) and Burgoyne, Cheek,

O'Sullivan et al. (1988) described the possible role of the cytoskeleton in secretion from adrenal chromaffin cells. Sontag, Aunis and Bader (1988) destabilized actin filaments in the subplasmalemmal cytoskeleton and found that secretion was potentiated. McKay, Cobianchi and Schneider (1987) suggested, to the contrary, that cytoskeletal elements such as microtubules do not play a pivotal role in stimulus-secretion coupling. The hypothesis that the cytoskeleton plays an inhibitory role in secretion (*i.e.*, chromaffin vesicles are suspended within the cytoskeleton) and, when the cytoskeleton is disrupted, secretion ensues, was explored by Cheek, Hesketh, Richards et al. (1986).

The importance of protein kinases in stimulus-secretion coupling was demonstrated by Wakade, Malhotra and Wakade (1986), Lee and Holz (1986), Brocklehurst, Lee and Pollard (1986), Burgoyne, Morgan and O'Sullivan (1988) and others. Increases in intracellular calcium may activate protein kinase C, setting in action the machinery for exocytosis.

Several laboratories are attempting to develop a cell-free model of exocytosis (Rosenheck and Plattner , 1986; Izumi, Yanagihara, Wada et al., 1986). Izumi et al. (1986) performed experiments that suggested a phospholipase may interact with calcium in events leading to chromaffin vesicle lysis. Winkler (1988) reviewed the evidence that exocytosis is the mechanism of secretion from adrenal chromaffin cells and sympathetic nerves. He noted that the molecular mechanisms of exocytosis are still a matter of speculation (cf. also Strittmatter, 1988).

Primary interest in adrenal medullary enzymes concerns those involved with catecholamine synthesis. For example, Dagerlind and Schalling (1988) used *in situ* hybridization histochemistry to localize the mRNAs for tyrosine hydroxylase, dopamine β-hydroxylase, and phenylethanolamine N-methyltransferase in the adrenal medulla and other tissues. Craig, Buckle, Lamouroux et al. (1986) demonstrated that the gene for human tyrosine hydroxylase is on chromosome 11p15. The same group demonstrated that the gene for human dopamine β-hydroxylase is on chromosome 9q34 (Craig, Buckle, Lamouroux et al., 1988), finally proving that tyrosine hydroxylase and dopamine β-hydroxylase are not genetically linked. Additional studies on the molecular biology of tyrosine hydroxylase were reported by O'Malley (1986), Grima, Lamouroux, Boni et al. (1987), D'Mello, Weisberg, Stachowiak et al. (1988), Saadat, Stehle, Lamouroux et al. (1988) and others. Acheson, Edgar, Timpl et al. (1986) and Saadat and Thoenen (1986, 1988) continued their earlier studies on the effect of the environment of the chromaffin cell on tyrosine hydroxylase activity. They have demonstrated the environment of cultured cells has a direct effect on enzyme activity.

Dopamine β-hydroxylase continues to be extensively studied. Dhawan, Duong, Ornberg et al. (1987) characterized this soluble enzyme at the molecular level. Work from the laboratory of Villafranca (e.g., Fitzpatrick and Villafranca, 1987) has expanded the understanding of the mechanisms by which dopamine β-hydroxylase converts dopamine to norepinephrine. Kruse, Kaiser, DeWolf et al. (1986a,b) developed more specific inhibitors of dopamine β-hydroxylase. An alternate strategy for the inhibition of dopamine β-hydroxylase was presented by May, Wimalasena, Herman et al. (1988). The primary structure of human dopamine β-hydroxylase was determined by Lamouroux, Vigny, Faucon Biguet et al. (1987). They also presented data which indicated the soluble and membrane-bound forms of the enzyme is from the same molecule. It appears that post-translational glypiation is responsible for modifying the membranous enzyme.

The role of ascorbic acid as an electron donor for the reaction catalyzed by dopamine β-hydroxylase was further characterized by Menniti, Knoth and Diliberto (1986). The electron transfer is mediated by cytochrome b_{561} and the biophysics of this transfer have been described by Kelley and Njus (1986). Cytochrome b_{561} isolated from chromaffin vesicles was partially sequenced by Perin, Fried, Slaughter et al. (1988).

Phenylethanolamine N-methyltransferase (PNMT) is the enzyme that converts norepinephrine to epinephrine. The complete nucleotide and deduced amino acid sequence of bovine PMNT was reported by Baetge, Suh and Joh (1986) and Batter, D'Mello, Turzai et al. (1988). The human enzyme has also been sequenced (Baetge, Behringer, Messing et al., 1988; Kaneda, Ichinose, Kobayashi et al., 1988).

Other enzymes in the adrenal medulla have been more completely characterized. For example, Percy and Apps (1986) and Moriyama and Nelson (1987, 1988a, b, c) contributed to our knowledge of the proton translocating ATPase of the adrenal chromaffin vesicle membrane.

The composition of the adrenal chromaffin cell, especially the chromaffin vesicle, is relatively well understood. Winkler, Apps and Fischer-Colbrie (1986) noted recent significant progress in the elucidation of the molecular composition of the chromaffin vesicle. Winkler, Fischer-Colbrie, Obendorf et al. (1988) demonstrated common antigens among adrenergic and cholinergic vesicles. Several previously undemonstrated compounds have been identified within the adrenal medulla, including corticotropin-releasing factor (Bruhn, Engeland, Anthony et al., 1987a; Minamino, Uehara and Arimura, 1988) and a dopamine-releasing factor (Chang and Ramirez, 1988).

Synaptin has been demonstrated in the adrenal medulla. Gaardsvoll, Obendorf, Winkler et al. (1988) showed that synaptophysin, synaptin, and p38 are immunochemically identical. Since a protein is normally designated by its original name, Gaardsvoll et al. suggested that synaptophysin/p38 (and probably also SVP38) be termed synaptin. Wiedenmann, Rehm, Knierim et al. (1988) and Schilling and Gratzl (1988) showed that adrenal medullary synaptin is associated with chromaffin vesicles. Obendorf, Schwarzenbrunner, Fischer-Colbrie et al. (1988b) reported synaptin in membranes of chromaffin vesicles, as well as other membranes of the adrenal chromaffin cell.

The major protein contained within the chromaffin vesicle is chromogranin A. The amino acid sequence for the bovine (Iacangelo, Affolter, Eiden et al., 1986; Benedum, Baeuerle, Konecki et al., 1986), human (Konecki, Benedum, Gerdes et al., 1988), porcine (Iacangelo, Fischer-Colbrie, Koller et al., 1988), and rat (Iacangelo, Okayama and Eiden, 1988) chromogranin A has been elucidated. While bovine chromogranin A was thought to have a molecular mass of about 75 kDa, its actual molecular mass is 48 kDa. Human chromogranin A has 439 protein residues preceded by an 18-residue signal peptide. Reports from several laboratories have shown that chromogranin A is widely distributed in both endocrine and nervous tissues. Rieker, Fischer-Colbrie, Eiden et al. (1988) demonstrated chromogranin A- and B-like proteins in a large series of species extending from arthropods to mammals.

As noted by Winkler, Fischer-Colbrie, Schober et al. (1988), there are two possible classes of functions for chromogranins (and other contents of the chromaffin vesicle) -- a function within the vesicle or a function after release. Within the vesicle, chromogranins may have a role in a storage complex with calcium, as shown by Reiffen and Gratzl (1986a,b). After release, peptides derived from chromogranin A have been shown to have a negative feedback on catecholamine secretion (Simon, Bader and Aunis, 1988). Such a mechanism could function *in vivo* as a delayed down-regulation of secretion to protect the chromaffin cell from exhaustion.

Regulation of the peptide and protein content of the chromaffin vesicle is better understood. For example, Siegel, Iacangelo, Park et al. (1988) used *in situ* hybridization histochemistry to demonstrate a heterogeneous distribution of mRNA for chromogranin A. Fischer-Colbrie, Iacangelo and Eiden (1988) showed that chromogranins, enkephalins, and neuropeptide Y are regulated separately by neural and humoral factors.

Interest in neuropeptide Y in the adrenal medulla has increased.

Kuramoto, Kondo and Fujita (1986) demonstrated that about half the chromaffin cells in the adrenal medulla exhibit neuropeptide Y immunoreactivity. Fischer-Colbrie, Diez-Guerra, Emson et al. (1986) and Pruss, Mezey, Forman et al. (1986) found neuropeptide Y to be present within chromaffin vesicles. Each chromaffin vesicle contained about 429 copies of the peptide. Bastiaensen, De Block and De Potter (1988) reported that neuropeptide Y is co-localized with enkephalins in chromaffin vesicles of the bovine adrenal medulla. The gene for human neuropeptide Y has been isolated by Minth, Andrews and Dixon (1986).

Possible physiological functions of peptides, including neuropeptide Y, were investigated by Gaumann, Yaksh, Tyce et al. (1988a) and Gaumann and Yaksh (1988). The concurrent secretion of peptides and catecholamines was measured in the effluent of the adrenal vein in anesthetized cats during hemorrhage. Under certain conditions, levels of [Met]enkephalin, neuropeptide Y, peptide YY, and neurotensin increased. However, opiate receptor antagonists did not change the neuropeptide levels, suggesting that opiate receptors do not mediate the release of peptides during hemorrhage. Additional evidence suggested that opioids are not involved in the response of the adrenal medulla during hemorrhage.

The regulation of peptide processing is becoming better understood. For example, Kanamatsu, Unsworth, Diliberto et al. (1986) demonstrated that splanchnic nerve activity has a direct effect on proenkephalin A biosynthesis. Other laboratories have shown that agonists stimulate the processing of proenkephalin in cultured adrenal chromaffin cells (Waschek, Dave, Eskay et al., 1987; Kley, Loeffler, Pittius et al., 1986). Eiden (1987b) presented a recent overview of enkephalin synthesis and secretion throughout the neuroendocrine system.

The adrenal medulla of the rat is often used as a neurobiological model. The morphology of this organ has been recently described by Tischler and DeLellis (1988a) and Tomlinson, Durbin and Coupland (1987).

The neural control of the adrenal medulla has been more completely elucidated. For example, Appel, Wessendorf and Elde (1986,1987) and Appel and Elde (1988) extended the neurochemical characterization of the preganglionic sympathetic neurons that innervate the adrenal medulla. Kesse, Parker and Coupland (1988) provided the first evidence of postganglionic innervation to the adrenal medulla from ganglia of the sympathetic chain.

A role for sensory fibers in the nerve supply to the adrenal medulla was suggested by Khalil, Livett and Marley (1986) and Amann and Lembeck (1986). They found that, when sensory fibers are eliminated by a specific neurotoxin, there was a change in the control of adrenal medullary secretion. This is the first time that a potential role for sensory fibers in adrenal medullary control has been indicated.

Central nervous system modulation of adrenal medullary activity was studied in several laboratories. For example, Bereiter, Engeland and Gann (1986) and Bereiter and Gann (1988) demonstrated that stimulation of sites in the caudal brainstem affects the adrenal medulla. The function of the adrenal medulla during aggressive behavior was studied by Stoddard, Bergdall, Conn et al. (1987). Plasma catecholamines were measured during naturally-elicited defense behavior in freely moving cats. This is the first study reporting the function of the adrenal medulla during natural behavior in unanesthetized animals.

Gold (1986, 1987) presented evidence that the adrenal medulla may have a direct effect on memory. Specifically, epinephrine released from the adrenal medulla liberated hepatic stores of glucose which enhanced memory storage. Black, Adler, Dreyfus et al. (1987) suggested that the adrenal medulla may be a useful model for investigating molecular mechanisms of memory in the sympathetic nervous system. Zhang, McGaugh, Juler et al. (1987) suggested that enkephalins secreted by the adrenal medulla may play a role in memory.

Seidler and Slotkin (1986) and Slotkin and Seidler (1988) reported that the development of functional innervation to the adrenal medulla is responsible for suppressing the neonatal ability of the organ to respond to non-neurogenic stimuli. This suggests that the regulation of the adrenal medulla varies at different ages.

McCarty (1986a) investigated the response of normotensive and hypertensive strains of rats to insulin and concluded that the pattern of development of sympathetic target tissues in early life may contribute to the hypertension which occurs in particular strains. Hendley, Cierpial and McCarty (1988), using new strains of rats, demonstrated that stress-induced sympathoadrenal hyperresponsiveness is associated with the hyperactive trait, but not the hypertensive trait, of spontaneously hypertensive rats.

Cultured chromaffin cells and their neoplastic derivative, PC12 cells, have been used extensively to understand the effects and mechanisms of growth factors, especially nerve growth factor. For example, Claude, Parada, Gordon et al. (1988) demonstrated that acidic fibroblast growth factor was as potent as nerve growth factor in stimulating proliferation of cultured rat adrenal chromaffin cells, but failed to support long-term survival. The effects of basic nerve growth factor were similar to those of the acidic nerve growth factor. Dahmer and Perlman (1988a,b) showed that insulin-like growth factor I has an effect on chromaffin cells and PC12 cells.

Imaging of the adrenal gland for clinical purposes has taken a quantum leap with magnetic resonance imaging. Several clinical laboratories described their observations using this new imaging technology. This procedure appears to be particularly useful for localizing tumors of the adrenal gland, including pheochromocytomas.

Weiss (1986) demonstrated involvement of the adrenal medulla in acquired immune deficiency syndrome (AIDS). She suggested that the apparent selective destruction of the adrenal medulla in a high proportion of autopsied patients is a result of infection with the retrovirus.

The question of why a neural structure, the adrenal medulla, is embedded within an endocrine organ, the adrenal cortex, is not yet fully answered. However, answers to this question relate to the synthesis of epinephrine and the shape of the adrenal chromaffin cell. Specifically, glucocorticoids enhance the activity of phenylethanolamine *N*-methyltransferase so that norepinephrine is readily converted to the more active hormone epinephrine. When cultured chromaffin cells are grown in the absence of glucocorticoids, they assume a more neuronal shape. Additionally, Sietzen, Schober, Fischer-Colbrie et al. (1987), La Gamma and Adler (1987), Yoburn, Franklin, Calvano et al. (1987), and Inturrisi, Branch, Robertson et al. (1988) recently reported that adrenal cortical steroids have a direct effect on peptides and chromogranins within the adrenal medulla. This effect apparently occurs at the transcriptional level (Stachowiak, Rigual, Lee et al., 1988; Stachowiak, Lee, Rigual et al., 1988) and indicates that the adrenal cortex has a more profound effect on the composition and function of the adrenal medulla than was previously believed.

Several general reviews of the function of the adrenal medulla have been published. This includes an edited collection by Heym (1987) that reviews the histochemistry of the adrenal medulla and related neural structures. Pearse (1986) reviewed his APUD (*a*mine *p*recursor *u*ptake and *d*ecarboxylation) theory that unites the adrenal medulla with other structures of similar origin and function. Johnson (1987) edited a major volume reviewing the cellular and molecular biology of the adrenal medulla and related tissues. An excellent review of paraneurons, including adrenal chromaffin cells, was published by Fujita, Kanno and Kobayashi (1988). The papers presented in the following chapters of **The Adrenal Medulla 1986-1988** not only summarize recent work on the adrenal medulla but also indicate a promising future for work with this model neurobiological system.

CHAPTER 2

MORPHOLOGY

<u>COMPARATIVE</u> <u>MORPHOLOGY</u>

In invertebrates and lower vertebrates, the adrenal medulla is not a discrete organ but rather is represented by dispersed chromaffin cells. This arrangement is present in birds. Ghosh (1986) reviewed the gross and light-microscopic morphology of the avian adrenal glands. He pointed out that birds are the only homoeothermic vertebrates that do not have a discrete adrenal medulla.

In a study of the ancestry of vertebrate chromaffin tissue, Sakharov and Zhuravlyova (1988) described a well-developed system of intensely fluorescent cells in the esophagus of five ascidian species (sea squirts) by using methods that detect cellular catecholamines. The catecholamine-containing cells form a strand in the connective tissue layer of the esophagus. Electron microscopic studies revealed numerous secretory vesicles 50 to 140 nm in diameter. The dense content of the vesicles could be depleted by reserpine. These cells appeared to form a primitive system that is a phylogenetic precursor to the adrenal medulla. This may be the most primitive animal in which an adrenal medulla-like organ has been described.

The morphology of the chromaffin tissue in the adrenal gland of the axolotl (*Amblystoma mexicanum*) was described by Lascar and Taxi (1986). Epinephrine- and norepinephrine-containing cells could be distinguished on the basis of size and shape of vesicles within the cells. The vesicles were smaller than reported for other anurans (frogs and toads), but the innervation of the chromaffin cells appeared to be similar to that of other anurans.

An extensive study of the morphology of the adrenal gland in 10 different species of anurans and 12 species of urodelans (salamanders) was published by Grassi Milano and Accordi (1986). They found two patterns in anurans: in one the gland had a more medial position on the ventral surface of the kidneys and was more diffuse, observed in frogs of the more primitive families; in the other there was more lateral tissue and it was aggregated, seen in the more advanced families. The urodelans had a large variability in adrenal glandular structure without a clear taxonomic pattern, although an increased compactness of the gland and mingling of chromaffin cells with steroidogenic cells was found only in the more advanced urodelans. They suggested that the adrenal gland becomes more compact and chromaffin and steroidogenic tissues aggregate in the transition from primitive to advanced families, both in anurans and in urodelans.

El-Maghraby (1988) compared the ability of the chromaffin reaction and formaldehyde-induced fluorescence to differentiate between norepinephrine- and epinephrine-containing cells in the adrenal medulla of the newborn rat. Both direct and indirect chromaffin reactions demonstrated catecholamines, but the low levels did not permit a clear distinction between the two cell types. In contrast, the fluorescence method could provide a clear distinction between norepinephrine- and epinephrine-containing cells.

A detailed histochemical and immunohistochemical study of the adrenal gland of the bullfrog (*Rana catesbeiana*) was performed by Kuramoto (1987). Using a combination of techniques, he identified three types of chromaffin cells in the adrenal gland: cells with epinephrine and serotonin; cells with norepinephrine and serotonin; and cells with epinephrine, enkephalin, and serotonin. Varicose nerve fibers with substance P-like, calcitonin gene-related peptide-like, peptide HI-like, and gastrin-releasing peptide-like immunoreactivities were found on solitary or small clusters of chromaffin cells. These peptides and neuropeptide Y also were localized around small arteries and arterioles supplying the adrenal gland. Additional experiments demonstrated the co-localization of these peptides on nerve fibers.

Immunohistochemical and biochemical evidence of the presence of serotonin in amphibian adrenal chromaffin cells was presented by Delarue, Leboulenger, Morra et al. (1988). Using antisera against serotonin and tyrosine hydroxylase on consecutive sections of the frog inter-renal gland, they found, by an indirect immunofluorescence technique, that a majority of chromaffin cells were also immunopositive for serotonin. Using antibodies to phenylethanolamine *N*-methyltransferase, they also demonstrated that serotonin reactivity was present in almost all epinephrine-producing cells which represented about 90% of the chromaffin cells. Using electron microscopic techniques, they localized serotonin immunoreactivity to secretory vesicles mostly located in the periphery of epinephrine-containing cells. Serotonin and its metabolite could be detected in extracts of the gland by high-pressure liquid chromatography. It was suggested that serotonin may be released with epinephrine under conditions of stress.

Golding and Pow (1987) examined nerve terminals that formed synapses with adrenal chromaffin tissues in the goldfish (*Carassius auratus*), frog (*Rana pipiens*), hamster, and rat (Wistar strain). They identified two types of presumptive secretory vesicles in the terminals: electron-lucent synaptic vesicles 30 to 50 nm in diameter, and secretory vesicles 60 to 100 nm in diameter. Sites of vesicle exocytosis were observed in each of the animals investigated. Their results indicated that dense-cored vesicles within synaptic terminals are involved in the discharge as well as the storage of neurochemical mediators.

The distribution of oxytocin and vasopressin in the adrenal glands of rat, cow, hamster, and guinea pig was studied with immunohistochemical techniques by Hawthorn, Nussey, Henderson et al. (1987). The adrenal cortex of all species studied contained both peptides, but the medulla showed considerable species variation. In the cow and the rat, both peptides appeared to be present in epinephrine- and norepinephrine-containing cells. In the hamster, oxytocin was present in epinephrine-containing cells whereas vasopressin was present in both cell types. The guinea pig adrenal medulla, which contains only epinephrine-secreting cells, was positive for both peptides.

Singh and Mathew (1987) reported on the ultrastructure of adrenal chromaffin cells in the rat. They identified epinephrine-containing vesicles as being moderately electron-dense whereas norepinephrine-containing vesicles were more electron-dense.

In a detailed study of the normal Wistar rat adrenal medulla performed by Tomlinson, Durbin and Coupland (1987), a morphometric analysis was correlated with catecholamine levels in fresh tissue. In addition to collecting quantitative information regarding various volume parameters in the adrenal medulla and chromaffin cells, these authors determined the catecholamine content of both cell types. They calculated the epinephrine content of a vesicle to be about 3.5×10^{-12} μmol and the norepinephrine content to be about 5×10^{-12} μmol.

Tischler and DeLellis (1988a) described the morphology and physiology of the normal rat adrenal medulla. They emphasized the value of the rat adrenal medulla as an experimental model and reviewed current concepts relating to the development of medullary cell lineages and to factors that affect synthesis, storage, and secretion of neurohormones in the adrenal medulla.

In a companion paper, Tischler and DeLellis (1988b) described proliferative lesions of the rat adrenal medulla. Such lesions develop spontaneously or can be drug-induced in many strains of rats. However, there often is ambiguity as to whether the changes observed are hyperplastic or neoplastic. Relationships between these lesions and lesions of the human adrenal medulla are thoroughly discussed, particularly with reference to drug testing in rodents.

Another morphologic study of the details of the venous drainage of the adrenal gland in man (Monkhouse and Khalique, 1986) is of interest because the secretory products of the adrenal medulla and cortex reach the systemic circulation via this route. These authors described several variations in this drainage pattern. In addition to the well-known veins draining directly or indirectly into the inferior vena cava, they found connections with the azygos and lumbar veins that may provide a route for venous adrenal blood to the superior vena cava.

The clinical anatomy of the suprarenal arteries was studied by Toni, Mosca, Favero et al. (1988) by using quantitative aortography. In their study of 100 patients without disease of the adrenal glands, they found that the site of origin varied, particularly for the superior and middle suprarenal vessels. Statistically significant differences were found in mean diameters of each arterial vessel in all subjects examined, thus supporting the concept of differential arterial supply to different portions of the gland.

DEVELOPMENTAL STUDIES

Le Douarin and Smith (1988) reviewed the development of the peripheral nervous system from the neural crest. This repeated some of their classic work with quail-chick chimeras. Using their system, they demonstrated that the adrenal medulla arises from the neural crest at somite levels 18 to 24.

The development of the vasculature in the suprarenal glands in the chick embryo was studied by Bertossi, Ribatti and Roncali (1987) and compared with that in the thyroid gland. The main process was the transformation of immature microvessels with thick walls and narrow lumens into fenestrated sinusoids with wide lumens and very thin endothelial linings. They observed that as the capillary wall matured the endothelial basement lamina formed.

Accordi and Corbellini (1986-1987) examined the morphogenesis of the adrenal gland of the urodele *Triturus cristatus* during its larval development and metamorphosis. During development, the adrenal gland is in the region between the pronephros and mesonephros and along the mesonephros. They observed differentiation of the chromaffin cells and steroidogenic cells in a cephalad-to-caudad direction. There also is a dorsal-to-ventral migration of the adrenal tissue.

von Dalnok and Menssen (1986) described the effects of glucocorticoids on the development of chromaffin cells in the adrenal medulla and paraneuronal cells in the carotid body of the rat. They documented the postnatal growth of adrenal chromaffin cells with quantitative ultrastructural data and ascertained that the concentration of dense cored vesicles had doubled by the end of postnatal week 1 in both epinephrine- and norepinephrine-containing cells. This was confirmed by biochemical measurements. In contrast to the carotid body chief cells, they found a 3-fold postnatal increase in the innervation density of adrenal chromaffin cells. They also found that treatment with dexamethasone favored the development of epinephrine-containing cells and inhibited the normal postnatal growth of axons and the development of synapses on adrenal chromaffin cells.

The effect of hydrocortisone administration *in vivo* on the labeling index and size of chromaffin tissue in the postnatal and adult mouse was studied by Monkhouse (1986). In the postnatal adrenal medulla at all ages studied, hydrocortisone had no effect on size of the medulla or on labeling indices of either epinephrine- or norepinephrine-containing cells, although it led to a marked diminution in adrenal cortical volume. The relative proportion of epinephrine-containing cells increased between age 8 days and adulthood; this was not affected by hydrocortisone. The cortico-medullary ratio remained unchanged from postnatal day 8 onward.

The development of the adrenal medulla in the opossum was described at the light (Carmichael, Spagnoli, Frederickson et al., 1987) and electron microscopic (Spagnoli, Frederickson, Robinson et al., 1987) levels. This is the first documentation of the development of adrenal chromaffin cells in a marsupial. It was determined that, at birth, the adrenal medulla is not formed but undergoes development from migrating pheochromoblasts to an aggregated adrenal medulla during the first 6 postnatal days. Chromaffin vesicles appeared to develop from both the Golgi complex and the Golgi/endoplasmic reticulum/lysosome system. The fact that the adrenal medulla develops postnatally while the animal is in the mother's marsupium is of interest because the organ then is relatively accessible for experimental study.

Gemmell and Nelson (1988) examined the ultrastructure of the pituitary and adrenal glands in newborns of two marsupial species, the brushtail possum (*Trichosurus bulpecula*) and the northern native cat (*Dasyurus hallucatus*). They were able to distinguish adrenal medullary cells from cortical cells on the basis of their appearance. It was suggested that the cells are functional at birth.

Ogawa, Ishikawa and Ohta (1986) studied the transdifferentiation of adrenal chromaffin cells into neuronal cells. Using cells from young rats, they showed by *in vitro* and *in vivo* methods that the phenotype of the cell is influenced more by the site of development than by the specific origin of the cell. They speculated that environmental signals interact with the cell genome to change the genes, although not irreversibly, from the neuronal to the endocrine type or vice versa.

Rohrer, Acheson and Thibault (1986) studied the developmental potential of quail dorsal root ganglion cells, using the quail-chick transplantation technique. They found that some cells were able to differentiate into adrenergic chromaffin cells. These potential precursors of adrenal chromaffin cells were present in the dorsal root ganglia in significant numbers only during early development. Le Douarin and Smith (1988) reviewed research on the development of the peripheral nervous system, including the adrenal medulla, from the neural crest.

The development of epinephrine, norepinephrine, dopamine, 5-hydroxytryptamine, and histamine was investigated in the adrenal glands and other organs of a desert rat (*Mastomys natalensis*) by Pandey, Dubey and Dhawan (1986). The predominant catecholamines in the adrenal glands were epinephrine and norepinephrine which reached peak levels at postnatal day 24. Dopamine levels were lower and peaked at about the same time. Levels of 5-hydroxytryptamine were much lower and showed an increase on postnatal day 12.

McMillen, Mulvogue, Coulter et al. (1988) used an immunocytochemical technique to investigate the development of the sheep adrenal medullary cells containing enkephalins and certain catecholamine-synthesizing enzymes. Positive staining with antibodies to dopamine β-hydroxylase was observed throughout the medulla in both adult and fetal (>80 days of gestation) adrenal glands. Positive staining with antibodies to phenylethanolamine *N*-methyltransferase was observed as early as 80 days of gestation; the staining with this antiserum was mainly confined to peripheral medullary cells. Enkephalin was found in peripheral and central portions of the adrenal medulla in fetuses between 80 and 120 days of gestation. At later dates, including adult glands, enkephalin was detected more uniformly in peripheral regions than in central regions.

MORPHOLOGIC DESCRIPTIONS AND ALTERATIONS

Rosenheck and Plattner (1986) characterized vesicles made from plasma membranes of adrenal chromaffin cells and examined their interaction with chromaffin vesicles. This study was designed to characterize morphologically an *in vitro* model for exocytosis. They found that the intramembrane structure of plasma membrane vesicles is distinct from that of other organelles, including chromaffin vesicles. They showed that most of the plasma membrane vesicles were oriented right-side-out. Unfortunately, inside-out vesicles are needed to mimic exocytosis. Isotonic ionic media caused vesicle aggregation in suspensions of plasma membrane vesicles and chromaffin vesicles. Analysis by freeze-fracture techniques revealed clusters of membrane-intercalated particles at the sites of contact between aggregated membranes.

Grant, Aunis and Bader (1987) examined the morphology and secretory activity of digitonin- and α-toxin-permeabilized bovine adrenal chromaffin cells. The toxin-treated cells closely resembled the control cultured cells whereas digitonin-treated cells showed gradations in cytoplasmic densities (suggesting extraction), some swelling of the endoplasmic reticulum, and occasional discontinuities in the plasma membrane and free vesicles in the extracellular medium. In both cell preparations, mitochondria were swollen. Although both cell systems secreted catecholamines in response to calcium, their calcium requirements and the kinetics of release were different. These results suggest that both exocytosis and vesicle translocation are operational in α-toxin-treated cells but that the translocation step, or the docking of vesicles at the plasma membrane, may be impaired in digitonin-treated cells.

In studies by Himmelhoch and Rossi (1986), plasma membranes from chromaffin cells of bovine adrenal medulla were isolated, frozen, and sectioned without previous chemical fixation. The nicotinic acetylcholine receptor was localized by using a monoclonal antibody prior to postfixation with a heavy metal. The binding sites were localized in patches on the surface of the vesicles. This indicated that the vesicles were right-side-out. Because the identification of binding sites was performed prior to chemical fixation, they concluded that the patchy distribution of binding sites corresponds to a similar distribution *in vivo*.

Ornberg, Duong and Pollard (1986) identified small vesicles within chromaffin vesicles isolated from bovine adrenal chromaffin cells when quick-freezing methods were used. One to five intravesicular vesicles were seen between the core of the chromaffin vesicle and the inner surface of the vesicle membrane. The intravesicular vesicles are released from the cell during exocytosis. Ornberg et al. (1986) speculated that the structural heterogeneity provided by the intravesicular vesicles may be related to the functional heterogeneity of vesicle contents.

Cryopreservation of human fetal adrenal medullary cells was performed by Silani, Pizzuti, Strada et al. (1988). The cells were dissociated, and a cryoprotectant was used. More than 90% of the thawed cultured clusters appeared morphologically to be intact, showing extensive neurite outgrowth and absence of appreciable cell loss. These and other findings indicate that human adrenal medullary cells maintain distinctive functional characteristics after cryopreservation.

Connections between mitochondria and chromaffin vesicles have been observed in thick sections of rat adrenal medulla. In an attempt to isolate these mitochondrion-chromaffin vesicle complexes, Pfeiffer, Brimijoin, Studelska et al. (1987) analyzed fractions of bovine adrenal medulla homogenates. Some complexes were seen in the part of the gradient between the two fractions richest in mitochondria and chromaffin vesicles. Complexes were not seen when the peak fractions of mitochondria and chromaffin vesicles were mixed and incubated before morphologic examination.

Coupland (1987) published a brief general review of the histochemistry of the chromaffin cell. This is an interesting historic perspective on the identity of epinephrine- and norepinephrine-containing cells.

Hearn (1987) evaluated the effect of specimen postfixation by osmium tetroxide on postembedding localization of chromogranin A. Using a protein A-gold technique, he detected chromogranin A-positive vesicles in adrenal medulla and other organs and tumors. It was determined that ultrastructural features were optimally preserved in tissues postfixed in osmium tetroxide, and the use of this fixative is compatible with postembedding localization of chromogranin A.

Pelto-Huikko (1987) described a method for correlative microscopy of catecholamines and neuropeptides in the adrenal medulla. The method is based on the histochemical demonstration of catecholamines by fluorescence microscopy and subsequent immunocytochemistry of neuropeptides in the same section. Double-labeling methods for studying the coexistence of catecholamines and catecholamine-synthesizing enzymes are also presented.

El-Maghraby (1988) compared the ability of the chromaffin reaction and formaldehyde-induced fluorescence to differentiate between norepinephrine- and epinephrine-containing cells in the adrenal medulla of the newborn rat. Both direct and indirect chromaffin reactions demonstrated catecholamines, but the low levels did not permit a clear distinction between the two cell types. In contrast, the fluorescence method could provide a clear distinction between norepinephrine- and epinephrine-containing cells.

Sulzer, Piscopo, Ungar et al. (1987) fixed several types of neuronal tissues in aldehydes and incubated them in solutions containing lead ions. They found that lead was associated with diverse neuronal vesicles, including chromaffin vesicles of adrenal medullary cells and most of the synaptic vesicles of presynaptic terminals in the adrenal medulla. They suggested that the lead deposits indicate the presence of sulfated macromolecules in these synaptic vesicles. These could be anionic binding sites which could play a significant role in calcium storage, stabilization of pH gradients, or control of osmotic phenomena.

Selenium is an essential trace element that is toxic if given in excess. Thorlacius-Ussing and Rasmussen (1986) performed a histochemical study to determine the cellular localization of selenium in the adrenals from rats exposed to sodium selenite; these organs are known to have the highest concentration in rats. Selenium was shown to accumulate in both norepinephrine- and epinephrine-containing cells in a dose-dependent fashion. At the ultrastructural level, selenium accumulated primarily in the chromaffin vesicles and, to a lesser extent, in the lysosomes.

Rungby (1986) localized silver in the peripheral nervous system following the administration, orally or via intraperitoneal injection, of silver salts. Silver deposits in neurons and adrenal chromaffin cells were located in the cytoplasm. Silver was present in all organs in large amounts in the connective tissue membranes, macrophages, vascular basal laminae, and supporting cells.

Rasmussen and Thorlacius-Ussing (1987) detected mercury in adrenals from rats exposed to methyl mercury. It was mainly within the adrenal cortex, although a slight increase in the amount of mercury within chromaffin cells was observed after long exposure times. Both epinephrine- and norepinephrine-containing cells had such deposits.

Stevens and Trogadis (1986) described a relatively simple method for reconstructive three-dimensional electron microscopy. For example, they described the three-dimensional morphology of the PC12 cell. One of their observations was a cytoplasmic channel that penetrated from one side of the nucleus to the other. They termed this channel the "nuclear tube."

Gupta, England and Notter (1987) studied the morphologic effects of antimitotic drugs on PC12 cells by scanning electron microscopy. Untreated and cells treated with an antimitotic drug had very small extensions and numerous fine processes on the cell surface. Cells treated with nerve growth factor with or without the drug exhibited many large processes that were longer when both the growth factor and the drug were used. With transmission electron microscopy, the morphologic features were similar in groups treated or not treated with the drug. Gupta et al. (1987) concluded that antimitotic agents inhibit PC12 cell growth without affecting important cellular features.

The targeting of secretory vesicles to cytoplasmic domains was examined by Matsuuchi, Buckley, Lowe et al. (1988) in PC12 cells. They suggested that cytoskeletal elements normally carry secretory vesicles to the surface. When the cytoskeletal elements coalesce in an extending process, vesicles are transported nonselectively to the tips of the cytoskeletal elements where they accumulate.

Dennerll, Joshi, Steel et al. (1988) assessed the mechanical properties of PC12 cells by applying a force with calibrated glass needles and measuring resulting changes in neurite length and deflection of the needle. They were able to determine neurite spring constants and rest tensions; these were decreased by treatment with cytochalasin D. Treatment with nocodazole increased neurite rest tensions but produced no change in spring constants. The observations suggest a specific type of complementary force interaction underlying axonal shape--a neurite actin network under tension and neurite microtubules under compression.

Cultured adrenal chromaffin cells from newborn rats were treated with inhibitors of RNA synthesis and then studied by electron microscopy by Magalhães, Serra and Magalhães (1987). The most striking changes observed with the inhibitors were localized in the nucleoli: segregation, fragmentation, and microsegregation of the nucleoli. The exact effect varied with the inhibitor used.

Bardakhch'yan and Kirichenko (1986, 1987) examined the ultrastructural changes in the adrenal medulla and cortex during endotoxin shock. Ultrastructural changes in the chromaffin cells during the intermediate stage of endotoxic shock correlated with severe microcirculatory disturbances leading to intravascular coagulation and hemorrhages and with changes in the innervation where signs of irreversible degeneration were found. These lesions indicated that the stress exceeded the adaptive capacity of the cells, the majority of which were exhausted and killed during the latter stages of endotoxemia.

The ultrastructural organization of chromaffin vesicles was

studied in the rat adrenal medulla, after varying periods of immobilization stress, by Arefolov, Malikova and Valdman (1987). Using morphometric techniques, they found that the overall volume of chromaffin vesicles was markedly changed during stress. The degree of such stress-induced reorganization depended on the stage of the stress reaction and involved first epinephrine- and then norepinephrine-containing cells.

Electron microscopic studies of the avian adrenal medulla were reviewed by Ghosh and Guha (1988). They pointed out that the information available on this organ in birds is meager compared with that on the mammalian adrenal medulla. For example, there is ample morphologic evidence of exocytosis as the means of secretion from the mammalian adrenal medulla, but this is not established in birds.

Carmichael, Banerjee, Mandal et al. (1985) described the ultrastructure of the adrenal medulla during different phases of the migratory cycle of the common snipe (*Capella gallinago*). The results suggest that the adrenal medulla is in a resting state during autumn. Evidence of increased cellular activity is seen during winter and early spring, and catecholamine levels are increased during this period. This suggests that there is increased stress during this time.

Seasonal changes in the adrenal medulla of male blossom-headed parakeets (*Psittacula cyanocephala*) in relation to environmental and reproductive cycles were described by Maitra (1987). Norepinephrine, but not epinephrine, concentrations and cells become significantly higher in the spring than in the fall. Maitra (1987) hypothesized that testicular activity is correlated with the periodic activity of the adrenal medulla in these animals.

The seasonal changes in the ultrastructure of the adrenal gland of the Canada goose (*Branta canadensis interior*) were examined by George and John (1988). Two types of adrenal medullary cells could be distinguished by the appearance of epinephrine- and norepinephrine-containing chromaffin vesicles. These vesicles were seen to accumulate in the spring premigratory phase and to be depleted in the spring post-migratory phase. Epinephrine-containing vesicles were of normal density during the breeding phase, but there was a depletion of norepinephrine-containing vesicles during the breeding and molting phases; these vesicles returned to normal in the fall pre-migratory phase.

The effect of insulin on the adrenal medulla in two species of birds was investigated by Carmichael, Banerjee, Mandal et al. (1987). Although a marked hypoglycemia was found in both species, significant depletion of catecholamine from the adrenal gland did not occur. Changes seen by electron microscopy were depletion of a minority of chromaffin cells and an increase in intracytoplasmic vacuolization in most cells. The structure of chromaffin vesicles was changed noticeably in one species (pigeon, *Columba livia*) but not in the other (bulbul, *Pycnontus cafer*).

Ramachandran and Patel (1986) studied the seasonal alterations and effects of pinealectomy on the relative weights and histologic appearance of adrenals in pigeons. Pinealectomy was noted to induce a decrease in weight of the adrenal during the breeding season and an increase in weight during the nonbreeding season. The histologic appearance of the adrenal gland suggested that it is actively involved during the breeding season.

Shaposhnikov (1986) examined the adrenal medulla of mature (6 months cld) and aging (26 months old) white rats, paying particular attention to the size of the nucleus. No changes were noticed in the ultrastructure of chromaffin cells or in the size of the nucleus with aging.

Guillot, Guillot, Martínez-Soriano et al. (1988) examined morphologic changes in the adrenal gland after soft-laser irradiation of the pineal gland. Morphologic signs of increasing activity were observed in all layers of the cortex and the medulla. The effect was similar to that found after removal of the pineal gland, demonstrating that the pineal gland exerts control over the adrenal gland through luminous stimuli.

Bezrukov and Shaposhnikov (1987) studied the ultrastructure of the adrenal glands of mature and old rats after electrical stimulation of the hypothalamic ventral medial nuclei. The changes included hypertrophy of the nuclei, increased amounts of polyribosomes, hypertrophied Golgi apparatus, and canaliculi of the endoplasmic reticulum. Catecholamine levels in the blood were increased. In the old rats, certain unequal ultrastructural and functional peculiarities were revealed. It was concluded that hypothalamic regulation of adrenal function is disturbed in older animals with a decreased adaptive capacity of adrenal secretory cells.

CHAPTER 3

STIMULUS-SECRETION COUPLING: RECEPTORS

For a comprehensive introduction to the topic of stimulus-secretion coupling in chromaffin cells, the impressive two-volume review edited by Rosenheck and Lelkes (1987) is highly recommended. This collection of 15 chapters covers a wide spectrum of topics relating to stimulus-secretion coupling including morphology, receptors, ionic events, intracellular proteins, the cytoskeleton, exocytosis, and physiologic aspects.

A brief review of receptors, particularly non-nicotinic receptors, on chromaffin cells was published by Marley (1987). He presented recent data to show the many ways in which function of the adrenal medulla can be regulated. He also pointed out that details of receptor action vary among species.

A difference between stimulus-secretion coupling in epinephrine-containing cells and in norepinephrine-containing cells was discovered by Marley and Livett (1987a). This is important because it is well-known that different physiologic stimuli result in epinephrine and norepinephrine being secreted from the adrenal medulla in different ratios. This is the first evidence of postsynaptic mechanisms that may contribute to this phenomenon. They exposed isolated bovine adrenal chromaffin cells first to nicotine and then to increased potassium levels. There was a marked difference in secretion of epinephrine and norepinephrine under these conditions. Their results suggest that stimulus-secretion coupling or the exocytotic machinery in epinephrine-containing cells is different from that in norepinephrine-containing cells.

Critchley, Ellis, Henderson et al. (1986) found that muscarinic and nicotinic agonists selectively released epinephrine and norepinephrine, respectively, from isolated perfused adrenal glands of the cat. The outputs of epinephrine and norepinephrine from isolated perfused adrenal glands of the dog remain in a fixed ratio under basal conditions and when stimulated by muscarinic or nicotinic agonists. In a whole-animal preparation (anesthetized dog), they found that a combination of muscarinic and nicotinic antagonists was needed to block reflex responses of the adrenal medulla. A nicotinic antagonist was more effective in blocking the baroreceptor reflex response than the chemoreceptor reflex response.

Livett, Marley, Mitchelhill et al. (1987) reviewed four techniques used to study stimulus-secretion coupling in chromaffin cells:

tritiated norepinephrine loading, trihydroxyindole fluorometric assay, high-performance liquid chromatography with electrochemical detection, and adenosine triphosphate (ATP) assay. The advantages and disadvantages of these and some related techniques were reviewed for studies of stimulus-secretion coupling and other applications in which adrenal medullary secretory products are measured.

Kumakura, Ohara and Satô (1986) described a system to measure the real-time dynamics of the secretory function in cultured adrenal chromaffin cells. A cell-bed perfusion technique with amperometric detection was used. Experiments using either acetylcholine or high potassium concentration to evoke secretion suggest that direct monitoring is useful for studies on the regulatory mechanisms of secretory function.

Boarder and McArdle (1986) described a procedure for establishing stable primary cultures of bovine chromaffin cells on microcarrier beads. Cells were grown on the beads and incorporated into a superfusion apparatus, permitting the efficient perfusion of cells at a high density. Among other things, they found that the time course of the responses of norepinephrine-containing and epinephrine-containing cells did not differ, although there consistently was a preferential stimulation of norepinephrine release, particularly with high potassium levels. It was suggested that this superfusion technique accurately reflects the time course of cellular events in stimulus-secretion coupling.

A new method for the continuous recording of catecholamine secretion from the adrenal gland in rats in response to splanchnic nerve stimulation was developed by Kumakura, Sato and Sazuki (1988). The method consists of *in situ* perfusion of the adrenal gland and direct continuous recording of released total catecholamines by an electrochemical detector. Splanchnic nerve stimulation evoked current responses in a stimulus frequency-dependent manner. The amount of catecholamines released was linearly related to peak amplitude and to the total integrated output current for various current responses, indicating that the electrochemical detector current response in this system represents a highly reliable characteristic of the secretory process in the adrenal medulla. This system has the advantage of monitoring total catecholamine secretion from the adrenal gland *in situ* directly and continuously.

The effect of continuous stimulation of splanchnic nerves at different frequencies on the secretion of catecholamines from the isolated rat adrenal gland was examined by Wakade (1988). Secretion evoked at 10 Hz decreased more than 60% in 1 hour and could be diminished by hexamethonium and atropine or naloxone. In contrast, secretion evoked at 1 Hz was well maintained for several hours and was not sensitive to hexamethonium and atropine. However, it was sensitive to naloxone. The major conclusion was that secretion of catecholamines can be maintained for several hours when splanchnic nerves are stimulated at low frequency. Under these conditions, the release of noncholinergic transmitter(s) from splanchnic nerves is mostly responsible for inducing the secretion of catecholamines.

NICOTINIC RECEPTORS

Two probes previously shown to identify the nicotinic acetylcholine receptor on neurons were shown by Higgins and Berg (1987) also to recognize a component on the surface of bovine adrenal chromaffin cells. The first probe is a monoclonal antibody, and the second is an α-neurotoxin. Their results indicated that the component on the bovine adrenal chromaffin cell recognized by the antibody and the neurotoxin is likely to be the nicotinic receptor, and this receptor is similar to that previously seen on neurons.

The extent of pharmacologic and immunologic homology between the authentic nicotinic acetylcholine receptor from the periphery and α-bungarotoxin binding sites of PC12 cells (derived from a rat adrenal medullary tumor) and neuronal tissues was ascertained by Lukas (1986). He showed that the binding sites on the PC12 cells are comparable to binding sites on other neural tissues.

The effects of neosurugatoxin, a neurotoxin from the Japanese ivory mollusk, on the nicotinic response of bovine adrenal chromaffin cells was examined by Bourke, Bunn, Marley et al. (1988). Neosurugatoxin inhibited acetylcholine- and nicotine-induced catecholamine secretion from the cultured cells. This inhibitory effect was reversible and independent of the presence of the agonists, but the toxin had no effect on catecholamine release evoked by high levels of potassium ions or histamine. Also, the neurotoxin had no effect on the stimulation of phosphatidylinositol metabolism evoked by muscarine. These results suggest that neosurugatoxin may be useful as a nicotinic receptor probe.

Whiting, Schoepfer, Swanson et al. (1987) raised a monoclonal antibody against acetylcholine receptors from chicken brain and used it to identify a functional nicotinic acetylcholine receptor on PC12 cells. This antibody has permitted localization, purification, and characterization of these receptors.

Quik, Fournier and Trifaró (1986) investigated the possibility that the binding site for α-bungarotoxin is separate from the specific binding site for acetylcholine. They found that a factor(s) derived from rat and bovine brain that affects the *in vitro* binding of α-bungarotoxin also affects the binding of the toxin to chromaffin cell membranes. Exposing cultured chromaffin cells to this factor(s) caused a marked decrease in the number of binding sites for α-bungarotoxin but did not change the affinity.

In a follow-up study, Quik, Geertsen and Trifaró (1987) showed that some, but not other, nicotinic antagonists can markedly enhance the number of α-bungarotoxin binding sites in adrenal chromaffin cells through an interaction at a nicotinic cholinergic receptor. Their results suggest that the nicotinic-like α-bungarotoxin binding site and the functional nicotinic receptor are distinct in this tissue. The two main lines of evidence are: the differential effects of the various nicotinic antagonists in increasing the number of toxin binding sites compared with the ability of these agents to inhibit the functional nicotinic response, and the observation that an 8-fold increase in α-bungarotoxin receptor number does not enhance responsiveness of the functional nicotinic receptor. These results also imply that the toxin receptors have a functional role in the adrenal medulla.

Kemp and Edge (1987) characterized α-bungarotoxin binding in PC12 cells. Using sensitive methods, they could not detect a flux of Na^+ that was sensitive to α-bungarotoxin. These and other results raise a question of whether the α-bungarotoxin binding protein can act as a functional nicotinic acetylcholine receptor in PC12 cells.

Acetylcholine receptor channels and their blockade by clonidine in cultured bovine adrenal chromaffin cells was studied by Cull-Candy, Mathie and Powis (1988). Using three variations of the patch-clamp recording technique (recording from whole cells, outside-out patches, and cell-attached patches), they showed that clonidine markedly decreased the size of the acetylcholine-induced whole-cell current and altered the shape of the noise spectrum. The kinetics of single acetylcholine channels were more complex than suggested by noise analysis. Clonidine decreased the time constants of both components of the burst-length distribution and also decreased the frequency of opening of acetylcholine channels. It was concluded that the action of clonidine on nicotinic acetylcholine receptors and chromaffin cells may have involved a complex blocking mechanism, perhaps related to agonist-induced desensitization.

The effects of prolonged depolarization on the nicotinic acetylcholine receptors of PC12 cells was studied by DeLorme and McGee (1988). They grew PC12 cells under conditions that produce chronic depolarization in order to examine the possibility that changes in membrane potential or subsequent increases in intracellular calcium levels are responsible for adaptation of neuronal nicotinic receptors. Among other things, they found that concurrent depolarization did not affect the functional down-regulation produced by chronic exposure of the cells to carbachol. Thus, neuronal nicotinic acetylcholine receptors on PC12 cells do not appear to be regulated by depolarization or a prolonged increase in the intracellular calcium level.

The characteristics of the inhibitory action of clonidine on catecholamine release from bovine adrenal chromaffin cells were investigated by Ohara-Imaizumi, Miyakawa and Kumakura (1988). Clonidine inhibited acetylcholine-evoked release of catecholamine but not the release evoked by high potassium concentration. The inhibition of acetylcholine-evoked release by clonidine was not reversed by α_2-adrenergic antagonists. Treatment of these cells with pertussis toxin reversed the inhibitory effect of clonidine, but it did not affect the inhibitory actions of hexamethonium and nifedipine. It was concluded that clonidine inhibition of catecholamine release is not mediated by the α_2-adrenoceptor but may be mediated by a specific receptor for clonidine.

Higgins and Berg (1988c) showed that a cAMP-dependent process enhances the function of adrenal chromaffin nicotinic acetylcholine receptors as a population in the plasma membrane. This was demonstrated by showing that cAMP analogues caused specific increases both in the level of nicotine-induced catecholamine release from the cells and in the level of the nicotine-induced conductance change occurring in the cells. The responsiveness of the receptors to regulation appears to depend on the length of time the receptors have been on the cell surface. Receptors that have been newly inserted into the plasma membrane generate a greater nicotinic response than do older receptors; and unlike older receptors, their response to agonists is not enhanced after treatment of the cells with cAMP analogues. Their findings indicated that nicotinic acetylcholine receptors or associated components undergo a maturation in the plasma membrane that alters their function and their regulation by second-messenger systems.

DESENSITIZATION

The phenomenon of desensitization of the nicotinic secretory response of adrenal chromaffin cells was reviewed by Marley (1988). The secretion of catecholamines in response to nicotinic stimulation is known to diminish over time, and this generally has been attributed to loss of nicotinic receptor responsiveness. However, it was pointed out that virtually every step in stimulus-secretion coupling, from nicotinic receptor activation to exocytosis of chromaffin vesicles, can be desensitized or inactivated. These findings suggest that loss of responsiveness to an agonist can result from more than one mechanism and that caution should be exercised before one assumes that it is due solely to desensitization of the receptor.

Marley and Livett (1987b) investigated whether adrenal opioid peptides acted like the tachykinins (including substance P) to prevent desensitization of the nicotinic secretory response of isolated bovine adrenal chromaffin cells. Dynorphin and metorphamide produced protection against desensitization of the nicotinic response at low concentrations whereas etorphine and morphine only produced this effect at higher concentrations. All of the opioid compounds tested were considerably more potent as inhibitors of nicotine-induced catecholamine secretion from the cells than as protectors against desensitization of this response. Marley and Livett (1987b) concluded that adrenal opioid peptides probably do not act on adrenal opioid binding sites characterized by ligand binding studies to prevent the desensitization of the nicotinic response. These peptides are probably not involved in a mechanism to maintain catecholamine secretion during stress.

The effects of a number of α- and β-adrenoceptor agonists and antagonists on the modulation of secretion from bovine adrenal chromaffin cells were investigated by Wan, Powis, Marley et al. (1988). The secretion of endogenous catecholamines and ATP were monitored. The degree of desensitization of the nicotinic response was dependent on the concentration of nicotine to which the cells had been exposed previously. Both clonidine and propranolol inhibited the catecholamine secretion induced by nicotine during the pre-incubation period. However, regardless of the inhibition of secretion, neither clonidine nor propranolol had an effect on the onset or the rate of nicotine-evoked desensitization subsequently observed. These and other data suggested that inhibition of the nicotinic response and desensitization of the nicotinic response are regulated independently. These findings do not support the existence of α- or β-adrenoreceptor modulation of secretion from isolated bovine chromaffin cells.

By directly monitoring the secretory process in a cell-bed perfusion apparatus, Ohara-Imaizumi and Kumakura (1986) used real-time monitoring to study the secretory process in cultured bovine adrenal chromaffin cells. They demonstrated that prolonged stimulation of the chromaffin cells with either acetylcholine or high potassium concentration produces an initial burst of a secretory response that lasts for a few seconds; this initial burst is followed by a progressive decline for inactivation of the response. When additional acetylcholine was added during the declining phase of the primary response, an additional response was evoked. The magnitude of this second response was decreased by extracellular calcium in a dose-dependent manner. However, neither the peak amplitude nor the decrease in the rate of the primary response was affected by calcium. This suggested that the transient nature of the acetylcholine-evoked response is mainly attributable to inactivation of transduction mechanisms occurring at later phases of stimulus-secretion coupling and not only to a desensitization of the nicotinic receptors.

The phenomenon of desensitization of catecholamine secretion in the perfused adrenal gland of the rat--using high potassium, nicotine, and muscarine as secretagogues--was investigated by Malhotra, Wakade and Wakade (1988a). The secretion of catecholamines could be depressed by continuous exposure to nicotine for 30 minutes but, even after this desensitization, the introduction of potassium at high level or as muscarine along with the nicotine still led to a large increase in the secretion of catecholamines. Omission of calcium ions from the perfusion medium containing either nicotine or potassium caused a rapid decline in the secretion of catecholamines, suggesting influx of external calcium. Muscarine-induced secretion was decreased only slightly by omission of calcium, suggesting mobilization of internal calcium stores. Addition of a calcium chelating agent to the calcium-free medium completely abolished the secretion evoked by all agents. Nicotine-evoked secretion was associated with an increase in the uptake of radioactive calcium. However, after continuous exposure to nicotine, the uptake of radioactive calcium did not increase significantly after 30-minutes. Potassium-evoked secretion was associated with an increase in the uptake of radioactive calcium, and this increase was detected even after one hour of continuous exposure. It appeared that the voltage-sensitive calcium channels opened by high levels of potassium did not undergo rapid inactivation and remained open to allow influx of extracellular calcium and secretion of catecholamines. This study showed that secretion is well maintained during prolonged stimulation of the rat adrenal medulla by high levels of potassium or muscarine.

Simasko, Soares and Weiland (1986) examined the desensitization of nicotinic acetylcholine receptors on PC12 cells to determine whether these cells exhibit a nonrecoverable component as reported in other systems. The uptake of radioactive sodium was used to quantitate receptor function. They found that, in addition to classically described desensitization, a second process occurs that results in a nonrecoverable loss of receptor activity. They called this second process "inactivation." The onset of inactivation was dependent on the concentration of the desensitizing ligand and was blocked by nicotinic antagonists. No inactivation was observed even after 2 hours of incubation in a recovery buffer. The mechanism underlying inactivation remains to be determined, but Simasko et al. (1986) suggested that it may be the first step in nicotinic receptor down-regulation.

In a follow-up study, Simasko, Durkin and Weiland (1987) examined the effects of substance P on activation and desensitization of the nicotinic receptor and PC12 cells. They found that substance P affects the desensitization phase but not inactivation. Their results support the conclusion that substance P allosterically modulates the nicotinic receptor but they also indicate that some degree of block of the receptor-linked channel occurs.

In an impressively complete series of studies, Boyd (1987) and Boyd and Leeman (1987) characterized desensitization of acetylcholine receptors on PC12 cells. Using the influx of radioactive sodium to measure receptor activation, Boyd (1987) found distinct kinetic phases of desensitization: a rapid phase that was characterized by rate constants dependent on the chemical nature and concentration of the agonist, and a slower phase that was characterized by rate constants less dependent on these. This second phase may correspond to the inactivation described by Simasko et al. (1986). However, Boyd (1987) found that after removal of the agonist the recovery from desensitization was reversible and occurred by two distinct kinetic phases. He proposed that the development of equilibrium desensitization may be a necessary component of the processes involved in irreversible receptor deactivation.

Boyd and Leeman (1987) evaluated the inhibition of cholinergic receptor function by substance P. In the absence of agonists, substance P converts the acetylcholine receptors in a time- and concentration-dependent manner to a state that is not responsive to agonists. This effect was reversed when substance P was removed. In the presence of agonist, substance P caused a strong enhancement of both the rate and the extent of agonist-mediated desensitization. In contrast, the second, slower, component of agonist-mediated desensitization is inhibited by substance P. Taken overall, their findings provide a basis for physiologic roles for substance P in the short- and long-term regulation of cholinergic transmission at the postsynaptic level.

Higgins and Berg (1988a) showed that tritiated nicotine binds with high affinity to a desensitized form of the acetylcholine receptor on bovine adrenal chromaffin cells. The receptor is the same as that previously identified by binding of the monoclonal antibody mAb35 and can be modulated by exposure to mAb35. Desensitization of the receptor is accompanied by an increase in agonist affinity with no apparent change in antagonist affinity. These and other results demonstrate that the bovine adrenal chromaffin acetylcholine receptor is similar to receptors from muscle and electric organ in undergoing an agonist-induced conversion to a desensitized state having increased affinity for agonists.

Higgins and Berg (1988b) also showed that chronic exposure of these cells to mAb35 causes loss of the surface acetylcholine receptors through antigenic modulation. They could then take advantage of the modulation and other manipulations to demonstrate that the normal metabolic stability of the receptors in the plasma membrane is similar to that previously described for receptors on skeletal muscle. Unexpectedly, they found a substantial discrepancy between the rate at which receptors reappear in the plasma membrane after antigenic modulation and the rate at which recovery of the nicotinic response is obtained. The results suggested a difference in the functional properties of new and old acetylcholine receptors in the plasma membrane.

MUSCARINIC RECEPTORS

Recent studies have emphasized that, although acetylcholine receptors are of primary importance for catecholamine release from the adrenal medulla, the predominance of nicotinic or muscarinic subtypes varies widely among species. Knight and Baker (1986) demonstrated that catecholamine secretion from the adrenal medulla of the chicken is activated almost entirely by acetylcholine acting on muscarinic receptors. They isolated cells from chicken adrenal glands and found that acetylcholine, carbamoylcholine, potassium, veratridine, metha- choline, muscarine, and oxotremorine all were effective secretagogues, whereas nicotine was not. The secretion evoked by acetylcholine was blocked by low concentrations of atropine but was relatively insensitive to hexamethonium. Further experiments suggested that muscarinic activation of these cells from the chicken facilitate tetrodotoxin-insensitive depolarization, thereby opening conventional voltage-sensitive calcium channels.

Nakaki, Sasakawa, Yamamoto et al. (1988) demonstrated a functional shift from muscarinic to nicotinic cholinergic receptors over time in cultures of bovine adrenal chromaffin cells. During periods of less than two days, muscarinic but not nicotinic stimulation led to an increase in inositol triphosphate generation, cGMP formation, and calcium ion mobilization. At periods of more than four days, nicotinic but not muscarinic receptors become preferentially coupled to these same parameters.

Calcium ion fluxes and catecholamine secretion induced by muscarinic stimulation were compared to those induced by nicotinic stimulation from chromaffin cells isolated from the guinea pig adrenal medulla by Mishbahuddin (1987). Muscarine was a more potent secretagogue than nicotine. Muscarine-induced catecholamine secretion was slow in onset and depended on the presence of extracellular calcium. Muscarine did not cause significant uptake of radioactive calcium, but experiments using a fluorescent indicator showed that it caused an increase in intracellular-free calcium that was sufficient to stimulate catecholamine secretion. An intracellular calcium antagonist inhibited both the catecholamine secretion and the increase in cytosolic calcium induced by muscarine. It was concluded that muscarine stimulates catecholamine secretion by mobilizing calcium from intracellular stores.

The differences between the mechanisms of muscarinic and nicotinic receptor-mediated catecholamine secretion, with respect to their dependence on voltage changes and extracellular calcium, were examined by Nakazato, Ohga, Oleshansky et al. (1988) in perfused adrenal glands of the guinea pig. Specifically, they examined the effects of atropine or hexamethonium and of the removal of extracellular calcium on the secretory responses to acetylcholine, pilocarpine, nicotine, and high potassium during perfusion with Locke's and isotonic potassium chloride solutions. Their experiments suggested that both nicotinic and muscarinic receptors are involved in acetylcholine-induced catecholamine secretion by guinea pig adrenal chromaffin cells. Activation of muscarinic or nicotinic receptors appeared to stimulate the catecholamine release through different mechanisms with respect to both voltage-dependence and calcium requirements.

Harish, Kao, Raffaniello et al. (1987) demonstrated that muscarine-evoked catecholamine secretion from perfused rat adrenal glands can occur in the absence of extracellular calcium, presumably by mobilization of intracellular calcium. They suggested that this may be due to muscarinic receptor-mediated generation of inositol triphosphate.

Forsberg, Rojas and Pollard (1986) found that muscarinic receptor stimulation in bovine adrenal chromaffin cells leads to enhanced metabolism of inositol phospholipid. They also found that the muscarinic antagonist atropine decreases acetylcholine-induced secretion, suggesting that muscarinic receptor stimulation primes the cells for nicotinic receptor-mediated secretion, perhaps by causing small nonstimulatory increases in cytosolic calcium levels mediated by inositol triphosphate.

Adrenal chromaffin cells isolated from the guinea pig adrenal medulla were used by Misbahuddin and Oka (1988) to examine the catecholamine secretion and calcium movement induced by muscarinic stimulation in comparison with those induced by nicotinic stimulation. Both agonists increased catecholamine secretion, but the stimulation by muscarine was greater. The secretion of catecholamines was dependent on the presence of calcium ions in the medium, but only nicotine was associated with a rapid increase in radioactive calcium uptake. Experiments using the calcium indicator Quin-2 showed that muscarine caused an increase in cytoplasmic free-calcium ion concentration. These and other results indicated that, in isolated guinea pig adrenal chromaffin cells, nicotine stimulated catecholamine secretion by increasing calcium uptake into the cells whereas muscarine stimulated catecholamine secretion by mobilizing calcium from intracellular storage sites.

Using chromaffin cells from the bovine adrenal medulla, Swilem, Yagisawa and Hawthorne (1986) found muscarinic receptors on the cells; but muscarinic agonists did not provoke catecholamine release. With Quin-2 as a fluorescent reagent to measure intracellular free calcium levels, they found that muscarinic agonists do not increase cytosolic free calcium concentrations but, unlike the nicotinic agonists, they cause phosphoinositide hydrolysis.

The presence of muscarinic binding sites in the bovine adrenal medulla was investigated by Barron, Murrin and Hexum (1986) using the specific muscarinic receptor antagonist tritiated quinuclidinyl benzilate. Their results demonstrate the presence of more than one type of muscarinic binding site in the bovine adrenal gland.

Barron and Hexum (1986a) further investigated the role of muscarinic receptors in the modulation of catecholamine release from the bovine adrenal medulla, and they expanded on this by monitoring the concomitant release of [Met]enkephalin-immunoreactive material. They found that the stimulation of the muscarinic receptor by itself is not sufficient to elicit the release of catecholamines and [Met]enkephalin, but it does appear to modulate secretion stimulated by nicotinic receptor activation.

Yamanaka, Kigoshi and Muramatsu (1986a) characterized muscarinic receptors in bovine adrenal medullary microsomes by using tritiated quinuclidinyl benzilate. They found that specific binding of the agonist to microsomes is rapid, reversible, saturable, and of high affinity. Whereas quinuclidinyl benzilate appeared to bind to a single class of sites, other muscarinic agonists (acetylcholine, carbamoylcholine, oxotremorine) appeared to occupy at least two sites. Apparently, the predominant site is the M_1 muscarinic site. In a follow-up study, Yamanaka, Kigoshi and Muramatsu (1986b) found that copper enhanced the affinity of carbamoylcholine at the low-affinity binding site, with a slight increase in the affinity at the high-affinity binding site. On the other hand, copper slightly decreased the binding affinity of the antagonist pirenzepine and of atropine. They concluded that low concentrations of copper may modulate the muscarinic receptors in the adrenal medulla by selectively increasing agonist affinity.

Investigating the M_1- and M_2-subtypes of muscarinic receptors, Wakade, Kahn, Malhotra et al. (1986) used a specific agonist of M_1 receptors (McN-A-343) in the perfused rat adrenal gland. Even at high concentrations, this specific agonist caused only a modest secretion of catecholamines compared with low concentrations of muscarine. In the presence of pirenzepine, an M_1 antagonist, the concentration-secretion curve for muscarine was shifted to the right by a factor of 10. Although the pirenzepine data suggest that M_1 receptors are responsible for the secretion of catecholamines from the rat adrenal medulla, this conclusion is not supported by the results obtained with the M_1-receptor agonist McN-A-343. They decided that it was not possible to reach a definitive conclusion regarding the subtypes of muscarinic receptors involved in the secretion of catecholamines in the rat adrenal medulla.

Borges, Ballesta and García (1987) provided evidence suggesting that the adrenal chromaffin cell from the cat preferentially secretes epinephrine in response to methacholine by activating calcium entry into the cell through an ionophoric channel associated with or controlled by an M_2-muscarinic receptor. It appeared that this channel is chemically operated and seems to be unrelated to voltage-sensitive calcium channels. They used specific antagonists and specific receptor binders in collecting their data.

In two studies, Pozzan, Di Virgilio, Vicentini et al. (1986) and Vicentini, Ambrosini, Di Virgilio et al. (1986) examined the activation of muscarinic receptors in PC12 cells (a cultured rat pheochromocytoma cell). Pozzan et al. (1986) focused on calcium homeostasis, with reference to internal calcium levels, influx, and redistribution. They compared undifferentiated PC12 cells with cells treated with nerve growth factor which induces a neuronal-like differentiation accompanied by a large increase in the number of muscarinic receptors. Differences in calcium homeostasis were noted between undifferentiated and differentiated cells. These effects were due to the activation of the muscarinic receptor because they were not affected by nicotinic blockers and were completely eliminated by low concentrations of muscarinic antagonists. Intracellular calcium levels could be elevated by a muscarinic receptor-dependent process that seemed to involve redistribution of calcium from cytoplasmic stores and influx of calcium through a channel other than the voltage-dependent calcium channel. The study by Vicentini et al. (1986) dealt with the correlation between receptor-coupled phosphoinositide and calcium responses. They analyzed phosphoinositide hydrolysis and calcium responses induced by muscarinic-receptor activation. They also studied the effects of a number of inhibitors and other drugs that were expected to interfere at various levels with the process initiated by receptor activation. The release of inositol triphosphate and its possible metabolites and the increase in calcium were found to occur with carbachol concentration curves that were very similar. These and other results suggested that phosphatidylinositol hydrolysis and increase in intracellular calcium occur as two successive events in the intracellular transduction cascade initiated by receptor activation.

Bunn, Marley and Livett (1988b) measured the accumulation of inositol phosphates as an index of muscarinic receptor activation to examine the influence of a number of different opioid compounds on muscarinic receptors in cultured bovine adrenal chromaffin cells. Muscarine produced a dose-dependent 1.5-fold increase in total inositol phosphates, and this could be inhibited by atropine. The 10 opioid compounds represented agonists for all of the identified opioid receptor subtypes. However, none of these opioids had any significant effect on either basal or muscarinic-induced total inositol phosphate accumulation. Bunn et al. (1988b) concluded that it is unlikely that opioid peptides can modulate the inositol phosphate second-messenger system within adrenal chromaffin cells.

Yamanaka, Muramatsu and Kigoshi (1988) examined muscarinic receptor binding in bovine adrenal medullary microsomes after exposure to tetranitromethane, a compound that modifies tyrosine and cysteine residues in proteins. Tetranitromethane caused a concentration-dependent irreversible decrease in the maximum number of muscarinic binding sites and a slight increase in the equilibrium association constant. These and other results indicated that tetranitromethane causes a loss of muscarinic binding sites and a decrease in the binding affinity of the receptors in the bovine adrenal medulla, possibly through modifications of functional groups such as tyrosine residues.

The effect of muscarine on neurosecretion from PC12 cells was studied by Rabe, DeLorme and Weight (1987). When PC12 cells were exposed to muscarine, the cells responded rapidly with increases in intracellular inositol triphosphate level and intracellular calcium and release of stored transmitter. These phenomena were blocked by a muscarinic antagonist but were unaffected by a nicotinic antagonist. These and other results suggest that the muscarine-stimulated release of neurotransmitter may be associated with an inositol triphosphate-induced mobilization of intracellular calcium.

ADRENERGIC RECEPTORS

There have been conflicting reports in the literature on the role of α-adrenergic receptors in regulating secretion from the adrenal medulla. Powis and Baker (1986) performed experiments with perfused bovine adrenal glands and isolated bovine chromaffin cells to investigate the suggestion that α-adrenergic receptors are present that regulate catecholamine secretion. They found that, although the specific α_2-adrenergic agonist clonidine inhibits catecholamine secretion evoked by acetylcholine, carbachol, and nicotine, other observations contradict a regulatory role for α-adrenergic receptors. Their accumulated data suggest that functional α_2-adrenergic receptors of the classical type are not present on bovine chromaffin cells and that clonidine must act on the cells in some other way. They suggested that clonidine may act at the nicotinic receptor.

Sharma, Wakade, Malhotra et al. (1986) showed that the secretion of catecholamines from the perfused rat adrenal gland is not regulated by α-adrenergic receptors. They found that various selective agonists and antagonists of α-adrenergic receptors did not modify the secretion of catecholamines evoked by splanchnic nerve stimulation or nicotine. They concluded that the negative feedback mechanism that controls the release of norepinephrine at central and peripheral adrenergic synapses does not operate in the adrenal medulla of the rat.

Nguyen and DeLéan (1987) investigated the pharmacologic characteristics of clonidine in the co-secretion of catecholamines and enkephalins in cultured bovine adrenal chromaffin cells. They observed that clonidine inhibits nicotine-stimulated secretion of catecholamines and [Leu]enkephalin, but neither the specific α_2-adrenergic antagonist yohimbine nor the α_1-adrenergic antagonist prazosin could fully reverse the inhibitory effect. They ruled out a role for α-adrenergic receptors in the mediation of clonidine inhibition of co-secretion of catecholamine and enkephalins in bovine chromaffin cells.

The lack of agreement in this area is illustrated by Foucart, Nadeau and de Champlain's (1987) suggestion that catecholamine secretion from the adrenal medulla of the dog appears to be partly modulated by a presynaptic inhibitory mechanism that involves α_2-adrenergic receptors. They tested the effect of several α-adrenergic agonists and antagonists on catecholamine secretion evoked by electrical stimulation of the splanchnic nerve at different frequencies and found that the agonist appeared to be more efficient at low frequencies whereas antagonists appeared to be more efficient at higher frequencies. They suggested that adrenal medullary α_2-adrenergic receptors would be saturated at higher frequencies of stimulation. The species specificity of these phenomena also need to be considered.

In vivo interactions between prejunctional α_2- and β_2-adrenoreceptors at the level of the adrenal medulla were studied by Foucart, de Champlain and Nadeau (1988). Electrical stimulation of the splanchnic nerve increased catecholamine release from the adrenal gland of the dog. Intravenous injection of a β_2-adrenergic antagonist significantly decreased this response, and a subsequent injection of an α_2-adrenergic agonist further decreased the release of catecholamines. However, if the agonist was injected first, the release was not different from that in controls; and the subsequent injection of the antagonist also did not modify the release in response to electrical stimulation. These results suggest that the blockade of presynaptic β_2-adrenergic receptors decreases the release of adrenal catecholamines without interfering with the activation of α_2-adrenoceptors. In contrast, the pretreatment with an α_2-adrenergic agonist, which does not modify the release of catecholamines, seems to interfere with the inhibitory effect of a β_2-adrenergic antagonist.

Orts, Orellana, Cantó et al. (1987) showed that adrenal medullary catecholamine release evoked by splanchnic nerve stimulation is not modulated by α- or β-adrenoceptors in the cat. Transmural electrical stimulation of the adrenal gland increased total catecholamine secretion which was inhibited by propranolol and other antagonists. It was suggested that propranolol may inhibit secretion by blocking ion fluxes through the acetylcholine receptor ionophore. Clonidine may inhibit secretion by this same mechanism or may interfere with some intracellular secretory event.

The effects of a new synthetic α_2-adrenergic receptor agonist (an imidazole derivative, DJ-7141) on catecholamine secretion from isolated bovine adrenal chromaffin cells were examined by Oka, Misbahuddin, Ishimura et al. (1987). The agonist did not affect basal catecholamine secretion but did inhibit the catecholamine secretion induced by nicotine. The inhibitory effect was less than that of clonidine. However, this inhibitory effect was not antagonized by α_2-adrenergic antagonists, suggesting that the effect of DJ-7141 is independent of its effect on α_2-adrenergic receptors.

Yanagihara, Wada and Izumi (1987) examined the effects of α_1- and α_2-adrenergic agonists on the synthesis of catecholamines in cultured bovine adrenal chromaffin cells. Clonidine inhibited carbachol-stimulated synthesis of catecholamines and uptake of calcium into the cells. Other α_2-adrenergic agonists also inhibited synthesis of catecholamines. α_1-Adrenergic agonists did not affect synthesis. Other results suggested that α_2-adrenergic agonists inhibit carbachol-stimulated synthesis of catecholamines by suppressing tyrosine hydroxylase activity, probably through the inhibition of calcium uptake.

Noronha-Blob, Marshall, Kinnier et al. (1986) examined the pharmacologic profile of α_2-adrenergic receptors in PC12 cells. It is known that stimulation of α_2-adrenergic receptors increases adenylate cyclase activity. These authors showed that PC12 cells are functionally deficient in the α_1-adrenergic receptor response linked to adenylate cyclase, but this response is efficiently coupled to α_2-adrenergic receptors.

The local modulation of adrenal catecholamine release by β_2-adrenoceptors in the anesthetized dog was studied by Foucart, Nadeau and de Champlain (1988). They evaluated the effect of a β-adrenoceptor agonist, isoprenaline, on the release of adrenal catecholamines elicited by splanchnic nerve stimulation. Subsequently, the administration of selected β_1- or β_2-adrenoceptor antagonists was used to study the reversibility of the potentiating effect of isoprenaline and to identify the subtype of β-adrenoceptor involved in this facilitatory mechanism. Their results suggested that the release of adrenal catecholamines is modulated locally by a positive feedback mechanism through activation of β_2-adrenoceptors.

The characteristics of the inhibitory action of clonidine on catecholamine release from bovine adrenal chromaffin cells were investigated by Ohara-Imaizumi, Miyakawa and Kumakura (1988). Clonidine inhibited acetylcholine-evoked release of catecholamine but not the release evoked by high potassium concentration. The inhibition of acetylcholine-evoked release by clonidine was not reversed by α_2-adrenergic antagonists. Treatment of these cells with pertussis toxin reversed the inhibitory effect of clonidine, but it did not affect the inhibitory actions of hexamethonium and nifedipine. It was concluded that clonidine inhibition of catecholamine release is not mediated by the α_2-adrenoceptor but may be mediated by a specific receptor for clonidine.

Liggett, Marker, Shah et al. (1988) correlated plasma catecholamine levels with intravascular α- and β-adrenoreceptors (on platelets and mononuclear lymphocytes, respectively) and extravascular α- and β-adrenoreceptors in lungs of patients undergoing pulmonary resection. In lung α_1-adrenergic receptor densities were positively correlated with plasma levels of norepinephrine and epinephrine. In contrast lung β-adrenergic receptor densitiies were inversely related to plasma catecholamine concentrations. No correlations were found between plasma catecholamines and adrenergic receptor densities in the intravascular system. The authors concluded that extravascular tissues should be used to study the regulation of adrenergic receptors by endogenous catecholamines in man.

The effects of α_1- and α_2-adrenergic agonists on synthesis of catecholamines in cultured bovine adrenal chromaffin cells was examined by Yanagihara, Wada and Izumi (1987). α_1-Adrenergic agonists did not affect catecholamine synthesis, but α_2-adrenergic agonists inhibited carbachol-stimulated synthesis of catecholamines in a concentration-dependent manner. Their results suggested that α_2-adrenergic agonists inhibit carbachol-stimulated synthesis of catecholamines by suppressing tyrosine hydroxylase activity, probably through inhibition of calcium uptake. However, the involvement of α_2-adrenoceptors in the inhibitory effects of α_2-adrenergic agonists on catecholamine synthesis is still unsettled.

DOPAMINERGIC RECEPTORS

González, Artalejo, Montiel et al. (1986) characterized a dopaminergic receptor that modulates catecholamine release from the bovine adrenal medulla in radioligand binding studies with nicotine-evoked catecholamine release monitored. Apomorphine markedly inhibited nicotine-evoked release; haloperidol reversed the inhibitory effect of apomorphine. These and other data suggest that the bovine adrenal chromaffin cell contains a dopaminergic receptor, apparently of the D_2 type, that modulates the catecholamine secretory response evoked by nicotine. González et al. (1986) suggested that selective dopaminergic agonists might prove to be useful clinically in the treatment of hypertension.

Lyon, Titeler, Bigornia et al. (1987) evaluated the interaction between dopamine receptor ligands and particulate membrane fractions from bovine chromaffin cells and adrenal medullary homogenates by using the D_2 dopamine receptor radioligand tritiated *N*-methylspiperone. They characterized these binding sites; for example, they detected about 1,000 receptors per cell. Other studies, including competition studies, suggested that dopamine may modulate adrenal medullary function through D_2 receptors.

Quik, Bergeron, Mount et al. (1987) also detected dopamine receptors in the bovine adrenal medulla by using tritiated spiperone as the radioligand. They found a high-affinity site with the characteristics of a D_2 receptor. This receptor was distinct from the spiperone binding site previously described in the adrenal cortex.

Shigetomi, Ueno, Tosaki, et al. (1987) superfused adrenal glands of rats with solutions containing dopamine and synthetic atrial natriuretic peptide. Dopamine remarkably decreased basal release of epinephrine and norepinephrine, and this inhibition was blocked by a dopamine D_2 antagonist. The polypeptide did not have such an effect. They concluded that dopamine, but not atrial natriuretic peptide, may play an important role in suppressing the activity of the adrenal medulla.

Bigornia, Suozzo, Ryan et al. (1988) provided evidence confirming a role of dopamine receptors as inhibitory modulators of catecholamine release from bovine adrenal chromaffin cells. They further showed that the mechanism of modulation involves inhibition of stimulated calcium uptake. Apomorphine caused a dose-dependent inhibition of the radioactive calcium uptake stimulated by either nicotine or high potassium concentration. This inhibition was reversed by a series of specific dopamine receptor antagonists. The combined results suggest that dopamine receptors on adrenal chromaffin cells alter calcium channel conductance which, in turn, modulates catecholamine release.

γ-AMINOBUTYRIC ACID (GABA) RECEPTORS

The evidence supporting GABA as an important modulator of secretion from the adrenal medulla was reviewed by Alho, Fujimoto, Guidotti et al. (1986), Guidotti and Hanbauer (1986), and Kitayama and Tsujimoto (1986). There is strong evidence that GABA is present in nerve terminals and chromaffin cells within the adrenal medulla in cow, dog, and rat. Several lines of evidence reviewed in these papers suggest that GABA evokes catecholamine secretion from the adrenal medulla through the activation of $GABA_A$ receptors located on adrenal chromaffin cells. This apparently works on chloride channels, resulting in chromaffin cell depolarization.

Kataoka, Fujimoto, Alho et al. (1986) documented that GABA modulates the spontaneous release of catecholamine and the release elicited by electrical stimulation of the splanchnic nerve in the dog. $GABA_A$ receptor agonists increase catecholamine secretion, even in denervated glands. They also showed that the extent of catecholamine release elicited by splanchnic nerve stimulation was decreased by a $GABA_A$ agonist and was increased by a $GABA_A$ receptor antagonist. They suggested that the release of catecholamines elicited by nicotinic receptor stimulation may be decreased by endogenous GABA.

Kataoka, Ohara-Imaizumi, Ueki et al. (1988) evaluated the role of GABA in the secretory function of cultured adrenal chromaffin cells by using the method of real-time monitoring. GABA evoked the secretion of catecholamines from the adrenal chromaffin cells in a dose-dependent manner. These and other findings provided additional evidence favoring the possibility that the membrane depolarization evoked by GABA is operative for both inhibitory and facilitatory regulation of the chromaffin cell responsiveness to acetylcholine.

The release of enkephalin-containing peptides was shown by Fujimoto, Kataoka, Guidotti et al. (1987) to be modulated by $GABA_A$ in experiments using the adrenal gland of the dog. They found that the release of enkephalin-containing peptides elicited by electrical stimulation is potentiated by a $GABA_A$ antagonist. This is in agreement with the concept that $GABA_A$ receptor activation causes bursts of chloride channel openings, with the outward chloride current eliciting membrane depolarization and thereby preventing subsequent responsiveness to nicotinic receptor activation.

In an interesting series of studies, Kitayama (1985), Kitayama, Morita, Dohi et al. (1986), and Kitayama, Kóyama, Morita et al. (1986) characterized the role of GABA in catecholamine secretion from the adrenal medulla. Kitayama (1985) and Kitayama, Morita, Dohi et al. (1986) found that GABA-evoked catecholamine release from the dog adrenal medulla is dependent on K^+. Kitayama, Kóyama, Morita et al. (1986) investigated the role of Ca^{2+} in GABA-evoked catecholamine release from bovine adrenal chromaffin cells. They suggested that the activation of GABA receptors on adrenal chromaffin cells facilitates the influx of Ca^{2+} through voltage-sensitive channels, leading to the release of catecholamines. Kitayama, Kóyama, Morita et al. (1986) showed that GABA facilitated the calcium uptake associated with the increase of catecholamine release in cultured bovine adrenal chromaffin cells. Both effects of GABA, on calcium uptake and on catecholamine release, were blocked by bicuculline and picrotoxin and decreased by, nifedipine, a calcium channel blocker. This supported their proposal that activation of the GABA receptor on adrenal chromaffin cells facilitates calcium influx through voltage-sensitive calcium channels, leading to the release of catecholamines.

The modulation of the $GABA_A$ receptor by alphaxalone was investigated by Cottrell, Lambert and Peters (1987) using voltage-clamp recordings from bovine adrenal chromaffin cells. This short-acting steroid general anesthetic reversibly potentiated, in a dose-dependent manner, the amplitude of membrane currents elicited by locally applied GABA. An isomer of alphaxalone did not potentiate GABA-induced currents but, at higher concentrations, the isomer suppressed the amplitude of currents elicited either by GABA or acetylcholine. Because the isomer is not an anesthetic, this suggests that the effect of alphaxalone on GABA-evoked membrane currents may not be important in producing anesthesia.

In a continuation of this line of research, Callachan, Cottrell, Hather et al. (1987) examined the modulation of the $GABA_A$ receptor on bovine chromaffin cells by progesterone metabolites. Their results suggest that certain naturally occurring steroids potentiate the actions of GABA and directly activate the $GABA_A$ receptor.

Martinez, Gimenez, Castro et al. (1987) suggested that both $GABA_A$ and $GABA_B$ receptors are present on bovine adrenal chromaffin cells. Under their experimental conditions, they found a displacement of tritiated GABA by baclofen.

Hamann, Desarmenien, Desaulles et al. (1988) investigated the effects of an arylaminopyridazine derivative of GABA on dose-response curves for GABA-induced depolarizations from neurons and excised membrane patches from bovine chromaffin cells. The derivative did not alter the mean open time of GABA-activated channels and did not introduce further short closing gaps within the bursts. These and other results might be of importance for further studies of presynaptic GABA actions on transmitter release.

The effect of several ligands and calcium ions on the binding of tritiated GABA to bovine adrenal medulla membranes was investigated by Castro, Oset-Gasque, Cañadas et al. (1988). With no blockade, the GABA binding showed two components: one of low affinity, and one of high affinity. Muscimol specifically blocked low-affinity sites, and baclofen blocked high-affinity sites. Calcium ions were necessary for maximum binding to high-affinity sites but did not significantly affect the low-affinity sites. These results indicate that bovine adrenal medulla has $GABA_A$ receptor population of low affinity together with a $GABA_B$ receptor population of high affinity.

The occurrence and distribution of specific tritiated-muscimol binding sites, most probably identical with $GABA_A$ receptors, were studied in sections of the rat adrenal gland by light microscopic autoradiography by Amenta, Collier, Erdö et al. (1988). Specific binding was found primarily in association with chromaffin cells. A limited number of binding sites also were found in the adrenal cortex. In anesthetized hexamethonium-pretreated rats, intravenous administration of GABA produced a series of excitatory cardiovascular effects that were mimicked by intravenous administration of muscimol but not baclofen. These findings indicated the presence on adrenal chromaffin cells, of GABA receptor sites whose excitation can produce changes in cardiovascular function.

Modulation of the $GABA_A$ receptor on bovine adrenal chromaffin cells by depressant barbiturates and pregnane steroids was examined by Peters, Kirkness, Callachan et al. (1988). Their data support the concept that the $GABA_A$-mimetic and potentiating actions of barbiturates and steroids are similar. However, their results obtained with combinations of steroids and barbiturates in the ligand-binding assay appear to be inconsistent with the two classes of compounds interacting with a common site to modulate the $GABA_A$ receptor activity.

OPIOID PEPTIDE RECEPTORS

Malhotra and Wakade (1987b) used nicotinic and muscarinic receptor antagonists to investigate noncholinergic receptors in the secretion of catecholamines evoked by stimulation of the splanchnic nerve in the rat. The antagonists had a strong inhibitory effect at high frequencies of nerve stimulation, but they were not as effective at lower frequencies. The secretion of catecholamines that remained after blockade of cholinergic receptors was almost completely inhibited by naloxone. It was concluded that neurally evoked secretion of catecholamines is mediated by acetylcholine and a noncholinergic substance(s); the contribution of noncholinergic substance(s) predominates at low neuronal activity whereas that of acetylcholine is maximal at high neuronal activity. Blockade of the noncholinergic component by naloxone suggests that an opioid peptide may be involved in the secretion of catecholamines in the rat adrenal medulla.

Critchley, MacLean and Ungar (1988) stimulated the distal end of the cut left splanchnic nerve in anesthetized dogs while collecting the venous affluent of the left adrenal gland for measurement of secreted catecholamines. The resting output of catecholamines was inhibited by low-frequency stimulation but was augmented at high frequencies. The inhibition of catecholamine output by low-frequency stimulation was reversed by opiate antagonists (naloxone and nalmefene) but enhanced by angiotensin-converting enzyme inhibitors (captopril and enalapril). Their results fit a model in which both resting output of catecholamines and neurally evoked release are regulated by a balance of excitatory and inhibitory modulators.

Kimura, Katoh and Satoh (1988) examined how opioid agonists and antagonists modify the splanchnic nerve stimulation-induced release of catecholamines from the dog adrenal medulla *in vivo* in an attempt to elucidate whether opioid receptors play a functional role in controlling catecholamine release. [Leu]enkephalin and morphine attenuated the increase in catecholamine output induced by low-frequency nerve stimulation without affecting the basal catecholamine output. Nerve stimulation-induced enhancement of catecholamine secretion was markedly increased by naloxone and by naltrexone. Basal catecholamine output was unaffected by both these agents. These results suggested that endogenously released opioid peptides inhibit the release of catecholamines by activating opioid receptors in the canine adrenal gland.

The modulation of bovine adrenal medullary secretion by the opiate agonist etorphine and the opiate antagonist diprenorphine was investigated by Barron and Hexum (1986b). Etorphine inhibited the spontaneous outflow of [Met]enkephalin by approximately 10% but had no significant effect on the spontaneous release of catecholamines. Etorphine significantly decreased the secretory effect of acetylcholine. Diprenorphine had no significant effect on the spontaneous release of either enkephalins or catecholamines, but it reversed the inhibition of release caused by etorphine. Barron and Hexum (1986b) concluded that this supported the contention that opiates modulate the secretion of catecholamines and enkephalins from the adrenal gland.

Livett and Marley (1986) studied the effect of opioid peptides and morphine on histamine-induced catecholamine secretion from cultured bovine adrenal chromaffin cells. Dynorphin, metorphamide, morphine, and diprenorphine each had no effect on the catecholamine secretion induced by histamine. The histamine-induced secretion was calcium-dependent and similar to that reported for other species. The results with opioid peptides and morphine suggest that endogenous adrenal opioid peptides do not act on the opioid binding sites found on adrenal chromaffin cells to modify their secretory response to histamine.

Marley and Bunn (1988) investigated the ability of opioid compounds to modify responses of cultured bovine adrenal chromaffin cells to angiotensin II. They used etorphine as the opioid agonist because it acts at several opioid receptor subtypes known to be on bovine adrenal cells. Although angiotensin II increased basal catecholamine secretion from cultured cells and stimulated the basal accumulation of inositol phosphates, opioid agonists had no effect on catecholamine secretion induced by angiotensin II and, at appropriate concentrations, had no effect on angiotensin II-induced inositol phosphate accumulation. It was pointed out that the functions of adrenal medullary opioid receptors remain to be elucidated.

Bunn, Marley and Livett (1988a) used autoradiography to examine the distribution of opioid binding subtypes in the bovine adrenal gland. Specific opioid binding sites were restricted to the adrenal medulla. Kappa sites were highly concentrated over nerve tracts but also were found over adrenal chromaffin tissue, mostly at the periphery of the medulla. Delta opioid sites were selectively localized in the central region of the medulla. New opioid sites were low in number and distributed throughout the adrenal medulla.

A technique in which nonmodified opioid peptides were photoactivated and bound in a nondissociable manner to opioid sites in the bovine adrenal medulla was reported by Cantau, Bourhim, Giraud et al. (1987). They found that, under certain conditions of ultraviolet irradiation and a specific concentration of proteins, etorphine could be cross-linked to opioid binding sites. Other opioid peptides tested could not be cross-linked. They suggested that covalent binding of nonmodified opioid peptides would be of interest for the biochemical characterization of binding sites.

Murase, Kamikubo, Murayama et al. (1987a, 1987b) characterized adrenal medullary opioid receptors. Both studies used a plasma membrane fraction obtained from the bovine adrenal medulla. In the first study, Murase et al. (1987a) found that a tritiated enkephalin and tritiated diprenorphine bound to the membranes with high affinities. The results from a large series of experiments suggest that bovine adrenal medullary membranes contain high-affinity, stereospecific opioid receptors and that the binding of opioids to the receptors is influenced by cations. They also found opioid receptors on human pheochromocytoma cells. Murase et al. (1987b) studied the possible coupling of opioid receptors to guanosine triphosphate-binding proteins to clarify the mechanism of opioid binding to bovine adrenal medullary membranes. Their results suggested that opioid receptors are coupled to pertussis toxin-sensitive guanosine triphosphate-binding protein which may not influence adenylate cyclase.

Marley, Mitchelhill and Livett (1986a) reported that the peptide metorphamide is 100 times more potent than [Met]enkephalin in inhibiting catecholamine secretion from isolated bovine adrenal chromaffin cells. Exocytosis was monitored by measuring adenosine triphosphate release or catecholamine release. Naloxone and diprenorphine failed to antagonize the inhibitory action of metorphamide on nicotine-induced catecholamine release. Marley et al. (1986a) suggested that the novel amidated COOH terminus of this peptide is important for its action.

Eighteen endogenous opioid peptides, all containing the sequence of either [Met^5]- or [Leu^5]enkephalin, were tested by Marley, Mitchelhill and Livett (1986b) for their ability to modify nicotine-induced secretion from bovine chromaffin cells. ATP released from suspensions of freshly isolated cells was measured as an index of secretion. They collected evidence that suggested that the actions of some opioid peptides are nonspecific or that there are still undescribed receptors on the adrenal chromaffin cell.

Carydakis, Bourhim, Giraud et al. (1986) reported on the interaction of three tricyclic antidepressant drugs (clomipramine, imipramine, and amitryptyline) with δ, μ, and κ opioid binding sites in the bovine adrenal medulla. Clomipramine was the only drug that interacted with δ and μ sites. All three drugs showed a significant interaction with subtypes of the κ binding site. These and other results led Carydakis et al. (1986) to suggest that an interaction between tricyclic antidepressants and opioid binding sites might be the basis of their analgesic action.

Liebisch, Bommer, Schimak et al. (1987) confirmed that the nicotine-induced release of catecholamines from bovine adrenal chromaffin cells is inhibited by amidorphin. They also found that the peptide inhibited the release of opioid peptides from the cells. They demonstrated the inhibitory potency of the COOH-terminal portion of amidorphin and also showed that the secretion of norepinephrine induced by histamine was not inhibited by amidorphin.

Lieber and Oehme (1986) showed that naloxone diminished the increase in the basal norepinephrine outflow in normotensive and spontaneously hypertensive rats. The data suggest an interaction between opioid peptides and the basal catecholamine outflow during maintenance of hypertension.

Mannelli, Maggi, DeFeo et al. (1986) studied the effects of placebo or naloxone administration on plasma catecholamine levels in normal patients and 15 hypertensive patients suspected of having a pheochromocytoma. In normal subjects, naloxone caused a significant increase of epinephrine secretion, whereas norepinephrine did not change. Similarly, in the group of hypertensive patients, epinephrine secretion increased after naloxone. These and other results suggest the catecholamine secretion from normal and pathological chromaffin tissue is modulated by endogenous opioids; this modulation seems particularly evident in patients with epinephrine-secreting pheochromocytomas.

Binding of human β-endorphin to bovine adrenal medullary membranes was characterized by Kamikubo, Murase, Murayama et al. (1986) as time-dependent, saturable, and stereospecific. Their overall results suggested that binding of β-endorphin to membranes of bovine adrenal medulla consists of a high-affinity opioid-sensitive component and a low-affinity non-opioid component. The non-opioid component of the binding may be related to the COOH terminus of the β-endorphin molecule.

By using specific agonists and antagonists, Kamikubo, Murase, Niwa et al. (1986, 1987) showed that adrenal medullary δ-opioid receptors apparently are linked to islet-activating protein-sensitive guanosine triphosphate-binding proteins that are not directly coupled to adenylate cyclase.

Kamikubo, Murase, Murayama et al. (1987) solubilized adrenal medullary opioid receptors with digitonin and characterized the solubilized receptors. The binding of tritiated diprenorphine to the solubilized material was rapid and saturable. Kinetic studies of this binding indicated that active opioid receptors can be solubilized with digitonin from bovine adrenal medullary membranes.

The opioid-binding activity of a digitonin extract of bovine adrenal medullary membranes was studied by Kamikubo, Murase, Murayama et al. (1987). Binding of tritiated diprenorphine to the solubilized material was rapid and saturable. Several opioids displaced dynorphin. The complex of diprenorphine and the solubilized binding sites could be eluted and showed an apparent molecular mass of 200 kDa. These results indicate that active opioid receptors can be solubilized with digitonin from adrenal medullary membranes.

An extensive review of opioid receptors and opioids in peripheral tissues was published by Hedner and Cassuto (1987). They pointed out that opioids are found in many peripheral organs, including the adrenal medulla. Opioid receptors are also found to be widely distributed, particularly in the digestive tract and on various cells in the immune system.

Evidence for disulfide bonds in membrane-bound and solubilized opioid receptors from bovine adrenal medulla was repeated by Kamikubo, Murase, Murayama et al. (1988). Pretreatment of membranes with dithiothreitol or mercaptoethanol inhibited tritiated diprenorphine binding by decreasing the number of binding sites. This inhibitory action was time- and dose-dependent. These and other results indicate that disulfide bonds are involved in opioid binding activity of the opioid receptor system.

ATRIAL NATRIURETIC FACTOR RECEPTORS

The cardiac atria of mammals contain peptides, referred to as atrial natriuretic factors, that cause vasodilation and natriuresis. Lynch, Braas and Snyder (1986) used autoradiography to localize atrial natriuretic factor receptors in rat adrenal gland, kidney, and brain. Binding sites were localized within the adrenal cortex but not within the adrenal medulla. Chai, Sexton, Allen et al. (1986), using autoradiography, also found receptors in the adrenal cortex of the rat but not within the adrenal medulla. Examining the bovine adrenal gland by an *in vitro* autoradiographic technique, Higuchi, Nawata, Kato et al. (1986) demonstrated binding of α human atrial natriuretic factor in the adrenal cortex but not the medulla.

In contrast, Fuchs, Shigematsu and Saavedra (1986) did localize atrial natriuretic factor receptors within the adrenal medulla of the tree shrew (*Tupaia belangeri*). They suggested a possible role for atrial natriuretic factor or related peptides in the control of catecholamine secretion from the adrenal medulla of this primate and possibly other primates.

Heisler and Morrier (1988) demonstrated that bovine adrenal chromaffin cells contain functional atrial natriuretic peptide receptors. In binding studies, they found that bovine adrenal medulla membranes had a single class of high-affinity binding sites. Addition of the peptide to cells in culture resulted in a concentration-dependent increase in cyclic guanidine monophosphate (cGMP) synthesis. They suggested that the atrial natriuretic peptide receptors have a function in adrenal chromaffin cells.

Maurer and Reubi (1986a) used autoradiography to localize receptors for atrial natriuretic factor, angiotensin II, and somatostatin within the adrenal glands of rat, mouse, hamster, rhesus monkey, guinea pig, and cow. Some binding of atrial natriuretic factor was seen in the adrenal medulla of the guinea pig but not of the other animals. Rat, mouse, and hamster had an accumulation of receptors for angiotensin II in the adrenal medulla. Somatostatin sites were localized within the adrenal medulla of the rat. Maurer and Reubi (1986) suggested that these three peptides have species variability and a less-interrelated and possibly heterogeneous mode of action in the adrenal medulla as opposed to the adrenal cortex where there is less species variability and an apparent functional relationship between the peptides.

Swithers, Stewart and McCarty (1987) and Stewart, Swithers and McCarty (1987) carried out a pair of studies of binding sites for atrial natriuretic factor in adrenal glands and kidneys of four groups of rats--Dahl hypertension-sensitive, hypertension-resistant, spontaneously hypertensive, and Wistar-Kyoto normotensive rats--by quantitative autoradiography. Almost no binding was noted in the adrenal medulla in any one of these rats.

Stewart, Swithers, Plunkett et al. (1988) reviewed the current literature regarding the distribution and binding characteristics of receptor sites for atrial natriuretic factor in peripheral and central target tissues. They pointed that, although binding sites do not appear to be present in the bovine adrenal medulla, they are present in the rat adrenal medulla and the adrenal medulla of the tree shrew.

High-affinity receptors for atrial natriuretic factor were identified by Rathinavelu and Isom (1988) in PC12 cells. Radioiodinated synthetic atrial natriuretic factor was bound to a single class of high-affinity binding sites that could be characterized. Photoaffinity labeling of the receptor specifically labeled two protein bands. Atrial natriuretic factor receptors on PC12 cells could provide a unique model for the study of this receptor.

Atrial natriuretic factors were tested for their effects on cGMP production in two neurally derived cell lines, including PC12 cells, by Fiscus, Robles, Waldman et al. (1987). They found that PC12 cells respond to four different analogs of atrial natriuretic factor with increases in cellular content of cGMP. Because the additional cGMP presumably is being generated at the internal surface of the cell membrane, they also predicted that this cGMP may be readily available for export from the cells. Indeed, both cell lines tested responded to an analog of atrial natriuretic factor with large increases in cGMP efflux.

OTHER PEPTIDE RECEPTORS

Although chromogranin A was found to have no direct effect on secretion from cultured bovine adrenal chromaffin cells, Simon, Bader and Aunis (1988) demonstrated that secretion from these was controlled by peptides derived from chromogranin A. This observation potentially provides information as to a function for chromogranin A. It was suggested that peptides derived from chromogranin A could exert a delayed feedback control on chromaffin cell secretory activity. This mechanism may be important during stress situations when desensitization could render the adrenal medulla functionless.

Schwartz, Sheikh and O'Hare (1987) identified receptors on PC12 cells for neuropeptide Y and pancreatic polypeptide. These two peptides have a distinct tertiary structure. They demonstrated that PC12 cells are suitable for structure-function studies of peptides with this tertiary structure and studies on the cellular events following binding of similar peptides.

Neuropeptide Y, which is known to be stored within chromaffin vesicles of adrenal chromaffin cells, was shown by Dahlöf, Persson, Lundberg et al. (1988) to inhibit the release of epinephrine and norepinephrine that is evoked by preganglionic nerve stimulation in the pithed rat. Systemic infusion of neuropeptide Y significantly decreased the stimulus-induced increase in plasma epinephrine. However, the increase in plasma norepinephrine was significantly decreased only when two periods of stimulation were administered. It was concluded that preganglionic nerve-stimulation-induced release of epinephrine and norepinephrine is under inhibitory control of neuropeptide Y. The inhibition of norepinephrine release from sympathetic nerve terminals is most probably due to a prejunctional site of action whereas the decrease in secretion of epinephrine could be due to a local effect of neuropeptide Y in the adrenal medulla.

The possible role of neuropeptide Y in catecholamine secretion was studied in bovine adrenal chromaffin cells by Higuchi, Costa and Yang (1988). Neuropeptide Y produced a concentration-dependent inhibition of nicotine-stimulated release of norepinephrine and epinephrine. A structurally related peptide, human pancreatic polypeptide, showed a similar inhibitory effect on catecholamine release, but other peptides had little or no effect. It was suggested that neuropeptide Y has a modulatory role on catecholamine secretion, possibly acting through a specific receptor.

Neuropeptide Y and pancreatic polypeptide belong to a family of regulatory peptides that have a distinct tertiary structure referred to as the PP-fold. Schwartz, Sheikh and O'Hare (1987) discovered high-affinity receptors specific for these two PP-fold peptides on PC12 cells. The affinity for pancreatic polypeptide was 100 times higher than that for neuropeptide Y. They concluded that PC12 cells are suitable for structure-function studies of PP-fold peptides and studies on the cellular events that occur after cells bind these peptides.

Nieber and Oehme (1987) examined the effect of substance P and an analog of substance P on presynaptic and postsynaptic release in rat adrenal gland slices. Substance P and its analog inhibited the evoked release of preloaded tritiated norepinephrine in a dose-dependent manner. They also inhibited the outflow of tritiated acetylcholine. The authors concluded that the NH_2 terminus of substance P has an important role in the effect of substance P during prolonged stimulation, such as during periods of stress.

Nieber, Oehme, Arefolov et al. (1988) noted that two structurally related tetrapeptides, the amino-terminal tetrapeptide of substance P and tuftsin, had some similarities in modifying immune reactions and normalizing stress-induced disorders in the adrenal medulla. Both peptides inhibited nicotine-evoked catecholamine outflow, the substance P-derived peptide being more effective. The substance P-derived peptide also inhibited the electrically stimulated release of acetylcholine whereas tuftsin did not. It was suggested that the regulation of the cholinergic-adrenergic interaction in adrenals is mediated by specific receptors that are different on the presynaptic and postsynaptic sides.

Richter and Grunwald (1987) examined the effect of substance P on the membrane potential of adrenal chromaffin cells in slices of rat adrenal medulla. Substance P caused a significant depolarization of the chromaffin cell membrane at relatively low concentrations. Their results suggest that substance P may have a direct excitatory effect on adrenal chromaffin cell membranes.

Khalil, Marley and Livett (1988b) studied the modulatory actions of two members of the tachykinin family, neurokinin A and B, on endogenous catecholamine secretion from cultured bovine adrenal chromaffin cells. Both neurokinin A and B were found to have two distinct actions, similar to substance P, on nicotine-induced catecholamine release. One was an inhibitory action at low nicotine concentrations and the other was a protective action against desensitization by high nicotine concentrations. On a molar basis, the efficacy of neurokinin A or B was 1/30th of that of substance P. When they tested the ability of a substance P antagonist to antagonize the modulatory actions of substance P on the nicotinic response, the results suggested the possibility that the actions of substance P on bovine adrenal chromaffin cells may be mediated through two receptor subtypes of two different affinities.

The effects of substance P and an amino-terminal analogue on inositol phospholipids in rat adrenal medulla slices were examined by Minenko, Tjulkova and Oehme (1988). The analogue had a substantially smaller effect on radioactive phosphorus incorporation into inositol phospholipids than did substance P. It was assumed that the moderate stimulation of the hydrolysis of inositol phospholipids by the substance P analogue is sufficient to exert an influence on the adrenal medulla.

Khalil, Marley and Livett (1988) reassessed substance P for its ability to modify nicotine-induced catecholamine secretion from cultured bovine adrenal chromaffin cells. Substance P exhibited biphasic effects in its inhibition of the nicotinic secretory response and protection against desensitization.

Malhotra and Wakade (1987a) examined the effects of [Met]enkephalin and vasoactive intestinal polypeptide (VIP) on the perfused adrenal gland of the rat. [Met]Enkephalin increased basal release of catecholamines and potentiated acetylcholine-evoked release. VIP caused a significant increase in the secretion of catecholamines that was not affected by antagonists of nicotinic and muscarinic receptors or by denervation. However, VIP-evoked release was calcium-dependent and could be decreased by naloxone. Malhotra and Wakade (1987a) suggested that VIP may be the noncholinergic excitatory substance present in the splanchnic nerves and released along with acetylcholine during stimulation of nerves to evoke secretion of catecholamines from the rat adrenal chromaffin cells.

Malhotra, Wakade and Wakade (1988b) investigated the molecular mechanism involved in the exocytotic secretion of catecholamines by VIP and the effects of this peptide on radioactive calcium uptake and phosphoinositide breakdown. The omission of calcium from the perfusion medium had almost no effect on vasoactive intestinal peptide-induced secretion; however, the addition of a calcium chelator abolished secretion. Stimulation with the peptide did not result in a net increase in calcium uptake, and uptake was not modified by a protein kinase C activator. All of these effects of vasoactive intestinal peptide were similar to those of muscarine. These and other results led to the conclusion that the inositol 1,4,5-triphosphate generated on activation of vasoactive intestinal peptide and muscarinic receptors is linked to exocytotic secretion of adrenal catecholamines through release of internal calcium ions.

The effects of VIP on catecholamine secretion from isolated guinea pig adrenal chromaffin cells were studied by Misbahuddin, Houchi, Nakanishi et al. (1986). VIP alone induced only a slight secretion of catecholamines but did potentiate acetylcholine-induced secretion. VIP appeared to stimulate muscarine-induced catecholamine secretion but not nicotine-induced secretion.

Misbahuddin, Oka, Nakanishi et al. (1988) studied the stimulatory effect of VIP on catecholamine secretion from isolated guinea pig adrenal chromaffin cells. The peptide induced a dose-dependent catecholamine secretion that was slow and continued for at least 30 minutes. This secretion was dependent on the presence of calcium in the medium, but no increase in calcium uptake by the cells could be measured. Acetylcholine and muscarine induced a marked increase in the free intracellular calcium level but that VIP induced only a slight increase. Apparently, vasoactive intestinal polypeptide induces catecholamine secretion by increasing the sensitivity of catecholamine secretion to calcium level.

In a subsequent study, Misbahuddin, Oka, Nakanishi et al. (1988) showed that VIP induced a dose-dependent catecholamine secretion that was slow and continued for at least 30 minutes. This secretion was dependent on the presence of calcium in the medium, but no increase in calcium uptake by the cells could be measured. Acetylcholine and muscarine induced a marked increase in the free intracellular calcium level but that VIP induced only a slight increase. Apparently, VIP induces catecholamine secretion by increasing the sensitivity of catecholamine secretion to calcium levels.

The characteristics of the bradykinin-evoked secretory response in the perfused rat adrenal medulla were studied by Warashina, Fujiwara, Hirano et al. (1988). Bradykinin increased catecholamine response, but this decreased over time due to desensitization. The stimulatory effect of bradykinin was removed by a washout period, but the desensitization persisted for longer periods. The secretory effect appeared to be only partially dependent on calcium in the external medium. A calcium channel blocker did not affect bradykinin-evoked secretion. These results suggest that bradykinin leads to catecholamine secretion by a pathway different from that for acetylcholine.

Using tritiated bradykinin, Kozlowski, Rosser and Hall (1988) identified binding sites in PC12 cells. The binding characteristics paralleled those of the B2 bradykinin receptor. Exposure to nerve growth factor significantly increased the number of bradykinin binding sites, but other characteristics of the binding remained unchanged.

Plevin and Boarder (1988) demonstrated that histamine, bradykinin, and angiotensin II stimulate the production of inositol phosphates in bovine adrenal chromaffin cells. The H1 receptor appeared to be implicated in the response to histamine, whereas the BK-2 receptor was implicated in the response to bradykinin. Bradykinin gave the greatest stimulation of inositol triphosphate production, whereas histamine gave a larger inositol monophosphate accumulation. Angiotensin II and carbachol also increased phosphoinositide turnover.

Somatostatin receptors have been visualized in the rat adrenal gland by Maurer and Reubi (1986b). They used autoradiography with an iodinated derivative of a somatostatin analog and found that somatostatin receptors are not exclusively restricted to the zona glomerulosa but also can be found in the adrenal medulla. The sites in the medulla have a higher affinity than those in the cortex. This phenomenon is species-specific because they found no binding sites for somatostatin in the cow and guinea pig adrenal medulla and binding sites only in the adrenal medulla in the hamster, mouse, and rhesus monkey.

Reubi, Maurer, von Werder et al. (1987) examined human endocrine tumors for somatostatin receptors. Although some tumors contained somatostatin receptors, pheochromocytoma, a tumor of the adrenal medulla, did not.

OTHER RECEPTORS

The possible interaction of tritiated glutamate binding sites with anion channels in rat neural tissues, including the adrenal medulla, was investigated by Yoneda, Ogita, Nakamuta et al. (1986). The adrenal was found to have low but measurable amounts of binding activity. In a study more specifically directed toward the adrenal medulla of the rat, Yoneda and Ogita (1986) localized tritiated glutamate binding sites. The binding to membranes from the adrenal medulla was 5 times greater than that to membranes from the adrenal cortex. The characteristics of binding in the adrenal medulla were different from those in the central nervous system.

The enhancement of tritiated glutamate binding by N-methyl-D-aspartic acid in the rat adrenal gland was investigated by Yoneda and Ogita (1987a). They found that addition of this amino acid, one of the agonists for central excitatory amino acid neurotransmitter receptors, induced a significant augmentation of the adrenal binding in a concentration-dependent manner. They performed a series of experiments that suggested that the rat adrenal may contain glutamate binding sites that are distinctly different from those in the central nervous system. They were able to solubilize these binding sites from rat adrenal glands by using various detergents (Yoneda and Ogita, 1987b).

Saavedra and Alexander (1986) quantified angiotensin II in several neural tissues including the adrenal medulla. They localized binding sites within the adrenal medulla and found that the number of binding sites was decreased in sino-aortic denervated rats. They suggested that peripheral mechanisms involving angiotensin II receptors may be associated with the pathophysiology of neurogenic hypertension.

Printz and Boyd (1986) examined differences in the angiotensin II receptors in a tissue from the central nervous system and in cultured bovine adrenal chromaffin cells. The receptor on the chromaffin cells was functional as demonstrated by both catecholamine and chromogranin A secretion studies. Studies with tissues from the central nervous system indicated that a complex system is operational.

Using radioautography, Bianchi, Gutkowska, Charbonneau et al. (1986) detected the presence of binding sites for angiotensin II in both norepinephrine- and epinephrine-containing cells in the rat adrenal medulla. Electron microscopy showed that angiotensin II binds to the cell surface of norepinephrine-containing cells, is progressively internalized, and is associated with lysosomes and Golgi complex within 20 minutes, whereas in epinephrine-containing cells the angiotensin II seems to be internalized earlier and recycled back to the cell surface within 5 minutes. Their results suggest different intracellular pathways for angiotensin II in the two cell types of the rat adrenal medulla.

Binding sites for angiotensin II and angiotensin III were detected in the rat adrenal medulla by Himeno, Nazarali and Saavedra (1988). Using quantitative *in vitro* autoradiographic techniques, they characterized these binding sites and suggested that angiotensin III and angiotensin II may share the same binding sites in the adrenal gland. Their results support the hypothesis that angiotensin III in the adrenal medullae has a role as well as in the zona glomerulosa of the adrenal cortex.

Brust and Printz (1988) examined whether the bovine adrenal chromaffin cell receptor for angiotensin II responds to angiotensin III. They showed that angiotensin III is nearly as active with angiotensin II as a secretagogue in these cultured cells.

Bommer, Liebisch, Kley et al. (1987) demonstrated that histamine is a potent secretagogue for opioid peptides in adrenal chromaffin cells *in vitro*. The release of the peptides was calcium-dependent. Moreover, histamine also produced a profound compensatory increase in cell peptide content after 48 hours of exposure. These histamine-induced effects were antagonized by an H_1-receptor antagonist whereas H_2-receptor antagonists were less effective.

Noble, Bommer, Liebisch et al. (1988) reported that histamine as well as other neuroactive substances stimulated the release of catecholamines from bovine adrenal chromaffin cells. Histamine stimulation was found to be dose-dependent and occurred through activation of H_1 histaminergic receptors. In contrast to cholinergic nicotinic activation, histamine initially induced a small release of catecholamines but, with prolonged exposure, it produced a much greater response than nicotine did. In addition, the H_1-receptor showed little desensitization over time. This histamine-induced release was found to be calcium-dependent and was attenuated by calcium channel blockers. These studies suggest that histamine as well as certain other neuroactive substances could play an important role in the physiology and biochemistry of adrenal chromaffin cells.

Macquin-Mavier, Clerici, Franco-Montoya et al. (1988) assessed the respective contributions of adrenal and extra-adrenal mechanisms to the increase in sympathetic activity elicited by histamine. Plasma epinephrine measurements were used as an index of sympathoadrenal activity. In intact guinea pigs, histamine infusion caused an increase in plasma epinephrine level. In guinea pigs that had undergone chemical sympathectomy with 6-hydroxydopamine, the increase in plasma epinephrine induced by histamine was distinctly less. This increase was blocked by the ganglion-blocking agent hexamethonium and by pithing in guinea pigs. Pretreatment with the H_2-receptor agonist cimetidine or the H_1-receptor antagonist mepyramine decreased the increase in plasma epinephrine. These results suggest that, in guinea pigs, epinephrine is released not only from the adrenal medulla but also from nerve endings and that histamine releases epinephrine by indirect action through central reflex pathways.

Ruff and Santais (1988) reviewed the H_1 and H_2 histamine receptors. Among many effects relating to allergic reactions, histamine increases secretion from the adrenal medulla.

The effect of benzodiazepines on evoked catecholamine release from cultured bovine adrenal chromaffin cells was investigated by Kitayama, Morita, Dohi et al. (1988). Midazolam inhibited the catecholamine release evoked by acetylcholine, high potassium, and veratridine but not that evoked by the ionophore A23187 or caffeine in calcium-free media. Other benzodiazepines also inhibited acetylcholine-evoked catecholamine release but only at high concentrations. Additional results suggested that benzodiazepines at high doses inhibit evoked catecholamine release from adrenal chromaffin cells, possibly through a blockade of calcium influx.

Negishi, Ito, Tanaka et al. (1987) found that prostaglandin E_2 bound to a specific fraction of membranes prepared from the bovine adrenal medulla. They were able to determine that the prostaglandin E_2 receptor is a glycoprotein with an approximate molecular mass of 110 kDa. Their results also suggest that the prostaglandin E_2 receptor may be functionally associated with a pertussis toxin- insensitive guanosine triphosphate-binding protein and is not coupled to the adenylate cyclase in the bovine adrenal medulla.

Negishi, Ito, Yokohama et al. (1988) described the purification of a novel pertussis toxin substrate G-protein, named G_{am}, from bovine adrenal medulla. They also reported the functional reconstruction of partly purified prostaglandin E receptor and pure G_{am}, G_i, or G_o and phospholipid vesicles and demonstrated that the interaction of receptor with G-protein can induce both a high-affinity agonist-binding state in the receptor and an agonist-stimulated GTPase activity of G-protein. These and other results indicate that the prostaglandin E receptor can couple functionally with G_{am}, G_i, or G_o in phospholipid vesicles and suggest that G_{am} may be involved in signal transduction of the prostaglandin E receptor in bovine adrenal medulla.

Yokohama, Tanaka, Ito et al. (1988) studied the possibility that the specific binding of prostaglandin E_2 to adrenal medullary membranes could be related to the regulation of catecholamine release by using cultured bovine adrenal chromaffin cells. Prostaglandin E_2 induced catecholamine release in the presence of ouabain, an inhibitor of Na^+, K^+-ATPase. This finding is somewhat surprising in light of reports that prostaglandin E_2 inhibits catecholamine release induced by various stimuli in adrenal glands. They found that the prostaglandin E_2 receptor, like the muscarinic receptor, is linked to phosphoinositide breakdown with the resultant formation of inositol triphosphate and mobilization of intracellular calcium. They also suggested that protein kinase C may be activated by diacylglycerol formed simultaneously with inositol triphosphate and is involved in secretion of catecholamines from chromaffin cells.

Yokohama, Negishi, Sugama et al. (1988) went on to show that prostaglandin E_2-induced phosphoinositide metabolism is blocked by pretreatment with a phorbol ester. Using intact cultured bovine chromaffin cells, they also examined the inhibitory effect of this phorbol ester on the individual steps of the activation process of phosphoinositide metabolism. This effect was seen with a phorbol ester that activated protein kinase C but not with an inactive phorbol ester. Their results suggested that protein kinase C serves as a feedback regulator for prostaglandin E_2-phosphoinositide metabolism. The site of action of phorbol ester appears to be distal to the coupling of the receptor to GTP-binding protein but on a component specific to the agonist-induced phosphoinositide metabolism.

The effects of prostaglandins on catecholamine secretion and calcium ion fluxes were studied in a primary culture of bovine chromaffin cells by Koyama, Kitayama, Dohi et al. (1988). Several forms of prostaglandin-induced catecholamine release from cultured cells as well as enhanced release induced by catecholamines, high level of potassium, veratridine, and A23187. This induction was calcium-dependent. Prostaglandin increased radioactive calcium uptake and showed an additive effective with acetylcholine on calcium uptake. These and other results suggest that prostaglandins enhance basal and stimulation-evoked catecholamine release from chromaffin cells, possibly through facilitation of calcium ion influx.

The effects of opioid compounds on catecholamine secretion and phosphatidylinositol turnover induced by prostaglandins E_1 and E_2 in cultured bovine adrenal chromaffin cells were studied by Marley, Bunn and Livett (1988). Prostaglandin E_2 was more potent than E_1 in inducing catecholamine secretion in a calcium-dependent manner. Neither etorphine nor diprenorphine affected the catecholamine secretion induced by either prostaglandin. Prostaglandin E_2 was more effective than E_1 in increasing phosphatidylinositol turnover, and etorphine and diprenorphine had no effect on this turnover. The results indicate that prostaglandins can facilitate catecholamine secretion independently of their effects on phosphatidylinositol metabolism. They also indicate that endogenous adrenal opioid peptides do not act on the opioid binding sites on adrenal chromaffin cells to modify their responses to prostaglandins.

Yamada, Morita, Dohi et al. (1988) demonstrated that prostaglandins E_1, E_2, F_2, and D_2 potentiated the acetylcholine-evoked catecholamine release from perfused dog adrenal glands in a dose-dependent manner. These prostaglandins alone slightly increased catecholamine release. Additional experiments suggested that the mechanism of action involves the effect of calcium ion flux across the plasma membrane. The authors suggested that prostaglandins formed endogenously in response to physiologic stimulation of chromaffin cells could play a facilitating modulatory role in catecholamine release.

Engeland, Lilly, Bruhn et al. (1987) assessed the possibility that corticotropin-releasing factor is important in the control of adrenal catecholamine secretion. They developed a preparation that permits measurement of adrenal venous output *in vivo* in response to arterial injection into the adrenal gland of the dog. Adrenal injection of corticotropin-releasing factor at relatively high concentrations stimulated epinephrine and norepinephrine secretion. The response was only 1% as sensitive as that to acetylcholine. They showed that the releasing factor has an effect on the adrenal medulla but is probably not a physiologically important secretagogue for catecholamines.

Chern, Kim, Slakey et al. (1988) showed that adenosine can exert a potent stimulatory effect on adenylate cyclase and can enhance secretion from bovine adrenal chromaffin cells in the presence of forskolin. Synergism between adenosine and forskolin is seen in both the stimulation of secretion and the increase in cAMP levels in the cells. However, an adenine agonist increases the cellular cAMP content in the presence of forskolin without having any positive effect on secretion. This finding suggests that the increase in cAMP level may not be the sole cause of the increase in secretion stimulated by adenosine.

Insulin binding was studied in subpopulations of bovine adrenal chromaffin cells by Serck-Hanssen and Søvik (1987). Using radioiodinated insulin, they showed that chromaffin cells have high-affinity binding sites for insulin and these binding sites are mainly on epinephrine-containing cells. In a following study, Serck-Hanssen, Søvik and Lie (1988) characterized insulin receptors in isolated bovine adrenal chromaffin cells. The cells were incubated with radioiodinated insulin, and specific binding was found to be linearly related to the number of cells. There apparently is a single class of noninteracting receptors with about 1,700 receptors per cell. The apparent molecular mass of the insulin-binding subunit of the receptor was found to be 135 kDa.

Serck-Hanssen, Søvik and Lie (1988) investigated insulin receptors in isolated bovine adrenal chromaffin cells. The cells were incubated with radioiodinated insulin, and specific binding was found to be linearly related to the number of cells. There apparently is a single class of noninteracting receptors with about 1,700 receptors per cell. The apparent molecular mass of the insulin-binding subunit of the receptor was found to be 135 kDa.

The effects of islet-activating protein on catecholamine release, calcium mobilization, and inositol triphosphate formation in cultured adrenal chromaffin cells were studied by Sasakawa,

Yamamoto, Nakaki et al. (1988). Their results suggest that islet-activating protein acts on the common process of the depolarization-mediated and receptor-mediated secretory mechanisms. They concluded that the GTP-binding protein may play a role in catecholamine secretion at a step distal to calcium mobilization.

The binding characteristics of tritiated saxitoxin and its binding site were examined in bovine adrenal chromaffin cells by Wada, Arita, Kobayashi et al. (1987). These cells showed a specific binding of saxitoxin that was saturable and reversible. Experiments indicated that saxitoxin binds to a specific site on voltage-dependent sodium channels and inhibits the influx of sodium. Tritiated saxitoxin is a useful compound for analysis of voltage-dependent sodium channels in adrenal chromaffin cells.

Schmid-Antomarchi, Hugues and Lazdunski (1986) demonstrated that PC12 cells produce a large number of apamin receptors. As measured with radioiodinated apamin, these cells have at least 50 times more of these receptors than do other cell types that are known to have apamin receptors and apamin-sensitive calcium-activated potassium channels in their membranes. Schmid-Antomarchi et al. (1986) concluded that PC12 cells are an excellent system for further study of these channels.

Clark, Stumpf, Bishop et al. (1986) demonstrated binding sites for 1,25-dihydroxyvitamin D_3 within the adrenal medulla of mice. Immunocytochemical staining for phenylethanolamine N-methyltransferase revealed that both epinephrine-containing and non-epinephrine-containing cells concentrate 1,25-dihydroxyvitamin D_3 in their nuclei. Clark et al. (1986) suggested that 1,25-dihydroxyvitamin D_3 may directly affect certain functions of these adrenal medullary cells.

Williams, Abreu, Jarvis et al. (1987) characterized adenosine receptors in PC12 cells. Their findings are consistent with the selective labeling of an A_2-type adenosine receptor on these cells.

Nakata and Fujisawa (1988) showed that PC12 cell membranes have A_2-like adenosine binding sites as assessed by using the specific ligand 5'-N-ethylcarboxamide[^{3}H]adenosine. Specific binding to PC12 cell membranes at 0°C was saturable and showed a monophasic saturation profile. In contrast, binding at 30°C exhibited a biphasic profile, suggesting the presence of two specific binding sites. When these adenosine binding sites were solubilized, they retained the same ligand binding characteristics as those of the membrane-bound form. These binding sites were partially characterized.

Watanabe, Kawada, Kurosawa et al. (1988) investigated the mechanism of the enhancing effect of capsaicin on adrenal catecholamine secretion and determined how adrenal sympathetic nerves participate in this. Intravenous administration of capsaicin in anesthetized rats caused a rapid and marked increase in adrenal sympathetic nerve activity. This was a dose-dependent phenomenon. Cholinergic blocking attenuated the adrenal epinephrine secretion caused by capsaicin. The direct action of capsaicin was examined by using retrograde perfusion of the left adrenal gland with capsaicin. Capsaicin did not enhance catecholamine secretion. These results suggest that the physiologic enhancement of catecholamine secretion by capsaicin is mainly through activation of the central nervous system.

Kawada, Sakabe, Watanabe et al. (1988) studied the effects of some pungent principles from spices on adrenal catecholamine secretion as compared with the effect of capsaicin. An increase in catecholamine secretion, especially of epinephrine, was observed not only during capsaicin infusion but also during infusion of piperine (a pungent principle from pepper) or zingerone (ginger). Sulfa-containing and volatile pungent principles, such as those from mustard and garlic, did not cause any catecholamine secretion. They suggested that some pungent principles from dietary spices can induce a warming action via adrenal catecholamine secretion.

The ability of platelet-activating factor to stimulate dopamine release and modify calcium homeostasis in PC12 cells was studied by Bussolino, Tessari, Turrini et al. (1988). The ability of platelet-activating factor to release dopamine is related to its molecular form, with only the R configuration being active. Nanomolar concentrations of this form induced a significant release of dopamine, and this was related to induced changes in calcium levels.

CHAPTER 4

STIMULUS-SECRETION COUPLING: IONS

It is well-established that an intermediate step in stimulus-secretion coupling is change in intracellular ion concentrations, particularly cations, after activation of a receptor on the cell surface. This chapter will describe different aspects of these ionic events.

A tribute to the late Peter Baker, who made many significant contributions to our understanding of stimulus-secretion coupling, was published by Knight (1988). Many of the contributions of Peter Baker were reviewed, including his work with the squid giant axon and permeabilized adrenal chromaffin cells.

An overview of stimulus-secretion coupling, highlighting the contributions of Peter Baker, was offered by Rink and Knight (1988). Among other things, they discussed the development of the electropermeabilized cell which allows control of the low molecular weight components of the cytosol while leaving the exocytotic apparatus and process intact.

In an excellent series of reviews, Baker and Knight (1987a, 1987b), Baker (1987), and Knight (1987) described the role of ions in stimulus-secretion coupling, particularly in adrenal chromaffin cells. Details of the experimental technique and the role of calcium in stimulus-secretion coupling were described. The interaction between calcium and the other aspects of stimulus-secretion coupling was thoroughly examined.

Ozawa and Sand (1986) reviewed the physiology of ionic events in excitable endocrine cells, including the literature on the adrenal chromaffin cell as well as pancreatic beta cells, adenohypophyseal cells, and calcitonin-secreting cells.

Techniques used to isolate the ionic events of stimulus-secretion coupling for experimental investigation were reviewed by Knight and Scrutton (1986), including methods used to gain access to the cytosol (mainly detergents and electropermeabilization). They concluded that the electropermeabilized cell preparation is versatile and has many advantages for studying stimulus-secretion coupling, not only in adrenal chromaffin cells but also in other cells.

The control of exocytosis from adrenal chromaffin cells was reviewed by Holz (1988). He summarized by pointing out that nicotinic stimulation results in calcium ion influx. The increase in cytosolic calcium has at least three effects which are important for secretion: 1) calcium directly activates the secretory pathway through as yet unknown mechanisms; 2) calcium activates phospholipase C and may generate diglyceride; and 3) perhaps acting together with diglyceride, calcium causes the translocation and activation of protein kinase C. The activation of protein kinase C enhances calcium-dependent secretion.

The use of specific toxins to block cationic channels as a means of studying stimulus-secretion coupling was stressed in articles by Wanke, Ferroni, Gattanini et al. (1986) and Cattaneo and Grasso (1986). In both of these papers, α-latrotoxin, derived from the black widow spider, was used to study ionic channels in PC12 cells. Wanke et al. (1986) identified a channel that is different from the classic voltage- and receptor-operated channels present in PC12 cells and from the large conductances induced by the toxin in artificial lipid membranes. Cattaneo and Grasso (1986) used antibodies against the α-latrotoxin molecule and identified an antibody that inhibited toxin-stimulated dopamine release from PC12 cells, prevented toxin-induced calcium accumulation, altered toxin-dependent phosphoinositide breakdown, and prevented toxin-induced channel formation in artificial lipid bilayers. Several lines of experimental evidence led them to suggest that the effects they described resulted primarily from the blockade of an event that immediately follows binding and is central for the full expression of toxin action.

The effects of hydrostatic pressure on the function of ion channels were examined by Heinemann, Conti, Stühmer et al. (1987). This involved a patch-clamp study under high hydrostatic pressure. Calcium occurrence in bovine adrenal chromaffin cells was found to be independent of pressure effects. A slight effect on the mean amplitude and the gating kinetics of sodium currents was measured at moderate pressures. However, when exocytosis was studied at high pressure by monitoring the cell capacitance, more drastic effects were seen. According to these results, the process of exocytosis is the most likely site at which hydrostatic pressure can act to produce nervous disorders. Furthermore, it was demonstrated that pressure can be a useful tool in the investigation of other cellular responses, since Hinemann et al. (1987) were able to separate different steps occurring during exocytosis owing to their different activation volumes.

Thieffry, Chich, Goldschmidt et al. (1987a, 1987b) described a large ionic channel in subcellular fractions of bovine adrenal medulla. In this electrophysiologic study of organelle membranes, they identified a channel that is voltage-sensitive and has four levels of conductance. The identity of the organelle(s) has not yet been determined.

Thieffry, Chich, Goldschmidt et al. (1988) were able to incorporate membranes from subcellular fractions of adrenal medulla into phospholipid bilayers formed at the tip of microelectrodes. Current fluctuations recorded in the presence of a transmembrane potential revealed the existence of a voltage-dependent channel of large conductance. They were able to partially characterize this channel and demonstrated that it differs from the voltage-dependent anion channel of outer mitochondrial membranes. This channel may be a candidate for a separate class of undescribed channels in the mitochondrial membrane from bovine adrenal medulla.

CALCIUM IONS

Several authors have considered the role of calcium in stimulus-secretion coupling. Pollard, Brocklehurst, Forsberg et al. (1987) reviewed the regulation of membrane fusion by calcium during secretion by adrenal chromaffin cells and beta cells of the pancreas. Among other things, they pointed out that between the intracellular changes in calcium and the final secretory events is a set of as yet poorly understood intermediate events. Knight (1986a) concluded that secretion of catecholamines from chromaffin cells and related secretory events require calcium and magnesium adenosine triphosphate (ATP). He stressed that there apparently are many interactions between calcium and guanine nucleotides. García, Artalejo, Borges et al. (1987) and Meldolesi and Pozzan (1987) reviewed the pharmacology of the calcium channel within the plasma membrane of the adrenal chromaffin and related cells. Both reviews presented a molecular model for the calcium channel.

Meldolesi, Malgaroli, Wollheim et al. (1987) reviewed fluorescence techniques for measuring intracellular calcium. They focused primarily on advantages, drawbacks, and limitation of procedures that use Quin-2 and Fura-2. The review included much of the literature on these types of measurements in adrenal chromaffin cells and PC12 cells. A briefer review was also proffered (Meldolesi, Volpe and Pozzan, 1988).

Capponi, Lew, Schlegel et al. (1986) also reviewed techniques for measuring calcium concentrations and membrane potentials within adrenal chromaffin cells and other isolated cells. Specifically, they critiqued the techniques based on Quin-2 for measuring calcium levels and bisoxonol and other potential-sensitive fluorescent probes for measuring average membrane potential. The use of a calcium clamp to study the intracellular environment during secretion was reviewed by Baker, Knight and Umbach (1985). They pointed out that microinjection of calcium into cells such as adrenal chromaffin cells is unreliable and that the microinjection of calcium buffers is much more satisfactory. Furthermore, they predicted that photolabile calcium chelators would become available and would allow experimenters to achieve rapid step changes in intracellular free calcium levels.

Ahnert-Hilger and Gratzl (1988) reviewed the current information on the use of pore-forming proteins in the plasma membrane of chromaffin cells as a tool to dissect stimulus-secretion coupling. Rather than detergents, they used bacterial toxins for the stable, selective, and prolonged permeabilization of the plasma membrane of cells. This method allows for the formation of a pore of controlled size in the plasma membrane with little likelihood of damage to intracellular structures. The current and future roles for pore-forming proteins in the study of exocytosis and related events was reviewed.

The fact that an increase in Ca^{2+} within secretory cells is not the universal trigger in stimulus-secretion coupling was reviewed by Gomperts (1986). This topic was taken up specifically with isolated bovine chromaffin cells by Cobbold, Cheek, Cuthbertson et al. (1987) who monitored calcium transients in response to nicotine and high potassium concentration in single adrenal chromaffin cells microinjected with the photoprotein aequorin. Both agonists produced a transient (60 to 90 sec) increase in Ca^{2+}. At certain levels, the treatment with potassium resulted in the same calcium transient as produced by certain levels of nicotine but less than one-third of the secretory response. Cobbold et al. (1987) suggested that nicotinic agonists generate an alternative second messenger in addition to the increase in Ca^{2+}.

Penner and Neher (1988) reviewed the role of calcium in stimulus-secretion coupling in excitable (including adrenal chromaffin cells) and nonexcitable cells. It is clear that in excitable cells an increase in intracellular calcium is the triggering event that induces secretion. This does not hold for nonexcitable cells. They pointed out the relative importance of intracellular calcium in the regulation of cellular functions, and it is not surprising that several mechanisms regulate intracellular calcium concentration. The major pathway for calcium ions in excitable cells is by voltage-activated calcium channels, but release of calcium from intracellular stores via second messengers predominates in nonexcitable cells and also may be important in excitable cells. In addition, receptor-operated channels and second-messenger-gated conductances may prove to be important. All of these pathways are subject to regulation by various interactive second-messenger systems which provide necessary tuning for an appropriate control of intracellular calcium.

To study the relationship between cell calcium concentration and exocytotic secretion, Kao and Schneider (1987) used Quin-2 to ascertain the dependence of catecholamine secretion on cytosolic Ca^{2+}. In isolated bovine adrenal chromaffin cells, they found a threshold of about 300 nM Ca^{2+} in the cytosol before detectable secretion occurred; half-maximal secretion occurred near 2 μM calcium. Muscarinic receptors mediated a smaller increase in cytosolic calcium but, unlike that elicited by nicotinic receptors, did not require extracellular calcium.

The regulation of cytosolic calcium was studied in digitonin-permeabilized chromaffin cells by Kao (1988). Accumulation of radioactive calcium ions by permeabilized cells was measured at various calcium concentrations. There was a small uptake of calcium in the absence of ATP. In the presence of ATP, permeabilized cells accumulated calcium into two pools: one that was sensitive to a proton ionophore, and one that was not. The ionophore-sensitive pool was mainly in mitochondria. Other results suggested that the ionophore-insensitive pool was the endoplasmic reticulum. At physiologic levels of calcium, the presumed endoplasmic reticulum pool was responsible for about 90% of the ATP-stimulated calcium uptake.

Artalejo, López, Moro et al. (1988) provided direct evidence of a functional coupling among dihydropyridine receptors, calcium channels, and secretion. They found that the inhibitory action of nitrendipine is considerably enhanced if, before calcium stimulation, the adrenal medullary tissue is impregnated with nitrendipine in depolarizing but not hyperpolarizing conditions. The binding of nitrendipine to adrenal medullary tissue was considerably enhanced in depolarization. Blockade of calcium uptake into and catecholamine secretion from the same glands was also enhanced in the depolarized tissues.

Sasakawa, Ishii and Kato (1986) found that, when cultured bovine adrenal chromaffin cells were stimulated by nicotinic agonists in a calcium-free medium containing a calcium chelator, intracellular Ca^{2+} concentration increased. This was not seen when the cells were stimulated with high potassium concentration or veratridine. The increase in Ca^{2+} induced by carbamoylcholine was blocked by hexamethonium but not by atropine. These and other results suggested to Sasakawa et al. (1986) that intracellular sites of calcium storage are able to release the ion in response to nicotinic receptor stimulation.

Kinetic properties of single calcium channels in bovine adrenal chromaffin cells were studied by using the patch-clamp technique by Hoshi and Smith (1987). They found three types of calcium channel openings with different mean durations: less than 1 msec, 3 to 6 msec (these were more frequently seen during depolarization), and >12 msec. These were observed in the presence of the calcium agonist BAY K 8644. All three types of openings were similar in the unitary current amplitude. The authors suggested that a kinetic model with at least two open states is required to explain activation of calcium channels in chromaffin cells.

Wakade, Malhotra, Sharma et al. (1986) showed that, when the isolated rat adrenal gland is perfused with hypertonic Krebs bicarbonate solution followed by normal Krebs solution, there is a dramatic increase in catecholamine secretion. The effect is even more dramatic when calcium-free hypertonic Krebs solution is used. Immediately after the switchover from hypertonic to isotonic solution, the accumulation of radioactive calcium increased 6-fold over the control accumulation. Wakade et al. (1986) suggested that the most likely explanation for these findings is that, once opened after the switchover, calcium channels become inactive very slowly and allow Ca^{2+} to enter the chromaffin cells. Experiments suggested that the calcium channels are opened by a relative change in the membrane potential.

The role of high cytoplasmic pH in the inhibitory action of high osmolarity on secretion from bovine adrenal chromaffin cells was investigated by O'Sullivan and Burgoyne (1988). Using fluorescent indicators, they monitored the effect of increased osmolarity on cytoplasmic pH and cytoplasmic free calcium ions. Increased osmolarity increased both pH and calcium but had no effect on the calcium increase elicited by either high potassium level or nicotine. Increasing the pH by other methods was shown to inhibit secretion from chromaffin cells. The increase in intracellular pH by hyperosmolar solutions was proposed as one of the mechanisms by which increased osmolarity inhibits secretion.

The relationship among calcium influx, cytosolic calcium levels, and catecholamine release in cultured bovine adrenal chromaffin cells was determined by Zimlichman, Pollard, Keiser et al. (1986). They found that secretagogues such as veratridine that caused an increase in calcium influx had the greatest effect on catecholamine secretion. Other agonists, such as the calcium ionophore A23187, increased cytosolic calcium concentration but did not produce significant catecholamine secretion. These and other experiments suggested that the rate at which calcium enters the cell, but not cytosolic calcium concentration, regulates the release of catecholamines.

Zimlichman, Goldstein, Zimlichman et al. (1987) found that angiotensin II increases cytosolic calcium concentration and stimulates catecholamine release from cultured bovine adrenal chromaffin cells. Their data suggest that the mechanism of activation by angiotensin II may include an increase in cytosolic calcium concentration due to opening of membrane calcium channels which may be unrelated to cholinergic receptor-mediated calcium channels.

Artalejo, García and Aunis (1987) were able to resolve potassium-evoked calcium uptake to a time scale of 1 sec. They demonstrated a linearity of calcium uptake during the first 5 sec of stimulation. Experiments with specific antagonists and agonists of calcium channels suggested that the calcium influx is mainly through specific voltage-dependent calcium channels. Uptake of strontium was faster than that of calcium, and barium uptake was even faster. Further experiments suggested that intracellular calcium concentration determines the rate of inactivation of the cationic channels.

Gandía, López, Fonteríz et al. (1987) found that potassium-evoked calcium uptake into cat adrenal medullary tissues was inhibited potently by the dihydropyridine calcium antagonist PN 200-110, with a concomitant decrease in catecholamine secretion. Verapamil, diltiazem, and flunarizine were much less potent antagonists. These authors also determined the order of potencies of a series of inorganic antagonists.

Kunze, Hamilton, Hawkes et al. (1987) correlated the binding of PN 200-110 and nitrendipine with their pharmacologic actions on voltage-dependent membrane calcium channels in PC12 cells. They found that binding was complicated by the presence of large numbers of low-affinity sites. However, there was good agreement between the number of high-affinity sites and the number of functional calcium channels. The electrophysiologic effects were also complicated and could be stimulatory or inhibitory, depending on the membrane potential. At the molecular level, the effects appeared to be on gating; changes in channel conductance did not occur.

Yamada, Teraoka, Nakazato et al. (1988) used the intracellular calcium antagonist TMB-8 to block the catecholamine secretion from perfused cat adrenal glands evoked by caffeine or acetylcholine. TMB-8 reversibly inhibited the catecholamine secretion evoked by caffeine and by acetylcholine in the presence of hexamethonium during perfusion with calcium-free medium containing a calcium chelator. In contrast to acetylcholine, caffeine was much more effective in releasing catecholamines from the cat adrenal glands in the absence of extracellular calcium than in its presence. These results support the view that muscarinic receptor activation causes catecholamine secretion by mobilizing calcium from an intracellular pool, just as caffeine does.

The effects of BAY K 8644 on the catecholamine secretory responses to the calcium ionophore A23187 or ouabain in the cat adrenal medulla were examined by Artalejo and García (1986). The secretory profile obtained with the ionophore was not modified in the presence of BAY K 8644, but it markedly potentiated the secretory effects of ouabain. They concluded that BAY K 8644 potentiates only those catecholamine secretory responses that are mediated through the activation of voltage-sensitive calcium channels; the drug does not seem to affect secretory responses by acting on the membrane sodium/calcium exchange system or at some intracellular calcium-dependent component. Also, it is likely that ouabain enhances the rate of adrenal catecholamine release by a dual mechanism: chromaffin cell depolarization, and activation of a membrane sodium/calcium exchange system.

Fonteríz, Gandía, López et al. (1987) found that both enantiomers of PN 200-110 inhibit potassium-evoked catecholamine release from cat adrenal glands. The (+) enantiomer of Sandoz 202-791 potentiates secretion, but the (-) enantiomer behaves as a potent inhibitor. They concluded that, because potassium-evoked radioactive calcium uptake also was potently inhibited by (+) PN 200-110, the chromaffin cell dihydropyridine receptor is associated with the voltage-dependent calcium channel that exhibits an exquisite stereoselectivity.

Ladona, Aunis, Gandía et al. (1987) tested whether or not BAY K 8644 enhanced the secretory responses associated with calcium channel activation rather than those not involving such channels. They examined the rates of catecholamine release in response to stimulation with nicotine, muscarine, or potassium in perfused cat adrenal glands and cultured bovine adrenal chromaffin cells. BAY K 8644 markedly potentiated the secretory responses to nicotine or potassium, and this was competitively antagonized by a dihydropyridine calcium channel antagonist. They concluded that BAY K 8644 selectively potentiates catecholamine secretion through the activation of voltage-sensitive calcium channels and that calcium gains access to the cell through a common pathway during stimulation by nicotine or high potassium concentration. They concluded that the dihydropyridine calcium agonist and antagonist act on a common site on the chromaffin cell membrane, perhaps through a specific receptor near the voltage-dependent calcium channel.

Sala, Fonteríz, Borges et al. (1986) described the effects of calcium or strontium and the concentration of the divalent calcium channel permeant cation on the kinetics of catecholamine release from the perfused cat adrenal gland upon sustained depolarization with high potassium concentration. By using several manipulations, including BAY K 8644, the role that calcium channels might play in this phenomenon also was evaluated. Their experiments suggest that calcium, but not strontium, modulates the inactivation of the late secretory response and that this modulation seems to be exerted at a step distal to the calcium channel.

Cárdenas, Montiel, Esteban et al. (1988) used PN200-110 and BAY K 8644 to examine secretion from epinephrine- and norepinephrine-containing adrenal chromaffin cells in the perfused cat adrenal gland. PN200-110 inhibited and BAY K 8644 potentiated the catecholamine release evoked by a cholinergic agonist or potassium stimulation. Epinephrine and norepinephrine secretions were affected equally. Because these two drugs act specifically on voltage-dependent calcium channels, it seems that the secretion of the two different catecholamines from their respective cell types is regulated by the same type of channel.

Moreland, Ushay, Kimball et al. (1988) examined the mechanism of action of BAY K 8644 by examining the possibility that the pressor response to BAY K 8644 may also be the result of indirect activation of vascular smooth muscle by the release of adrenal catecholamines. Intravenous administration of BAY K 8644 increased mean arterial pressure in conscious rats. This pressor response was blocked by calcium channel blockers at low doses. α-Adrenoceptor antagonists completely blocked the BAY K 8644-induced pressor responses and converted them to depressor responses. Adrenalectomy did not alter the inhibitory effect of α-adrenoceptor antagonists on the pressor response, but it did prevent the reversal of the BAY K 8644 pressor response to a depressor response. In addition, adrenalectomy did not affect the ability of an α-adrenoceptor antagonist to reverse the pressor response to exogenous epinephrine administration. These and other data suggest that the pressor response to BAY K 8644 may involve both direct activation of vascular smooth muscle cells and indirect activation of the muscle cells by adrenal catecholamines.

Sorimachi, Nishimura and Yano (1985) found that BAY K 8644 increased the rate of catecholamine secretion from the perfused cat adrenal gland. The magnitude of response was inversely proportional to the extracellular concentration of calcium. They supported the assumption that the calcium influx triggered by a decrease in extracellular calcium concentration becomes substantial enough to increase secretion significantly in the presence of BAY K 8644.

Sorimachi, Nishimura and Yano (1986) investigated the effects of divalent cations on catecholamine release in response to potassium depolarization and to substitution of sucrose or sodium in order to study the role of surplus potential in the gating mechanism of calcium channels. The results in a number of experiments with various cations are in accord with the concept that a high concentration of permeant divalent cations decreases the number of activated calcium channels by decreasing the surface potential. Also, calcium appears to be more effective than strontium in decreasing surface potential by binding to negative charges on the surface of the cell. The increase in calcium secretion induced by BAY K 8644 was greater in the presence of strontium than with calcium.

In a related study, Yamagani and Sorimachi (1987) investigated the possible involvement of change in surface potentials in the secretory mechanism in the rabbit adrenal gland. Specifically, they investigated the effects of nitrophenol compounds, uncouplers of oxidative phosphorylation. The results suggest that nitrophenols stimulate secretion by two independent mechanisms: one is related to an effect on surface potentials of the chromaffin cell membrane, and the other appears to be dependent on sodium and calcium.

In PC12 cells, Johnson, Conroy and Isom (1987) found that addition of potassium cyanide led to a gradual increase in cytosolic calcium concentrations. These and other results demonstrate that cyanide induces an accumulation of cytosolic calcium and this additional calcium load appears to originate primarily from the extracellular compartment. Johnson et al. (1987) supported the implication of calcium as an intracellular mediator of cyanide toxicity.

The effect of cyanide on the release of catecholamines was re-evaluated by Borowitz, Born and Isom (1988) in isolated bovine adrenal glands stimulated with four different agonists. Cyanide increased the catecholamine release induced by barium or cadmium and, to a lesser extent, the secretion induced by acetylcholine or potassium. These data suggest that cyanide acts by multiple mechanisms to enhance evoked catecholamine release. It was suggested that changes in plasma membrane permeability may be crucial in the alterations of ion flux and evoked catecholamine release caused by cyanide.

The mechanism of the inhibitory effect of thiopentone on calcium uptake and secretion from cultured bovine adrenal chromaffin cells was studied by Matsumoto, Sumikawa and Kashimoto (1986). Thiopentone inhibited carbachol-induced calcium uptake into and catecholamine release from the cells in a concentration-dependent manner. The results suggest that thiopentone blocks the stimulus-secretion coupling in chromaffin cells as a result of inhibiting calcium uptake through nicotinic receptor-linked channels. The linkage between receptor stimulation and calcium channel activation seems to be the process most susceptible to inhibition by thiopentone.

The action of pentobarbitone on stimulus-secretion coupling in bovine adrenal chromaffin cells was investigated by Pocock and Richards (1987). They examined in detail the action of pentobarbitone on secretion induced by the activation of nicotinic receptors and by high potassium concentration and the relationship between intracellular Ca^{2+} and exocytosis. They concluded that the inhibitory effect of pentobarbitone results from inhibition of calcium movement through the voltage-sensitive calcium channel and from inhibition of ion movements through the channels gated by the nicotinic receptor.

Greenberg, Carpenter and Messing (1987) compared the properties of depolarization-dependent radioactive calcium uptake in PC12 cells grown with and without ethanol. Ethanol exposure increased calcium uptake, but the ethanol-induced component of uptake retained properties of calcium flux through voltage-dependent calcium channels, including sensitivity to calcium channel-modulating drugs. The authors suggested that such drugs may have a role in counteracting ionic events underlying ethanol dependence and withdrawal.

The effects of metalloendoproteinase inhibitors on secretion and intracellular Ca^{2+} concentration in bovine adrenal chromaffin cells was investigated by Harris, Cheek and Burgoyne (1986). It is known that the inhibitor 1,10-phenanthroline blocks the catecholamine release elicited by nicotine or potassium; but the inhibitor carbobenzoxy-Gly-Phe-NH_2 inhibits only the catecholamine release in response to nicotine and enhances that due to potassium. Both inhibit the increase in intracellular Ca^{2+} induced by both agonists. These and other results show that metalloendoproteinase inhibitors have complex effects on chromaffin cells, including effects on the regulation of intracellular calcium concentration, but do not inhibit calcium-activated exocytosis itself.

Lelkes and Pollard (1987) found that oligopeptide metalloendoproteinase inhibitors inhibit catecholamine release from isolated bovine adrenal chromaffin cells. In contrast, catecholamine release from digitonin-permeabilized cells stimulated with calcium was virtually unaffected. Lelkes and Pollard (1987) presented evidence that the actual site of action of these inhibitors is transmembrane calcium movement and the maintenance of intracellular calcium concentration. They concluded that metalloendoproteinase activity is not directly involved in a fusion process as a proteinase *per se*. Instead, they suggest that the more likely mechanism of action of these inhibitors, and hence presumably of the cellular metalloendoproteinases, may be at the level of intracellular calcium homeostasis.

Peppers and Holz (1986) showed that digitonin-permeabilized PC12 cells release catecholamines in the absence of a secretagogue in a calcium and magnesium ATP-dependent manner, consistent with exocytosis. In addition, they showed that exogenous diacylglycerol and phorbol esters that activate protein kinase C enhance secretion from permeabilized PC12 cells.

Di Virgilio, Milani, Leon et al. (1987) studied the detailed kinetics of the responses induced in PC12 cells by depolarization with high potassium concentration. They found that voltage-gated calcium channels are inactivated within a few seconds after depolarization, and this inactivation is both voltage- and intracellular calcium-dependent. They suggest that localized intracellular calcium gradients form close to the plasma membrane shortly after depolarization and that the calcium concentration reached in these regions is the relevant variable in the regulation of secretion.

Inoue and Kenimer (1988) investigated muscarinic receptor-stimulated norepinephrine release in the influx of calcium into PC12 cells. They showed that the muscarinic agonist methacholine stimulates calcium influx and norepinephrine release in a dose-dependent manner. Experiments performed in a sodium-free medium or with inhibitors of voltage-dependent calcium channels suggested the involvement of a receptor-activated calcium channel that differs significantly from the voltage-dependent calcium channel involved in nicotinic receptor-stimulated release. These effects were inhibited by pertussis toxin. These experiments provide the first evidence that muscarinic stimulation evokes neurotransmitter secretion by opening a receptor-activated calcium channel that is controlled by a pertussis toxin-sensitive protein.

To study the regulation of voltage-sensitive calcium channels, DeLorme, Rabe and McGee (1988) used PC12 cells and the binding of tritiated nitrendipine to determine the number of dihydropyridine-sensitive channels. The uptake of radioactive calcium ions was used to determine the functional state of the channels on the cell surface. Prolonged depolarization by increased extracellular calcium ion concentration caused concomitant time- and concentration-dependent decreases in nitrendipine binding and depolarization-dependent uptake of calcium. These and other results suggest that the number of functional voltage-sensitive calcium channels can be regulated by changes in intracellular calcium, such as those associated with prolonged depolarization. However, because calcium channel number remained decreased while intracellular free calcium concentration returned to normal, other mechanisms controlling channel number also must be involved.

Artalejo, Bader, Aunis et al. (1986) found that, on stimulation of cultured bovine adrenal chromaffin cells with potassium, radioactive calcium uptake is quickly enhanced and reaches a plateau within 1 min. Using techniques with a temporal resolution of 10 sec, they compared calcium fluxes with electrophysiologic measurements of chromaffin cell calcium currents. They showed that, upon sustained depolarization with high potassium concentration, the rates of both calcium uptake and norepinephrine release decreased in parallel, suggesting that the voltage-dependent calcium channel activity modulates the kinetics of the early secretory response.

Di Virgilio, Milani, Leon et al. (1987) used fluorescent indicators to study the kinetics of the responses induced by depolarization with high potassium in PC12 cells. Their results clearly demonstrated that the L-type calcium channels are inactivated within a few seconds after depolarization, and this inactivation is both voltage- and calcium-dependent. Moreover, the correlation of changes in calcium and secretion kinetics revealed interesting clues as to the roles of calcium transients in the regulation of exocytosis.

SODIUM

Nakazato, Ohga and Yamada (1986) used a perfused isolated splanchnic nerve and adrenal gland preparation from the guinea pig to determine whether potassium deprivation and ouabain exerted different effects on the catecholamine secretion evoked by splanchnic nerve stimulation, by acetylcholine, and by excess potassium and whether the effect of ouabain on these evoked responses was influenced by varying the extracellular concentration of sodium, magnesium, or cobalt. Their results suggest that ouabain enhances the catecholamine secretion evoked by splanchnic nerve stimulation, acetylcholine, and excess potassium by increasing the rate of calcium influx through the acetylcholine receptor-linked calcium channel or the voltage-dependent calcium channels in a sodium-dependent manner.

Wada, Takara, Yanagihara et al. (1986) examined the effects of ouabain and extracellular potassium deprivation on carbachol-induced influx of radioactive potassium and calcium and the secretion of catecholamines in cultured bovine adrenal chromaffin cells. The treatment remarkably potentiated sodium and calcium influx and catecholamine secretion; this potentiation was not observed in a sodium-free medium. These and other results suggest that sodium influx via nicotinic receptor-associated sodium channels increases the activity of Na^+, K^+-ATPase, and the inhibition of this ATPase augments carbachol-induced calcium influx and catecholamine secretion by potentiating cellular accumulation of sodium. It appeared to Wada et al. (1986) that both nicotinic receptor-associated sodium channels and Na^+, K^+-ATPase both modulate the influx of calcium and the secretion of catecholamines by modulating cellular concentrations of sodium.

Wada, Arita, Yanagihara et al. (1988) demonstrated that, in bovine adrenal medullary cells, phencyclidine inhibited carbachol-induced influx of radioactive sodium and radioactive calcium as well as the secretion of catecholamines in a concentration-dependent manner. Tritiated phencyclidine bound specifically to adrenal medullary cells, and the binding characteristics were elucidated. It was suggested that phencyclidine does not inhibit voltage-dependent calcium channels and calcium-dependent channels. Furthermore, phencyclidine binds to two populations of sites, each of which is functionally linked to nicotinic receptor-ion channel complex and to voltage-dependent sodium channels. Inhibition of sodium influx by phencyclidine decreases calcium influx, potassium efflux, and catecholamine secretion.

Boarder, Marriott and Adams (1987) investigated the dependency of stimulus-secretion coupling in cultured bovine adrenal chromaffin cells on external sodium concentration under conditions of continuous perfusion of the cells and the susceptibility of this coupling to inhibition by stereoisomers of nicardipine, a calcium channel blocker. In the absence of external sodium, secretion of catecholamines in response to nicotine is not impaired nor do the isomers of nicardipine inhibit this response. They suggested that, primarily, calcium enters the cell directly through an acetylcholine receptor-linked channel and entry is not necessarily dependent on external sodium.

Yamamoto (1986) investigated the dynamics of strychnine block of single sodium channels in isolated bovine adrenal chromaffin cells by using the patch-clamp technique. Sodium current inactivation was eliminated by treatment with *N*-bromoacetamide. When strychnine was applied to the cytoplasmic face of the sodium channels, repetitive rapid transitions between open and blocked states within single openings of sodium channels were recorded. Many other features of strychnine block of the channels were measured and were consistent with a sequential model in which strychnine molecules block open sodium channels and then the blocked channels cannot close until strychnine molecules leave the blocking site.

Takara, Wada, Arita et al. (1986) examined the effects of ketamine, an intravenous anesthetic, on radioactive sodium influx, radioactive calcium influx, and catecholamine secretion in cultured bovine adrenal chromaffin cells. The influx of sodium caused by carbachol or veratridine was depressed by ketamine with a concentration-inhibition curve similar to that of calcium influx and catecholamine secretion. They found that, at clinical concentrations, ketamine can inhibit nicotinic receptor-associated ionic channels and inhibit sodium influx via the receptor-associated ionic channel and that this is responsible for the inhibition of catecholamine secretion. At higher concentrations, the anesthetic also inhibits voltage-dependent sodium channels but has no effect on voltage-dependent calcium channels.

Izumi, Wada, Yanagihara et al. (1986) studied the mechanism of catecholamine release by monensin, a sodium ionophore. In cultured bovine adrenal chromaffin cells, monensin caused the release of catecholamines simultaneously with the influx of radioactive sodium into the cells. The release of catecholamines was dependent on extracellular sodium but not calcium. In isolated chromaffin vesicles, monensin caused a sodium-dependent release of catecholamines simultaneously with the influx of sodium into the vesicles. Monensin did not cause the release of dopamine β-hydroxylase, leading Izumi et al. (1986) to the conclusion that monensin causes a nonexocytotic release of catecholamines.

Wada, Arita, Kobayashi et al. (1987) examined the binding characteristics of saxitoxin and its binding site in bovine adrenal chromaffin cells. These cells showed a specific binding of saxitoxin that was saturable and reversible. Saxitoxin inhibited veratridine-induced sodium influx with a potency similar to that of tetrodotoxin. However, veratridine, aconitine, and scorpion venom did not inhibit saxitoxin binding at concentrations that increased sodium influx. These results indicate that saxitoxin binds to a specific site on voltage-dependent sodium channels and inhibits the influx of sodium.

Arita, Wada, Takara et al. (1987) investigated the effects of antidepressants on ionic channels and secretion of catecholamines in bovine adrenal chromaffin cells. Tricyclic and tetracyclic antidepressants inhibited carbachol-induced influx of radioactive sodium and calcium and the secretion of catecholamines. Their experiments suggested that these antidepressants bind to two populations of binding sites that are functionally associated with nicotinic receptor-associated ionic channels and with voltage-dependent sodium channels. This leads to inhibition of sodium influx which in turn decreases the calcium influx and catecholamine secretion caused by carbachol or veratridine.

Takahashi, Sugino and Kudo (1986) found that monensin induced a profound release of radiolabeled materials from a subclone of PC12 cells preloaded with tritiated norepinephrine. The release was suppressed by the absence of external sodium but not external calcium. Monensin caused a slight increase in intracellular calcium. Takahashi et al. (1986) suggested that monensin expels norepinephrine from storage vesicles, the amine is metabolized by cytoplasmic monoamine oxidase, and the metabolites are released from the cell in a nonexocytotic manner.

Cárdenas, Montiel, Artlejo et al. (1988) used PN200-110, a dihydropyridine calcium channel blocker, to clarify further the role of external sodium ions in triggering catecholamine release by nicotinic stimulation. Their experiments demonstrated a strong sodium dependency of the blocking effects of PN200-110 on nicotinic catecholamine release from cat adrenal glands. This suggested a prominent role of sodium ions in cholinergic-receptor-mediated secretion.

Gordon, Merrick, Auld et al. (1987) prepared antipeptide antibodies that distinguish between two subtypes of sodium channels. The antibodies were used to localize the subtypes of sodium channels in central and peripheral tissues. The channels are primarily expressed in the central nervous system and are not detected in significant numbers in adrenal medulla, skeletal or cardiac muscle, sympathetic ganglia, sciatic nerve, or the cauda equina.

Gordon, Merrick, Auld et al. (1987) used anti-peptide antibodies to distinguish between two sodium channel subtypes in neurons from the rat brain. These two subtypes were primarily in the central nervous system and were not detected in significant number in the adrenal medulla or other peripheral organs.

In two studies, Uvnäs and Åborg (1987a, 1987b) found that the contents of chromaffin vesicles may be released by a mechanism related to ion exchange. The release of vesicle contents when isolated vesicles were superfused with sodium- or potassium-containing solutions was similar to that of a weak synthetic anion exchanger.

OTHER IONS

Lopatin, Kondratiev, Longinov et al. (1988) used patch clamp techniques to examine potassium channels of PC12 cells. Under physiologic conditions, inward sodium and calcium channels were not observed; calcium-dependent and potential-activated potassium channels were detected. There were two types of channels which differed not only in their kinetics, but also in their voltage-current characteristics.

A role of an anionic channel in the membrane of secretory vesicles has been proposed as part of the mechanism of secretion from adrenal chromaffin cells as well as other cells. Stanley, Ehrenstein and Russell (1988) presented evidence for anion channels in secretory vesicles. Using vesicles from the bovine neurohypophysis, they detected an anion channel that could provide a pathway for anion transport that has previously been described for chromaffin vesicles. The secretory vesicle anion channel may play a role in calcium-induced secretion.

Rogawski, Pieniek, Suzuki et al. (1988) investigated the effects of phencyclidine, a psychotomimetic dissociative anesthetic, and several related drugs on voltage-dependent potassium currents in PC12 cells. Whole-cell voltage clamp recordings demonstrated two kinetically distinct voltage-dependent outward potassium current components in these cells. These and other results demonstrated that, because phencyclidine and related drugs are powerful selective blockers of a sustained component of the potassium channel in PC12 cells, blockade of potassium channels is unlikely to be responsible for the psychotomimetic or anticonvulsive properties of phencyclidine but could account for the convulsant potential of the drugs.

The electrical activity in chromaffin cells of the intact mouse adrenal gland was measured by Nassar-Gentina, Pollard and Rojas (1988). They studied the dependence of the resting membrane potential on extracellular calcium ions and concluded that it is controlled by potassium channels. Measurement of input resistance in medullary cells in situ suggested that the chromaffin cells were electrically coupled. Furthermore, acetylcholine induced a substantial decrease in input resistance that may be due, at least in part, to a decrease in cell-to-cell junctional resistance. Because the effects of acetylcholine on various electrical characteristics of mouse medullary chromaffin cells are mediated by muscarinic receptor activation, Nassar-Gentina et al.

(1988) suggested that acetylcholine-evoked catecholamine secretion in this preparation is also mediated by muscarinic receptor activation.

Hoshi, Garber and Aldrich (1988) examined the effect of forskolin on voltage-gated potassium channels and found that it is independent of adenylate cyclase activation. Forskolin directly altered the gating of a single class of voltage-dependent potassium channels in PC12 cells, and this alteration occurred in isolated cell-free patches, independently of soluble cytoplasmic enzymes. Hoshi et al. (1988) pointed out that this direct action of forskolin can lead to misinterpretation of results in experiments in which forskolin is assumed to activate adenylate cyclase selectively.

Quinine blockade of currents through calcium-activated potassium channels was studied by Glavinovic and Trifaró (1988) using excised inside-out patch recordings from cultured bovine adrenal chromaffin cells. Application of quinine to the intracellular side of the membrane in micromolar concentrations chopped the unitary potassium currents into bursts of brief openings. Characterization of these and related phenomena demonstrated that quinine is clearly a poor blocker.

Wada, Kobayashi, Arita et al. (1987) loaded cultured bovine adrenal chromaffin cells with radioactive rubidium and examined the effects of high potassium concentration, veratridine, and carbachol on the efflux of rubidium as well as some properties of the potassium permeability mechanisms. An extensive series of experiments led to the suggestion that adrenal chromaffin cells have at least three distinct types of potassium permeability mechanisms: basal potassium efflux, calcium-dependent efflux, and sodium-dependent potassium efflux. It seems that nicotinic receptors mediate potassium efflux by increasing sodium influx via nicotinic receptor-associated ionic channels rather than calcium influx via voltage-dependent calcium channels.

In isolated bovine adrenal chromaffin cells investigated by Pocock and Simons (1987), lead ions did not affect basal secretion but did inhibit catecholamine secretion in response to carbachol and high potassium concentration. Calcium influx and sodium influx were also inhibited by lead. In a companion paper, Simons and Pocock (1987) reported that lead enters chromaffin cells through calcium channels. The evidence for this is that lead uptake is antagonized by calcium, inhibited by a calcium channel blocker, and stimulated by BAY K 8644. The permeability of the channels to lead appears to be at least 10 times their permeability to calcium.

Izumi, Toyohira, Yanagihara et al. (1986) investigated barium-evoked release of catecholamines from digitonin-permeabilized bovine adrenal chromaffin cells. They found that the maximal release of catecholamines by barium was greater than that caused by calcium, suggesting that barium directly influences secretion rather than mobilizes intracellular calcium stores. The catecholamines released by barium were accompanied by dopamine β-hydroxylase, showing that barium causes an exocytotic release of catecholamines. They concluded that barium can substitute for calcium in triggering exocytosis from adrenal chromaffin cells.

Hirano, Kidokoro and Ohmori (1987) used cesium ions to eliminate current flowing through various potassium channels in a study of acetylcholine dose-response relationships in cultured rat adrenal chromaffin cells. Their results in this technically difficult voltage-clamp study suggest that cesium may slowly block acetylcholine receptor channels from the outside in the resting state.

Abajo, Castro, Garijo et al. (1987) studied the effect of lithium on catecholamine release from the perfused cat adrenal gland. Replacement of sodium by lithium evoked a progressive increase in the spontaneous release of catecholamines that reached a maximum within 45 minutes and was calcium-dependent. This was not seen when sucrose or choline was substituted for sodium. Their results indicated that lithium accumulates in the cells and partially substitutes for sodium in the sodium-calcium countertransport system at the plasma membrane.

Volonté and Racker (1988) used lithium chloride to stimulate the formation of inositol monophosphate in PC12 cells that had been exposed to nerve growth factor. This was also seen in membranes isolated from PC12 cells that had differentiated in response to nerve growth factor. Although observations with intact cells are difficult to interpret without ambiguity, the results obtained with isolated membranes support the interpretation of the stimulatory action of lithium in PC12 cells.

CHAPTER 5

STIMULUS-SECRETION COUPLING:
INTRACELLULAR PROTEINS AND NUCLEOTIDES

As indicated in the previous two chapters, secretion in chromaffin cells is mediated by the influx of calcium into the cell through channels in the plasma membrane or by the mobilization of intracellular calcium stores (or by both mechanisms). The effect of this increase in intracellular calcium levels on secretion is believed to be mediated by one or more proteins within the cell.

<u>CALCIUM-BINDING</u> <u>PROTEINS</u>

Proteins isolated from the cytosol of adrenal chromaffin cells by calcium-dependent affinity chromatography have been called "chromobindins." The roles of these calcium-dependent membrane-binding proteins in the regulation and mechanism of exocytosis were reviewed by Creutz, Zaks, Hamman et al. (1987a). Another general review of calcium-binding proteins and secretion was published by Hutton (1986).

Creutz, Zaks, Hamman et al. (1987b) compared 23 chromobindins with other known molecules. Chromobindin 4 was identified as a 32 kDa protein called "calelectrin" or "indonexin." Chromobindin 20 was identified as a 67 kDa variant of calelectrin. Chromobindin 8 was identified as p36, a substrate for a tyrosine-specific kinase. Chromobindin 6 was identified as p35, a substrate for the tyrosine kinase activity associated with the epidermal growth factor receptor. Chromobindin 9, which is known to be a substrate for protein kinase C, was found to be related to p35 and may be a precursor of chromobindin 6.

Martin and Creutz (1987) examined the properties of seven chromobindins. These proteins bind to vesicle membranes in the presence of calcium; however, they are not released from the membrane when the calcium is removed unless adenosine triphosphate (ATP) is present. The proteins range from 53 kDa to 59 kDa and form a multisubunit complex of about 800 kDa. Martin and Creutz (1987) named this complex "chromobindin A." The binding of chromobindin A to membranes is stimulated by calcium, strontium, and barium. The release is stimulated by various nucleotides. It was suggested that chromobindin A binds to a protease-sensitive receptor on the vesicle membrane, and this complex is involved in exocytosis, being partially responsible for the dependence of this process on ATP.

Hamman, Gaffey, Lynch et al. (1988) cloned and characterized the cDNA that encodes chromobindin 4. The translated amino acid sequence of chromobindin 4 (also known as endonexin, 32 kDa calelectrin, or P II) shows the four-domain structure characteristic of proteins in this class. The nucleotide sequence is 55% to 61% identical to that of related membrane-binding proteins. Southern blot analysis suggests the presence of a single gene for this protein.

Chromobindin A and τ-protein both consist of a small group of closely related proteins that can form aggregates. Sternberg, Baudier, Akizuki et al. (1988) confirmed that τ-protein is present in bovine adrenal medullary tissue and showed the similarities between it and chromobindin A. They also demonstrated differences in immunoreactivity and susceptibility to phosphorylation.

Geisow, Fritsche, Hexham et al. (1986) reported that calelectrin and p36 and p32.5 contain a 17-amino acid consensus sequence that is conserved and is present in multiple copies. They suggested that this common sequence may be present in other calcium-binding proteins and may be related to their ability to bind to biologic membranes.

A procedure was devised by Michener, Dawson and Creutz (1986) to determine whether, in the stimulated chromaffin cell, phosphate is incorporated into chromobindins in a calcium-dependent manner. Stimulation by several secretagogues led to the incorporation of phosphate into chromobindin 9. Incorporation of phosphate into this protein was dependent on extracellular calcium and had a time course that paralleled the secretion of catecholamines, returning to baseline levels after 30 minutes when secretion terminated.

Okumura-Noji, Kato and Tanaka (1986) examined the calcium-dependent phosphorylation of proteins in PC12 cells. They found that a group of proteins ranging from 50 kDa to 55 kDa and a 95 kDa protein were phosphorylated. Calcium/calmodulin-dependent protein kinase appeared to be responsible for the phosphorylation of the 50 kDa to 55 kDa proteins and, partly, of the 95 kDa protein. Depolarization of intact PC12 cells by potassium induced phosphorylation of the 95 kDa protein.

Synexin was the first chromobindin to be identified and characterized. The effect of synexin on the aggregation and fusion of chromaffin vesicle ghosts (empty vesicles) at pH 6 was investigated by Nir, Stutzin and Pollard (1987). Their two assays used fluorescence intensity increases to monitor continuous mixing of aqueous contents or membrane mixing as an approach to study of the kinetics and extent of fusion. The results indicated that the initial aggregation and fusion process is extremely fast. A small increase in volume was found to accompany the fusion between chromaffin vesicle ghosts. Their results also indicated that the main activity of synexin was to enhance the rate of aggregation.

Rojas and Pollard (1987) measured synexin-induced changes in the electrical phosphatidylserine bilayers. Such changes can come about as a consequence of insertion of protein dipoles into the substances of the bilayer. They interpreted their data to indicate that, in the presence of calcium, synexin can change from a water-soluble form to one that can penetrate into the substance of the bilayer, thus modifying the dielectric properties of the membrane.

Creutz, Snyder, Husted et al. (1988) examined the pattern of repeating aromatic residues in synexin and found that they were similar to the cytoplasmic domain of synaptophysin. Twenty percent of the synexin sequence was found in one contiguous sequence of 61 residues and a nonoverlapping sequence of 20 residues. They found a repeat of six peptides occurring eight times in the series. This pattern of periodic aromatic residues suggested the presence of a novel secondary structure. This is similar to repeats present in synaptophysin, gliadin, and type II keratin.

The effect of pH on synexin-mediated fusion of chromaffin vesicle ghosts was examined by Stutzin, Cabantchik, Lelkes et al. (1987). They used a freeze-thaw technique to prepare ghosts loaded with a self-quenching concentration of the fluorescent probe FITC-dextran. When the loaded ghosts were mixed with empty ghosts in the presence of synexin, the two compartments fused, resulting in dilution of the probe and a concomitant increase in fluorescence. A pH profile of fusion revealed an apparent midpoint of activation at pH 5.2.

A molecular basis for synexin-driven calcium-dependent membrane fusion was presented by Pollard, Burns and Rojas (1988). Examination of cDNA clones for synexin showed that the molecule is similar to some other vesicle-aggregating proteins; but it does contain a unique, long, highly hydrophobic amino-terminal leader sequence followed by a characteristic 4-fold repeat that is homologous with repeats found in other members of the synexin gene family. The highly hydrophobic character of synexin seems consistent with information previously obtained that synexin is able to insert directly into interior bilayers prepared not only from purified phosphatidyl serine but also from biologic membranes. There also was evidence that synexin forms calcium-selective channels when the protein is applied to the cytosolic side of the plasma membrane. Therefore, the synexin molecule spans the membrane. From these and other data, Pollard et al. (1988) developed the concept that the fusion process may involve synexin forming a "hydrophobic bridge" between two fusing membranes. Lipid movement across this bridge may then be the material basis for fusion.

The role of calmodulin in calcium signal transduction was reviewed by Tanaka (1988). Studies with calmodulin antagonists, hydrophobic fluorescent probes, hydrophobic chromatography, and alternate activators of calcium-calmodulin-dependent enzyme revealed that calcium ion induces conformational changes in calmodulin that expose hydrophobic regions on the surface of the molecule; these regions may act as sites of interaction with target enzymes and calmodulin antagonists. Also, a similar molecular mechanism of calcium signal transduction has also been reported with other calcium-modulated proteins such as troponin C and S-100 protein.

The presence of calmodulin-binding proteins in three neurosecretory vesicles (bovine adrenal chromaffin vesicles, bovine posterior pituitary secretory vesicles, and rat brain synaptic vesicles) was investigated by Fournier and Trifaró (1988a). When detergent-solubilized membrane proteins from each type of organelle were applied to calmodulin-affinity columns in the presence of calcium, several calmodulin-binding proteins were retained. In all three membranes, proteins of about 65 kDa and 53 kDa were found consistently. Two monoclonal antibodies previously shown to react with a cholinergic vesicle membrane protein were reactive with the 65 kDa protein present in chromaffin vesicle membranes. These and other results clearly indicate that an immunologically identical calmodulin-binding protein is expressed in at least three different neurosecretory vesicle types, thus suggesting a common role for this protein in secretory vesicle function.

Fournier and Trifaró (1988b) described calmodulin-binding proteins in chromaffin cell membrane. They also showed that the 65 kDa calmodulin-binding protein, previously shown to be common to several neurosecretory vesicle membranes, is also present in the plasma membrane of chromaffin cells. This suggests that the protein may serve as a link between vesicle and plasma membranes and may direct the fusion process during exocytosis.

Several cytosolic proteins bind to secretory vesicle membranes in a calcium-dependent manner and thus may be involved in the mediation of membrane interactions during exocytosis. One of these proteins, calpactin, was investigated by Drust and Creutz

(1988). Calpactin was shown to be a tetramer consisting of two heavy (36 kDa) chains and two light (10 kDa) chains. They found that calpactin promotes the calcium-dependent aggregation and fatty acid-dependent fusion of chromaffin vesicle membranes at a lower calcium concentration than reported for other vesicle-aggregating proteins. The same calcium concentration is required for secretion from permeabilized chromaffin cells. They also presented data to suggest that the amino-terminal portion of the 36 kDa protein modulates the calcium/lipid binding sites in the core portion of the protein.

Commenting on the report by Drust and Creutz (1988), Burgoyne (1988) agreed that, at least in an *in vitro* model of membrane interaction, calpactin can mediate membrane-membrane association at physiologically relevant levels of calcium ions. He pointed out that there is not yet any direct evidence that such a mechanism occurs during exocytosis. Permeabilized cells may be appropriate for finally determining the importance of calpactin in exocytosis.

CYTOSKELETON AND CONTRACTILE PROTEINS

The involvement of cytoskeletal and associated proteins in stimulus-secretion coupling is unclear. McKay, Cobianchi and Schneider (1987) examined the possible role of microtubules in cholinergic nicotinic receptor-related events, using the microtubule-disrupting agent colchicine and its nondisruptive isomer β-lumicolchicine. Both compounds inhibited acetylcholine-induced secretion and their potencies were similar. β-Lumicolchicine, but not colchicine, inhibited potassium-induced secretion. Because both compounds disrupted receptor-mediated secretion, the evidence does not support a role for microtubules in receptor-mediated events. The difference in potassium-evoked secretion suggests that these structural isomers have different modes of action on chromaffin cell function.

Burgoyne and Cheek (1987a) presented a lucid account of the role of the cytoskeleton in secretion from adrenal chromaffin cells. This more general account included evidence that reorgani-zation of the cortical actin network is necessary to allow vesicles to reach sites of exocytosis in stimulated cells. This reorgani-zation may involve changes in actin filament cross-linking, assembly, and interactions with secretory vesicle and plasma membranes. The role of specific proteins also was discussed.

Drubin, Kobayashi and Kirschner (1986) reviewed the evidence for the association of τ protein with microtubules in various cells and tissues ("τ protein" is a term used to describe a subgroup of microtubule-associated proteins that co-assemble with microtubules and promote the assembly of purified tubulin). They found large amounts of the 200 kDa protein in the adrenal gland and low amounts of the 55 kDa to 68 kDa and 125 kDa species in undifferentiated PC12 cells. The levels were higher in differentiated PC12 cells.

Kotani, Murofushi, Maekawa et al. (1986) investigated the biochemical characteristics of microtubule-associated proteins of the bovine adrenal medulla and cortex. They found that the medulla was rich in high molecular weight microtubule-associated proteins. The τ protein appeared to be a minor species in the adrenal gland.

Keith (1987) investigated the slow transport of tubulin in differentiated PC12 cells by measuring the fluorescence recovery after photobleaching of a fluorescent tubulin analog. He found that in these cells, as in peripheral axons, tubulin is transported in coherent nondiffusing waves at two different slow rates. It appears that most, if not all, of the tubulin is moving out into the neurites of the PC12 cells. He suggested that these cultured cells to represent a good model of axonal transport.

Cheek, Hesketh, Richards et al. (1986) examined the hypothesis that chromaffin vesicles are suspended within the cytoskeletal matrix; when the intracellular calcium level increases, the vesicles are released from the matrix, leading to exocytosis. They pointed out that considerable reorganization of many of the cytoskeletal elements would have to take place upon stimulation of the cells, including solation of the filaments that restrain the vesicles in the unstimulated state. In an investigation of the association of adrenal medullary actin-binding proteins with endogenous actin in a model cytoskeletal system using supramolecular gels obtained by incubation of bovine adrenal medullary cytoplasmic extracts, they found that a multi-component three-dimensional gelation system consisting of both actin-dependent and actin-independent elements was formed. The gels are composed exclusively of cytoskeletal elements: microtubules, microfilaments, and intermediate filament proteins. They also reported that the actin-based system was calcium-dependent.

Aunis and Bader (1988) reviewed the evidence that the cytoskeleton acts as barrier to exocytosis in secretory cells such as the adrenal chromaffin cell. Chromaffin vesicles appear to be entrapped in a highly organized cytoskeletal network under the plasma membrane. When cells are stimulated, molecular rearrangements of this subplasma lemmel cytoskeleton take place allowing vesicles to reach the cell surface.

Cheek and Burgoyne (1986) used special stains to demonstrate actin filaments concentrated in the peripheral portion (cortex) of resting bovine adrenal chromaffin cells. Cortical actin filaments were disassembled 15 sec after stimulation by nicotine and had reassembled 30 sec later. Disassembly was independent of external calcium, insensitive to the inhibitor trifluoperazine, and not elicited by high potassium concentration, muscarinic agonists, or phorbol ester. They suggested that disassembly of cortical actin filaments may allow access of chromaffin vesicles to exocytotic sites and act in conjunction with an increase in intracellular Ca^{2+} to bring about the secretory response to nicotinic agonists.

The control of the cytoskeleton during secretion was reviewed by Burgoyne, Cheek, O'Sullivan et al. (1988). Among other things, they presented evidence from their studies on the bovine adrenal chromaffin cell that exocytosis normally may be prevented by the subplasmalemmal cytoskeleton and that the cytoskeleton is a target for second messenger control during the process of exocytosis.

Sontag, Aunis and Bader (1988) used streptolysin O, a streptococcal cytotoxin, to permeabilize the chromaffin cell plasma membrane for examining the role of the subplasmalemmal cytoskeleton in the secretory response of these permeabilized chromaffin cells. Specifically, they examined the effect of the introduction of agents that destabilize actin filaments in the permeabilized cells. They found that changes in actin filament organization potentiate secretion evoked by calcium, an observation that is consistent with the concept that disassembly of filamentous actin is required for maximal secretory response.

Burgoyne, Cheek and Norman (1986) demonstrated that one of the cytosolic vesicle-binding proteins, molecular mass 70 kDa, is a form of caldesmon, the calmodulin-regulated actin-binding protein. Cytoplasmic gels assembled from an adrenal medullary extract in the absence of calcium contained actin and the 70 kDa protein. The association of both of these proteins with the cytoplasmic gel was inhibited by low concentrations of calcium. Also, they demonstrated that the 70 kDa protein is localized at the periphery of chromaffin cells. These results are consistent with the concept that caldesmon has a role in regulating the organization of actin filaments at the cell periphery during secretion.

Cheek and Burgoyne (1987) demonstrated that cyclic adenosine monophosphate (AMP) inhibits both nicotine-induced actin disassembly and catecholamine secretion from bovine adrenal chromaffin cells. Increased intracellular cyclic AMP concentration, achieved by three independent methods, did not inhibit potassium-induced secretion but did inhibit nicotine-induced secretion. One such agent, forskolin, inhibited actin disassembly but did not affect the increase in intracellular calcium concentration. This is further evidence that actin disassembly is required in addition to the increase in intracellular calcium in order to obtain a maximal secretory response in chromaffin cells. Their results also point to a role for cyclic AMP in the regulation of stimulus-induced actin disassembly.

Lelkes, Friedman, Rosenheck et al. (1986) used agents that destabilize or stabilize F-actin to examine the role of actin filaments in stimulus-secretion coupling. They found that destabilizing agents promoted calcium-stimulated secretion in digitonin-permeabilized bovine adrenal chromaffin cells. In contrast, stabilizers produced the opposite effect. They concluded that stimulus-secretion coupling in chromaffin cells may require the reorganization of actin for modulating both ion transport across the plasma membrane and exocytosis *per se*.

Sarafian, Aunis and Bader (1987) examined the role of certain soluble cytoplasmic proteins in the secretory process. Digitonin-permeabilized adrenal chromaffin cells progressively lose their capacity to secrete catecholamines in response to calcium. They provided direct evidence that one or more of the cytoplasmic proteins that diffuse from the cell after digitonin treatment are responsible for this loss of secretory activity. Soluble proteins that leaked from permeabilized cells were collected, dialyzed, and concentrated. When these proteins were added back to permeabilized cells that were unable to secrete, catecholamine release was fully restored. One of the release proteins was characterized as calmodulin. However, addition of calmodulin alone was ineffective in restoring secretory activity.

The distribution, structural organization, and state of phosphorylation of neurofilaments were examined in bovine adrenal chromaffin cells by Grant, Demeneix, Aunis et al. (1988). They cultured cells under various conditions and used a series of monoclonal antibodies directed against phosphorylated and nonphosphorylated epitopes. Nonphosphorylated neurofilament epitopes were detected immunocytochemically to various degrees. Staining was usually limited to a paranuclear region from which fine filaments sometimes appeared to radiate. In marked contrast, none of the antibodies directed against phosphorylated neurofilament epitopes stained the structures. They also showed that, under conditions that provoked neurite outgrowth from chromaffin cells, phosphorylation from neurofilaments in the neurites could be demonstrated.

Chang, Toloza and Bulinski (1986) examined actin isoform expression during differentiation of PC12 cells. They found that the ratio of β to γ isoforms of actin decreased from 1.3 to 0.99 after 6 days of treatment with nerve growth factor. This appeared to be related to neurite outgrowth and may be mediated by cyclic AMP.

Nagatsu, Suzuki, Kiuchi et al. (1987) found that the release of dopamine from PC12h cells by high potassium concentration was inhibited by a specific inhibitor of myosin light-chain kinase in a dose-dependent manner. The inhibitor also specifically inhibited the phosphorylation of a 20 kDa protein. They suggested that myosin light-chain kinase may play a stimulatory role in the reaction that releases catecholamines from rat pheochromo-cytoma cells.

Tropomyosins were isolated from the bovine adrenal medulla by Côté, Doucet and Trifaró (1986a). They found three polypeptides with molecular masses of 32 kDa, 35.5 kDa, and 38 kDa. The molar ratio of the two major polypeptides, 32 kDa and 38 kDa, was 2:1. The 38 kDa polypeptide exhibits a stronger affinity for F-actin than do the other forms. Peptide profiles obtained after limited proteolytic digestion show some similarity between the two predominant tropomyosins of the bovine adrenal medulla and also between these and the α and β forms of bovine skeletal muscle tropomyosin.

Friedman, Lelkes, Rosenheck et al. (1986) used a technique whereby, via liposome fusion, macromolecules can be injected into the cytoplasm of bovine adrenal chromaffin cells to introduce heavy meromyosin and its subfragment, in both intact and poisoned forms, into chromaffin cells in an effort to determine whether or not cellular actin and myosin are capable of playing a role in stimulus-secretion coupling. Heavy meromyosin was found to stimulate secretion, cause depolarization, and increase sodium uptake. A sodium/proton antiporter, activated by the interaction of heavy meromyosin with microfilaments within the cell, seems to be involved in the action of heavy meromyosin and also could be a component of stimulus-secretion coupling induced by agonists.

The effects of purified myosin light-chain kinase on myosin light-chain phosphorylation and catecholamine secretion in digitonin-permeabilized chromaffin cells were investigated by Lee, Holz and Hathaway (1987). In the absence of exogenous myosin light-chain kinase, calcium enhanced phosphorylation of the myosin light-chain. In the presence of calcium, myosin light-chain kinase caused an approximately 2-fold increase in myosin light-chain phosphorylation. Under the same conditions, secretion was

unaltered by myosin light-chain kinase. These experiments
indicate that the phosphorylation of myosin light-chain by myosin
light-chain kinase is not a limiting factor in secretion in digitonin-
treated chromaffin cells. Lee et al. (1987) suggested that the
activation of myosin is not directly involved in secretion from the
cells.

Georges, Trifaró and Mushynski (1987) studied the
phosphorylation of neurofilament proteins in cultures of adrenal
chromaffin cells and the distribution of neurofilament subunits
between soluble and cytoskeletal fractions. They found that
neurofilament proteins in chromaffin cells are in a
hypophosphorylated state, as determined by the comigration of
the 160 kDa and 210 kDa subunits with *in vitro* dephosphorylated
bovine brain subunits on SDS gels. Pulse-chase experiments with
radioactive phosphate showed that neurofilament proteins rapidly
attained maximal phosphorylation levels. They also found
differences between the phosphopeptide maps of cytoskeleton-
associated and soluble middle molecular weight neurofilament
subunit, suggesting that the localization of phosphate moieties
rather than extent of phosphorylation influences the association
of the subunit with neurofilaments.

Parysek and Goldman (1987) characterized intermediate
filaments in PC12 cells. They found that a 57 kDa protein, not
vimentin, is the major component of them. Neurofilament triplet
proteins appear to be minor components of the total intermediate
filament component of the cytoskeleton.

Franke, Grund and Achtstätter (1986) found that intermediate
filaments from PC12 cells differ profoundly from those of neurons
and adrenal medullary cells but are similar to those of certain
neuroendocrine tumors of epithelial origin. As determined by
electron microscopy, immunolocalization, and biochemical
analyses, they found that, in addition to neurofilaments, the PC12
cells contain an extended meshwork of bundles of intermediate-
sized filaments of cytokeratin comprising cytokeratins A and D,
equivalent to human cytokeratin polypeptides no. 8 and no. 18,
irrespective of whether they are grown in the presence or absence
of nerve growth factor. They concluded that PC12 cells
permanently produce intermediate filaments of both the
epithelial and the neuronal type and thus contain a combination
of intermediate filaments different from those of adrenal
medullary cells.

PROTEIN KINASES

It has been demonstrated that certain phorbol esters, which are also tumor promoters, mimic the effects of endogenous diacylglycerol in activating protein kinase C. Wakade, Malhotra and Wakade (1986) demonstrated that the secretion of catecholamines from the rat adrenal gland evoked by stimulation of splanchnic nerves, high potassium concentration, or nicotine is facilitated by a phorbol ester. Polymyxin B, an inhibitor of protein kinase C, produced concentration-dependent inhibition of the evoked secretion, and the effect was reversed by the phorbol ester. Furthermore, they showed that an increase in the accumulation of radioactive calcium in the adrenal medulla after stimulation with nicotinic agonists and high potassium concentration is further enhanced by the phorbol ester. They suggested that protein kinase C is involved in the exocytotic secretion of catecholamines by regulating the influx of calcium through voltage-sensitive and nicotine receptor-linked calcium channels of rat chromaffin cells.

Using digitonin-permeabilization to permit manipulation of the intracellular composition of chromaffin cells, Lee and Holz (1986) altered levels of protein kinase C within bovine adrenal chromaffin cells with phorbol esters and dioctanoylglycerol. Only those phorbol esters that activated protein kinase C enhanced catecholamine secretion as well as protein phosphorylation. Dioctanoylglycerol had similar effects. They concluded that the phorbol esters and dioctanoylglycerol enhance calcium-dependent catecholamine secretion, and this is associated with enhanced protein phosphorylation that probably is mediated by protein kinase C.

In a follow-up study, TerBush and Holz (1986) investigated the abilities of activators of protein kinase C to increase the membrane-bound form of the enzyme in intact and digitonin-permeabilized bovine chromaffin cells. They found direct evidence that the enhanced secretion and phosphorylation induced by phorbol esters and diglyceride occur through activation of protein kinase C. Their results also suggested that micromolar concentrations of calcium, in the absence of phorbol esters or exogenous diglyceride, activate protein kinase C in permeabilized and intact cells.

Burgoyne, Morgan and O'Sullivan (1988) examined the role of endogenously activated protein kinase C in calcium-activated exocytosis from digitonin-permeabilized adrenal chromaffin cells. Protein kinase C activity was decreased by down-regulation after long-term treatment with a phorbol ester or by the inhibitor sphingosine. Both treatments resulted in a substantial decrease in the catecholamine secretion elicited by micromolar amounts as calcium, indicating that endogenous activation of protein kinase C is a major requirement for calcium-activated exocytosis in chromaffin cells.

TerBush, Bittner, and Holz (1988) investigated protein kinase C translocation from cytosol to membrane in response to a nicotinic agonist and to depolarization with high potassium concentration. The significant and unique aspects of their study included: 1) the demonstration that calcium ion influx causes an extremely rapid (within 2 seconds) translocation of protein kinase C to membranes which correlate well with the secretory response; and 2) the development of a method to quantitate the relationship between membrane-bound protein kinase C and the enhancement of calcium-dependent secretion. This quantitation leads to the conclusion that secretagogue-induced translocation of protein kinase C to membranes is sufficient to enhance the secretory response.

Knight, Sugden and Baker (1988) presented evidence implicating protein kinase C in exocytosis from electropermeabilized bovine chromaffin cells. The calcium sensitivity of exocytosis was increased by activators of protein kinase C. Putative inhibitors of protein kinase C blocked both the phorbol ester-sensitive component of secretion and the underlying insensitive component. These inhibitors also inhibited protein kinase C activity *in vitro*; protein kinase C activity was also irreversibly inhibited by high potassium concentration.

Brocklehurst, Lee and Pollard (1986) purified protein kinase C from the bovine adrenal medulla by a rapid procedure which resulted in an approximate 1,500-fold increase in specific activity. They found that the properties of protein kinase C in crude and pure preparations are very similar. The calcium concentrations required for half-maximal activation of the enzyme are much higher than the calcium concentrations required for half-maximal secretion from permeabilized cells.

Brocklehurst and Pollard (1986) found that, under certain conditions, the phorbol ester 4B-phorbol 12-myristate 13-acetate can potentiate the calcium-dependent effect of high potassium concentration on catecholamine secretion. In contrast, this phorbol ester does not produce the same enhancement of secretion when the cells are stimulated by acetylcholine, nicotine, or veratridine.

Sasakawa, Ishii, Yamamoto et al. (1986) examined the effects of protein kinase C activators on the dynamics of internal calcium levels in cultured adrenal chromaffin cells. These activators partially inhibited the carbachol-induced increase in intracellular calcium concentration but had no effect on the potassium-induced increase. They suggested that protein kinase C activation causes an uncoupling of signal transduction between nicotinic receptors and calcium channels.

Tachikawa, Takahashi, Shimizu et al. (1987) examined the effect of the specific protein kinase C inhibitor polymyxin B on secretagogue-evoked secretion of catecholamines from cultured bovine adrenal chromaffin cells. Polymyxin B inhibited the phorbol ester-induced secretion of catecholamines and also inhibited secretion induced by the calcium ionophore ionomycin and acetylcholine. Polymyxin B blocked the increase in intracellular calcium concentration induced by acetylcholine or potassium. In contrast, it did not affect the ionomycin-induced increase in calcium. They strongly suggested that catecholamine secretion induced by phorbol esters or ionomycin is mediated via activation of protein kinase C. They further indicated that, in potassium- or acetylcholine-evoked secretion, polymyxin B inhibits secretion by blocking calcium influx into the cells.

The effects of tetanus toxin on catecholamine release from intact and digitonin-permeabilized chromaffin cells were examined by Bittner and Holz (1988). They observed an inhibitory effect that was specific for tetanus exotoxin and the B fragment of tetanus toxin; the C fragment had no effect. The toxin inhibited calcium-evoked secretion from permeabilized cells and could not be overcome with increasing calcium concentrations. Tetanus toxin also inhibited the catecholamine secretion enhanced by phorbol ester-induced activation of protein kinase C. Thus, the toxin or a proteolytic fragment of the toxin apparently can enter digitonin-permeabilized cells to interact with a component of the calcium-dependent exocytotic pathway to inhibit secretion.

DiVirgilio, Pozzan, Wollheim et al. (1986) reported that phorbol ester inhibits the degree of cytoplasmic calcium increase induced by depolarizing agents in PC12 cells. Plasma membrane potential and calcium efflux are not affected. They raised the possibility that the voltage-gated calcium channel is under inhibitory control by protein kinase C.

The roles of various intracellular signals and their possible interactions in the control of neurotransmitter release were investigated in PC12 cells by Meldolesi, Gatti, Ambrosini et al. (1988). Agents that affect primarily the cytosolic concentration of calcium ions and activators of protein kinase C were applied, alone or in combination, to growing chromaffin-like PC12 cells or to neuron-like PC12 cells differentiated by exposure to nerve growth factor. Their results demonstrated that release from PC12 cells can be elicited by both increasing the calcium level and activating protein kinases. Also, these various control pathways interact extensively. Activation of muscarinic receptors by catecholamine carbachol induced appreciable release responses which appeared to be due to a synergistic interplay between calcium and protein kinase C activation. The increasing calcium levels stimulated release in both types of PC12 cells, whereas the activation of protein kinase C was more active in the chromaffin-like PC12 cells. These and other results suggest that release of neurotransmitter from PC12 cells is changed by differentiation, with a diminished role of the mechanism mediated by protein kinase C.

Harris, Kongsamut and Miller (1986) also suggested a role for protein kinase C in the inhibitory regulation of calcium influx and catecholamine release from PC12 cells. They found that, although a phorbol ester had little effect on the cells by itself, it augmented the potassium-evoked release of catecholamines and blocked the influx of calcium.

Messing, Carpenter and Greenberg (1986) examined the effect of phorbol esters on calcium channel function in PC12 cells. They found that these activators of protein kinase C decreased potassium-evoked uptake of calcium and decreased the binding of a calcium channel antagonist to intact cells. Inhibition of binding was markedly decreased in PC12 membranes but was restored by reconstituting membranes with protein kinase C activity. They concluded that protein kinase C may participate in endogenous regulation of voltage-dependent calcium channels.

Hollingsworth, Ukena and Daly (1986) found that a protein kinase C-activating phorbol ester enhanced the responsiveness of the cyclic AMP generating systems to forskolin and 2-chloroadenosine in PC12 cells. Because the phorbol ester mimics the effect of diacylglycerols that form during the turnover of the membrane lipid phosphatidylinositol, the results suggest an interrelationship between the systems involved in phosphatidylinositol turnover and cyclic AMP generation in PC12 cells.

Matthies, Palfrey and Miller (1988) used permeabilized PC12 cells to investigate whether the basic mechanism of secretion also requires the participation of one or more protein kinases or whether regulated protein phosphorylation acts solely to modulate this process. In mutant PC12 cells lacking cAMP-dependent protein kinase and having little or no functional protein kinase C, calcium-dependent secretion occurred normally but was not enhanced by cAMP or phorbol esters. Inhibitors of calmodulin also failed to block calcium-triggered catecholamine release in digitonin-permeabilized PC12 cells. These and other results suggested that the basic mechanism of exocytosis does not involve an event mediated by cAMP-dependent protein kinase, protein kinase C, or calmodulin.

<u>NUCLEOTIDES</u>

The adenine nucleotide stores of cultured bovine adrenal chromaffin cells were radiolabeled by incubating the cells with ^{32}P-labeled phosphate and adenosine, and the turnover, subcellular distribution, and secretion of the nucleotides were examined by Corcoran, Wilson and Kirshner (1986). ATP represented 84% to 88% of the labeled adenine nucleotides; ADP, 11% to 13%; and AMP, 1% to 3%. By using both labeled phosphate and labeled adenosine, the turnover of nucleotides was found to be biphasic with a fast phase (half-life, 4 hours) and a slow phase (half-life, 7 to 17 days). The slow phase appeared to be made up of at least two pools, one being within chromaffin vesicles. Upon stimulation, catecholamines were released in 2 to 3 times the quantities of labeled nucleotides.

To study the role of cyclic nucleotides in the release of catecholamines, Tsujimoto, Morita, Kitayama et al. (1986) determined the effect of cyclic nucleotides on stimulation-evoked catecholamine release and on calcium flux in isolated perfused canine adrenal glands. A soluble form of cyclic AMP (cAMP) substantially enhanced the catecholamine release evoked by acetylcholine and slightly increased basal release. A soluble form of cyclic guanosine monophosphate (cGMP) increased the basal catecholamine release slightly. The calcium efflux evoked by acetylcholine or caffeine was enhanced significantly by soluble cAMP. Tsujimoto et al. (1986) suggested that cAMP may function as a facilitatory modulator in catecholamine release from the adrenal medulla via the effect on calcium flux.

Dohi, Morita, Kitayama et al. (1986) showed that unstimulated efflux of cAMP from perfused dog adrenal glands was slightly increased by a relatively high concentration of forskolin. The acetylcholine-evoked efflux of cAMP that preceded catecholamine release was enhanced by forskolin in a dose-dependent manner. Forskolin did not affect basal catecholamine release but enhanced acetylcholine-evoked release. The catecholamine release evoked by potassium or caffeine also was potentiated by forskolin. They suggested that cAMP generation may increase in response to stimulation of adrenal chromaffin cells and that the resulting increase in this nucleotide may function in modulating catecholamine release.

Morita, Dohi, Kitayama et al. (1987a) examined the effect of forskolin on stimulation-evoked cAMP increase and catecholamine release in isolated bovine adrenal chromaffin cells.

Acetylcholine increased the cAMP concentration with a peak effect at 1 minute after the addition. Pretreatment with forskolin enhanced this increase. The catecholamine release induced by acetylcholine was enhanced by forskolin, but forskolin alone did not enhance release. These and other results suggest that cAMP may play a role in the modulation of catecholamine release from chromaffin cells.

In a follow-up study, Morita, Dohi, Kitayama et al. (1987b) found that stimulation-evoked calcium fluxes in cultured bovine adrenal chromaffin cells are enhanced by forskolin. Forskolin increased acetylcholine-induced uptake and efflux of radioactive calcium. It was suggested that cAMP increases stimulation-induced catecholamine release by enhancing calcium uptake across the plasma membrane or altering calcium flux in an intracellular calcium storage site.

A study of the effects of vasoactive intestinal peptide on chromaffin cell cAMP level and catecholamine secretion was performed by Wilson (1988). He showed that vasoactive intestinal peptide increases the cAMP in primary cultures of bovine adrenal chromaffin cells, an effect that is potentiated by inhibitors of cyclic nucleotide phosphodiesterases. This increase occurs within minutes and lasts for as long as 24 hours. The peptide alone failed to evoke catecholamine secretion from chromaffin cells but potentiated potassium-, veratridine-, and nicotine-evoked secretion. Because vasoactive intestinal peptide is found in nerve terminals innervating the adrenal medulla, this neuropeptide appears to be a modulator of chromaffin cell function, possibly via regulation of cAMP levels.

Chern, Herrera, Kao et al. (1987) found that ATP, ADP, and adenosine inhibit acetylcholine-stimulated secretion from bovine adrenal chromaffin cells. The cells must be incubated with ATP for about 90 sec for maximal inhibition, but inhibition by adenosine occurs much faster. This suggests that ATP and ADP exert their effect after being hydrolyzed to adenosine. They also found that external ATP can diminish the increase in cytosolic calcium that is induced by acetylcholine.

McHugh and McGee (1986) examined the effects of acute treatment of PC12 cells with forskolin. They presented evidence that forskolin has a direct, anesthetic-like, inhibitory effect on carbachol-stimulated uptake of rubidium through the nicotinic acetylcholine receptors, which is not mediated by AMP.

Marriot, Adams and Boarder (1988) showed that treating adrenal chromaffin cells with forskolin increased cAMP level, decreased the maximal stimulation of release of norepinephrine by nicotine, and increased release in response to increased external potassium and the calcium ionophore A23187. The presence of the phosphodiesterase inhibitor Ro 20-17-24 with forskolin potentiated both the stimulation of cAMP and the inhibition of nicotine-induced norepinephrine release. Dideoxyforskolin, an analogue of forskolin that does not stimulate adenylate cyclase, inhibited both potassium- and nicotine-stimulated release. Both forskolin and dideoxyforskolin decreased the calcium transient level in response to nicotine as measured by a fluorescence indicator. These results support a model in which an increase in cAMP inhibits the activation of nicotinic receptors but augments stimulus-secretion coupling downstream from calcium entry. Marriot et al. (1988) pointed out that the data do not indicate a simple relationship between total intracellular cAMP level and the attenuation of nicotinic stimulation of release.

Gatti, Madeddu, Pandiella et al. (1988) studied changes in cAMP concentrations in intact PC12 cells exposed to various treatments. They further characterized the spectrum and the neutral interrelationships of the intracellular signals induced by receptor activation in PC12 cells. This included interactions with cytosolic calcium levels and relationships to nerve growth factor-induced differentiation.

GTP-BINDING PROTEIN

Burgoyne (1987) presented a general review of the role of guanosine triphosphate (GTP)-binding protein (G protein) in the control of exocytosis. After reviewing several related presentations at a recent meeting, he suggested that a new G protein functions in signal transduction leading to the production of an as yet unidentified second messenger that controls exocytosis and perhaps other cellular functions.

Bittner, Holz and Neubig (1986) examined the effect of guanine nucleotides in digitonin-permeabilized bovine chromaffin cells. A nonhydrolyzable analog of GTP produced an ATP-dependent but calcium-independent stimulation of norepinephrine release from the cells. A similar effect was produced by other nonhydrolyzable analogs, but no effect was seen with various nucleotides including GTP. These and other data suggest that a guanine nucleotide-dependent process interacts in some fashion with one or more components of the normal calcium-dependent secretory pathway. However, it may not be an intrinsic part of the mechanism underlying calcium-dependent secretion.

Ahnert-Hilger, Bräutigam and Gratzl (1987) used α-toxin-permeabilized PC12 cells to examine the role of G proteins and protein kinase C in exocytosis. They measured the metabolism of catecholamines in the cytoplasm as an indication of whether or not the secretory products are exposed to the cytoplasm. Catecholamine metabolites formed in the cytoplasm were neither formed nor released from the permeabilized cells in the presence of calcium. Calcium-induced catecholamine release from the permeabilized cells was enhanced by a phorbol ester and inhibited by guanosine 5'-O-(3-thiotriphosphate). The latter effect was abolished by pretreatment of the cells with pertussis toxin but not cholera toxin. Thus, it appears that calcium-induced exocytosis can be modulated via the protein kinase C system as well as by G proteins.

PHOSPHOLIPASES

Phospholipases appear to have a role in stimulus-secretion coupling by mechanisms such as catalysis of the formation of inositol lipids including phosphatidyl-inositol and its phosphorylated relatives. These and related topics were reviewed at length by Hawthorne (1986).

Bartlof and Franson (1987) described the modulation of acid-active vesicle-associated phospholipase A_2 by bovine adrenal medullary cytosol. They demonstrated that various cytosolic and commercially available proteins influence lysosomal phospholipase A_2 activity. Both stimulation and inhibition occur through ionic strength-dependent interaction with substrate. They suggest that such interactions may be important in the *in vivo* regulation of acid-active phospholipases A_2.

Using permeabilized chromaffin cells and the fluorescent probe Quin 2, Stoehr, Smolen, Holz et al. (1986) found that inositol triphosphate specifically triggered an immediate and dose-dependent release of calcium from intracellular stores. Although representing only a small fraction of total cellular calcium, the amount released by inositol triphosphate could significantly enhance cytosolic Ca^{2+} and may account for the effects of muscarine on calcium metabolism in chromaffin cells.

Eberhard and Holz (1987) examined the ability of cholinergic agonists to activate phospholipase C in bovine adrenal chromaffin cells by assaying the production of inositol phosphates in cells prelabeled with tritiated inositol. They found that both nicotinic and muscarinic agonists increased the accumulation of labeled inositol phosphates and that the effects mediated by the two types of receptors were independent of each other. Nicotinic receptor stimulation and increased potassium probably increase the accumulation of inositol phosphates through calcium influx and an increase in cytosolic calcium. This did not appear to happen with muscarinic agonists. They suggested that derivatives of inositol phosphates produced during cholinergic stimulation of chromaffin cells may modulate secretion and other cellular processes by activating protein kinase C or releasing calcium from intracellular stores.

Eberhard and Holz (1988) discussed the evidence for the activation of phospholipase C by intracellular calcium ions. This may be a direct effect rather than the result of calcium-dependent release of neurotransmitters that activate phospholipase C through a receptor-mediated mechanism. Calcium-activated phospholipase C may represent a positive feedback system for calcium: small increases in cytosolic calcium induced by calcium influx across the plasma membrane may produce higher cytosolic calcium concentrations due to inositol triphosphate-induced release of calcium from intracellular stores. The activation of phospholipase C by calcium also may provide a mechanism for diacylglycerol generation and protein kinase C activation after calcium influx. Thus, the regulation of phospholipase C activity by calcium may be physiologically important in regulating cytosolic calcium and protein kinase C in excitable tissues.

Forsberg, Feuerstein, Shohami et al. (1987) pointed out the relationship between adrenal chromaffin cells and adjacent endothelial cells. They measured the effect of several secretory products of chromaffin cells on cultured bovine adrenal medullary endothelial cells and found that only ATP stimulates prostacyclin formation. They suggested that the increase in prostacyclin formation may be secondary to mobilization of intracellular calcium by inositol triphosphate, leading to activation of phospholipase A_2, liberation of arachidonic acid, and conversion of arachidonic acid to prostacyclin. They proposed that the function of ATP, known to be localized with catecholamines in the chromaffin vesicle, may be to regulate blood flow in the adrenal medulla by interacting with adjacent endothelial cells.

Noble, Bommer, Sincini et al. (1986) prelabeled bovine adrenal chromaffin cells with tritiated inositol and then measured the accumulation of labeled inositol phosphate after stimulation with various pharmacologic agents. Carbachol, bradykinin, and histamine produced significant accumulation of labeled inositol phosphate over basal levels; histamine produced the greatest effect. Specific antagonists of H_1 receptors but not H_2 receptors blocked this effect. The question remains as to the role of H_1 receptor activation in the metabolism of chromaffin cells.

Minenko, Kiselev, Tulkova et al. (1987) examined the effects of acetylcholine, after atropine blockade, on the kinetics of orthophosphate incorporation into phosphatidyl phosphate and phosphatidyl bisphosphate as well as on the hydrolysis of inositol phospholipids in rat adrenal medulla slices prelabeled with tritiated *myo*-inositol. They found an increased accumulation of phosphate in the phospholipids; other data indicated a decrease in turnover of phosphatidylinositol bisphosphate.

Sasakawa, Nakaki, Yamamoto et al. (1987) examined the accumulation of inositol triphosphate in cultured bovine adrenal chromaffin cells. Stimulation with potassium increased inositol triphosphate accumulation within the cells.

The effect of substance P on the incorporation of labeled phosphate into inositol phospholipids from rat adrenal medulla slices was examined by Minenko and Oehme (1987). After 20 sec, incorporation of phosphate into all inositol phospholipids was decreased without changes in the phosphatidylcholine/ phosphatidylserine ratio. These and other data suggested a possible mechanism for regulation of catecholamine secretion by substance P related to changes in inositol phospholipids.

The effects of hormonal stimulation in the presence and absence of lithium ions on the formation of inositol mono- and polyphosphates in PC12 cells was investigated by van Calker, Assmann and Greil (1987). Lithium potentiated the agonist-induced increase in amounts of inositol mono-, di-, and triphosphate but not the increase in amount of inositol tetraphosphate.

In the technique-oriented paper by Horwitz and Perlman (1987), inositol phospholipid metabolism in PC12 cells was described. They found that muscarine increases the accumulation of labeled inositol phosphate to the same extent in cells in monolayer and in suspension. Thus, either method can be used to study metabolism in PC12 cells, although each type of cell preparation has certain advantages under specific conditions.

Boarder, Plevin and Marriott (1988) showed that angiotensin II has a powerful synergistic action on prostaglandin E_1-stimulated cAMP accumulation in cultured bovine adrenal chromaffin cells. The response was similar to that for stimulation of phosphoinositide breakdown by angiotensin II in these cultures. The potentiation of stimulated cAMP levels was seen with a protein kinase C-activating phorbol ester; pretreatment with active phorbol ester, which would be expected to diminish protein kinase C levels, attenuated the angiotensin II potentiation of cAMP. Using digitonin-permeabilized cells, they showed that adenylate cyclase activity was stimulated by prostaglandin E_1 in a dose-response relationship as was cAMP accumulation in intact cells, but the permeabilized cells showed no response to angiotensin II. This may be related to the diacylglycerol-mediated activation of protein kinase C by angiotensin II which enhances the stimulatory effect that prostaglandin E_1 has on adenylate cyclase.

Fouchier, Bastiani, Baltz et al. (1988) demonstrated that glycosylphosphatidylinositol is involved in the membrane attachment of proteins in chromaffin vesicles. Incubation at 37°C for treatment of chromaffin vesicle membranes with a certain phospholipase C converted 82- and 68 kDa molecules from amphiphilic to hydrophilic forms. They were immunoreactive with an antibody known to be revealed only after removal of a diacylglycerol anchor. Further experiments suggested the presence of a nonacetylated glucosamine residue in the determinant. This is one of the first reports suggesting that a glycosylphosphatidylinositol anchor might exist in membranes other than a plasma membrane.

The effects of substance P on inositol phospholipids of adrenal medulla slices from spontaneously hypertensive and normotensive rat were investigated by Minenko, Gabrysiak and Oehme (1988). Substance P decreased radioactive phosphate incorporation into inositol phospholipids of both rat strains. This effect was stronger in hypertensive rats. Substance P caused a less potent diesteratic hydrolysis of inositol phospholipids prelabeled with tritium in spontaneously hypertensive rats.

OTHER PROTEINS

Besides the many proteins and related compounds referred to above, several other proteins have been implicated in stimulus-secretion coupling. One of many methods used to elucidate the role of these proteins is microinjection. Trifaró and Seward (1987) reviewed some of the microinjection techniques used to study secretion.

Côté, Doucet and Trifaró (1986b) found that stimulation of bovine adrenal chromaffin cells changed the phosphorylation state of several proteins as examined by labeled phosphate incorporation. Enhanced phosphorylation of 22 protein bands as well as increased dephosphorylation of a 20.4 kDa protein band was observed when extracts of cells stimulated by either acetylcholine or potassium were subjected to gel electrophoresis; acetylcholine and potassium did not always give the same degree of phosphorylation. These and many additional findings demonstrate that protein phosphorylation and dephosphorylation are stimulated during catecholamine secretion from chromaffin cells and that the phosphorylation state of these proteins is affected by calcium deprivation and an antagonist of calmodulin. Côté et al. (1986) pointed out that the relationship between the level of protein phosphorylation and events of stimulus-secretion coupling remains to be established.

The phosphorylation and thiophosphorylation of saponin-permeabilized bovine chromaffin cells was examined by Brooks and Brooks (1987a). The rationale for using ATP-thiophosphate was that it could fix phosphorylation-dependent reactions in the thiophosphorylated state because the resultant thiophosphoproteins were resistant to the action of cell phosphatases. Thiophosphate is incorporated primarily into two cellular proteins, 54 kDa and 47 kDa. Calcium enhanced thiophosphorylation of the 47 kDa but not the 54 kDa protein. They found different electrophoretic banding patterns for thiophosphorylation and phosphorylation. This is likely due to the irreversibility of the thiophosphorylation reaction and the reversibility of the phosphorylation reaction. Brooks and Brooks (1987a) suggested that the inability to turn over thiophosphate groups in association with changes in secretion may permit identification of phosphoproteins that are putatively involved in secretion.

Brooks and Brooks (1987b) examined the modification of protein thiophosphorylation in saponin-permeabilized bovine chromaffin cells by various compounds known to influence chromaffin vesicles. The role of a 47 kDa protein was specifically investigated. Their data suggest that thiophosphorylation of this protein may be associated with nucleotide translocation across the vesicle membrane.

Ehrlich and Kornecki (1987) reviewed the role of extracellular protein phosphorylation systems in the regulation of cellular responsiveness. They concluded that cells which store ATP within secretory vesicles and release it by exocytosis, such as the adrenal chromaffin cell, extracellular protein phosphorylation may be an important mechanism in cellular activation and intercellular communication. Extracellular protein phosphorylation systems have been identified on adrenal chromaffin cells.

The cholinergic regulation of protein phosphorylation in bovine adrenal chromaffin cells was studied by Haycock, Browning and Greengard (1988). After exposure of cultured chromaffin cells prelabeled with radioactive phosphate to acetylcholine, the phosphorylation of 100 kDa and 60 kDa proteins increased. The latter protein was identified as tyrosine hydroxylase. Immunoprecipitation with antibodies to three known phosphoproteins revealed an acetylcholine-dependent phosphorylation of these proteins. These three proteins were also shown to be present in bovine adrenal chromaffin cells by immunolabeling techniques. These were protein IIIa and IIIb and an 87 kDa protein selectively phosphorylated by protein kinase C. These and other data demonstrate that cholinergic activation of chromaffin cells increases the phosphorylation of several proteins and that several protein kinase systems may be involved in these effects.

Using immunoprecipitation to isolate protein III from extracts of total chromaffin cell proteins, Haycock, Greengard and Browning (1988) demonstrated the presence of the subgroups IIIa (74 kDa) and protein IIIb (55 kDa) in chromaffin cells isolated from the bovine adrenal medulla. They also characterized the regulation of the phosphorylation in intact cells by secretagogues. Treatment of chromaffin cells with acetylcholine produced calcium-dependent increases in both the phosphorylation of protein III and the release of catecholamines. These effects of acetylcholine were mimicked by nicotine but not by muscarine. These and other results suggested that protein III phosphorylation was associated more directly with an increase in intracellular calcium than with secretion *per se*.

Phosphorylated proteins of bovine chromaffin cells labeled with radioactive orthophosphate have been analyzed by Gutierrez, Ballesta, Hidalgo et al. (1988) by two-dimensional polyacrylamide gel electrophoresis and autoradiography. Computer-assisted image analysis was used to assemble a data base map of phosphorus-labeled proteins. About 500 polypeptides were detected, numbered, and characterized according to their intensity of labeling, molecular weight, and isoelectric point. The effect of various manipulations on these proteins was also studied. This data base will allow the location of specific phosphoproteins and serve as a reference for future studies.

Sarafian, Aunis and Bader (1987) demonstrated that the loss of cytosolic proteins inhibits catecholamine secretion from cultured bovine adrenal chromaffin cells. Cells were permeabilized with digitonin, and it was noted that the secretory response to calcium diminished with time. Proteins were collected from the medium and were able to restore the secretory response. This demonstrates that, among the proteins leaked from digitonin-permeabilized cells, there are specific proteins crucial to the exocytotic mechanism. Two of these proteins were identified as calmodulin and protein kinase C.

The effects of trypsin on the secretion stimulated by micromolar amounts of calcium ions and phorbol ester in digitonin-permeabilized adrenal chromaffin cells were investigated by Holz and Senter (1988). Addition of trypsin to the cells after digitonin treatment completely inhibited subsequent calcium-dependent catecholamine secretion. The same concentrations of trypsin did not inhibit secretion from permeabilized cells if trypsin was present only prior to cell permeabilization. This indicated that trypsin entered permeabilized cells and that an intracellular trypsin-sensitive protein is involved in secretion. A phorbol ester that activates protein kinase C enhanced the calcium-dependent secretion from permeabilized cells. Trypsin inhibited this enhancement.

Adam-Vizi, Rosener, Aktories et al. (1988) directly tested the hypothesis that inhibition of secretion by botulinum neurotoxin type D occurs by an intracellular process involving ADP-ribosylation. They measured the extents of inhibition of secretion and of ADP-ribosylation in the same cells. Although the inhibitory effect of unpurified toxin closely paralleled intracellular ribosylation, the two events clearly are unrelated because, by using purified D and C-3 toxins together with their antibodies, each of these events could be either stimulated or inhibited independently of each other.

Brocklehurst and Pollard (1988) found that pertussis toxin stimulated catecholamine release from bovine adrenal chromaffin cells in a calcium-dependent manner. This occurred in the absence of any stimulatory or inhibitory agonists. The release of catecholamines was associated with the ADP-ribosylation of an approximately 40 kDa protein present in the total membrane fraction. These results are consistent with the existence of an exocytosis-linked G-protein.

Grandori and Hanafusa (1988) localized the proto-oncogene product p60c-*src* in chromaffin vesicle membranes. It is part of a complex with a 38 kDa protein, and enrichment of kinase activity is due to a parallel enrichment of p60c-*src* protein. It was suggested that the 36 kDa protein along with p60c-*src* may be related to secretory activities.

An 80 kDa protein that binds to G-actin at an equimolar ratio in a calcium-dependent manner was purified chromatographically from bovine adrenal medulla by Ashino, Sobue, Seino et al. (1987). The protein is a monomer and under certain conditions causes fragmentation of actin filaments. Considering the mode of action on actin filaments, Ashino et al. (1987) suggested that this 80 kDa protein may be a gelsolin-like protein. It is also probably a calcium-binding protein.

Bader, Trifaró, Langley et al. (1986) identified a gelsolin-like protein in adrenal chromaffin cells obtained from cow and pig.

Actin- and calcium-sensitive actin-binding proteins were purified on affinity columns using DNase I as ligand. A 91 kDa actin-binding protein was eluted and shown to be reactive against an antibody to gelsolin. A 93 kDa protein, found to be similar to brevin, was shown to be a component of blood but not chromaffin cells. An 85 kDa protein also was found but it apparently is not related to gelsolin. They suggested that calcium may activate the 91 kDa protein, and this may be important for the movement of chromaffin vesicles to the plasma membrane.

The localization of fodrin, also known as spectrin, was investigated at the ultrastructural level in the rat adrenal gland by Langley, Perrin and Aunis (1986). By use of an affinity-purified antibody directed against the α-fodrin subunit, all chromaffin cells, cortical cells, and nerve fibers and their surrounding Schwann cells were found to be labeled close to the cytoplasmic side of their plasma membranes. The labeling appeared to be more intense in chromaffin cells. It was suggested that fodrin may play a role in the positioning of chromaffin vesicles near the plasma membrane.

Perrin, Langley and Aunis (1987) showed that anti-α-fodrin inhibits secretion from permeabilized chromaffin cells. The antibody, or its Fab fragments, did not modify basal release but did specifically inhibit calcium-induced catecholamine release. Their observations indicate that fodrin and the cytoskeleton participate in the release mechanism. In commenting on this article, Burgoyne and Cheek (1987b) suggested that fodrin is directly responsible for the initial interaction of secretory vesicle and plasma membranes leading to membrane fusion or that the calcium-activated proteolysis of fodrin may be part of the process by which the peripheral cytoskeleton is cleared to allow secretory vesicles access to exocytotic sites. Commenting on these two articles, Siegel (1987) suggested that synapsin-like molecule may interact with fodrin and other cytoskeletal proteins in the secretory response. In a follow-up comment, Burgoyne and Baines (1987) pointed out that the interaction of actin filaments with fodrin and membranes may be a mechanism common to chromaffin cells, nerve terminals, and other cells.

Tanaka, Yokohama, Negishi et al. (1987) showed that pretreatment of cultured bovine adrenal chromaffin cells with pertussis toxin facilitated nicotine-induced catecholamine release. This facilitation was correlated with the ability of the toxin to catalyze the ADP-ribosylation of an approximately 40 kDa membrane protein. When ADP-ribosylation was inhibited, the response to the toxin was reversed. They suggested that this ADP-ribosylation facilitates catecholamine release, apparently without affecting calcium uptake.

Knight (1986b) demonstrated that evoked catecholamine secretion from cultured bovine adrenal chromaffin cells is inhibited by commercially available botulinum toxins types A, B, and D. Basal secretion is also inhibited. Calcium influxes are unaffected. Knight (1986b) suggested that this evidence supports the idea that botulinum toxins block secretion by acting downstream from the calcium transient, at or near the site of exocytosis.

In a report of a meeting on exocytosis, Burgoyne (1987) suggested that the protein p21 (a product of the *ras* oncogene) could be the substrate for botulinum toxin type D. Adam-Vizi, Knight and Hall (1987) clarified this by pointing out that the membrane protein that is ADP-ribosylated by botulinum toxin type D is not p21. They showed that, although botulinum toxin type D ribosylates proteins from bovine adrenal medulla homogenate, the toxin fails to ADP-ribosylate purified N- and HA-*ras* gene products under similar incubation conditions.

Matsuoka, Syuto, Kurihara et al. (1987) showed that botulinum toxin types C_1 and D caused ADP-ribosylation of a 24 kDa protein in membranes prepared from PC12 cells. The ribosylation reaction was dependent on the presence of magnesium chloride and GTP. Matsuoka et al. (1987) suggested that the ADP-ribosylation reaction is responsible for the development of the biologic activity of the botulinum neurotoxins and that the target of this reaction may be novel GTP-binding proteins localized on cell membranes.

To characterize dopamine release by permeabilized PC12 cells, Ahnert-Hilger and Gratzl (1987) depleted cells of endogenous ATP before stimulation with calcium. They found that magnesium ATP, calmodulin, or proteins containing sulfhydryl or hydroxyl groups are not necessary for exocytosis in permeabilized PC12 cells.

Morita, Ishii, Uda et al. (1988) leached ATP from digitonin-permeabilized adrenal chromaffin cells in an effort to examine the role of ATP in exocytosis. Catecholamine release from these pretreated cells was studied in the absence and presence of exogenous ATP. In an ATP-free medium, exocytosis induced by calcium was greatly diminished; this could be restored by the presence of ATP but not other adenine nucleotides. This suggested that cytoplasmic ATP may be involved in the exocytotic mechanism of catecholamine secretion.

The effect of the metabolic inhibitor 2,4-dinitrophenol on the concentration of cytoplasmic ATP and the activity of catecholamine secretion was studied by Nakanaishi, Morita and

Oka (1988) in cultured bovine adrenal chromaffin cells. Pretreatment of the cells with 2,4-dinitrophenol resulted in a decrease in the concentration of cytoplasmic ATP. Catecholamine secretion evoked by either carbachol or high potassium also was decreased by this pretreatment. These and other results suggested that ATP in the cell cytoplasm may play an important role in the regulation of catecholamine secretion as a factor modulating the activity of exocytosis in the adrenal chromaffin cell.

The secretory proteins from adrenal chromaffin vesicles, specifically chromogranins, were found by Veeraragavan, Coulombe and Gagnon (1988) to be excellent methyl acceptor proteins *in vitro*. Dopamine β-hydroxylase, another secretory protein within chromaffin vesicles, was poorly methylated in their experimental system. These chromogranins are the first secretory proteins to be methylated stoichiometrically *in vitro* without prior deamidation. The role of chromogranin methylation remains to be determined.

Gagnon, Veeraragavan and Coulomb (1988) examined protein carboxymethylation in adrenal chromaffin cells. Under physiologic conditions, they found that stimulation of a splanchnic nerve results in an increase in adrenal medullary protein methyl ester formation as well as in augmented methanol production. With adrenal medullary cells in culture, carboxymethylated chromogranin A is detected in mature chromaffin vesicles after labeling. These and other data suggest that methylation of proteins may occur in nascent chains before the peptides are injected into the rough endoplasmic reticulum.

Mundy, Hermann and Strittmatter (1987) noted that chromaffin cells have metal-dependent endoproteinases in both the plasma membrane and the soluble fraction of homogenized cells. They demonstrated that the metal-dependent endoproteinases in these two subcellular fractions can be differentiated by selective inhibitors. In both intact and permeabilized cells, the inhibitor phosphoramidon did not block exocytosis and did not inhibit the soluble proteinases but did inhibit the plasma membrane metalloproteinase. Because soluble metalloproteinase activity is inhibited by other proteinase inhibitors at concentrations that block exocytosis, a soluble metalloproteinase, but not the plasma membrane one, appears to be required in exocytosis.

Farber, Wilson and Wolff (1987) found calmodulin-dependent phosphoprotein phosphatase activity in PC12 cells as well as in other cultured cell lines. This phosphatase apparently is related to that found in the brain.

CHAPTER 6

STIMULUS-SECRETION COUPLING: EXOCYTOSIS

The ultimate event in stimulus-secretion coupling is the release of secretory products into the extracellular space. This occurs chiefly, if not exclusively, via the mechanism of exocytosis. The membrane fusion events associated with exocytosis have been reviewed by DeLisle and Williams (1986), Green (1987) and Lindstedt and Kelly (1987). These reviews cover evidence from studies on the chromaffin cell and other systems. Green (1987) discussed the opposing views as to whether swelling of a secretory vesicle precedes membrane fusion or follows it. It was concluded that a role for local swelling and membrane fusion may exist in systems such as the chromaffin vesicle. Lindstedt and Kelly (1987) pointed out that we have a better understanding of how exocytosis is triggered than of the exocytotic event itself.

The evidence that exocytosis is the mechanism of secretion from the adrenal medulla and sympathetic nerves was reviewed by Winkler (1988). Among other things, he pointed out that exocytosis generally has been accepted as a mechanism of secretion for some time, but the molecular mechanisms involved in exocytosis are still a matter of speculation.

Strittmatter (1988) reviewed the molecular mechanisms of exocytosis by using the adrenal chromaffin cell as a model system. This review emphasized new insights in the molecular mechanisms of exocytosis.

Permeabilized cells have been shown to be a valuable model for study of stimulus-secretion coupling. Bader, Thiersé, Aunis et al. (1986) used chromaffin cells that were permeabilized with a toxin from *Staphylococcus aureus*. α-Toxin was shown to permeabilize the plasma membrane but not the membrane of chromaffin vesicles. In α-toxin-treated cells, released proteins could not be sedimented and lactate dehydrogenase was still associated within cells, which provides direct evidence that the secretory product is liberated by exocytosis. Bader et al. (1986) pointed out certain advantages of permeabilizing cells with this α-toxin compared with digitonin.

Direct morphologic evidence of exocytosis from cultured bovine adrenal chromaffin cells was presented by Brooks and Carmichael (1987). Stimulation of the cells in the presence of tannic acid resulted in precipitation of secretory proteins during the process of exocytosis. In this fashion, exocytotic profiles were integrated over a 10-minute time period. They also demonstrated holes within the plasma membrane of saponin-permeabilized cells.

Using immunolabeling methods for detecting the successful fusion of chromaffin vesicle membranes with the plasma membrane, Schäfer, Karli, Gratwohl et al. (1987) provided evidence that digitonin-permeabilized PC12 cells retained their capacity to release catecholamines via exocytosis. They also demonstrated that it is possible to introduce macromolecules as large as antibodies into exocytosis-competent chromaffin cells and into PC12 cells. This led to the conclusion that pores formed by digitonin do not impair the process of exocytosis, although they are big enough to allow macromolecules to pass in both directions.

Seidler and Slotkin (1986a) pointed out that neonatal rats can secrete adrenal catecholamines by a mechanism that does not require nerve stimulation. Using two different stimuli, hypoxia and nicotine, they found that depletion of catecholamines was accompanied by a decrease in uptake, indicating a loss of chromaffin vesicle integrity, which is characteristic of exocytosis. They found a significant proportion of adrenal catecholamines to be extravesicular, but only the vesicular pool was involved in the secretory response.

Forsberg and Pollard (1988) measured the exocytotic secretion of ATP from adrenal medullary cells by an on-line method. They found that barium ions induced ATP release and that this required the entry of the ions through either voltage- or receptor-gated calcium channels. This conclusion is based on the observations that short preincubations with low concentrations of either nicotine or potassium ions greatly enhanced barium-induced ATP release and that this augmentation could be blocked with the nicotinic receptor antagonist hexamethonium or the calcium antagonist nifedipine, respectively. Moreover, both nicotine and potassium stimulated uptake of radioactive barium. These results support the hypothesis that the cellular events leading to barium-induced secretion coincide, at least in part, with the events leading to calcium-dependent exocytosis.

The functional integrity of adrenal chromaffin vesicles was studied in the perfused rat adrenal gland by Wakade, Wakade and Malhotra (1988). Continuous perfusion with a fluid at high potassium concentration resulted in a large release of catecholamines, representing almost 30% of the total tissue catecholamine content, within 45 minutes. Despite such a large secretory response, the catecholamine content of the potassium-stimulated adrenal medulla was comparable to that in unstimulated controls, suggesting an enhanced resynthesis to maintain normal levels. When tyrosine hydroxylase was inhibited, tissue catecholamine content decreased. Depleted catecholamines could be restored by adding tyrosine, dopa, or dopamine (not norepinephrine or epinephrine) to the perfusion medium. This repletion could be blocked by a tyrosine hydroxylase inhibitor but not by calcium deprivation. It was suggested that chromaffin vesicles may be reutilized for the purpose of synthesis, storage, and secretion of catecholamines.

The effect of free fatty acids on the cation-induced fusion of large liposomes was investigated by Meers, Hong and Papahadjopoulos (1988). Arachidonic acid was found to act synergistically with promoters of liposomal aggregation such as magnesium ions, spermine, and synexin. The overall rate of liposome fusion was enhanced, as would be expected from actions at separate kinetic steps.

<h2 style="text-align:center"><u>TECHNIQUES</u></h2>

During exocytosis, the entire contents of the chromaffin vesicles are thought to be liberated into the extracellular space or into the medium in the case of cultured cells. This includes the release of ATP. Two methods for the detection of ATP secreted from chromaffin cells were described (White, Bourke and Livett, 1987; Rojas, Forsberg and Pollard, 1987). White et al. (1987) developed a method for direct, continuous detection of secretion of ATP by using the firefly luciferin-luciferase assay with detection by a photomultiplier directly below the culture well. They were able to achieve a time resolution of secretion on the order of seconds. They could detect desensitization of the nicotinic receptor and demonstrated a biphasic secretory response. Rojas et al. (1987) used a similar assay system and also achieved good time resolution. They included an interesting discussion of the biophysics of the assay and of exocytosis in general.

Catecholamine release from perfused cat adrenal glands was continuously monitored on-line by running the perfusion fluid through an electrochemical detector in a method developed by Borges, Sala and García (1986). They pointed out several advantages of their system, including the ability to examine catecholamine released by low concentrations of acetylcholine, nicotine, or potassium.

As mentioned in Chapter 3, Boarder and McArdle (1986) and Kumakura, Ohara and Satô (1986) reported methods for real-time monitoring of catecholamines from cultured chromaffin cells.

Development of a cell-free model of exocytosis is a high-priority goal in many laboratories. Such a reliable model would be extremely valuable in studying exocytosis. Commentaries by DeBlock and DePotter (1987a; 1987b) and Konings (1987) discussed an earlier model of this type involving fusion of chromaffin vesicles with vesicles made of bovine adrenal medullary plasma membranes. There apparently are difficulties with this system; and many of the earlier reported results apparently were artifacts.

Izumi, Yanagihara, Wada et al. (1986) used isolated chromaffin vesicles and plasma membranes from adrenal medullary cells to study the possible involvement of phospholipase A_2 in stimulus-secretion coupling. Treatment of plasma membranes with the phospholipase in the presence of calcium caused lysis of chromaffin vesicles in a calcium-dependent manner. They suggested that calcium acts in two steps of exocytosis: transformation of plasma membranes into a lytic form accomplished by phospholipase A_2 and interaction of chromaffin vesicles with plasma membranes. This method needs to be examined very carefully because of artifacts, as mentioned above.

Stutzin (1986) developed a technique for studying fusion of biologic membrane vesicles. Empty bovine chromaffin vesicles (ghosts) were loaded with a fluorescent compound at self-quenching concentrations and then made to fuse with empty ghosts. The fusion was monitored by measuring the unquenching of the fluorescence. Stutzin (1986) suggested that this method may be of value for studying fusion processes in many membrane systems.

Kolchinskaya, Chaika and Kravets (1988) examined the interaction of chromaffin cell membrane fragments with artificial phospholipid membranes. Membrane fragments obtained from bovine adrenal medulla were shown to bind tritiated nitrendipine. The interaction of these fragments with an artificial phospholipid bilayer induced preferable conductance for calcium and barium ions. This effect was blocked by cadmium and stimulated by a calcium channel agonist, BAY K 8644. Kolchinskaya et al. (1988) suggested that fragments of chromaffin cell membrane containing functional calcium channels can be incorporated into an artificial membrane.

<u>OSMOTIC FORCES</u>

The role of osmotic forces in exocytosis from adrenal chromaffin cells was reviewed by Holz (1986). In a study of calcium-dependent secretion from digitonin-permeabilized bovine chromaffin cells, Holz and Senter (1986) found that shrunken vesicles can indeed undergo exocytosis and that, with intact cells, increased ionic strength rather than chromaffin vesicle shrinkage contributes to the inhibition of secretion by hyperosmotic solutions.

Wakade, Malhotra, Sharma et al. (1986) showed that perfusion of the rat adrenal medulla with a hypertonic solution followed by an isotonic solution resulted in an explosive secretion of catecholamines; this was due to prolonged opening of calcium channels.

Ladona, Bader and Aunis (1987) used cultured bovine adrenal chromaffin cells to study the effects of hyperosmolarity on the nicotine- and high potassium-induced secretory response. They also used cells permeabilized with digitonin or α-toxin. Hyperosmolarity does not affect the spontaneous release of catecholamines from either intact or permeabilized cells. The nicotine- and potassium-induced secretions from intact cells are inhibited by hypertonic solution; 100% inhibition of net release was observed at 660 mosM. This inhibition was reversible. They also showed that hyperosmolarity has intracellular effects on calcium release from permeabilized cells evoked by calcium. They concluded that hyperosmolarity has multiple effects on the cell membrane and the protein constituents associated with it but also has a significant effect on intracellular events associated with exocytosis.

In a range of intracellular calcium levels similar to that seen during secretion, Miyamoto and Fujime (1988) noted a drastic decrease in the elastic modulus of bovine chromaffin vesicle membrane. These measurements were made by a combination of osmotic swelling and dynamic light-scattering methods. This result suggests that the vesicle membrane becomes extremely flexible as a prelude to exocytosis.

<u>MEMBRANE</u> <u>RETRIEVAL</u>

Exocytosis involves the fusion of the secretory vesicle membrane with the plasma membrane. Obviously, if this were to continue for too long a time, the surface area of the plasma membrane would increase to an untenable size. It is obvious that retrieval of membrane must take place in order to maintain a more or less constant cell surface. The question has been, "Is the retrieved membrane specifically the vesicle membrane that was inserted or is it any random membrane?" This question was answered by Patzak and Winkler (1986). They used antibodies to a specific glycoprotein known to be localized on the inner surface of the chromaffin vesicle membrane. This glycoprotein would be exposed on the cell surface after exocytosis. Using studies at both the light and electron microscope levels, they identified patches of membrane labeled with the antibody and these patches were internalized within 45 minutes after the cessation of stimulation. In this fashion, they were able to follow the specific retrieval of vesicle membrane from the surface and identify its intracellular pathway. The vesicles are partly recycled to newly formed chromaffin vesicles.

In a follow-up study, Patzak, Aunis and Langley (1987) used morphometric methods to examine at the ultrastructural level the events that occur after retrieval. They found that the labeling was most pronounced in prelysosomes and lysosomes. The retrieved membrane did not appear to recycle into new vesicles. They suggested that the fate of vesicle membrane retrieved after cell stimulation may be influenced by external conditions.

Lotshaw, Ye and Edwards (1986) used antibodies against another glycoprotein, dopamine β-hydroxylase, to follow the specific retrieval of chromaffin vesicle membranes after exocytosis. At the light microscope level, they found anti-dopamine β-hydroxylase fluorescence distributed as patches. Thirty minutes after labeling, the fluorescence appeared to be almost completely internalized. Other experiments showed that the half-life of the label on the cell surface was approximately 7 minutes. These results demonstrate that dopamine β-hydroxylase is rapidly and selectively retrieved from the cell surface, probably from the site of exocytosis.

vonGrafenstein, Roberts and Baker (1986) used horseradish peroxidase as a fluid phase marker to monitor the time course of endocytosis and membrane retrieval in a quantitative manner. They developed a kinetic model which assumes that exocytotic membrane fusion and endocytotic membrane retrieval are consecutive events. They found that extracellular calcium is not essential for membrane retrieval.

OTHER CONSIDERATIONS

Exocytosis is widely considered to be the basic mechanism for release of chromaffin vesicle contents. However, Knoth, Viveros and Diliberto (1987) presented evidence of release of newly acquired ascorbate and α-aminoisobutyric acid from adrenal chromaffin cells through specific transporter mechanisms. They demonstrated that these substances are taken up by the cells into the cytosol and are released in a fashion different from that of endogenous catecholamines when stimulated under various conditions. For example, in permeabilized cells, catecholamines are released only in the presence of external calcium whereas ascorbate and α-aminoisobutyric acid are released in a calcium-independent fashion. The release of these two compounds is sensitive to inhibitors of metabolism and transport.

Penner, Neher and Dreyer (1986) noted that injection of botulinum neurotoxin type A into chromaffin cells strongly inhibited secretion as revealed by the measurement of cell capacitance. Their results indicate that such toxins, which normally are ineffective in chromaffin cells because they are not bound and internalized, are ineffective because they do not reach their site of action. Furthermore, they localized the secretion-blocking effects of the toxin to a fragment representing the light chain covalently linked to part of the heavy chain, suggesting that this part of the molecule contains the active site.

Schäfer, Karli, Schweizer et al. (1987) examined chromaffin vesicles in chromaffin cells and PC12 cells that were aligned underneath the plasma membrane. This alignment, which is followed by the recognition of an attachment to fusion sites in the plasma membrane, is referred to as "docking." Docking was easier to observe with the electron microscope in permeabilized PC12 cells than in chromaffin cells. Transport of chromaffin vesicles from the Golgi to the plasma membrane docking sites seems to depend on a mechanism sensitive to permeabilization.

Tapparelli, Grob and Burger (1987) studied membrane events in exocytosis by examining the effect of different detergents on potassium-stimulated release of norepinephrine from PC12 cells. They found that Triton X-100, having a chain length of 16 carbon atoms, was effective in inhibiting secretion. Other data suggest that the site of action of Triton X-100 is at the level of altering the movement of ions during the stimulatory phase of secretion.

The effects of protease inhibitors on the secretion of catecholamines were studied in cultured bovine adrenal chromaffin cells by Nakanishi, Morita, Oka et al. (1986). Although inhibitors of serine proteases could inhibit the carbachol-induced secretion, they failed to inhibit secretion evoked by either high potassium concentration or the calcium ionophore A23187. A thiol protease inhibitor had no effect on secretion. The results indicate that serine protease inhibitors may inhibit receptor-mediated secretion, probably through their effects on the plasma membrane, thus suggesting that involvement of the serine and thiol proteases in exocytosis is unlikely.

Citovisky, Laster, Schuldiner et al. (1987) demonstrated that

Sendai virus particles are able to fuse with vesicles obtained from bovine chromaffin vesicles. This was demonstrated by electron microscope studies or by fluorescence dequenching in appropriate preparations. Their results indicate that, under hypotonic conditions, fusion between Senai virions and biologic membranes does not require the presence of specific receptors. Such fusion is characterized by the same features as fusion of living cultured cells with and infection by, Sendai virions.

CHAPTER 7

EFFECTS OF DRUGS

MPTP

Di Paolo, Bédard, Daigle et al. (1986) studied the long-term effects of 1-methyl-4-phenyl-1,2,3,6-tetrahydropyridine (MPTP) on catecholamine concentrations in monkeys. They found that, at 5 months after the start of MPTP treatment, epinephrine and dopamine levels were decreased in the adrenal medulla but norepinephrine concentration remained unchanged.

Brooks, Brooks and Carmichael (1987) examined the effect of chronic and acute MPTP administration on the function and structure of cultured bovine adrenal chromaffin cells. They found that MPTP causes a profound depletion of catecholamines in chromaffin cells that is evident both biochemically and morphologically by electron microscopy. The biochemical changes are not prevented by inhibitors of monoamine oxidase or desipramine.

Reinhard, Diliberto, Viveros et al. (1987) examined the subcellular compartmentalization of MPP^+ (a metabolite of MPTP) in cultured bovine adrenal chromaffin cells. Cells accumulated MPP^+ in a time- and concentration-dependent manner, and the compound was localized predominantly in chromaffin vesicles. Negligible amounts of MPP^+ were detected within mitochondria or in the cytosol. MPP^+ was co-secreted with catecholamines. Incubation of cells with MPP^+ in the presence of an inhibitor of vesicular uptake potentiated the toxicity of MPP^+ to the cells. Reinhard et al. (1987) suggested that the ability of chromaffin cells to take up and store MPP^+ in the chromaffin vesicle prevents the interaction of the toxin with other structures and thereby prevents damage to the cell.

Daniels, Reinhard and Painter (1988) used nuclear magnetic resonance to study the accumulation of MPP^+ into bovine chromaffin vesicles. MPP^+ labeled with [13]carbon is taken up by chromaffin vesicles; the linewidth of this nuclear magnetic resonance signal was significantly increased, suggesting a restriction in the molecular motion of the molecule. This was confirmed by additional studies. Nuclear magnetic resonance data on the interaction between MPP^+ and ATP suggest that there is not a specific interaction between intravesicular ATP and MPP^+ but rather the restriction of the molecule is due to the viscosity of the vesicular matrix.

Scherman, Darchen, Desnos et al. (1988) demonstrated that MPP$^+$ is a substrate of the vesicular monoamine uptake system of chromaffin vesicles. Chromaffin vesicle ghosts accumulated tritiated MPP$^+$ in a manner that was time-dependent and dependent on ATP and magnesium. They presented additional data to establish that MPP$^+$ is a substrate of the vesicular monoamine transporter. This illustrates the broad specificity of this transporter.

The metabolism of MPTP in bovine chromaffin cells was examined by Wilson and Beeler (1988). They also examined the loss of catecholamines induced by MPTP and MPP$^+$ and the accumulation of these toxic molecules in chromaffin cells. MPP$^+$ appeared to be costored with catecholamines in chromaffin vesicles, whereas MPTP was found in soluble and particulate fractions but not in chromaffin vesicles. Although chromaffin cells form MPP$^+$ from MPTP and store MPP$^+$, these studies suggested that catecholamine depletion in these cells results from the actions of MPTP and MPP$^+$ that are not stored in chromaffin vesicles. MPTP evokes catecholamine secretion from chromaffin cells via nicotinic receptors. MPTP and free MPP$^+$ are toxic to the cells, apparently through blockade of metabolic processes. These and other findings suggest that cells with a high capacity for MPP$^+$ uptake but limited vesicular storage capacity would be susceptible to the toxic actions of this drug.

Hadjiconstantinou, Tjioe, Alho et al. (1987) presented morphologic and biochemical evidence that MPTP causes the accumulation of lipofuscin in the cortex of the mouse adrenal gland. Catecholamine contents in the medulla were significantly diminished.

The uptake and metabolism of MPTP were examined in PC12h cells by Naoi, Takahashi and Nagatsu (1987). They found that MPTP was oxidized to MPP$^+$, although this is generally thought to be catalyzed by type B monoamine oxidase and these cells contain only the type A enzyme. Their data show that the uptake and oxidative conversion of MPTP may take place in cells with monoamine oxidase type A and that the neurotoxicity of MPP$^+$ may not be due directly to its storage in subcellular compartments.

The short-term effects of MPTP on catecholamine levels in the heart, adrenal gland, retina, and caudate nucleus of the cat were studied by Ambrosio, Blesa, Mintenig et al. (1988). The content of epinephrine, but not norepinephrine, was significantly lower in MPTP-treated cats than in controls. Decreases in catecholamines in other organs were also noted.

The distribution of radioactivity in mice after a single injection of tritiated MPTP was studied by quantitative whole-body autoradiography and liquid scintillation counting by Lydén-Sokolowski, Larsson and Lindquist (1988). A high uptake of radioactivity was observed in the adrenal medulla and some specific regions of the brain. Inhibitors of monoamine oxidase B decreased this uptake, but an inhibitor of monoamine oxidase A or a catecholamine reuptake inhibitor did not cause a dimunition in the MPTP uptake.

Naoi, Takahashi, Ichinose et al. (1988a) demonstrated that aromatic L-amino acid decarboxylase activity was inhibited in PC12h cells by MPP^+. The enzyme activity was enhanced by the addition of the cofactor pyridoxal phosphate, and MPP^+ inhibited this enhancement. MPP^+ did not affect the binding of the enzyme to substrates. In further experiments, Naoi, Ichinose, Takahashi et al. (1988b) showed that the intracellular concentration of the decarboxylase protein was not decreased. These and other data indicate that MPP^+ may induce conformational changes in aromatic L-amino acid decarboxylase protein and decrease its affinity to the cofactor pyridoxal 5-phosphate.

In a follow-up study, Naoi, Takahashi and Nagatsu (1988) showed that MPP^+ markedly decreased the dopamine content in PC12h cells after 6 days in culture. MPP^+ was found to decrease the activity of tyrosine hydroxylase, monoamine oxidase, and aromatic L-amino acid decarboxylase. These results indicate that MPP^+ not only inhibits the synthesis of catecholamines but also that of an enzyme participating in their catabolism in cells and thus may perturb catecholamine levels in various neuronal systems.

The effects of MPTP and its metabolite MPP^+ were studied in specific neuronal and glial cell lines in vitro by Notter, Irwin, Langston et al. (1988). One neuronal line was a subclone of a neuroblastoma line. MPTP had no morphologic effect on actively growing neuroblastoma cells. However, a low dose of MPP^+ was cytotoxic to the neuroblastoma cells, inducing vacuole formation, cell lysis, and cell growth inhibition over a 3-day period. Protein synthesis was inhibited by MPTP in a dose-dependent manner. These and other findings suggested that this in vitro system may be a useful model for assessing the role of individual cell populations in response to neurotoxic agents.

ETHANOL

Skattebol and Rabin (1987) examined the effects of ethanol on calcium uptake by PC12 cells. Their results indicate that ethanol acts primarily on the voltage-dependent calcium channel, and the direct effect of chronic ethanol exposure is an increase in calcium channel density and in sensitivity to ethanol.

The effect of chronic alcohol administration on the adrenal glands, thyroid gland, pancreas, and liver of rats was investigated by Markotko and Pankow (1987a). Morphometric and histochemical findings in the adrenal medulla indicate increased secretion and synthesis of catecholamines.

Harper, Pagonis and Littleton (1987) showed that ethanol inhibited catecholamine release from cultured adrenal chromaffin cells. The inhibition appeared to be a consequence of increased functional activity of dihydropyridine-sensitive calcium channels.

The effect of ethanol on muscarine-stimulated release of tritiated norepinephrine was studied by Rabe and Weight (1988) in PC12 cells. Above a threshold level, ethanol produced a dose-dependent inhibition of muscarine-stimulated release. Ethanol also inhibited the muscarine-stimulated increase in intracellular free calcium that corresponded with the inhibition of transmitter release. These and other results suggested that the effects of ethanol on neurotransmitter release are associated with the effects of ethanol on intracellular free calcium levels.

ANESTHETICS

Several laboratories have investigated the effects of anesthetic agents on the adrenal medulla. Yashima, Wada and Izumi (1986) showed that halothane inhibits the cholinergic receptor-mediated influx of calcium in cultured bovine adrenal chromaffin cells. Halothane did not inhibit the influx of calcium caused by depolarization with potassium.

The effects of halothane anesthesia on sympatho-adrenal function in rats were studied by Kawate, Kumekawa, Meguro et al. (1987). The activity of the nerve going to the adrenal gland decreased when clinically relevant doses of halothane were given. These and other results led to the conclusion that halothane decreases adrenal catecholamine secretion by depressing adrenal nerve activity as well as by acting directly on adrenal chromaffin cells. Since the effect was greater on epinephrine than norepinephrine secretion, it was also suggested that the two adrenal catecholamines are controlled by different nerve fibers.

Maggi and Meli (1986) reviewed the suitability of urethan anesthesia for physiopharmacologic investigations. They called attention to evidence that doses of urethan higher than clinically used lead to activation of the adrenal medulla; however, they pointed out that the information is insufficient to permit concluding that this effect is necessarily a consequence of urethan anesthesia.

The effects of four volatile anesthetics (halothane, isoflurane, influrane, and methoxyflurane) on the potassium-evoked increase in intracellular calcium in PC12 cells were investigated by Kress, Eckhardt-Wallasch, Tas et al. (1987). They found that the increase in intracellular calcium concentration was depressed in a similar fashion by all four anesthetics at clinical concentrations. They suggested that the anesthetics interfere with calcium fluxes through voltage-gated channels.

The effect of the local anesthetics tetracaine and bupivacaine on enzyme activities in the rat adrenal medulla was examined by Rosenberg, Nissinen, Mannisto et al. (1987). Tyrosine hydroxylase, dopamine β-hydroxylase, and catecholamine-O-methyltransferase were not inhibited by either anesthetic.

The action of volatile anesthetics on stimulus-secretion coupling in bovine adrenal chromaffin cells was studied by Pocock and Richards (1988). Concentrations of ethrane, halothane, isoflurane, and methoxyflurane within the anesthetic range inhibited the secretion of epinephrine and norepinephrine evoked by carbachol. Catecholamine secretion induced by stimulation with potassium was also inhibited but at higher concentrations. All four agents inhibited calcium influx, and this appears to be sufficient to account for the inhibition of secretion.

The sympathoadrenal medullary functions in aged rats under anesthesia were examined by Kurosawa, Sato, Sato et al. (1988). Among other things, they observed that both spontaneous ongoing sympathetic nerve activity and catecholamine secretion rates from the adrenal gland increased during aging in anesthetized and resting rats. The increases in sympathoadrenal medullary function during aging were reflected by the increased concentrations of catecholamines in plasma.

OTHER AGENTS

Hexum and Russett (1987) showed that administration of nicotine via osmotic mini-pumps to guinea pigs results in a significant increase in plasma concentrations of epinephrine and enkephalin-like peptides but not norepinephrine. The increase in plasma epinephrine and peptides was not accompanied by corresponding alterations in either adrenal medullary synthesis or blood pressure and heart rate. They suggested that this may partially explain the observation that chronic smoking does not induce hypertension.

The effects of insulin and insulin-like growth factors on the replication of PC12 cells were investigated by Dahmer and Perlman (1988a). Addition of insulin to cells in a low serum medium increased the incorporation of tritiated thymidine into the cells, increased the number of cells in the cultures, and decreased the percentage of cells in the G_0/G_1 phase of the cell cycle. Insulin-like growth factors also increased thymidine incorporation. These data suggest that insulin and insulin-like growth factors are growth factors for PC12 cells and that the growth-promoting effects of these agents may be mediated by a type I insulin-like growth factor receptor on these cells.

Ekker, Sourkes and Gabor (1988) used immunotitration to examine the changes in *S*-adenosylmethionine decarboxylase content after administration of piribedil or 2-deoxyglucose. They determined that there was no difference in the half-life of the decarboxylase activity in adrenal glands of untreated rats and of rats receiving piribedil. These and other experiments suggested that the decrease in adrenal *S*-adenosylmethionine decarboxylase activity and protein content caused by stress is due to a decrease in the rate of synthesis of the enzyme.

Gibbs (1986) showed that oxytocin inhibits peripheral catecholamine secretion in the urethan-anesthetized rat. Plasma levels of epinephrine were inhibited by 53% and norepinephrine by 43%. Also, the corticotropin (ACTH) level was decreased. He suggested that peripheral catecholamines at times may be directly involved in the control of ACTH secretion and also that oxytocin may have important functions in the regulation of adrenal catecholamine secretion.

Alho, Tähti, Koistinaho et al. (1986) examined the effect of toluene inhalation on catecholamines in the peripheral nervous system. Using the formaldehyde-induced fluorescence technique for histochemical demonstration of catecholamines, they found no change in the catecholamines in the adrenal medulla or sympathetic ganglia. In electron microscope studies, no clear pathologic changes were detected after exposure to toluene.

Gusovsky, Daly, Yasamuto et al. (1988) demonstrated the inhibitory effects of nifedipine, a dihydropyridine, on the maitotoxin-evoked release of ATP from PC12 cells. The maitotoxin-induced calcium-dependent phosphoinositide breakdown in the same cells was not affected by nifedipine or other calcium channel blockers. Lower concentrations of maitotoxin were required to stimulate phosphoinositide breakdown than to evoke ATP release from PC12 cells. The differences in concentration range required to elicit these responses and the differences in sensitivity to calcium channel blockers indicate that maitotoxin-evoked secretion and phosphoinositide breakdown are independent phenomena in PC12 cells.

The role of the sympatho-adrenal system in the tachycardia that accompanies the hypotensive response to hydralazine was studied in urethan-anesthetized rats by Vidrio and García-Márquez (1986). Using animals that had been demedullated or pretreated with 6-hydroxydopamine, they obtained results to indicate that hydralazine tachycardia is mediated by sympatho-adrenal activation. It was further suggested that the tachycardia is facilitated by a muscarinic mechanism that modulates sympathetic influences on cardiovascular function.

Mukherjee (1985) administered exogenous catecholamines to male weaver birds. The adrenal medulla in all cases remained virtually unaffected, but changes were noted in the adrenal cortex and testes.

The effect of ascorbic acid (vitamin C) on adrenal medullary hormones in birds was reviewed by Ghosh and Chatterjee (1986). Ascorbic acid acted as a co-factor to facilitate the conversion of dopamine to norepinephrine, and had a positive effect on adrenal corticosteroid production which in turn activated the enzyme that converts norepinephrine to epinephrine. In a more specific study, Chatterjee, Pal and Ghosh (1986) examined the effect of vitamin C in castrated juvenile pigeons. A fall in adrenal epinephrine level, but no significant change in norepinephrine was noted.

The effects of clonidine on adrenal medullary catecholamine levels were studied in normotensive rats by Gaillard, Tran, Rostin et al. (1987). Intraperitoneal injections of clonidine caused a dose-dependent decrease in epinephrine content of the adrenal gland. This effect was suppressed by denervation of the gland, demonstrating that clonidine decreases the catecholamine content of the adrenal medulla through a central action. It was suggested that the adrenal medulla is involved in the hypotensive effect of clonidine.

Anglade, Dang Tran, De Saint Blanquat et al. (1987) studied the action of clonidine on secretion from the adrenal medulla in dogs. Intravenous administration of clonidine induced a decrease in both adrenal catecholamine secretion and cardiovascular variables. When administered intracisternally, clonidine also decreased catecholamine release from the adrenal gland. Clonidine failed to modify the adrenal catecholamine release evoked by electrical stimulation of the splanchnic nerve. This demonstrated that in dogs clonidine decreases epinephrine release from the adrenal gland through a central and not a peripheral mechanism. This action may contribute to its antihypertensive effects.

Adrenal catecholamine concentration was measured in rats after 10 days of treatment with bromocriptine and haloperidol in a study by Baksi, Hughes and Strahlendorf (1986). They found that the bromocriptine treatment significantly decreased dopamine, norepinephrine, and epinephrine contents in a dose-dependent manner. On the other hand, haloperidol treatment had no significant effect on dopamine content and had a biphasic effect on epinephrine content. Norepinephrine concentration increased only with the lowest dose of the drug. They suggested that their observation may represent a "rebound" phenomenon for bromocriptine which has a shorter half-life than haloperidol.

Boksa (1986a) studied the effects of propranolol on catecholamines in cultured bovine adrenal chromaffin cells. She showed that *l*-propranolol was taken up into the cells in a nonstereoselective manner. *l*-Propranolol is not taken up by the same transporter as catecholamines. Propranolol inhibited epinephrine uptake but did not have an effect on the basal release of norepinephrine from the cells. However, *l*-propranolol inhibited carbachol release.

Gaillard, Rascol and Montastruc (1987) examined the effect of the dopaminergic antagonist domperidone on the concentrations of catecholamines in the adrenal medulla in rats. They found that this anti-parkinsonism drug increased the catecholamine levels in the adrenal medulla. This supported the suggestion that this drug also could be used in the management of orthostatic hypotension, and its usefulness may be in increasing sympathetic tone by acting on the sympathoadrenal axis. Also, because domperidone is considered to be a relatively specific D_2 antagonist, these data suggest a role for the D_2 dopaminergic receptor in modulating catecholamine release from chromaffin cells.

In an attempt to explain the mode of action of lithium on the treatment of bipolar affective disorders, Grof, Brown, Grof et al. (1986) examined the effect of lithium on peripheral catecholamines in man. After administration of therapeutic doses of lithium for 3 weeks to healthy volunteers, there was a differential response of catecholamines to insulin stimulation. The response of plasma norepinephrine remained unchanged, but the epinephrine response was dramatically decreased. This demonstrates that the adrenal medulla was directly affected, providing additional evidence for the separate neuroregulation of the adrenal medulla and sympathetic nerve endings.

Fujii, Yamamoto and Sato (1987) demonstrated that adrenal enucleation abolishes the development of hypersensitivity to haloperidol in rats. The loss of hypersensitivity could also be accomplished by denervation of the adrenal gland. These and other findings suggest that adrenal medullary catecholamines may play a modulating role in the response of the central nervous system to haloperidol.

Nakanishi, Morita, Murakumo et al. (1986) demonstrated that hydralazine causes a concentration-dependent inhibition of the catecholamine secretion induced by carbachol or potassium in cultured bovine adrenal chromaffin cells. It also significantly inhibits calcium uptake. This drug may inhibit catecholamine secretion by blocking calcium uptake into the cells.

In an exchange of letters, Makinen and Hamalainen (1986) and Boelsterli and Zbinden (1986) discussed the effect of xylitol on the adrenal medulla of rats. Makinen and Hamalainen (1986) concluded that this sugar substitute does not have a significant effect on the metabolism of the adrenal medulla. On the other hand, Boelsterli and Zbinden (1986) thought that long-term administration of xylitol led to a high incidence of adrenal medullary hyperplasia and probably also pheochromocytoma. The final assessment of human risk from long-term ingestion of xylitol and related compounds remains to be determined.

The effects of lactose and various polyalcohols on the general metabolism and pathology of the adrenal medulla were reviewed by Baer (1988). No conclusion could be drawn about how changes in calcium homeostasis affect adrenal medullary activity and growth. The proliferative changes in the adrenal medulla of rats fed various polyalcohols provided no implications regarding man.

The influence of antiepileptics administered to postnatal rats on histologic characteristics of the adrenal glands was examined by Novíc, Mijatov-Ukropina, Marinkovic et al. (1985). The most distinct change medulla was a decrease of the volume of the adrenal medulla and a narrowing of its vascular bed. There were parallel changes in chromaffin cells and sympathetic ganglion cells within the medulla.

Marzotko and Pankow (1987b) examined the effect of a single dose of dichloromethane on the adrenal medulla of rats. They noted increased urinary catecholamine excretion, morphologic changes, and a distinct decrease in the chromaffin reaction.

The effect of dantrolene sodium, a drug used as treatment for malignant hyperthermia, on catecholamine release from the perfused dog adrenal medulla was investigated by Sumikawa, Hayashi, Fukumitsu et al. (1987). Dantrolene did not affect the acetylcholine-induced catecholamine release but did inhibit the caffeine-induced release. Therefore, the therapeutic action of dantrolene on malignant hyperthermia could be due to inhibition of abnormal release of calcium in adrenal chromaffin cells.

Wernert, Antalffy and Dhom (1986) examined the effect of estradiol on the adrenal glands of rats in a morphometric study. They found that it causes a significant dilation of sinusoids in the adrenal medulla.

The effects of prolactin on the adrenal medulla in mice and hamsters were studied by Fernández-Ruiz, Esquifino, Steger et al. (1988). In the mouse, prolactin-secreting pituitary transplants produced increases in adrenal medullary weight and tyrosine hydroxylase activity and depletion of epinephrine, resembling results obtained previously in the rat. In the golden hamster, pituitary grafts increased the concentrations of epinephrine and norepinephrine in the adrenal and produced an apparent, but not significant, increase of tyrosine hydroxylase activity. It was concluded that prolactin exerts a stimulatory influence on tyrosine hydroxylase activity in the adrenal medulla of mice, hamsters, and rats but its actions on adrenal medullary weight and catecholamine stores are species-specific.

Monkhouse and Fussell (1988) used the fraction of labeled mitotic figures to derive basic data concerning the cell cycle of adrenal chromaffin cells and the effect of hydrocortisone on this. Hydrocortisone was administered to gravid mice from day 13 of pregnancy until term. Hydrocortisone administration did not affect the total cell cycle time nor the timing of specific phases. These results indicate that the cycle of these cells is not susceptible to the influence of corticosteroids at this stage of development.

Simonyi, Kanyicska, Szentendrei et al. (1988) investigated the effect of chronic morphine treatment on epinephrine biosynthesis. After 13 days of morphine treatment, the activity of phenylethanolamine N-methyltransferase and the epinephrine content were increased in the adrenal medulla of male rats. This effect was abolished by hypophysectomy. This led to the conclusion that the morphine-induced increase in epinephrine biosynthesis is not fully dependent on an intact pituitary-adrenal axis and may be mediated by a neural mechanism or by a direction action of morphine.

The effects of adrenalectomy on the pharmacokinetics and the antinociceptive activity of morphine were investigated by

Miyamoto, Ozaki and Yamamoto (1988) in an attempt to elucidate the mechanism of adrenalectomy-induced potentiation of morphine antinociception in rats. The plasma half-life of morphine given intravenously was significantly prolonged by adrenalectomy. A similar effect was seen with a high dose, but not a low dose, of morphine given subcutaneously. The antinociceptive potency of both doses of morphine administered subcutaneously was enhanced by adrenalectomy, but the plasma morphine concentrations were not equivalent. These and other results suggested that the enhancement of morphine antinociception by adrenalectomy cannot be explained by the increased morphine level alone.

Mannelli, DeFeo, Maggi et al. (1986) infused verapamil, a calcium channel blocker, into four patients with pheochromocytoma. The infusion did not cause changes in plasma epinephrine or norepinephrine levels that differed significantly from those induced by saline infusion. Blood pressure was decreased in these patients but apparently this was not due to inhibition of the catecholamine secretion from the tumor.

Ozutsumi, Sugimoto and Matsuda (1987) showed that *Clostridium perfringens* type A enterotoxin induces the release of norepinephrine from PC12 cells in a calcium-dependent manner. They suggested that neuronal release of catecholamines may underlie the acute systemic effects of this toxin.

Jacobs and Stevens (1986b) examined the effect of the microtubule-depolymerizing drug Nocodazole on the form of neurites extending from PC12 cells. Microtubule polymerization led to the formation of varicosities due to a clustering of membranous organelles in young neurites. Neurites older than 7 days appeared to be resistant to microtubule depolymerization. The authors were able to predict the volume and shape of a neurite quantitatively.

The effects of cytoskeleton-disrupting drugs on ouabain-stimulated catecholamine secretion from cultured bovine adrenal chromaffin cells were studied by Morita, Oka and Hamano (1988). Catecholamine secretion induced by ouabain was markedly enhanced by pretreatment of the cells with cytochalasin B. In contrast, neither colchicine nor vinblastine had any significant influence on the secretion. These results indicate that cytochalasin B enhance the catecholamine secretion induced by ouabain through its disrupting action on microfilaments within the cells, thus providing evidence of involvement of the microfilament system in the mechanism of the secretory action of ouabain on the chromaffin cell.

The effects of the anthracycline cytotoxic drug doxorubicin (Adriamycin) on the release of catecholamines from the bovine adrenal medulla were investigated by Pinto, Politi and Fernandez (1987). At low concentrations, doxorubicin facilitated the secretory response induced by acetylcholine and high potassium but did not affect the spontaneous catecholamine output. At higher concentrations of doxorubicin, significant and irreversible inhibition of the basal release of catecholamines, as well as that caused by acetylcholine or high potassium, was noted. These results suggest that doxorubicin effects could be mediated at the plasma membrane of the chromaffin cells. This study is compatible with the idea that increased adrenal medullary catecholamine release is involved in the cardiotoxic action of relatively low doses of doxorubicin used clinically.

The teratogenic effects of the herbicide nitrofen on the rat lung were examined by Lau, Cameron, Irsula et al. (1988). Among other observations, they found that adrenal catecholamines, which play an important role in surfactant reduction and fluid resorption in the lung during the transition to air breathing, were markedly decreased in treated animals.

Mosher and Kircher (1988) examined proliferative lesions of the adrenal medulla in rats treated with zomepirac sodium. This nonsteroidal anti-inflammatory drug caused an increase in focal hyperplasia and benign medullary tumors in 24-month-old male rats. Zomepirac sodium can be added to the list of substances that have been associated with increased proliferative lesions in the adrenal medulla of the aged rat.

Smoliakova and Rodivoz (1988) assessed the curative action of dopha-phosphaden on dystrophic changes in the retina induced by disturbances in the function of the sympathoadrenal system. Dystrophic changes in the retina were delayed by the administration of dopha-phosphaden which stimulates the activity of the sympathoadrenal system.

The effect of an external electrostatic field on catecholamine secretion by rat adrenals was studied by Artsruni, Zil'fian, Azgaldian et al. (1987). The effect of exposure to an electrostatic field included time-related modifications in structure and function of adrenal chromaffin cells and the concentration of catecholamines.

CHAPTER 8

ENZYMES

This chapter will deal primarily with enzymes involved in the synthesis of catecholamines. Tyrosine is converted to epinephrine in a four-step process that is catalyzed, in order, by tyrosine hydroxylase (TH), aromatic-L-amino-acid decarboxylase, dopamine β-hydroxylase (DβH), and phenylethanolamine N-methyltransferase (PNMT). These enzymes also are found in other parts of the nervous system, but those in the adrenal medulla have been studied extensively because that organ is composed of a large population of catecholamine-containing cells. Other enzymes that have been recently studied also will be reviewed.

The location of the gene for human TH was shown by Craig, Buckle, Lamouroux et al. (1986) to be on chromosome 11p15. The gene for human Dβh is on chromosome 9q34 (Craig, Buckle, Lamouroux et al., 1988). There has previously been suggestions that catecholamine synthesizing enzymes are derived from a "gene family," but this new information demonstrates that these two key enzymes are genetically unrelated.

Slotkin (1986) reviewed studies on the development of the sympatho-adrenal axis, including development of the catecholamine-synthesizing enzymes. His review stressed interrelationships among specific cellular events occurring during the maturation process and the physiologic function of the adrenal medulla as a catecholamine-secreting organ that mediates vital processes. Among other things, he pointed out that, prior to innervation of the adrenal medulla, chromaffin cells develop largely according to intrinsic genetic information and are subject to modification by hormonal and other metabolic inputs. After the establishment of neural connections, further development of the tissue occurs under transsynaptic control.

Wilburn, Goldsmith, Chang et al. (1986) examined the ontogeny of catecholamine-synthesizing enzymes and enkephalins in the primate fetal adrenal medulla. Using immunocytochemistry, they localized enzymes and enkephalin in adrenals from fetal rhesus and human adrenal glands. In the older fetuses, positive immunologic reactions were seen to the enzymes DβH and PNMT as well as [Leu]enkephalin. These were localized in the same cells. In fetuses, DβH could be detected at 15 weeks gestation, followed by PNMT and [Leu]enkephalin at 18 to 19 weeks. They suggested that the primate fetal adrenal medulla may be capable of co-secretion of catecholamines and enkephalins by the end of the second trimester of gestation.

Dagerlind and Schalling (1988) used the cDNA clone for human DβH for *in situ* hybridization to localize DβH mRNA in the central nervous system and the adrenal gland of the gray monkey. They also localized mRNA for TH, PNMT, and neuropeptide Y. In the adrenal gland, all chromaffin cells were TH- and DβH-positive. However, only cells in the peripheral part of the medulla were positive for PNMT. Only a few of the adrenal chromaffin cells were positive for neuropeptide Y. Whether the low amount of neuropeptide Y mRNA in the adrenal medulla of the monkey as compared with those previously reported for the rat reflects different translational processing or low amounts of peptide production remains to be determined.

Milner and Wurtman (1986) reviewed the coupling of catecholamine synthesis to precursor supply. They pointed out that, among other things, it is necessary for the animal to consume pure tyrosine, separate from the other amino acids in dietary proteins, in order to obtain major increases in tyrosine levels within tissues.

The effect of vasoactive intestinal polypeptide (VIP) on the synthesis of catecholamines from tyrosine in isolated bovine adrenal chromaffin cells was examined by Houchi, Oka, Misbahuddin et al. (1987). They found that VIP stimulated catecholamine synthesis from tyrosine but not from dopa. This stimulatory effect was not dependent on extracellular calcium. The effect was additive with that of carbachol but not with that of phorbol esters which activated protein kinase C. They suggested that VIP stimulated catecholamine synthesis by activation of TH and that protein kinase C was involved in this mechanism.

Regunathan and Sourkes (1988) assessed the effect of arginine vasopressin on adrenal TH and PNMT activities. Both enzymes showed marked increases after administration of arginine vasopressin to rats. Hypophysectomized animals showed induction of TH but not PNMT. This indicates that the pituitary influences the increase in PNMT caused by arginine vasopressin. Administration of arginine vasopressin via the lateral ventricle led to an increase in adrenal TH activity, suggesting that the peptide acts centrally to induce the enzyme.

Houchi, Morita, Minakuchi et al. (1986) showed that hydralazine inhibited catecholamine synthesis in cultured bovine chromaffin cells. Their results also suggested that this drug, used in the treatment of hypertension, may be able to suppress the hydroxylation of tyrosine to dopa and the conversion of dopamine to norepinephrine by inhibiting TH and DβH. In a follow-up study, Morita, Houchi, Nakanishi et al. (1986) showed that hydralazine results in allosteric alterations in the molecular structures of TH and DβH. They suggested that the inhibitory action of hydralazine on these enzymes is not totally based on the chelating activity of the drug.

Féty, Misére, Lambás-Señas et al. (1986) examined the effect of the neurotoxin DSP-4 on the catecholamine-synthesizing enzymes in the adrenal medulla and brain. They found slight increases in the activities of TH and DβH with relatively small changes in the activity of PNMT. They suggested that these changes were due to a reflex activation after denervation of sympathetic terminals by the neurotoxin.

Yanagihara, Yokota, Wada et al. (1987) studied the effects of intracellular pH and the ionic environment on catecholamine synthesis in cultured bovine adrenal chromaffin cells. Incubation of the cells in the absence of sodium led to a marked enhancement of catecholamine synthesis. Removal of the sodium led to a decrease in pH. Under conditions in which the pH was decreased further, catecholamine synthesis was enhanced further. Their results suggest that removal of extracellular sodium increases the synthesis of catecholamines, at least in part, by shifting the intracellular pH toward the optimal pH of TH.

The effects of protonophores on the synthesis of catecholamines and the intracellular pH in cultured bovine adrenal chromaffin cells were studied by Yokota, Yanagihara, Izumi et al. (1988). Two protonophores stimulated the synthesis of catecholamines and the cells. This was accompanied by a decrease in intracellular pH. Because the optimal pH of TH is around 6.0 in adrenal chromaffin cells, lowering the pH may increase the synthesis of catecholamines by means of this mechanism.

Ziegler, Kennedy and Elayan (1988) developed a sensitive radioenzymatic assay for epinephrine-synthesizing enzymes. This assay produces tritiated epinephrine from norepinephrine and tritiated S-adenosylmethionine. Therefore, this assay will detect the synthesis of epinephrine by enzymes other than PNMT. Epinephrine was shown to be formed in the adrenal medulla and several other tissues. Denervated tissues in rats without an adrenal medulla contained very little norepinephrine but still had epinephrine and epinephrine-forming enzymes present.

The production and regulation of TH, PNMT, and neuropeptide Y were studied in the rat and bovine adrenal medulla by Schalling, Dagerlind, Brené et al. (1988). By using both immunohistochemical and *in situ* hybridization methods, they detected cells positive for PNMT and neuropeptide Y; these exhibited close overlap in the bovine adrenal medulla and were preferentially located in the outer two-thirds of the medulla. Although TH and its mRNA were observed in virtually all chromaffin cells, TH mRNA levels were much higher in the cells positive for PNMT and neuropeptide Y. After administration of the catecholamine-depleting drug reserpine to rats, a brief increase followed by a dramatic decrease in the level of the mRNA for PNMT was observed. In contrast, mRNA for both TH and neuropeptide Y exhibited only an increase. Different regulatory mechanisms appear to operate for these three compounds coexisting in the adrenal medulla.

Stachowiak, Rigual, Lee et al. (1988) explored the molecular mechanisms by which the pituitary-adrenal cortex axis regulates the levels of enzymes in the catecholamine biosynthetic pathway. Specifically, they examined the effect of hypophysectomy and dexamethasone on TH and PNMT mRNA levels in the adrenal medulla and superior cervical ganglia of rats. The RNA levels in the adrenal medulla were significantly decreased after following hypophysectomy. This reduction could be reversed by the administration of dexamethasone. Dexamethasone did not affect mRNA levels in sham-operated rats. These and other results led to the conclusions that the regulation of enzyme levels in the adrenal medulla involves the induction of TH mRNA by increased plasma glucocorticoid levels and the maintenance of steady-state straight levels of PNMT mRNA by glucocorticoid-dependent mechanisms.

The relationship between catecholamine synthesis and release as a function of the cell cycle in PC12 cells was studied by Koike and Takashima (1986). The activities of TH and dopa decarboxylase were independent of the cell cycle, whereas the activity of DβH was modulated during the cell cycle. Both high-potassium-induced and carbachol-induced release varied with the phase of the cell cycle.

TYROSINE HYDROXYLASE

Tyrosine hydroxylase (Enzyme Commission number 1.14.16.2; also known as tyrosine 3-monooxygenase or L-tyrosine, tetrahydropteridine:oxygen oxidoreductase [3-hydroxylating]) is generally considered to be the rate-limiting enzyme in catecholamine synthesis. For this reason, TH has been studied extensively. Perhaps the enzyme in the adrenal medulla has been the one studied most thoroughly, but TH also is present in catecholaminergic neurons throughout the nervous system.

Biochemical and molecular biology properties, regulation,

purification procedures, and methods for determination of TH activity were reviewed by Joh, Hwang and Abate (1986). Tyrosine hydroxylase was compared with phenylalanine hydroxylase and tryptophan hydroxylase.

The properties of TH, as well as methods for assay and purification, have been reviewed for the enzyme isolated from bovine (Nagatsu and Oka, 1987) and rat (Fujisawa and Okuno, 1987) adrenal medulla as well as that from PC12 cells (Tank and Weiner, 1987). Nagatsu and Oka (1987) pointed out that purification of the enzyme from tissues of large animals has been difficult due to polymerization during the purification procedures. A method reviewed by Fujisawa and Okuno (1987) yielded less than 0.2 mg of protein from 600 rat adrenal glands. Tank and Weiner (1987) pointed out that the advantages of using cultured cells as a source of the enzyme are that the starting specific activity is relatively high and the final product is more homogeneous. The disadvantage of using cultured cells is that the yield of enzyme is much less than that obtained from tumor.

A simplified micro-assay for TH activity based on the trapping of radioactively labeled carbon dioxide was described by Shen, Hamilton-Byrd, Vulliet et al. (1986). This assay was claimed to be simple, rapid, sensitive, and convenient for performing a large number of determinations. Less than 300 μg of protein can be assayed.

Lee, Nohta, Umegae et al. (1987) described an assay for TH by high-performance liquid chromatography with fluorescence detection. The assay is based on the determination of L-dopa formed from the substrate L-tyrosine under optimal conditions for the enzyme reaction. The detection limit for the L-dopa formed enzymatically is 2 pmol per assay tube.

Hayashi, Miwa, Lee et al. (1988) presented a rapid and reliable method for the determination of TH activity *in vivo* in the rat adrenal gland. This method involves measuring the rate of accumulation of dopa in the gland after decarboxylase inhibition. By using the appropriate purification procedures, they could achieve rapid and accurate determination of dopa. With this method, they investigated *in vivo* the effects of immobilization stress, reserpine, and hypoxia on activity of TH in the adrenal gland.

Kuhn and Billingsley (1987) purified TH from PC12 cells and produced antibodies. The purified enzyme protein migrated as a single band on SDS gels, with a molecular mass of 60 kDa. Other characteristics of the molecule were determined. Polyclonal antibodies were produced in rabbits against pure TH and were judged to be monospecific by Western blot analysis. The IgG antibody fraction was used to purify the enzyme from crude extracts in a single step.

Nelson and Kaufman (1987) noted that TH from the bovine adrenal medulla was activated up to 4-fold by incubation with low concentrations of ribonucleic acids (RNA). This interaction with RNA was exploited along with a specialized chromatographic method to effect rapid purification of a stable form of the enzyme. The purified enzyme also contained RNA which represented about 10% of the total mass and appeared to be important for full activity.

As reported in **The Adrenal Medulla, Volume 4,** the amino acid sequence of TH has been elucidated. Several laboratories are gathering more information on the molecular structure of the enzyme. O'Malley (1986) isolated a rat adrenal medullary recombinant clone by cross-hybridization with TH complementary DNA (cDNA). The new clone has a mRNA size of 5.6 kilobases (kb) and also hybridizes to the 1.9-kb TH mRNA. There are several hybridizing bands in common between the TH cDNA and the adrenal medullary clone. These results demonstrate that the adrenal clone has sequences in common with the TH gene or is closely linked to it in the genome. Hybrid-selected mRNA translation products of the adrenal clone can be immunoprecipitated with anti-DβH antiserum. This suggests that the adrenal medullary cDNA may code for another catecholamine-synthesizing enzyme.

The organization and evolution of the rat TH gene was described by Brown, Coker and O'Malley (1987). Their results suggested that the genes for TH and phenylalanine hydroxylase are members of a gene family that has a common evolutionary origin.

Similar results were reported by O'Malley, Anhalt, Martin et al. (1987). They isolated a full-length genomic clone for human TH. A human genomic library was constructed and screened by using a rat cDNA for TH as a probe. With various techniques, they found that, in contrast to the rat gene, the human gene for TH undergoes alternative RNA processing to generate at least three distinct mRNAs. They also compared the human TH and phenylalanine hydroxylase genes and suggested that both evolved from a common ancestral gene and that major changes in the size of introns have occurred since their divergence.

Grima, Lamouroux, Boni et al. (1987) described a single human gene that encoded multiple THs with different predicted functional characteristics. Apparently, TH molecules are encoded by at least three distinct mRNAs. The expression of these mRNAs varies in different parts of the nervous system. They found that the "type 2" mRNA predominated in the adrenal medulla and pheochromocytoma as determined by S_1 nuclease mapping experiments.

Human TH cDNA was isolated by molecular cloning by Kaneda, Kobayashi, Ichinose et al. (1987). They found a novel type of cDNA clone whose NH_2-terminal sequence is similar to but really distinct from that of each of the three types of TH cDNA previously reported. Southern blot analysis of human genomic DNA indicated that TH is encoded by a single gene. This suggests that the four different forms of TH mRNA are produced by alternative RNA splicing from a single primary transcript.

Genomic DNA encoding the rat TH gene was isolated by Harrington, Lewis, Krzemien et al. (1987) from a lambda phage library by using a nick-translated fragment from a cDNA clone for rat TH. They determined the initiation site for TH mRNA synthesis. Examination of the 5' flanking region suggested that this region of the gene has features that confer tissue-restricted expression on the TH promoter.

Saadat, Stehle, Lamouroux et al. (1988) examined the cDNA that codes for bovine TH mRNA and determined the entire nucleotide sequence. The mRNA consists of 1,706 nucleotides. The predicted amino acid sequence of bovine TH shows a similarity of 66% with the rat enzyme at the amino-terminus and 91% at the carboxyl-terminus. The carboxyl-terminus region has been shown to include the catalytic site; therefore, it is conserved to a greater extent in different species. Other differences between the rat and bovine enzyme were noted.

A cDNA clone containing the entire coding region of quail TH was isolated and analyzed by Fauquet, Grima, Lamouroux et al. (1988). Comparison with rat and human TH reveals several highly conserved domains. One such domain contains the putative site of phosphorylation and is implicated in the posttranslational regulation of the enzyme.

Semenenko, Cuello, Goldstein et al. (1986) produced a monoclonal antibody directed against TH. This antibody was selected specifically for use in immunocytochemistry, and its application in immunocytochemical localization of the enzyme by immunofluorescence and the peroxidase-antiperoxidase (PAP) technique was described. They also demonstrated that the antibody will immunoprecipitate phosphorylated TH.

Rat adrenal gland and brain were analyzed by hybridization histochemical methods using a RNA probe complementary to mRNA, by immunohistochemical methods using anti-TH antiserum, and by retrograde tracing using the fluorescent compound Fast blue, by Schalling, Hökfelt, Wallace et al. (1986). This technique represents a potentially valuable tool in the further analysis of peripheral and central catecholaminergic systems.

Han, Snouwert, Towle et al. (1987) synthesized an oligodeoxyribonucleotide probe that is specific for TH mRNA and used this to study the expression of the TH gene and its regulation in adult and developing neural tissues. By using Northern blot hybridization of polyadenylated RNAs from adrenal gland and other tissues, a single 1.9-kb TH mRNA was detected in chromaffin cells. After reserpine administration, the intensity of the hybridization signal increased 3-fold. Use of this specific TH probe in *in situ* hybridization procedures represents a powerful approach to the study of regulation of TH gene expression.

Coker, Vinnedge and O'Malley (1988) described the structure of the promoter and intron I of the human and rat TH genes. The 5' flanking regions of the two genes were 74% identical and contained a completely conserved cAMP response element. Although both genes were single copies, multiple splicing events of the human transcript led to multiple mRNAs. Based on several lines of evidence, alternative forms of mRNA of rat TH analogous to those in the human were not present. Both rat and human promoters directed the transcription of reporter genes when introduced into rat pheochromocytoma and fibroblast cells.

D'Mello, Weisberg, Stachowiak et al. (1988) isolated a cDNA clone encoding bovine adrenal TH. They compared the amino acid sequence deduced from this cDNA with the sequences for rat and human THs. The bovine enzyme has 85% and 84% amino acid sequence identity with rat and human THs, respectively. Seventy-nine percent of the residues were identical in all three species, indicating a strong evolutionary conservation of the enzyme structure. Moreover, three of the four putative phosphorylation sites were conserved.

Kelner, Morita, Rossen et al. (1986) used digitonin-permeabilized bovine chromaffin cells to investigate the subcellular localization of TH and PNMT. Digitonin titration of the release of proteins and catecholamines revealed the existence of at least three subcellular compartments distinguished by their sensitivity to digitonin: soluble proteins, a "digitonin-sensitive" cytoplasmic protein pool, and the chromaffin vesicle. These and other data suggest that the two enzymes are in a detergent-labile association in the cell. Kelner et al. (1986) suggested that TH and PNMT may be localized on the surface of the chromaffin vesicle.

Faucon Biguet, Buda, Lamouroux et al. (1986) found that a single injection of reserpine caused a long-lasting enhancement of TH activity. They developed a sensitive method for assaying both TH mRNA levels and enzyme activity in tissue from a single rat. Reserpine caused a 4.2-fold increase in tyrosine mRNA in the adrenal medulla which was maximal 2 days after administration. This increase is about twice that in enzyme activity. This result suggests that induction of TH results from enhanced transcription of the TH gene. Comparison of these events in the adrenal medulla with those in selected regions in the brain suggests that there is a difference in the stability of the TH mRNA in different regions.

<u>Activation</u>
Tyrosine hydroxylase apparently exists within the chromaffin cell in an inactive or active form. The next few paragraphs will discuss general aspects of the activation, followed by a discussion of specific mechanisms of activation, primarily phosphorylation.

Morita, Teraoka and Oka (1987) examined the effects of isolated chromaffin vesicle membranes on the activity of soluble TH and also studied the association of the soluble enzyme with the vesicle membranes. Incubation of vesicle membrane with soluble fraction in a mixture intended to approximate the intracellular environment of the resting cell resulted in a marked increase in the enzyme activity. The association of the soluble enzyme with the vesicle membranes was reversible and specific. Their results seem to indicate that activity of the soluble enzyme may be modulated by the vesicle through association of the soluble enzyme with the surface of the vesicle, suggesting that interaction between TH and the chromaffin vesicle may play a role in regulation of catecholamine synthesis.

The effects of basic polypeptides on the activation of TH by adenosine triphosphate (ATP) were investigated by Morita, Nakanishi, Houchi et al. (1986) in a study of the involvement of macromolecular cell components in the regulation of the enzyme activity. Basic polypeptides enhanced the activation of TH by low concentrations of ATP, and the potentiating effects of these polypeptides were concentration-dependent. They suggested that basic polypeptides convert the enzyme from a nonsusceptible form to a form susceptible to ATP, thus potentiating the ATP-induced activation.

Morita, Houchi, Nakanishi et al. (1986) studied the effects of various inositol phospholipids on TH activity, using the isolated enzyme prepared from bovine adrenal medulla. They found that inositol phospholipids caused the direct activation of TH through electrostatic effects on the enzyme. The potency of this activation was dependent on the number of phosphate groups in the lipid molecule. Morita et al. (1986) suggested a possible involvement of inositol phospholipids in the regulation of TH activity *in vivo*.

The effect of spermine on TH activity was examined before and after phosphorylation of the catalytic subunit of cAMP-dependent protein kinase by Kiuchi, Kiuchi, Togari et al. (1987). Before phosphorylation, spermine inhibited the enzymatic activity, but phosphorylation abolished the ability of spermine to inhibit TH. These and other results suggest that spermine may inhibit TH activity by interacting with the pterin binding site of the enzyme molecule in a manner of negative cooperativity.

Tank, Ham and Curella (1986) further characterized the induction of TH by either cyclic adenosine monophosphate (cAMP) or glucocorticoids in PC18 cells, a subclone of PC12 cells. They demonstrated that the induction is elicited by cAMP analogs or compounds that increase intracellular cAMP levels and by glucocorticoids. They also showed that, when cells are incubated in the presence of both cAMP and glucocorticoid, the increase in TH activity is approximately equal to the sum of the increases in activity observed in the presence of either inducing agent alone. Furthermore, the increases in TH activity elicited by these inducing agents alone or in combination are due to increases in enzyme protein; and these increases are associated with increases in the rate of synthesis of the enzyme. In an accompanying paper, Tank, Curella and Ham (1986) related the changes in these variables to changes in TH mRNA and showed that the increase in mRNA caused by cAMP and dexamethasone is not additive. Their results suggest that more than one mechanism is involved in the regulation of TH synthesis in PC18 cells exposed to both cAMP and glucocorticoids.

Lewis, Harrington and Chikaraishi (1987) demonstrated that both glucocorticoid and cAMP stimulate the transcriptional activity of the TH gene in PC7e cells, another derivative of PC12 cells. Both inducers effect transcriptional changes within 10 minutes after treatment; the changes are maximal within 30 to 60 minutes. They also reported that the 5' flanking sequences of the TH gene contain the *cis* information necessary for induction of TH by cAMP. Sequences sufficient for glucocorticoid stimulation of transcription do not appear to be present in this 5' flanking region.

The environment of the cell appears to have a direct effect on TH activity. Acheson, Edgar, Timpl et al. (1986) investigated the effects of substrate-bound laminin on levels of enzyme activity in cultured calf adrenal chromaffin cells. Laminin increased the levels of TH, DβH, and PNMT. The effect was selective, and the levels of other enzymes were not increased. The increase in TH was preceded by activation of the enzyme. The effects of laminin appear to be developmentally regulated because neither activation nor increased levels of TH occur in adult adrenal chromaffin cells exposed to laminin.

As a first step toward the identification of molecules involved in cell contact-mediated enzyme induction, Saadat and Thoenen (1986) investigated whether plasma membranes of adrenal chromaffin cells can mimic the inductive effect of cell contact. Plasma membranes prepared from bovine adrenal chromaffin cells induced TH in a manner similar to that observed in high-density cultures. Acetylcholinesterase was not induced in a similar manner, suggesting that the induction of TH was not a result of a general increase in protein synthesis. Other experiments suggested that the molecule(s) involved in this induction appear to be intrinsic membrane proteins.

Saadat, Stehle, Lamouroux et al. (1987) studied the regulation of TH by cell-cell contact in cultured bovine adrenal chromaffin cells. Dissociation of cells from the adrenal medulla resulted in a rapid decrease in the TH mRNA. An enhanced rate of TH enzyme degradation was observed in chromaffin cells when brought into low-contact cultures. Tyrosine hydroxylase mRNA increased 4-fold in high-contact cultures within 1 day after plating. Similarly, the rate of synthesis of TH molecules was maximal after 1 day, although the increase in the absolute amount of TH occurred slowly.

The regulation of TH mRNA levels in PC12 cells by cell-to-cell contact was examined by Saadat and Thoenen (1988). They presented evidence that TH is regulated at the transcriptional level during cell-to-cell contact. When PC12 cells were plated at low density, the mRNA levels and enzyme activity decreased within several hours. The decrease in TH resulting from the loss of cell-to-cell contact appeared to be an active process because it occurred much more rapidly than would be expected from the turnover rate of the protein. In contrast, in cells replated at high density, TH activity and mRNA decreased transiently and then increased.

Changes in rat adrenal TH mRNA levels following chronic amphetamine administration was measured by Fontenot, Cass and Vulliet (1987). The TH activity of adrenals collected after 7 days of treatment was decreased when compared to controls. This was shown to be due to a decrease in mRNA for TH. The decreased rate of transcription may account for the lowered activity of TH.

To test if the peptide substance P might produce more long- term biochemical changes in chromaffin cells, Boksa (1986b) incubated bovine adrenal chromaffin cells for several days with substance P and measured the effects on TH activity and on the changes in TH activity and catecholamine levels induced by carbachol. The results indicated that, in addition to modulating the short-term release of catecholamines, substance P also may be involved in the long-term regulation of catecholamine levels in the adrenal medulla. Substance P inhibited the increase in TH activity normally induced by carbachol. Long-term stimulation with carbachol depleted endogenous catecholamines, and substance P prevented this depletion. Substance P alone had only a slight effect on TH activity. Other experiments indicated that the effect of substance P is relatively specific for carbachol-induced increases in enzyme activity.

The role of nicotinic cholinergic transmission in cold stress-induced alterations in rat adrenal medullary TH mRNA was investigated by Stachowiak, Stricker, Zigmond et al. (1988) by using RNA dot-blot hybridization and a cloned TH cDNA probe. Chlorisondamine, a ganglionic blocker, greatly attenuated the induction of TH mRNA levels caused by cold exposure, whereas carbachol and nicotine, cholinergic agonists, increased TH mRNA levels. These results suggest that cholinergic nicotinic receptors play a key role in the transsynaptic induction of adrenal TH gene expression.

Togari, Ichikawa and Nagatsu (1986) found that two molecular species of the calcium-dependent neutral protease, calpains I and II, activated TH that had been purified from bovine adrenal medulla. This activation was inhibited by the endogenous inhibitor of calpain, calpastatin. It remains to be seen whether or not this mechanism of proteolytic activation of TH operates *in vivo*.

Yanagihara, Uezono, Koda et al. (1987) found that, in digitonin-permeabilized bovine adrenal chromaffin cells, calcium caused an activation of TH that was dependent on the presence of ATP. This calcium-induced activation occurred in the presence of cAMP and activators of protein kinase C. Inhibitors of calmodulin had little effect on this activation. Yanagihara et al. (1987) suggest that micromolar concentrations of calcium stimulate the activity of TH. This apparently is mediated by a calcium-dependent phosphorylation.

The effects of polyamines on TH activity in the rat adrenal medulla were investigated by Togari, Murakami, Oshima et al. (1986). The polyamines putrescine, spermidine, and spermine showed biphasic effects--inhibitory at low concentrations and stimulatory at high concentrations. The evidence suggests that TH molecules could interact with polyamines or related substances at physiologic concentrations, causing an inhibition of TH activity.

Houchi, Masserano and Weiner (1988) demonstrated that bradykinin activates TH in PC12 cells. This stimulation is due to an increase in the affinity of TH for its pterin cofactor and can be blocked by a specific bradykinin receptor antagonist.

The mechanism of oxygen activation by TH isolated from PC12 cells was investigated by Dix, Kuhn and Benkovic (1987). The enzyme was considered to contain one atom of ferrous iron per subunit and catalyzes the conversion of its tetrahydropterin cofactor to a 4a-carbinolamine concomitant with substrate hydroxylation. They compared these mechanisms with those of the enzyme phenylalanine hydroxylase. Tyrosine hydroxylase also can use hydrogen peroxide as a cofactor for substrate hydroxylation, unlike phenylalanine hydroxylase. This suggests a subtle difference between the two enzymes in the iron ligand field. Dix et al. (1987) proposed a detailed mechanism for the reaction.

A review of the literature regarding the activation of TH by phosphorylation by Goldstein and Greene (1987) showed that a number of extracellular signals produce activation and phosphorylation via several different protein kinases in response to various extracellular signals. They compared the kinetic properties of the enzyme phosphorylated *in vitro* by different protein kinases.

Pocotte, Holz and Ueda (1986) identified a 56 kDa protein in cultured bovine adrenal chromaffin cells that is phosphorylated when catecholamine secretion is stimulated. Using immunologic techniques, they identified this protein as TH and identified multiple subunits of the phosphorylated enzyme. They examined the cholinergic pharmacology of *in situ* TH phosphorylation and activation and the effects of high potassium and barium concentrations. The data indicate that, under various conditions, TH can be phosphorylated by calcium-dependent and cAMP-dependent protein kinases. The phosphorylation is associated with an increase in the activity of TH *in situ*.

Pocotte and Holz (1986) used phorbol esters to investigate whether protein kinase C can regulate catecholamine synthesis within bovine adrenal chromaffin cells by phosphorylation and activation of TH. They found that only those phorbol esters that activate protein kinase C induced phosphorylation of a 56 kDa protein, presumably TH, and increased the production of dopa.

This phosphorylation and activation did not depend on extracellular calcium or an increase in cAMP.

Yanagihara, Tank, Langan et al. (1986) used peptide mapping analysis to examine the relationship between activation and phosphorylation of TH by high potassium concentration and dibutyryl cAMP. High potassium increased the incorporation of phosphate into two peptides, whereas dibutyryl cAMP increased phosphate incorporation into only one of these peptides. The phosphorylation of TH in PC12 cells occurs exclusively on serine residues. Their results suggest that TH in PC12 cells is phosphorylated on serine residues at two or more distinct sites after potassium-induced depolarization. Because only one of these sites is phosphorylated by cAMP-dependent protein kinase, activation of TH by potassium may involve phosphorylation by multiple protein kinases.

In a related study, Tachikawa, Tank, Weiner et al. (1987) examined the effects of a phorbol ester, diacylglycerols, and forskolin on TH activation and phosphorylation in PC12 cells. The phorbol ester (an activator of protein kinase C) and forskolin increased the activity and enhanced the phosphorylation of TH but apparently not by identical mechanisms. Some, but not all, diacylglycerols also activated the enzyme. Their results indicated that TH is activated and phosphorylated at different sites in PC12 cells exposed to phorbol ester and forskolin and that phosphorylation of either of these sites is associated with activation of TH. Furthermore, cAMP-dependent and calcium/phospholipid-dependent protein kinases may play a role in the regulation of TH.

In another study, Tachikawa, Tank, Yanagihara et al. (1986) further characterized the activation and phosphorylation of TH in PC12 cells treated with high potassium concentration, dibutyryl cAMP, or the calcium ionophore A23187. They defined the sites on the enzyme phosphorylated by these treatments by separation of the tryptic phosphopeptides derived from the enzyme. They also compared the tryptic phosphorylated peptides from the intact cells with those phosphorylated *in vitro*. The results suggested that the *in situ* phosphorylation at multiple sites on TH is catalyzed by calcium/calmodulin-dependent protein kinase; however, phosphorylation by this protein kinase is not sufficient to activate the enzyme. Cyclic AMP-dependent protein kinase also can phosphorylate TH *in situ*, and phosphorylation by this protein kinase is sufficient to activate the enzyme.

Campbell, Hardie and Vulliet (1986) identified four phosphorylation sites in the NH_2-terminal region of TH purified from PC12 cells. They described the isolation and sequence analysis of tryptic peptides phosphorylated by four protein kinases and by a fifth protein kinase that was detected as a trace contaminant in purified TH. From inspection of the cDNA sequence, the phosphorylation sites can be assigned to four serine residues in the NH$_2$-terminal region of the protein.

Waymire, Johnston, Hummer-Lickteig et al. (1988) reported that tryptic digests of TH isolated from chromaffin cells with prelabeled radioactive phosphate can be separated into seven peaks by reversed-phase high-pressure liquid chromatography. Acetylcholine or high-potassium-induced depolarization caused an increased incorporation of phosphate into five peptides at two different rates; three peptides were maximally phosphorylated within 15 seconds but maximal phosphorylation of the other two peptides was not achieved until 4 minutes. Comparison of the time course for TH phosphorylation with that for TH activation and for *in situ* conversion of tyrosine to dopamine has demonstrated that the rapid phosphorylation correlates with increased enzyme activity. The peptides phosphorylated more slowly appeared to correlate with a sustained increase in dopamine synthesis capacity induced by the stimulation of the cells.

Cyclic guanosine monophosphate (cGMP)-dependent protein kinase phosphorylates and activates TH, as shown by Roskoski, Vulliet and Glass (1987). The TH purified from PC12 cells also was activated by the catayltic subunit of cAMP-dependent protein kinase. The extent of activation was correlated with degree of phosphate incorporation into the enzyme. After trypsin digestion, the peptides phosphorylated by cGMP- and cAMP-dependent protein kinases showed identical elution profiles. This supports the concept that the protein kinases are phosphorylating the same residue of TH *in vitro*.

Pigeon, Ferrara, Gros et al. (1987) purified and determined the sequence of a tryptic peptide of TH carrying the site phosphorylated by the rat PC12 TH-associated kinase activity. A 1.5 kDa peptide (about 12 residues), carrying the phosphorylation site, was released from phosphate-labeled TH by limited proteolysis with trypsin. The sequence was found to correspond to residues 38 to 45 in the TH primary structure. Thus, the associated kinase phosphorylated a known serine molecule, one of the phosphorylation sites for the cAMP-dependent protein kinase. This apparently is a different serine from the one reported by Campbell et al. (1986). Pigeon et al. (1987) suggested that the discrepancy may be due to the contamination of TH by different kinase activities.

Roskoski and Roskoski (1987) reported that sodium nitroprusside, an activator of guanylate cyclase, produced an increase in TH activity in PC12 cells. The dibutyrylform of cGMP, as well as increased cAMP levels, also produced an increase in enzyme activity. The cAMP-dependent activation was greater than that of the cGMP-dependent messenger system. These results indicate that both cGMP and cAMP and their cognate protein kinases activate TH activity in PC12 cells.

Biopterins

Tyrosine hydroxylase is in a class of mammalian enzymes in which the catalytic activity is dependent on tetrahydrobiopterin as an electron donor. The importance of the role of tetrahydrobiopterin in the regulation of TH was reviewed in detail by Levine and Galloway (1987).

Smith and Nichol (1986) studied the synthesis, utilization, and structure of the tetrahydrobiopterin intermediates in the *de novo* biosynthesis of tetrahydrobiopterin in bovine adrenal chromaffin tissue. Their studies were made possible by the development of an improved high-performance liquid chromatography procedure for the resolution, quantitation, and isolation of the intermediates and isomers of tetrahydrobiopterin. They were able to determine the structure of some crucial intermediates. Based on these structures, on the sequence of biosynthesis of these compounds, on the bioconversion of the compounds to tetrahydrobiopterin, and on earlier work in their laboratory, they were able to describe in detail the synthetic reaction sequence from a previously known intermediate to tetrahydrobiopterin.

An extensive review of the regulation of tetrahydrobiopterin biosynthesis and cofactor replacement by tetrahydropterins was published by Nichol, Smith, Reinhard et al. (1988). Among other things, they pointed out that the regulation of tetrahydrobiopterin biosynthesis is best illustrated in cells containing TH, such as adrenal chromaffin cells. Changes in tissue levels of the cofactor appear to be correlated with the rate of catecholamine synthesis. The evidence that some degree of biopterin cofactor deficiency is associated with certain psychiatric and neurologic disorders indicates the need for a more detailed understanding of these complex relationships.

Abou-Donia, Wilson, Zimmerman et al. (1986) pointed out that the intracellular level of tetrahydrobiopterin normally may limit the rate of tyrosine hydroxylation. They presented further evidence that tetrahydrobiopterin is a regulatory factor for this reaction, that under certain conditions tetrahydrobiopterin content is regulated by GTP cyclohydrolase activity, and that the synthesis of tetrahydrobiopterin appears to be mediated by cAMP-dependent and -independent processes in a manner analogous to the regulation of TH synthesis.

Bigham, Smith, Reinhard et al. (1987) examined synthetic analogues of tetrahydrobiopterin that had cofactor activity for aromatic amino acid hydroxylases. They found that one of these compounds is an excellent cofactor for TH and other hydroxylases, does not destabilize the binding of substrate, and is recycled by dihydropteridine reductase. These compounds are being evaluated as cofactor replacements in biopterin-deficiency diseases.

Fitzpatrick (1988) studied the inhibition of purified bovine adrenal TH by several product and substrate analogues. Examination of the kinetic mechanism indicated that tetrahydrobiopterin bound first to the free enzyme followed by binding of tyrosine. Binding and inhibition were sensitive to pH.

Mechanisms
The catalytic mechanism of TH isolated from the cytosol of bovine adrenal medulla was studied by Haavik and Flatmark (1987) using new techniques of product isolation and characterization. They were able to identify and isolate the tetrahydrobiopterin oxidation products formed during catalytic turnover in the TH-catalyzed reaction by micro stopped-flow spectroscopy and fast high-performance liquid chromatography, combined with photodiode array depletion. It was found that the 4a-hydroxy-tetrahydrobiopterins proceeded the formation of the quinonoid forms with both L-tyrosine and L-phenylalanine as the substrate. The formation of 4-hydroxy-tetrohydrobiopterins and hydroxylated amino acids were highly coupled.

Haavik, Andersson, Peterson et al. (1988) used a new procedure that permits large-scale purification of TH from the bovine adrenal medulla to study physiochemical properties of the enzyme. The homogeneous enzyme had a 60 kDa subunit. The amounts of iron, zinc, and phosphate were measured. A broad light absorption band with its maximum around 700 nm explains the blue-green color. EPR spectra of 3.6K revealed high-spin iron (Fe III) in an environment of nearly axial symmetry. Additional physicochemical studies suggested that TH does not undergo major substrate- or cofactor-induced conformational changes as have been reported for a related enzyme, phenylalanine hydroxylase.

The catalytic domain of bovine adrenal TH was characterized by Abate, Smith and Joh (1988). Mild proteolysis of the enzyme produced a 34 kDa fragment that was catalytically active. They determined that the catalytic region is contained within the central region of TH, approximately 17 kDa from the amino-terminus. This region of TH shares a high degree of homology with some other enzymes and appears to be highly conserved, whereas the region toward the amino-terminus regulates cofactor binding and directs substrate specificity.

Andersson, Cox, Que et al. (1988) examined an unusual blue-green chromophore present in TH. Resonance Raman spectroscopy indicated the presence of a bidentate complex of catecholamine and ferric iron in the enzyme. The energies of the catecholamine-to-iron charge transfer transitions suggested a mixture of histidines and carboxylate coordinated to the iron center in the enzyme. This may be related to the potent feedback inhibition of TH by catecholamines.

Laschinski, Kittner and Bräutigam (1986) reexamined the inhibition of TH by catechols *in vitro*. They performed studies at pH 7.2 (as opposed to the pH optimum for TH, 6.0) and used the natural cofactor tetrahydrobiopterin in an attempt to attain assay conditions that were more physiologic. Several natural and synthetic catechols were compared for their potencies in inhibiting TH prepared from PC12 cells. Their results demonstrated that natural catechols and certain drugs are more effective as direct blockers of TH than generally assumed, provided that appropriate assay conditions are used. In the case of dopamine and norepinephrine, these findings suggest a reevaluation of their role for feedback control of TH *in vivo*.

Burke and Joh (1988) examined the effect of dopamine on TH in cultured rat adrenal medulla. The enzyme activity was inversely proportional to the concentration of dopamine in the medium in which the explants were cultured. This was interpreted to reflect a loss of the TH protein which was due to a decrease in the rate of synthesis. Metabolites of dopamine did not affect TH activity.

The effects of ascorbate and its congeners on the activity of TH purified to near homogeneity from PC12 cells were examined by Wilgus and Roskoski (1988). Ascorbate inactivates the enzyme under phosphorylating conditions. Under nonphosphorylating conditions, even much higher concentrations of ascorbate failed to activate the enzyme. Isoascorbate and dehydroascorbate also inactivated TH under phosphorylating conditions but not under nonphosphorylating conditions. Pretreatment of PC12 cells with all three compounds also decreased TH activity.

Nagatsu and Hirata (1987) found that MPTP, MPP^+, and a related compound inhibited TH. Apparently, both the pyridinium and the phenyl group are required for the inhibition.

The effect of MPP^+ on the phosphorylation of TH in PC12h cells was examined by Kiuchi, Hagiwara, Hidaka et al. (1988). After 3 days of treatment of MPP$^+$, TH activity was decreased to less than 50% of control. Phosphorylation of the TH molecule was also decreased by 50%. These results suggest that MPP$^+$ acutely inhibits the phosphorylation of TH, resulting in decreased enzyme activity.

AROMATIC L-AMINO ACID DECARBOXYLASE

Aromatic L-amino acid decarboxylase (also called dopa decarboxylase; Enzyme Commission number 4.1.1.28) catalyzes the conversion of dopa to dopamine and other reactions. This is a ubiquitous enzyme that is in no way rate-limiting in the synthesis of catecholamines. For these and other reasons, there is relatively little current work on this enzyme. Methods for the assay and purification of aromatic L-amino acid decarboxylase were reviewed by Sourkes (1987). He also described the biochemical properties of this enzyme.

The biochemistry and functional significance of aromatic L-amino acid decarboxylase were reviewed in detail by Bowsher and Henry (1986). They pointed out that, despite being the first reported enzyme in the synthesis of monoamines, major questions remain regarding the biochemical and functional aspects of this important enzyme.

Albert, Allen and Joh (1987) reported that a single gene codes for aromatic L-amino acid decarboxylase in both neuronal and non-neuronal tissues. They extracted the enzyme from the bovine adrenal medulla and purified it to homogeneity. They then produced highly specific antibodies and isolated the cDNA clone complementary to bovine adrenal enzyme mRNA. They used biochemical, immunochemical, and molecular biology techniques to determine if this enzyme can be distinguished from one that plays a presumably more general role in organs such as the liver and kidney. A single form of the enzyme was detected in rat and bovine tissue. Aromatic L-amino acid decarboxylase proteins from brain, liver, kidney, and adrenal medulla are biochemically and immunochemically indistinguishable.

Aromatic L-amino acid decarboxylase was purified from the bovine adrenal medulla and brain, with a monoclonal antibody by Nishigaki, Ichinose, Tamai et al. (1988). The molecular masses of the enzymes from both organs were estimated to be approximately 100 kDa. Western immunoblot analysis showed that the antibody recognized each enzyme. The enzymes from both organs were shown to be similar with regard to several characteristics.

The effect of carcinogenic heterocyclic amines on TH and aromatic L-amino acid decarboxylase in PC12h cells was examined by Naoi, Takahashi, Ichinose et al. (1988b). Some heterocyclic amines were found to decrease the activity of these two enzymes greatly. These and other results led to the conclusion that heterocyclic amines can be neurotoxins specific for dopaminergic neurons.

DOPAMINE β-HYDROXYLASE

Dopamine β-hydroxylase (also known as dopamine β-monooxygenase; Enzyme Commission number 1.14.17.1) converts dopamine to norepinephrine. A full length cDNA clone for DβH was isolated from a human pheochromocytoma by Lamouroux, Vigny, Faucon Biguet et al. (1987). The polypeptide chain of DβH is comprised of 578 amino acids, corresponding to an unmodified protein of 64,862 Da. It is apparent that only one form of DβH is coded for, although it exists in the chromaffin vesicle in soluble and membranous forms. It was suggested by Lamouroux et al. (1987) that the membranous form is modified by posttranslational glypiation.

A detailed review of DβH was published by Nagatsu (1986). Assay methods, purification procedures, and properties of DβH from human plasma (Frigon, 1987), bovine adrenal medulla (Ljones, 1987), and birds (Weppelman, 1987) were described.

The biochemistry and molecular biology of DβH were reviewed by Joh and Hwang (1987). They pointed out that the structure of DβH appears to be more complex than originally considered.

Stewart and Klinman (1988a) reviewed the structure and function of DβH of the adrenal chromaffin vesicle. In this extensive review, the kinetic and chemical mechanisms, the role of copper, mechanism-based inhibition, and other aspects of DβH were considered.

Kuhn, Hadman and Sabban (1986) used monensin to ascertain the location in the biosynthetic pathway where the 77 kDa membrane-bound subunit form of DβH is converted post-translationally to the 73 kDa soluble form. Treatment with low concentrations of monensin completely depleted the cells of norepinephrine and dopamine, had a small effect on protein synthesis, and enhanced post-translational processing of only DβH which was previously synthesized and presumably within neurosecretory vesicles. Treatment of PC12 cells for relatively short periods of time enhanced the conversion between the subunit forms of DβH. Their results suggested that the post-translational processing of DβH can occur prior to the exit from the Golgi apparatus and perhaps also in immature secretory vesicles prior to their acidification.

In view of their previous finding that soluble DβH exists in a tetramer/dimer equilibrium, Dhawan, Hensley, Osborne et al. (1986) investigated whether or not adenosine diphosphate (ADP) or fumarate stimulated the activity of the enzyme by affecting its state of dissociation. They reported that ADP binds to DβH and favors dissociation of the tetrameric form to the dimeric form and that this dissociation is correlated with ADP-dependent changes in the K_m of the enzyme. In contrast, fumarate had no significant effect on the state of DβH subunit dissociation despite its dramatic effect on the V_{max} of the enzyme.

Dhawan, Duong, Ornberg et al. (1987) extended their work on the soluble DβH to investigate the composition and dynamic association of subunits in the membranous form of the enzyme. They showed that the purified DβH from bovine membrane is a tetramer composed of nonidentical subunits. They reconstituted the membrane-bound enzyme into phospholipid vesicles and demonstrated the effect of pH and ADP on the molecular state of the enzyme. They found that both low pH and ADP binding promoted dissociation, resulting in release of soluble catalytic subunits. Furthermore, the data demonstrated that hydrophilic subunits from purified soluble hydroxylase associate with hydrophobic subunits of purified reconstituted membranous hydroxylase. This suggests that equilibrium exchange of subunits between soluble and membranous hydroxylase may occur in the chromaffin vesicle.

Stewart and Klinman (1988b) investigated the mode of attachment of the membrane-bound DβH. Specifically, they examined the possibility of a covalently attached glycosyl-phosphatidylinositol anchor. Incubation of fragmented chromaffin vesicle membranes with phosphatidylinositol-specific phospholipase C did not lead to solubilization of DβH activity. Furthermore, analyses of the inositol and phosphate contents of the soluble form of the enzyme indicated no covalently attached phosphate or inositol. These results demonstrate that anchoring of membrane-bound DβH does not occur by a glypiation mechanism. The results also argue against a post-translation covalent modification of the enzyme as the mechanism of membrane attachment.

McCafferty and Angeletti (1987) designed procedures directed toward isolation of peptides from the DβH molecule containing amino acids with low-redundancy codons. They reported the amino acid sequences from tryptic peptides and cyanogen bromide fragments of the enzyme molecule. The amino acid sequence presented should be helpful to investigators planning to clone the enzyme molecule.

Richard, Buda and Legay (1988) examined the biochemical and immunochemical characteristics of DβH from the rat adrenal medulla. First, they purified the soluble and membrane-bound forms of the enzyme by using high-pressure chromatography. Then, they purified the soluble enzyme which allowed them to obtain sufficient amounts of the pure enzymes to raise antibodies. The resulting antibody was characterized and used to investigate the immunologic properties of the two forms of the rat enzyme. Their results showed that the two enzymes contained similar epitopes, suggesting a close structural relationship.

Fitzpatrick and Villafranca (1987) reviewed advances in our understanding of the mechanism of DβH gained through the use of mechanism-based inhibitors. They also considered studies of the steady-state kinetics of the enzyme and the insights into the mechanism they have provided. Also, they reviewed substrate specificity because it provides information on the chemical reactivity of the actual hydroxylating species.

Fitzpatrick and Villafranca (1986) reported that several hydrazines are irreversible inhibitors of DβH and that the mechanism of inactivation appears to involve generation of a carbon-centered radical at the enzyme active site. With phenylhydrazine, inactivation results in the attachment of one phenyl moiety per enzyme subunit.

Fitzpatrick, Harpel and Villafranca (1986) used alternative substrates to probe the order of substrate addition to DβH. In addition to several different amines and substrates, they used ferrocyanide as a reductant in place of ascorbate. Their results were consistent with a ping-pong mechanism in which tyramine binds to the enzyme after the release of oxidized ascorbate. Subsequently, oxygen binds to form a ternary complex.

Colombo, Papadopoulos, Ash et al. (1987) developed a modified purification procedure for DβH isolated from the bovine adrenal medulla. Starting with a suspension of either chromaffin vesicles or adrenal medulla tissue, catalase is used in the homogenization step. With this precaution, DβH remains stable for subsequent purification. They isolated an enzyme of high stability and high specific activity. They also presented data dealing with ascorbic acid inactivation and the preparation of highly purified DβH apoenzyme.

Hamos, Desai and Villafranca (1987) completely removed the

oligosaccharide chains from DβH with glycosidic enzymes, measured the amount removed, and studied the effects of sugar removal on the activity of the enzyme. Adrenal medullary DβH is a glycoprotein containing approximately 5% carbohydrate by weight. Steady-state kinetic data with deglycosylated DβH showed minor differences between the native and the deglycosylated protein. Their data indicated that the oligosaccharide moieties present on DβH do not play a role in the catalytic activity of the enzyme.

Kruse, Kaiser, DeWolf et al. (1986) pointed out the therapeutic value of a specific inhibitor of DβH. They reported the synthesis and pharmacologic evaluation of several 1-benzylimidazole-2-thiols which are extraordinarily potent inhibitors of DβH. One of the compounds binds DβH approximately 10^6-fold more tightly than the phenethylamine substrates do. Several of the compounds were potent when administered orally. Kruse et al. (1986) also noted that the nature of the substituent in position 1 of the imidazole-2-thiol moiety plays an important role in determining inhibitory activity.

The synthesis and kinetic characterization of a new class of DβH inhibitors, 1-(4-hydroxybenzyl)imidazole- 2-thiol, was reported by Kruse, DeWolf, Chambers et al. (1986). These inhibitors, which incorporate a phenethylamine substrate mimic and an oxygen mimic in a single molecule, exhibit both the kinetic properties and the potency expected for a multisubstrate inhibitor and therefore are classified as such. Results of analyses suggested an extremely short intersite distance between the phenethylamine binding site and the active-site copper atom(s).

In an impressive series of studies, Kruse et al. investigated multisubstrate inhibitors of DβH. Kruse, Kaiser, DeWolf et al. (1986) synthesized and characterized some 1-(phenylalkyl)imidazole-2-thiones as a novel class of multisubstrate inhibitors of DβH. These inhibitors incorporate structural features that resemble both tyramine and oxygen substrates; and, as evidenced by steady-state kinetics, they appear to bind to both the phenethylamine binding site and the active site copper atom(s) in the enzyme. Kruse, Kaiser, DeWolf et al. (1987) prepared a series of 1-benzylimidazole-2-thiones to explore the effects of substitution in the benzole ring on the inhibition of DβH. A detailed structure-activity relationship for *in vitro* activity was discovered and was shown to correlate with four key structural features of the benzyl ring. Ross, Kruse, Ohlstein (1987) extended their structure-activity studies to 1-(pyridylmethyl)- and 1-(oxypyridylmethyl)imidazole-2-thiones in an attempt to exploit the pH differential that exists across the chromaffin vesicle membrane. They hypothesized that the weakly basic pyridyl compounds would diffuse into the acidic vesicles in their neutral forms where protonation would occur to enhance their effectiveness as inhibitors. This apparently is the case because the compounds inhibited DβH weakly *in vitro* but showed significant inhibition *in vivo*.

Goodhart, DeWolf and Kruse (1987) reported the mechanism-based inhibition of DβH by *p*-cresol (4-methylphenol) and other simple structural analogs of dopamine that lack a basic side-chain nitrogen. *p*-Cresol binds to DβH by a mechanism that is kinetically indistinguishable from normal dopamine substrate binding. The inhibitors studied are postulated to modify DβH covalently by direct insertion of an aberrant substrate-derived benzylic radical into an active-site residue.

DeWolf, Carr, Varrichio et al. (1988) reported the isolation and characterization of two modified tryptic peptides from DβH inactivated by *p*-cresol. Using a combination of techniques, they determined the sequence of the putative active site peptides, identified the site of attachment of *p*-cresol, and defined the chemical nature of the adduct formed. They presented additional data that indicated that a (4-hydroxyphenyl)methyl radical is generated during catalysis.

Bossard and Klinman (1986) tested a postulated mechanism for β-chlorophenethylamine inhibition of DβH in which enzyme-bound α-aminoacetophenone is generated, followed by an intramolecular redox reaction to yield ketone-derived radical cations as the enzyme inhibitor species. Phenylacetaldehyde was chosen to test this model because β-hydroxyphenyl acid aldehyde is expected to function as a reductant in a manner analogous to α-aminoacetophenone. Phenyl acid aldehyde exhibits the properties of a mechanism-based inhibitor. Some other amides were also found to be mechanism-based inhibitors--the product of hydroxylation is redox inactive. Therefore, they suggested that mechanism-based inhibitors are divided into two types: one type which undergoes hydroxylation prior to inactivation, and one type which only requires hydrogen atom abstraction.

The interaction of DβH with substrate analogs having either imidazole or pyrazole functions at the alkyl chain terminus was investigated by Sirimanne, Herman and May (1987). They demonstrated a novel activity of DβH with 1-(4-hydroxybenzyl)imidazole--namely, ready enzymic oxygenation leading to oxidative dealkylation. On the other hand, they reported that 1-(4-hydroxybenzyl)pyrazole is a potent reversible inhibitor of DβH. This represents the first report on the interaction of this important enzyme with heterocyclic compounds of this type and establishes the basis for new types of DβH substrates and inhibitors with moieties other than an amino function at the alkyl chain terminus.

The competence of DβH to process selenite substrates was investigated by May, Herman, Roberts et al. (1987). They anticipated that DβH would carry out selenoxidation with high facility comparable to that exhibited with sulfide substrates. They reported the first demonstration of facile selenoxidation activity for DβH and the ability of enzymatically generated selenoxide products to undergo recycling back to the selenide at the expense of reduced ascorbate. Because reduced ascorbate is the physiologic reductant essential to DβH activity in the chromaffin vesicle, these results establish the basis for a novel approach to modulating norepinephrine levels and may allow design of novel therapeutic agents.

Ascorbic Acid and Electron Transfer
It is well known that ascorbic acid is the most effective electron donor for the reaction catalyzed by DβH. Excellent reviews relating to ascorbic acid were published by Diliberto, Menniti, Knoth et al. (1987), Levine (1986a), and Levine and Hartzell (1987). The review by Diliberto et al. (1987) stressed the value of adrenal chromaffin cells as a model for study of the neurobiologic role of ascorbic acid. Levine (1986a) and Levine and Hartzell (1987) presented more general reviews on the requirement of the organism for ascorbic acid and the concept of optimal requirements.

The role of ascorbic acid in dopamine β-hydroxylation was investigated by Menniti, Knoth and Diliberto (1986). They presented direct evidence that intravesicular ascorbate donates electrons to DβH in bovine adrenal chromaffin cells in culture and that the cofactor is recycled during this process. The intravesicular ascorbate levels are maintained during β-hydroxylation by an extravesicular electron source which may be cytosolic ascorbate.

In a follow-up study, Menniti, Knoth, Peterson et al. (1987) recalculated the apparent K_m of DβH with respect to ascorbate in intact chromaffin cells as about 15 mM. Studies in isolated chromaffin vesicles and chromaffin vesicle ghosts indicate that the factor(s) that affects the interaction between the enzyme and its cofactor *in situ* resides within the chromaffin vesicles. These results raised the possibility that the availability of ascorbate may regulate the biosynthesis of norepinephrine *in vivo*.

Beers, Johnson and Scarpa (1986) pointed out that adrenal chromaffin vesicles must shuttle reducing equivalents from the cytosol inward to reduce ascorbic acid oxidized during norepinephrine biosynthesis by DβH within the vesicle. They provided evidence that incubation of chromaffin vesicles with dopamine or tyramine results in oxidation of the intravesicular ascorbate due to turnover of DβH and that reduction of the oxidized intravesicular ascorbate can be achieved when ascorbate is present in the extravesicular space. Evidently, chromaffin vesicles shuttle reducing equivalents inwardly from an extra- to an intravesicular ascorbate pool and the cytosolic ascorbate is the source of the intravesicular reducing equivalents required during norepinephrine synthesis.

The regulatory role of ascorbic acid in norepinephrine biosynthesis was studied in digitonin-permeabilized bovine adrenal chromaffin cells by Morita, Levine and Pollard (1986). They found that the permeabilization of chromaffin cells does not disrupt the functions of chromaffin vesicles, including dopamine uptake, norepinephrine synthesis, and storage of these amines. Dopamine uptake was not affected by the presence or absence of ascorbic acid in the medium, but the formation of norepinephrine was stimulated by ascorbic acid. These findings provide evidence that ascorbic acid may stimulate the conversion of dopamine to norepinephrine by increasing DβH activity rather than by increasing the substrate supply of dopamine. Morita et al. (1986) also suggested that the rate of norepinephrine synthesis may be regulated by the concentration of ascorbic acid within the cytosol.

Levine (1986b) characterized in detail the enhancement by ascorbic acid of norepinephrine formation from tyrosine in cultured bovine adrenal chromaffin cells as a model system for determining ascorbate requirements. His data indicated that ascorbic acid enhances only DβH activity without changing TH activity. This effect is specific for ascorbic acid, and ascorbic acid enhancement of DβH activity is maintained when chromaffin cells are incubated with various secretagogues.

Levine, Hartzell and Bdolah (1988) examined the roles of magnesium ATP and external ascorbic acid in norepinephrine formation in chromaffin vesicles. They presented data to indicate that external ascorbic acid and magnesium ATP are required together for maximal norepinephrine biosynthesis, independently of dopamine uptake. The effects of external ascorbic acid and magnesium ATP on norepinephrine were specific and synergistic but may be unrelated to maintenance of membrane potential or direct nucleotide regulation of DβH. Although the mechanism is not known, the data were consistent with the possibility that the ATPase of the chromaffin vesicle mediates these effects.

Herman, Wimalasena, Fowler et al. (1988) demonstrated the ascorbic acid dependence of membrane-bound DβH in adrenal chromaffin vesicle ghosts. This enzyme was demonstrated to use the intravesicular reduced ascorbate as an electron donor with great facility. In addition, they showed that the production of norepinephrine within the vesicle reaches high levels if reducing equivalents from externally applied ascorbate are provided. This process was shown to be dependent on magnesium-ATP and to be reserpine-sensitive, leading to the conclusion that the ATP-dependent proton pump, the catecholamine transporter, and cytochrome b_{561}-mediated electron shuttle are operative in chromaffin vesicle ghosts.

Öğüs (1987) showed that glutathione stimulates DβH. A possible role of the enzyme dehydroascorbate reductase in cofactor regeneration of DβH was proposed.

It is apparent from the above discussion that, for dopamine hydroxylation within the chromaffin vesicle, donation of electrons is required from a source outside the vesicle. Ahn and Klinman (1987) studied the ATP-dependent hydroxylation of dopamine in chromaffin vesicle ghosts preloaded with potassium ferrocyanide at various concentrations and subsequently exposed to either ascorbate or ferrocyanide in the assay medium. Their results demonstrated that both internal and external reductants activate hydroxylation and that they appear to act independently of each other. Ahn and Klinman (1987) suggested that electron donors may reduce membrane-bound enzyme on the outside surface of the membrane.

Njus, Kusnetz, Pacquing et al. (1988) reviewed the role of cytochrome b_{561} in maintaining the redox poise within chromaffin vesicles. Among other things, they pointed out that the reduction of intravesicular semidehydroascorbate occurs about at 2 orders of magnitude faster than decay of semidehydroascorbate by dysproportionation. This supports the hypothesis that cytochrome b_{561} functions *in vivo* to reduce intravesicular semidehydroascorbate and to regenerate intravesicular ascorbic acid. They also presented evidence that the cytochrome was accessible to substrates at either surface of the vesicular membrane. It was also reasoned that cytochrome b_{561} may mediate transmembrane electron transfer by itself.

Stewart and Klinman (1987) examined the relationship between structure and reactivity at the reductant site, examining the geometric constraints placed on the substrate site by the stereochemistry and mechanism of the hydroxylation reaction. Using ascorbate ferrocyanide and catecholamines as reductants, they showed that the structural requirements at the reductant site provide support for a binding site distinct from the substrate. Despite this finding, comparison of kinetics between slow reductants (catecholamines) and ascorbate unambiguously eliminates reaction mechanisms involving the reduction of enzyme forms leading from the binary enzyme-dopamine complex to enzyme-bound product.

A kinetic analysis of electron transport across chromaffin vesicle membranes was conducted by Kelley and Njus (1988). They examined the rate of electron transfer from ascorbate trapped within chromaffin vesicle ghosts to external ferricyanide. The rate of ferricyanide reduction becomes saturated at high ferricyanide concentration. The reciprocal of the rate is linearly related to the reciprocal of the ferricyanide concentration. These and other observations indicated that the phenomenon is associated with a rate constant for the oxidation of cytochrome b_{561} by ferricyanide. Their results support the hypothesis that cytochrome b_{561} functions *in vivo* to reduce intravesicular semidehydroascorbate, thereby maintaining the intravesicular ascorbic acid content.

Kelley and Njus (1986) confirmed the involvement of cytochrome b_{561} in transmembrane electron transfer by monitoring changes in the absorbance of cytochrome b_{561} in adrenal chromaffin vesicle membranes concurrently with electron transfer. The changes in absorbance show that the cytochrome does participate in electron transfer reactions on both sides of the membrane. Moreover, cytochrome b_{561} does appear to be capable of reducing dehydroascorbate to ascorbate.

Kent and Fleming (1987) purified cytochrome b_{561} from bovine adrenal chromaffin vesicles by fast protein liquid chromatography chromatofocusing. They demonstrated that, when reconstituted into ascorbate-loaded phosphatidylcholine vesicles, purified cytochrome b_{561} can supply transmembrane electrons for both the soluble and the membranous form of DβH. This electron transfer is dependent on the presence of a redox mediator which, *in vivo*, is most likely to be the ascorbate/semidehydroascorbate redox pair.

The structure of cytochrome b_{561} was examined by Perin, Fried, Slaughter et al. (1988). They purified cytochrome b_{561} from bovine adrenal chromaffin vesicles and used this to identify two cDNA clones. The structure predicted by the sequences of these cDNAs suggests a highly hydrophobic protein of 273 amino acids which spans the membrane six times with little extramembranous sequence. Cytochrome b_{561} is not homologous with any other cytochrome and thus represents a new class of electron carriers.

The role of the ascorbate shuttle in catecholamine formation was reviewed (Nutrition Reviews, 1986). Research in this area has broadened our understanding of vitamin C function.

Wakefield, Cass and Radda (1986a) undertook a detailed examination of the interactions among the various redox active components of the chromaffin vesicle, using optical difference spectroscopy to monitor the redox state of cytochrome b_{561} and quantitative electron paramagnetic resonance to determine the concentration of ascorbate-free radical. The data provide support for a model in which cytochrome b_{561} catalyzes transmembrane electron transfer from cytosolic ascorbate to intravesicular ascorbate free radical, thereby coupling the mitochondrial NADH:ascorbate free radical oxidoreductase to the intravesicular DβH. Electron flow into the vesicle is enhanced by the action of the proton-pumping ATPase in the vesicle membrane which also replenishes protons consumed by the turnover of DβH.

In an accompanying article, Wakefield, Cass and Radda (1986b) described an experimental system using hydrated nickel ion as a membrane-impermeable spin probe to selectively eliminate the electron paramagnetic resonance signal of extravesicular ascorbate free radical, thereby allowing distinction between ascorbate free radical inside and outside closed vesicles. Using this method, they found that ascorbate free radical inside the chromaffin vesicle indeed can be reduced by the action of the NADH:ascorbate free radical oxidoreductase on the topologically distant mitochondrion.

Similarities of electron transfer across chromaffin vesicles and neurosecretory vesicles from the bovine posterior pituitary gland were noted by Russell, Levine and Njus (1985). Two different tests showed that the ascorbic acid contained in posterior pituitary neurosecretory vesicles will reduce an external electron acceptor. It was suggested that, as in chromaffin vesicles, this electron transfer is mediated by cytochrome b_{561}.

Using spectrophotometric methods, Gonzalez, Caorsi and Berrios (1986) demonstrated the presence and quantified the amount of cytochrome b_{561} in purified secretory vesicles. The cytochrome is present in chromaffin vesicle membranes and other neurosecretory vesicles.

Mechanisms
Dopamine β-hydroxylase is a copper-containing enzyme. The enzyme-bound copper is essential for activity and alternates between the cuprous and cupric states during catalysis. Syvertsen, Gaustad, Schrøder et al. (1986) used a cupric-selective electrode to investigate the binding of the cupric ion to native and copper-free DβH. The results establish a stoichiometry of four high-affinity binding sites for copper per enzyme tetramer and more binding sites of lower affinity. The data for the four high-affinity sites indicate interaction in the binding to the sites.

van der Meer, Jongejan and Duine (1988) reported the presence of a pyrroloquinoline quinone within DβH from the bovine adrenal medulla. By using a specific inhibitor, the absorption spectrum of DβH was altered; and a product with this absorption spectrum could be detached. This fragment contained a hydrazone of pyrroloquinoline quinone. Dopamine β-hydroxylase is the first copper-quinoprotein hydroxylase yet described. This structure may have implications for the mechanism of action of DβH and its inhibition by hydrazines.

Syvertsen, Melø and Ljones (1987) studied the conformation of DβH by monitoring the protein fluorescence under different conditions. Protein fluorescence is a useful method because minor changes in the conformation of a protein may result in large effects on the quantum yield and the wavelength dependency of the emission. They determined changes in the conformation of the enzyme induced by copper, metal-chelating agents, pH changes, and anions. Three distinct conformations were characterized. For the copper-free apoenzyme, a unique binding site for the first copper atom was demonstrated by larger quenching of the protein fluorescence than for binding of additional copper atoms.

Magnetic resonance studies on the copper site of DβH in the presence of cyanide and azide anions were performed by Obata, Tanaka and Kawazura (1987). Their results suggested that two water molecules are ligated to the copper site of DβH.

Blackburn, Concannon, Shahiyan et al. (1988) reported detailed EPR spectral comparisons of DβH enzyme samples containing one and two copper atoms per subunit. Although small differences in binding constants were apparent between the two sites in the subunit, they were unable to detect significant spectral differences or any evidence of short-range magnetic interactions between them. The results pose interesting questions as to the mode of copper binding and to the mechanism of electron transfer to oxygen, not the least of which is the relatively weak binding of the cuprous ion to the native enzyme.

Markossian, Melkonyan, Paitian et al. (1988) studied the transfer of copper from copper thionein to apodopamine β-hydroxylase and in the opposite direction from DβH saturated with copper to thionein lacking copper. They demonstrated that copper can be transferred in both directions, restoring enzyme activity to the acceptor enzyme.

The active site of the cuprous ion in DβH isolated from the bovine adrenal medulla was studied by McCracken, Desai, Papadopoulos et al. (1988) using pulsed electron paramagnetic resonance spectroscopy. Examination of the stimulated electron spin-echo envelope revealed certain frequency component characteristics. It was shown that there are three, or more likely four, imidazole ligands bound to the copper.

Scott, Sullivan, DeWolf et al. (1988) reported the results of x-ray absorption spectroscopy experiments on highly concentrated samples of the cupric and cuprous oxidation states of bovine DβH with a 2:1 copper:unit stoichiometry. A significant change in the structure of the copper sites was occurred upon ascorbate-mediated reduction of cupric DβH to the cuprous form. Because the cupric form of the enzyme is inactive and may not be physiologically important, the data of Scott et al. (1988) indicate that a major structural rearrangement of the copper-containing active sites may be responsible for the reductive activation of DβH.

Other Considerations

Improved methods for the detection of DβH have been presented in two publications. Dillen, Claeys and De Potter (1986) described an improved method for the measurement of DβH activity in cerebrospinal fluid. The assay is based on incubation with dopamine at a saturating substrate concentration and quantitation of the reaction product norepinephrine by high-performance liquid chromatography with electrochemical detection. Racz, Kuchel, Debinski et al. (1986) described an assay for DβH activity in human and rat plasma and rat tissues that used reverse-phase high-performance liquid chromatography with electrochemical detection. Enzyme activity in the plasma was measured directly, without extraction of the enzyme. This method allows determination of DβH in small volumes of plasma and tissues.

Lima and Sourkes (1986) investigated the mechanism by which reserpine induces the activity of adrenal DβH when administered to rats. From this detailed pharmacologic study, they concluded that reserpine decreases cerebral serotonin; this, in turn, induces DβH via monoaminergic pathways.

Lima and Sourkes (1987) demonstrated that the continuous administration of corticotropin-releasing factor intraventricularly results in significant increases in DβH and PNMT activities in the adrenal glands of rats. The pattern of increase in DβH does not correspond to the effects observed on plasma corticosterone, a result that suggests that corticotropin-releasing factor is acting to increase the adrenal enzyme by means other than the pituitary-adrenal axis. In contrast, PNMT responds to the releasing factor in a manner indicating a correlation with glucocorticoid availability.

Dahlström, Bööj, Goldstein et al. (1987) applied a newly developed cytofluorometric scanning technique in a pharmacologic study of the influence of reserpine on the axonal transport of norepinephrine, DβH, TH, and neuropeptide Y. They suspected that these components of neurosecretory vesicles are transported in increased amounts after administration of reserpine. Surprisingly, the results obtained indicate that this might not be the case but suggest that the syntheses of DβH and neuropeptide Y are regulated independently after reserpine administration.

Shorr, Minnich, Varrichio et al. (1987) examined the homology that exists between catecholamine receptors and biosynthesis enzymes. Specifically, they used immunologic cross-reactivity to examine the structural relationship between β-adrenergic receptors and DβH. They reported evidence of a structural homology between these two entities.

Gripois and Valens (1986a) studied the induction of DβH activity in the adrenals of 2-week-old control and hypothyroid rats that were stimulated by insulin-induced hypoglycemia. In control rats, DβH induction was maximal at 48 hours after insulin administration. Hypothyroidism completely suppressed enzyme induction.

Markossian, Paitian, Mikaelyan et al. (1986) examined the effects of different forms of neurocuprein on DβH. The apo form of neurocuprein, an acidic copper protein in chromaffin vesicles, was found to be a potent inhibitor of the enzyme, whereas the holo form of neurocuprein had no effect on enzyme activity. The inhibitory capacity of neurocuprein may be due to the ability of the apoprotein to chelate copper. They suggested a role of neurocuprein as an endogenous protein that regulates DβH activity.

Dopamine β-hydroxylase is of clinical interest because of its possible relationship to the development of psychiatric disorders, hypertension, congestive heart failure, Lesch-Nyhan syndrome, and other diseases. As part of an overall study dealing with genetic factors involved in heart disease, Asamoah, Wilson, Elston et al. (1987) studied members of a multigeneration family to investigate the possible segregation of a major gene for DβH and its linkage to another locus. Individuals in this pedigree with a history of heart attack had significantly lower levels of DβH. Pedigree segregation analysis showed evidence of a codominant gene for DβH segregating in the family.

May, Wimalasena, Herman et al. (1988) reported a novel strategy for designing antihypertensives targeted at DβH. The strategy entails design of an alternate substrate such that it will be readily converted by the target enzyme to a product that is then capable of causing local depletion of an essential cofactor for the target enzyme. Such a strategy of "turnover-dependent cofactor depletion" is conceptually quite distinct from direct inhibition of the enzyme of interest and might represent an especially useful approach to modulating compartmentalized cofactor-dependent enzymes such as DβH for therapeutic purposes.

Man in 't Veld, Boomsma, Moleman et al. (1987) described a patient with a congenital deficiency of DβH. In this patient, norepinephrine and epinephrine were undetectable in plasma, urine, and cerebrospinal fluid; but the dopamine concentration was increased in all of these fluids. Dopamine β-hydroxylase was undetectable in plasma and cerebrospinal fluid. Physiologic and pharmacologic stimuli of sympathetic neurotransmitter release caused increases in plasma dopamine rather than in norepinephrine. In a comment on this report, Superti-Furga, Royce and Steinmann (1987) suggested that the impairment of β-hydroxylation of dopamine could be due to decreased availability of copper.

Fraeyman, Van de Velde and De Smet (1988) examined the molecular forms of DβH in rat adrenal gland and superior cervical ganglion. The adrenal gland had a tetrameric form of the enzyme with a molecular mass of 294 kDa and a novel molecular form with a molecular mass of 125 kDa. Pretreatment of the rat with cycloheximide markedly decreased enzyme activity without altering the molecular heterogeneity.

Palatini (1988) developed a kinetic model to examine the activation of DβH by anions. A model was presented that accounts for the numerous different effects of activating anions on the enzyme kinetics. It was claimed that the model presented has a general validity because it holds for any of the kinetic mechanisms thus far proposed for DβH.

PHENYLETHANOLAMINE *N*-METHYLTRANSFERASE

Phenylethanolamine *N*-methyltransferase (Enzyme Commission number 2.1.1.28, also known as S-adenosyl-L-methionine: phenylethanolamine *N*-methyltransferase) catalyzes the conversion of norepinephrine to epinephrine. The methyl donor for the *N*-methylation of norepinephrine is S-adenosylmethionine. General reviews of the properties of PNMT were published by Park (1986) and Fuller (1987).

The complete nucleotide sequence of the DNA encoding bovine

PNMT and the amino acid sequence of it were reported by Baetge, Suh and Joh (1986). They isolated a complementary DNA (cDNA) clone containing the full coding region for the enzyme and transfected cultured cells with an expression vector containing this cDNA. The transfected cells produced high levels of PNMT activity. Comparison of the sequences of PNMT and TH revealed a significant homology which supports previous protein and immunologic data suggesting that the catecholamine biosynthesis enzymes are structurally related.

Batter, D'Mello, Turzai et al. (1988) isolated the entire gene for bovine adrenal PNMT. Examination of the gene revealed that it is 1,594 base pairs in length and consists of three exons and two introns. Further information on the structure of the gene was revealed, including potential binding sites for glucocorticoid receptors. It was also revealed that there were no significant homologies with other catecholamine-synthesizing enzymes, indicating that PNMT is not a member of a multigene family of these enzymes.

Baetge, Behringer, Messing et al. (1988) cloned the human gene for PNMT and determined the complete nucleotide sequence. The structural gene consists of three exons and two introns spanning about 2,100 base pairs. A portion of the 5' flanking sequence was incorporated into transgenic mice. Antigen mRNA expression was detected in the adrenal glands and eyes. These results indicate that the enhancer(s) for appropriate expression of the human gene for PNMT is in the 5' flanking region.

Kaneda, Ichinose, Kobayashi et al. (1988) described the complete nucleotide sequence of the cDNA for human PNMT and deduced the amino acid sequence of the enzyme. They first isolated a cDNA clone from a bovine adrenal medulla cDNA library and used this cDNA fragment as a probe to screen a human pheochromocytoma cDNA library. They isolated a clone with an insert of about 1,000 base pairs which contained the complete coding region of the enzyme. Northern blot analysis indicated that this clone is a full-length cDNA. Determination of the nucleotide sequence revealed that human PNMT consists of 282 amino acid residues with a predicted molecular weight of 30,853. The amino acid sequence of the human enzyme was highly homologous (88%) to that of the bovine enzyme. Chromosomal assignment studies assigned the PNMT gene to chromosome 17.

Weisberg, Batter, Brown et al. (1988) reported the purification and partial determination of amino acid sequence of bovine adrenal PNMT. The amino acid sequences of several peptides derived from the purified protein were identical to those deduced from the exons of the genomic DNA clone.

Wong, Yamasaki and Ciaranello (1987) characterized the isozymes of bovine PNMT. The enzymatically active monomer has a molecular mass of 30 kDa and can be separated into at least four active charged isozymes. Kinetic characteristics were determined for each isozyme. Treatment of the isozymes with exoglycosidases indicated that carbohydrate substitution must be minimal. The kinetic and glycosylation data suggest that the isozymes of PNMT may be primary structural variants.

Park, Ehrlich, Evinger et al. (1986) examined differences in distribution of PNMT activity in four strains of rats. They found strain differences in the enzyme activity in the adrenal medulla and brain. By immunochemical titration, these variations were not due to presence of an inactive form of enzyme but to the amount of the enzyme protein.

In a follow-up study, Evinger, Park, Baetge et al. (1986) examined the biochemical basis for the strain difference at the level of the PNMT protein and mRNA production. One strain of rat had approximately 5 times more protein than another strain. The first strain also had 2 to 4 times more mRNA for PNMT. The strain specificity in the production of PNMT reflects differences in the expression of the gene for the enzyme. Thus, an inherited capacity for PNMT expression may provide the intrinsic determinants responsible for neurotransmitter production. The data of Evinger et al. (1986) provide a direct link between regulation of catecholamine enzyme biosynthesis at the genomic level and the availability of specific catecholamines for neurotransmitter and hormonal functions.

Bohn (1986) reviewed the expression and development of PNMT, specifically examining the role of glucocorticoids. She examined information provided by studies on the adrenal medulla and how it applied to epinephrine-containing neurons throughout the nervous system.

Bohn, Goldstein and Black (1986) examined the initial appearance and ontogeny of PNMT. They found that this enzyme is produced earlier in the brain than in the adrenal medulla and sympathetic ganglia. The development of the enzyme in the periphery, but not in the brain, is dependent on maintenance of physiologic levels of glucocorticoids.

In situ hybridization was used by Schalling, Dagerlind, Brene et al. (1987) to localize mRNA for PNMT. Hybridization of sections of rat and bovine adrenal glands resulted in a strong activity over numerous cells in the medulla. Some cells in the medulla were not reactive, and the pattern was somewhat different between rat and cow.

Banerji, Callas, Meyer et al. (1986) measured PNMT and DβH activities in the adrenal medulla of young rats after the administration of adrenocorticotropic hormone. Adrenal PNMT activity was increased significantly, whereas DβH activity remained unchanged. These results indicate that the pituitary-adrenocortical-adrenomedullary axis is functional in the neonatal rat. Apparently, glucocorticoids play a large role in stimulating adrenal catecholamine synthesis in the neonatal rat.

Lima and Sourkes (1986) used pharmacologic agents that affect monoaminergic functions to examine the regulation of PNMT in rats. The administration of reserpine brought about an increase in enzyme activity. The administration of dopamine agonists, a treatment that increases adrenal TH activity, did not modify adrenal PNMT activity. These and other data led to the conclusion that the induction of PNMT by reserpine involves depletion of catecholamines and serotonin, the depletion of serotonin having the more powerful effect. A serotonergic inhibitory pathway is involved in the central regulation of adrenal PNMT activity.

Gripois and Valens (1986b) showed that the postnatal development of adrenal PNMT activity is slightly accelerated by hypothyroidism and slowed by hyperthyroidism. Stimulation of the adrenals in rats by insulin- induced hypoglycemia does not lead to a change in enzymatic activity. If the stimulation is repeated for 4 days, it leads to a net increase in enzyme activity in control rats whereas no increase occurs in hypo- or hyperthyroid animals.

Byrd, Hadjiconstantinou and Cavalla (1986) found that PC12 cells synthesize small amounts of epinephrine and that dexamethasone increases both epinephrine content and PNMT activity. They suggested that PC12 cells may be useful in the investigation of the regulation of PNMT.

A new assay technique for PNMT activity by high-performance liquid chromatography with on-line radiochemical detection was described by Nissinen (1986). The method is based on the measurement of ^{14}C-labeled products of the substrate normetanephrine. The method is suitable for assaying PNMT activity in the adrenal and brain tissue.

OTHER ENZYMES

The enzymes discussed in the preceding sections are sequentially involved in the synthesis of epinephrine. The enzymes discussed in this section are related to other functions within the adrenal chromaffin cell. The enzymes that will be discussed are adenosine triphosphatase (ATPase), monoamine oxidase, acetylcholinesterase, ornithine decarboxylase, *S*-adenosyl-methionine decarboxylase, angiotensin-converting enzyme, enzymes involved in glucose utilization, lipases, phospho-diesterases, kinases, sphingomyelinase, and transglutaminase.

ATPase is present within the chromaffin vesicle membrane and is responsible for pumping proteins into the interior of the vesicle. A general review of proton-translocating ATPases was presented by Al-Awqati (1986). A general review of proton-pumping ATPases of secretory vesicles was published by Nelson (1987).

Wang, Moriyama, Mandel et al. (1988) cloned a cDNA that coded for an accessory polypeptide of the proton ATPase from chromaffin vesicles. This cDNA encoded a 32 kDa polypeptide that has an apparent molecular weight of 39 which was previously denoted as subunit IV of the ATPase from chromaffin vesicles. Northern blots revealed the presence of a single mRNA in bovine adrenal medulla.

Percy and Apps (1986) used chemical labeling by tritiated *N*-ethylmaleimide to identify the subunit of chromaffin vesicle membrane ATPase as the site of ATP hydrolysis. After treatment of the membranes with the inhibitor, only one polypeptide, of the at least five different polypeptides, was strongly radiolabeled; this was the largest (70 kDa) subunit of proton-translocating ATPase. Two-dimensional electrophoresis revealed heterogeneity in this polypeptide.

An ATPase sensitive to *N*-ethylmaleimide was purified 100-fold from chromaffin vesicle membranes by Cidon and Nelson (1986). The purified preparation contained four major polypeptides with molecular masses of about 115, 72, 57, and 39 kDa which co-purified with the ATPase activity. In contrast to the findings of Percy and Apps (1986), Cidon and Nelson (1986) found that the 39 kDa protein bound to radiolabeled *N*-ethylmaleimide. They concluded that the proton-translocating ATPase of chromaffin vesicle membranes contains at least four subunits, with the 115 kDa polypeptide being the main subunit and having the active site for the ATPase activity of the enzyme.

Moriyama and Nelson (1987a) purified the proton-translocating ATPase of chromaffin vesicles in a manner that maintained its proton-pumping activity when reconstituted into phospholipid vesicles. The enzyme was purified at least 50-fold with respect to both ATPase and proton-pumping activity. The ATP-dependent proton uptake activity of the reconstituted enzyme was absolutely dependent on the presence of chloride or bromide outside the vesicles, whereas sulfate, acetate, formate, nitrate, and thiocyanate were inhibitory. Moriyama and Nelson (1987a) suggested that membrane potential and anions are the main factors involved in the regulation of the pH gradient across the chromaffin vesicle membrane. Anion transporters and the muscarinic acetylcholine receptor may modulate the chloride concentrations in chromaffin cells and, in so doing, control the acidification of chromaffin vesicles.

In order to obtain more information on the various binding sites on the ATPase and to study the nucleotide binding sites as a first step for understanding the mechanisms of action of chromaffin vesicle ATPases, Moriyama and Nelson (1987b) investigated binding of N-ethylmaleimide and nucleotides to individual subunits of the enzyme. N-Ethylmaleimide bound to three of the five subunits, but inhibition of both ATPase and proton-pumping activity correlated with binding to the 72 kDa polypeptide. In the presence of ADP, the saturation curve of ATP changed shape, suggesting that the proton-translocating ATPase is an allosteric enzyme. This and other evidence led to the conclusion that a tightly bound ADP on the 72 kDa subunit is necessary for the activity of the enzyme.

Detailed methods of purification of the chromaffin vesicle H^+-ATPase were published by Nelson, Cidon and Moriyama (1988). Several different methods for purifying the enzyme were presented.

Moriyama and Nelson (1988a) addressed the question of how the pH difference created by H^+-ATPase is regulated. The pH difference across membranes such as the chromaffin vesicle membrane is achieved by modulation of anion concentrations and the sensitivity of the proton pumping activity to membrane potential and the pH difference. Modulators have only slight effect on the ATPase activity of this ATPase but proton pumping activity is strongly affected. This kind of control can be achieved by two distinct mechanisms: a spill in which the excess protons leak outside the membrane by a carrier evolved specifically for this function, or a slip in which the excess protons are not conducted across membrane even though the system is still utilizing energy.

Moriyama and Nelson (1988b) isolated an ATPase enzyme from chromaffin vesicles with an apparent molecular mass of about 115 kDa. This polypeptide had no relationship to the 115 kDa subunit of the chromaffin vesicle proton pump. It is possible that this ATPase is related to a single subunit of another ATPase. This ATPase was sensitive to vanadate and N-ethylmaleimide.

In a search for reversible and nontoxic inhibitors of H$^+$-ATPases, Moriyama and Nelson (1988c) examined the effect of fusidic acid and suramin on chromaffin vesicle ATPase and other ATPases. Fusidic acid inhibited the enzymes, apparently by a combination of uncoupling the proton pump and inhibition of the enzyme activity. Suramin was also a potent inhibitor of the H$^+$-ATPase from chromaffin vesicles.

Mandel, Moriyama, Hulmes et al. (1988) cloned and determined the sequence of the gene encoding the 16 kDa proteolipid of chromaffin vesicles. Sequence homology with other H$^+$-ATPases was detected. These findings suggest that the proteolipids of the vacuolar H$^+$-ATPases were evolved in parallel with the eubacterial proteolipid from a common ancestral gene that underwent gene duplication.

The magnesium ATPase activities of bovine adrenal chromaffin vesicles were studied in highly purified preparations of vesicle ghosts and in intact vesicles by Grønberg and Flatmark (1987). An inhibitor specific for mitochondrial ATPase caused a small inhibition that could be accounted for by a very minor contamination with mitochondria. Their experimental results support the conclusion that the intrinsic proton-translocating ATPase accounts for the major fraction of the overall magnesium ATPase activity of native chromaffin vesicle ghosts. This was demonstrated by the use of established ATPase inhibitors and spectroscopic probes for proton transport.

Dean, Nelson and Rudnick (1986) reconstituted the solubilized proton pump from adrenal chromaffin vesicles into proteoliposomes and compared its catalytic properties to those of native ATPase. Studies with selective inhibitors indicated that the chromaffin vesicle ATPase belongs to a specific class of ATP-dependent ion pumps. Comparisons of ATP hydrolysis with ATP-dependent serotonin transport suggest that approximately 80% of the ATPase activity in purified chromaffin vesicle membranes is coupled to proton pumping. Most of the remaining ATPase activity is due to contamination. These and other data demonstrate that the predominant ATP hydrolase of the chromaffin vesicle membrane is also responsible for ATP-driven amine transport and vesicle acidification in both native and reconstituted membranes.

Dean, Nelson, Agnew et al. (1987) determined the hydrodynamic properties of detergent-solubilized ATPase, which is coupled to proton pumping in bovine adrenal chromaffin vesicles, by sedimentation, equilibrium centrifugation, and gel permeation chromatography. The protein solubilized with detergent-containing phosphatidylserine sedimented as a particle of 264 kDa. By using various parameters, they calculated that the protein component has a mass of 134 kDa. The particle has an apparent Stokes radius of 4.3 nm. They concluded that the ATPase is an intrinsic membrane protein with a structure distinct from that of mitochondrial ATPase.

Examining the energetics of catecholamine uptake, Scherman, Soumaron and Henry (1986) looked for an acylphosphate derivative in chromaffin vesicle membrane preparations and investigated the possibility that the proton pump synthesizes a phosphorylated intermediate similar to that described for the proton pump in gastric mucosa. An acylphosphate was detected but was not inhibited by vanadate, indicating that the acylphosphate is not associated with the proton pump. Scherman et al. (1986) suggested that it is associated with another ATPase present in chromaffin vesicle membranes whose function is unknown.

Grønberg and Flatmark (1988) found that diethylstilbestrol reversibly inhibited the hydrolysis of magnesium ATP and proton pump activity in chromaffin vesicle ghosts. The parallel inhibition suggested a tight kinetic coupling between the two activities. The noncompetitive type of inhibition showed that the action of diethylstilbestrol is distal to the site of ATP binding and hydrolysis. Although nonspecific, the interaction of diethyl-stilbestrol with chromaffin vesicle membrane seems to affect primarily the H^+-ATPase.

Cuppoletti, Strasser and Dean (1988) identified an immunoreactive 8-azido-ATP-labeled protein common to the lysosomal and chromaffin vesicle membranes. An antibody raised against a chromaffin vesicle proton pump protein reacted on Western blots with a 70- to 80 kDa protein from the lysosomal membrane. They found that photolysis with 8-azido-ATP leads to inhibition of chromaffin vesicle proton pump function and pump-related ATP hydrolysis. An anti-chromaffin vesicle antibody reacted with an approximately 70 kDa protein of the chromaffin vesicle and lysosome. This raises the possibility that this protein, or similar proteins, may play a related role in pump function such as ATP binding or hydrolysis in these organelles.

Tamura, Lam and Inagami (1987) purified an endogenous inhibitor of Na, K-ATPase from the bovine adrenal gland. It was not determined if it was in the medulla or cortex. The inhibitor behaved like ouabain, and it was suggested that the inhibitor may mediate salt-induced high blood pressure.

Monoamine oxidase oxidizes various monoamine neuro-transmitters as well as exogenous bioactive monoamines. Various aspects of this biologically important enzyme were reviewed by Yu (1986). Monoamine oxidase was characterized in homogenates from various tissues by Lenzen, Freisinger-Treichel and Panten (1987). The contribution of the type B enzyme was 20% for rat adrenal medulla and 60% for bovine adrenal medulla. PC12 cells contained predominantly (90%) the type A monoamine oxidase; however, certain characteristics of the monoamine oxidase indicated abnormal behavior in this cell line.

Naoi, Suzuki, Takahashi et al. (1987) examined the effects of ganglioside supplementation of culture medium on type A and type B monoamine oxidase activities in PC12h cells. The activity was found to be mainly type A enzyme; and type B activity was negligible. After supplementation of the culture medium with a ganglioside, the cells produced type B monoamine oxidase activity. This suggested that gangliosides may be involved in the production of monoamine oxidase type B in neurons and in the regulation of levels of biogenic amines.

The effects of the neurotoxin N-methyl-4-phenylpyridinium ion (MPP$^+$) on monoamine oxidase in PC12h cells was investigated by Naoi, Suzuki, Kiuchi et al. (1987). MPP$^+$ was found to accumulate in PC12h cells, and monoamine oxidase activity in the cells was inhibited in a dose-dependent fashion within a certain range. On the other hand, TH activity and the concentration of dopamine were not changed.

Youdim, Heldman, Pollard et al. (1986) investigated the mechanisms of oxidative deamination of catecholamines by PC12 and bovine adrenal chromaffin cells. PC12 cells have a monoamine oxidase activity that oxidizes type A and type A-B substrates and is selectively inhibited by a type A inhibitor. In contrast, a type B substrate and type B inhibitors are ineffective. By these criteria, it is apparent that the monoamine oxidase in PC12 cells is solely type A. On the other hand, isolated chromaffin cells have type B monoamine oxidase. Youdim et al. (1986) pointed out that, although PC12 cells have some functional and biochemical properties in common with chromaffin cells and adrenergic neurons, they have a close resemblance to adrenergic neurons with regard to monoamine oxidase activity.

The conclusions of Youdim et al. (1986) were confirmed in a histochemical study by Carmichael and Pfeiffer (1987) that demonstrated type B monoamine oxidase in chromaffin cells of the human adrenal medulla and type A in vascular and other neural elements. Human pheochromocytoma apparently does not contain type B enzyme.

The distribution of acetylcholinesterase and pseudocholinesterases in the chick adrenal gland during the early phases of organogenesis was studied by Mastrolia, Bichi, Arizzi et al. (1986). In this histochemical study, acetylcholinesterase was demonstrated on the plasma membrane, in the perinuclear cisterna, and in some cisternae of the rough endoplasmic reticulum in chromaffin interrenal cells. Pseudocholinesterase activity was localized in the paranuclear space and rough endoplasmic reticulum. It was suggested that these enzymatic activities may be implicated in morphogenetic mechanisms.

Mastrolia, Manelli, Gallo et al. (1986) studied the cytochemical distribution of acetylcholinesterase activity in adrenal chromaffin cells of three amphibian species. Evidence of enzyme activity was found in both epinephrine- and norepinephrine-containing cells and neural elements present in the adrenal gland. In the chromaffin cells, the enzymatic activity was localized in the endoplasmic reticulum, the nuclear envelope, the cellular membrane, and the membrane of chromaffin vesicles. Generally, norepinephrine-containing cells appeared to be more reactive than epinephrine-containing cells. The intensity of the reaction appeared to vary among the species examined.

Gallo, Civini and Mastrolia (1987) studied the cytochemical localization of acetylcholinesterase activity in the adrenal chromaffin cells of the amphibian *Discoglossus pictus*. Evidence of enzyme activity was associated with all types of chromaffin cells including small vesicle chromaffin cells. On the whole, the epinephrine-containing cells (both normal and small-granule chromaffin types) were more reactive than the norepinephrine-containing cells. Gallo et al. (1987) suggested that small-granule chromaffin cells are "true" chromaffin cells in a different functional state.

The role of ornithine decarboxylase and the polyamines in nervous system development was reviewed by Slotkin and Bartolome (1986). The developmental pattern of ornithine decarboxylase activity, polyamines, and S-adenosylmethionine decarboxylase of rat adrenal gland was examined by Ekker and Sourkes (1987a). The activity of adrenal ornithine decarboxylase increases after birth, attains a peak at 17 days of age, and thereafter declines sharply. Induction of adrenal ornithine decarboxylase can be observed after immobilization stress at 4 days of age, but the induction caused by other methods is obtained only at 17 days. The results suggest that the developmental pattern of ornithine decarboxylase activity is independent of its responsivity to various stressors.

An enzyme catalyzing phosphoryl-group transfer from nucleoside triphosphates to ADP was isolated from the soluble protein fraction of bovine adrenal chromaffin vesicles by Taugner, Heym, Kummer et al. (1988). The enzyme was purified and had an apparent molecular mass of 58 kDa. This protein could be recognized by specific polyclonal antibody. The specificity of the enzyme was broad to nucleoside triphosphates but was narrow to nucleoside diphosphates, favoring adenosine diphosphate. Therefore, the name "nucleoside triphosphate-ADP-phosphorotransferase" was proposed. Immunocytochemical studies demonstrated that this enzyme is confined to chromaffin vesicles.

Ekker and Sourkes (1987b) investigated the effects of dopamine agonists, insulin, 2-deoxyglucose, and immobilization on *S*-adenosylmethionine decarboxylase activity in the rat adrenal medulla and cortex. These treatments are known to increase ornithine decarboxylase activity in this organ. Generally, the treatments decreased adrenal *S*-adenosylmethionine decarboxylase activity. Other experiments indicated that the activity of *S*-adenosylmethionine decarboxylase in the adrenal gland is under hormonal control.

Strittmatter, De Souza, Lynch et al. (1986) used autoradiography with tritiated captopril to localize angiotensin-converting enzyme in rat adrenal and pituitary glands. The maximum quantity of captopril binding sites in the adrenal medulla was 480 fmol/mg protein. The distribution of binding sites in the adrenal medulla is homogeneous. Subcellular fractionation of the bovine adrenal medulla revealed enrichment of angiotensin-converting enzyme in plasma membrane fractions but not in chromaffin vesicles.

Angiotensin-converting enzyme was characterized in the rat adrenal medulla by Israel, Barbella and Saavedra (1986). They found that the enzyme derived from the adrenal medulla and that from lung have similar properties. After unilateral adrenalectomy, there was a significant increase in adrenal medullary angiotensin-converting enzyme activity. This change was due not to a modified affinity of the enzyme for substrate but to an alteration in maximum velocity. The increase in activity was blocked by denervation of the adrenal glands. The results support the existence of a functional angiotensin-converting enzyme in the adrenal medulla that is under neuronal control.

Laliberte, Laliberte, Alhenc-Gelas et al. (1987) used immuno-cytochemical methods to localize angiotensin-converting enzyme in the rat adrenal gland. Enzyme immunoreactivity was found mainly, but not exclusively, on the luminal side of all types of vessels in the adrenal medulla. In addition, angiotensin-converting enzyme was localized on the plasma membrane of chromaffin cells but never in Schwann cells or in nerve fibers.

The results provide evidence of local production of angiotensin II at the vascular level in the adrenal medulla. Angiotensin II or other peptides, such as bradykinin, could be metabolized at the surface of chromaffin cells.

Wilson, Lynch and Snyder (1987) used autoradiography with tritiated captopril to localize and quantitate angiotensin-converting enzyme in various tissues after induction of hypertension. Elevated levels of the enzyme were noted in the adrenal medulla and other tissues following the induction of hypertension.

The enzymes and pathways of glucose utilization in the bovine adrenal medulla have been studied by Millaruelo, Sagarra, Delicado et al. (1986). They pointed out that scarce attention has been paid to energy metabolism in the adrenal medulla, a topic that must be of interest because this tissue consumes a great deal of energy, including the loss of ATP during exocytosis. Activities of six different metabolic enzymes were determined in extracts of the adrenal medulla. The isoenzyme patterns of four of those enzymes were also determined. The results pointed to a striking similarity between brain and adrenal medulla in terms of glycolytic isoenzyme patterns, with the exception of enolase. Nevertheless, there are notable differences in the distribution of glucose by different metabolic routes. In addition, adrenal chromaffin cells have the capacity to survive in anaerobic conditions, in distinct contrast to the brain.

González, Oset-Gasque, Gimenez Solves et al. (1987) reported the presence of succinic semialdehyde dehydrogenase in bovine adrenal medulla and in blood platelets. Both enzymes have similarities with the brain enzyme in terms of cofactor requirements, optimal pH, mitochondrial localization, and inhibition by AMP. The presence of succinic semialdehyde dehydrogenase in the adrenal medulla confirms the presence of a complete γ-aminobutyric acid bypass in this tissue.

Husebye and Flatmark (1987) characterized phospholipase activities of chromaffin vesicle membranes isolated from the bovine adrenal medulla. The vesicles contain a phospholipase A_2 activity. The preparations also revealed a phospholipase A_1 activity. The membranes also contain a lysophospholipase activity that accounts for the major part of the deacylation of membrane phospholipids.

Husebye and Flatmark (1988a) studied the kinetics and regulatory properties of phosphatidylinositol kinase in chromaffin vesicle ghosts. Phosphatidylinositol kinase activity appeared to be independent of pH but highly dependent on magnesium ions. By contrast, potassium and sodium had a slight inhibitory effect, and calcium ions reversibly inhibited the enzyme. In a follow-up study, Husebye and Flatmark (1988b) studied the effects of hydrophilic and amphiphilic cations on the activity of phosphatidylinositol kinase. They showed that some cations induced a biphasic (stimulation and inhibition) response in the enzyme, and other cations had a selective stimulatory effect. They concluded that the cations tested stimulate phosphatidylinositol kinase activity unspecifically by binding the positively charged groups to a membrane component, probably the kinase itself. This site appears to be different from that mediating the specific inhibition by calcium.

The plasma membranes of bovine adrenal chromaffin cells were isolated and the activity of enzymes involved in arachidonic acid liberation were investigated by Zahler, Reist, Pilarska et al. (1986). They demonstrated the generation of arachidonate by a membrane-bound diacylglycerol lipase as a result of treatment of the plasma membranes by phospholipase C from bacteria and also by the phosphatidylinositol-specific phospholipase C from chromaffin cells. The phospholipase C co-purified with plasma membranes.

Rindlisbacher, Reist and Zahler (1987) demonstrated that diacylglycerol lipase activity in the membranes of chromaffin cells consists of two enzymes working in series. First, a predominantly saturated fatty acid is split by a diacylglycerol lipase; the resulting molecule then is split by a monoacylglycerol lipase. Both enzymes are active at pH 6.0, but only the diacylglycerol lipase is active at pH 4.0. This enzyme activity is similar to that seen in platelets.

Tirrell and Coffee (1986) demonstrated that the bovine adrenal medulla contains an enzyme that catalyzes the hydrolysis of cyclic 2',3'-AMP to 2'-AMP. This cAMP phosphodiesterase appears to be similar to the phosphodiesterase found in the brain. The apparent molecular mass of the enzyme is 102.5 kDa; its function in the adrenal medulla or the brain is unknown at present.

Using monoclonal antibodies specific for the p60c-*src*, Parsons and Creutz (1986) detected high levels of this kinase in adrenal medullary tissue and in highly purified chromaffin vesicle membranes. An immune complex kinase assay was applied to fractions of adrenal medulla resolved on sucrose-density gradients. Thirty-seven percent of the total tissue p60c-*src* activity was found in association with chromaffin vesicles. Localization of a significant fraction of total cellular kinase activity to this secretory vesicle membrane suggests that the kinase may function in the regulation of neurotransmitter release.

Rainey, Mason, Cochet et al. (1988) examined the distribution of protein kinase C in the human fetal adrenal gland. Changes were seen in this enzyme under different conditions in the adrenal cortex, but changes in the enzyme within the adrenal medulla were not noted.

Serventi and Coffee (1986) characterized myosin light-chain kinase from the bovine adrenal medulla. The enzyme catalyzes the phosphorylation of the isolated light chain of skeletal muscle myosin and the light chain of intact adrenal medullary myosin. Several features suggested that adrenal medullary myosin light-chain kinase is similar in most, but not all, of its physical and kinetic properties to kinases isolated from other sources. The kinase may serve to regulate actin-myosin contractile activity in the adrenal medulla.

Sabatine and Coffee (1986) identified and characterized two cyclic nucleotide phosphodiesterases from the bovine adrenal medulla whose properties differ in some instances from those of the adrenal medullary phosphodiesterase activities previously described. One of these recently isolated phosphodiesterases is stimulated by cAMP.

Kecorius, Small and Livett (1988) described a method for the partial purification and characterization of a dipeptidyl aminopeptidase from the adrenal medulla. This enzyme was in the supernatant fraction of bovine adrenal medulla and had an apparent molecular mass of 68.1 kDa with a pH optimum of 9.5. Several peptides could be cleaved by this enzyme. This action involved the sequential removal of dipeptides from the amino-terminus. The ability of this dipeptidyl aminopeptidase to degrade certain neural peptides suggested that it could be involved in neuropeptide degradation.

An endopeptidase was isolated from bovine chromaffin vesicles by Maret and Fauchere (1988). The apparent molecular mass of this monomeric protein was 68 kDa and its pH optimum was 5.6, in agreement with the internal pH of chromaffin vesicles. This endopeptidase cleaved peptides at paired, but not single, basic residues. It was suggested that this endopeptidase may act as a maturation enzyme *in vivo*.

Neutral sphingomyelinase was characterized and localized in the bovine adrenal medulla by Bartolf and Franson (1986). The kinetic characteristics of the enzyme were elucidated. Sphingomyelinase activity was associated with a plasma membrane-microsomal fraction. The activity of this neutral sphingomyelinase in the adrenal medulla is similar in magnitude to that observed in other non-neural bovine tissues. It was suggested that this enzyme may be involved in membrane fusion and lysis during exocytosis through its ability to alter membrane composition.

Markossian, Paitian and Nalbandyan (1988) demonstrated that vanadyl ions may be used for reactivation of apodopamine β-monooxygenase. Maximal activity of the enzyme was achieved at about 400-fold molar excess of vanadyl ions whereas, for maximal reconstitution with copper, a 10-fold molar excess was necessary. At higher concentrations of vanadyl and of copper ions, inhibition of the enzyme was observed.

Byrd and Lichti (1987) found that the differentiation promoter sodium butyrate is able to cause a marked increase in transglutaminase activity in PC12 cells in a time- and dose-dependent manner. This increased transglutaminase activity is associated with growth arrest as well as with striking morphologic changes including increased cell adhesion. They hypothesized that transglutaminase may be involved in certain morphologic changes accompanying cellular differentiation and neoplastic transformation rather than in growth regulation *per se*.

The interaction of bovine and pancreatic hormones in the control hepatic enzymes during development was studied by Bohme, Belay, Dettmer et al. (1987). The hormones of the adrenal medulla were observed to antagonize the effects of glucagon in inducing liver enzymes.

Rabin (1988) determined the direct effect of chronic ethanol exposure on adenylate cyclase, Mg^+-ATPase, and Na^+,K-ATPase activities in PC12 cells. Exposure of the cells to ethanol for 4 days caused a dose-dependent increase in the stimulation of adenylate cyclase by ethanol *in vitro*. Conversely, 4-day treatment with ethanol increased the ATPase activities without altering the inhibitory effects of ethanol *in vitro*. Although ethanol slowed PC12 cell growth, the observed changes were not due to an ethanol-induced decrease in cellular density.

CHAPTER 9

COMPOSITION OF THE CHROMAFFIN CELL

The chromaffin vesicle is the characteristic organelle of the adrenal chromaffin cell. Probably because this was the first secretory vesicle to be isolated, it is the best characterized. This chapter will consider the composition of this model secretory vesicle and then the characterizations of other aspects of the chromaffin cell.

A review of the molecular composition of the chromaffin vesicle was published by Winkler, Apps and Fischer-Colbrie (1986). They pointed out the significant and clear-cut progress that has been made during recent years. They listed the relative numbers of the following molecules that are known to be localized within the chromaffin vesicle: catecholamines, nucleotides, calcium, ascorbic acid, chromogranins A, B, and C, enkephalins and related peptides, neuropeptide Y, dynorphin, neurotensin, substance P, dopamine β-hydroxylase, and cytochrome b_{561}. Much is known of the synthesis and function of these molecules, with the notable exception of a function for the chromogranins (see below). Winkler et al. (1986) pointed out that many of the controversies with regard to chromaffin vesicle composition that have raged in the past are now settled. They expressed optimism that the current questions and controversies will be settled in the next few years.

Winkler, Fischer-Colbrie, Schober et al. (1988) reviewed the contents of the chromaffin vesicle, referring to them as the "secretory cocktail." This mixture of components could exist to function within the organelle during storage or to function in concert after release. Of particular interest is chromogranin A. Although it may have a role in events such as calcium storage within the chromaffin vesicle, it generally is thought that this peptide acts after release. Winkler et al. (1988) speculated that chromogranin A, or peptides within chromogranin A, may have a presynaptic site of action to provide a local feedback mechanism.

Winkler, Fischer-Colbrie, Obendorf et al. (1988) asked the question, "Are there common antigens and common properties of adrenergic and cholinergic vesicles?" From comparing large dense-cored vesicles and small dense-cored vesicles, the answer was "Yes." This was true not only for morphologic similarities but also for biochemical and functional properties of these vesicles. One crucial question remains: Are small dense-cored vesicles formed from large dense-cored vesicles after exocytosis, or do they represent a completely separate vesicle population?

Historically, the most significant constituents of the adrenal chromaffin vesicle are the catecholamines--epinephrine, norepinephrine, and dopamine--generally in that order of concentration. Detecting and measuring these and other amines has long been of interest and importance. Kilpatrick, Jones and Phillipson (1986) presented a reliable semiautomated method using high-performance liquid chromatography (HPLC) coupled to colorimetric detection for the determination of monoamines and some of their metabolites. One notable feature of their method is the facile sample preparation which ensures rapidity, reproducibility, and prolonged sample storage capability. Using their method, one can assay amines in tissue weighing as little as 0.5 mg.

Bauersfeld, Ratge, Knoll et al. (1986) compared the analytical methods for catecholamines based on HPLC with amperometric detection with a radioenzymatic method. With a 1 ml plasma sample, they obtained detection limits of 25 ng/l and 18 ng/l for norepinephrine and epinephrine, respectively. They demonstrated an excellent correlation with the radioenzymatic method. They detailed the modifications needed to obtain these fine detection limits and pointed out that this method is relatively simple compared with radioenzymatic methods.

An improved radioenzymatic assay for plasma norepinephrine that uses purified phenylethanolamine *N*-methyltransferase was published by Henry and Bowsher (1986). They reported a sensitivity of less than 0.5 pg. They also pointed out that their method is reproducible and requires less manipulation than do previous radioenzymatic assays.

A new method for estimating the cerebrospinal fluid concentration of monoamine metabolites in the lateral ventricle of freely moving rats by the use of *in vivo* microdialysis was described by Becker, Adams and Robinson (1988). Among other observations, they noted that, in animals with adrenal medulla grafted into the lateral ventricle, there was an increase in the concentration of dihydroxyphenylacetic acid compared with pre-graft values or with animals with control grafts.

The peripheral distribution of free dopamine and its metabolites in the rat was determined by Favre, deHaut, Dalmaz et al. (1986). The highest level of dopamine, on a pg/mg basis, was in the carotid body. However, both the adrenal medulla and superior cervical ganglion were noted to have approximately equal concentrations of dopamine.

Coulter, McMillen and Browne (1988) measured epinephrine, norepinephrine, and dopamine in individual adrenal glands from adult rats and from rats at ages between day 16 of gestation and day 12 after birth. Epinephrine was detected at day 16 of gestation, and there was a 700-fold increase in epinephrine content of the rat adrenal by postnatal day 10. Norepinephrine was detected at day 16 of gestation, and there was a 26-fold increase by day 10. At all ages studied, the dopamine content of the rat adrenal was significantly lower than the epinephrine or norepinephrine content. At day 16 of gestation, there was more norepinephrine than epinephrine; but this changed such that the epinephrine/norepinephrine ratio was highest at day 10 (7.43:1), whereas it was 4.2:1 in the adult rat.

Catecholamines and their metabolites were measured in the urine of the insectivore *Suncus murinus* by Maruoka, Saito, Tanaka et al. (1988). There were no significant differences between *Suncus* and rats in regard to urinary content of epinephrine, norepinephrine, dihydroxyphenylacetic acid, and vanillylmandelic acid. In *Suncus*, however, the contents of dopamine and its metabolites were markedly lower and the contents of normetanephrine and metanephrine were higher than in rats. These results suggest that some tissues produce and excrete a larger amount of norepinephrine, epinephrine, and their metabolites and a smaller amount of dopamine and its metabolites in *Suncus*.

A general historical review of the relevance of dopamine to psychiatry was presented by O'Grady (1987). It was pointed out that the current thinking is that an imbalance of this and other amines plays a central role in mental disease. Previous trends also were reviewed.

<u>CHROMOGRANINS</u>

The first report of the isolation and sequence of a complementary DNA (cDNA) encoding bovine chromogranin A was by Iacangelo, Affolter, Eiden et al. (1986). This was followed quickly by a similar report by Benedum, Baeuerle, Konecki et al. (1986). Benedum, Lamouroux, Konecki et al. (1987) reported the primary structure of chromogranin B. Iacangelo et al. (1986) pointed out that, although chromogranin A has an apparent relative molecular mass of 75 kDa, its actual molecular mass is 48 kDa. Benedum et al. (1986) also identified a 48 kDa polypeptide chain. Chromogranin A was shown to be composed of 431 amino acid residues; chromogranin B was shown to be a 76 kDa polypeptide consisting of 657 amino acid residues (Benedum et al., 1987). Both Iacangelo et al. (1986) and Benedum et al. (1986, 1987) pointed out that the chromogranins are similar, if not identical, to secretory proteins found elsewhere. Ahn, Cohn, Gorr et al. (1987) also determined the nucleotide sequence for chromogranin A. They identified a 449-amino acid protein with a calculated mass of 50 kDa.

The confusion that exists because of the similarities between chromogranins and other secretory proteins led to a clarification published by Eiden, Huttner, Mallet et al. (1987) as the outcome of a meeting of the heads of involved laboratories throughout the United States and Europe. They proposed that the protein called "chromogranin A" retain its name on the basis of historical reasons and the fact that this protein is at its highest concentration in the adrenal medulla, a chromaffin tissue. Chromogranin B is identical to a protein called secretogranin I. It was proposed that the name "chromogranin B" be retained because this protein was first characterized in chromaffin tissue and is the major acidic protein of chromaffin vesicles in certain species. Another chromogranin, chromogranin C, is identical to secretogranin II. Eiden et al. (1987) proposed that the name "secretogranin II" be applied to this protein which was first characterized in a non-chromaffin endocrine organ, the pituitary gland, and it is at its highest concentration there. In chromaffin vesicles, it is only a minor component. The affix "II" should be retained because it seems quite possible that additional related acidic secretory proteins will be discovered. However, for the purposes of this review, the term "chromogranin C" will be used.

Preceding the work described above, Deftos, Murray, Burton et al. (1986) cloned the cDNA for chromogranin A. They demonstrated that this cDNA detects an mRNA in several different neuroendocrine tissues.

Also in an earlier report, Hamilton, Chu, Rouse et al. (1986) pointed out the structural similarities between chromogranin A and secretory protein I from the parathyroid. As pointed out above, these are now thought to be the same protein.

Another protein to enter the picture is pancreastatin, a molecule isolated from the pancreas that is known to inhibit glucose-induced insulin release. In commenting on pancreastatin and chromogranin A, Huttner and Benedum (1987) and Eiden (1987a) pointed out the striking similarity between these two molecules and argued that the similarity is too great to be coincidental. Both reports suggest that pancreastatin is produced from chromogranin A by limited proteolysis. Commenting on these two articles, Hutton, Davidson and Peshavaria (1987) raised the possibility that chromogranin A-related peptides may act locally on pancreatic endocrine cells or adjacent vasculature.

They suggested that chromogranin A-derived peptides should be seriously considered as an important part of the autoregulatory cycle of pancreatic endocrine cells.

The primary structures of human chromogranin A and pancreastatin were determined by Konecki, Benedum, Gerdes et al. (1987). The cDNA was obtained from a human pheochromocytoma. The nucleotide sequence reveals chromogranin A to be a 439-residue protein preceded by an 18-residue signal peptide. Comparison of the protein sequence of human chromogranin A with that of the bovine molecule shows high conservation of the NH$_2$-terminal and COOH-terminal domains as well as the potential dibasic cleavage sites; the middle portions show remarkable sequence variation. A portion of the human chromogranin A contains a sequence homologous to porcine pancreastatin. These and other observations suggest that human chromogranin A may be the precursor for human pancreastatin and possibly for other as yet unidentified biologically active peptides.

Iacangelo, Fischer-Colbrie, Koller et al. (1988) used specific oligonucleotide priming of double-stranded DNA to determine the sequence of a porcine chromogranin A adrenal medullary cDNA. Many similarities between porcine chromogranin A and that from human, cow, and rat were noted. The amino acid sequence of porcine pancreastatin was found within the sequence of chromogranin A. Thus, porcine chromogranin A could serve as a precursor for pancreastatin.

Hutton, Nielsen and Kastern (1988) obtained a cDNA that encoded the precursor form of the chromogranin A-related proteins β-granin and pancreastatin. The cDNA was obtained by immune screening of rat insulinoma and pancreatic islet cDNA libraries. Its sequence was virtually identical to that of rat adrenal chromogranin A, suggesting that the different molecular forms of chromogranin A immunoreactivity found in the adrenal medulla and endocrine pancreas are related to differences in post-translational proteolytic processing.

Iacangelo, Okayama and Eiden (1988) deduced the primary structure of rat chromogranin A from a rat adrenal cDNA clone. Comparison of rat and bovine chromogranin A revealed several similar features and some unique features of rat chromogranin A. Chromogranin A mRNA was detected in the adrenal medulla and other neural structures as well as in tumor cell lines derived from pancreas, pituitary, and adrenal medulla.

Helman, Ahn, Levine et al. (1988) isolated and characterized a cDNA clone specific for the mRNA that encodes human chromogranin A. The human cDNA has an overall nucleic acid identity to bovine chromogranin A cDNA of 86%. There were similarities with porcine pancreastatin that led to the suggestion that chromogranin A and pancreastatin both may be members of a larger family of calcium-binding proteins.

Murray, Deaven, Burton et al. (1987) assigned the gene for chromogranin A to human chromosome 14. They hybridized a chromogranin A cDNA probe cloned from a cDNA library of human medullary thyroid carcinoma cells to spots of individual human chromosomes flow-sorted onto nitrocellulose filters. Southern analysis of human genomic DNA with the same probe revealed only one to three restriction bands. These studies indicate that the chromogranin A gene is probably a single copy and not a member of a dispersed multigene family.

Ismael, Millar, Small et al. (1986) demonstrated that chromogranins can be degraded to enkephalin-like immunoreactive peptides by acetylcholinesterase. They incubated a homogeneous preparation of the enzyme with chromogranins isolated from bovine chromaffin vesicles and identified smaller proteins. The degraded proteins demonstrated enkephalin-like immunoreactivity. Ismael et al. (1986) proposed that acetyl-cholinesterase may play a physiologic role in the intracellular processing of peptides.

In a follow-up study, Small, Ismael and Chubb (1986) investigated the breakdown of chromogranin A by acetylcholinesterase and compared the peptidase activity of acetylcholinesterase with that of trypsin because trypsin-like enzymes have been proposed to be constituents of chromaffin vesicles. A number of peptidase inhibitors that strongly inhibited tryptic digestion of chromogranin A also inhibited the acetylcholinesterase digestion, although they were less potent. These and other experiments indicated that acetylcholinesterase has a peptidase activity that is similar, but not identical, to that of trypsin and suggest that a second nontryptic activity is also present. Furthermore, acetyl-cholinesterase may process chromogranin A to smaller chromogranins in bovine chromaffin cells.

Seidah, Hendy, Hamelin et al. (1987) demonstrated that chromogranin A can act as a reversible processing-enzyme inhibitor. Specifically, they demonstrated that it is a reversible competitive inhibitor of a newly characterized serine protease. This protease cleaves between pairs of basic amino acids in a number of prohormones, and chromogranin A appears to inhibit this step.

Wilson, Phan and Lloyd (1986) presented a method for the

purification of human chromogranin from adrenal glands obtained at autopsy. They also presented evidence suggesting NH_2-terminal processing of the peptide.

Eiden, Iacangelo, Hsu et al. (1987) described details of chromogranin A synthesis and secretion in bovine chromaffin cells. They raised an antiserum against chromogranin A and used it as a specific probe in cell-free extracts of the adrenal medulla and chromaffin cells. Chromogranin A was found to make up about 10% of the total protein of the chromaffin cell. The synthesis of chromogranin A and [Met]enkephalin appeared to be differentially regulated within the chromaffin cell because chronic treatment of cells with nicotine and forskolin caused an increase in [Met]enkephalin but not in chromogranin A. Eiden et al. (1987) determined that chromogranin A and [Met]enkephalin are excreted from functionally identical cellular compartments, presumably the chromaffin vesicle.

Fischer-Colbrie, Hagn, Kilpatrick et al. (1986) characterized a group of acidic proteins from bovine chromaffin vesicles with an antiserum raised against a protein extracted from the pituitary gland. The properties of these proteins are very similar to those of chromogranins A and B; however, there is no immunologic cross-reaction between these protein groups. Fischer-Colbrie et al. (1986) suggested that this third group of acidic proteins of chromaffin vesicles be named "chromogranins C." As mentioned above, Eiden et al. (1987) have agreed to refer to this protein as "secretogranin II."

Fischer-Colbrie and Schober (1987) developed methods, based mainly on HPLC, to isolate chromogranins A, B, and C. Amino acid analysis of the chromogranins revealed a similar composition for all three proteins, with glutamic acid being the most prevalent amino acid. The methods developed for isolating the proteins from bovine chromaffin vesicles also proved suitable for isolating chromogranins A and B from a pheochromocytoma, although chromogranin C was not present in sufficient amounts to be isolated. The chromogranins purified by these methods were used to raise specific antibodies.

Wohlfarter, Fischer-Colbrie, Hogue-Angeletti et al. (1988) raised specific antisera against synthetic peptide fragments of bovine chromogranin A. Soluble proteins of bovine chromaffin vesicles were subjected to two-dimensional immunoblotting with these antisera. The endogenous breakdown products of chromogranin A gave distinct patterns of immunostaining which enabled Wohlfarter et al. (1988) to correlate these peptides with defined regions of the chromogranin A molecule. The results established that, within chromaffin vesicles, degradation of chromogranin A by endogenous proteases can start either at the carboxyl- or the amino-terminus.

Chromogranin A and other secretory proteins were found to be carboxyl-methylated within secretory vesicles by Nguyen, Harbour and Gagnon (1987). It required 3 to 6 hours before methylated chromogranin A could be detected in mature chromaffin vesicles, a time consistent with the synthesis and storage of secretory proteins *in vivo*. Carboxyl-methylated chromogranin A was secreted from chromaffin cells by exocytosis stimulated by acetylcholine. Because protein carboxyl-methylase is a cytosolic enzyme, these results suggest that methylation of secretory proteins is a co-translational phenomenon.

Iguchi, Natori, Kato et al. (1988) examined the processing of chromogranin B into specific fragments in the bovine adrenal medulla and the pituitary gland. Using antibodies raised to these fragments, they demonstrated fragments in chromaffin vesicles as well as in the anterior pituitary. These fragments could be released from cultured bovine chromaffin cells by stimulation with nicotine or high potassium. They presented other results to suggest that chromogranin B is processed into small fragments and that this processing is tissue-specific.

The distribution of chromogranins within the body and in different organisms has been the subject of intense investigations. Hagn, Schmid, Fischer-Colbrie et al. (1986) used antisera against bovine chromogranins A, B, and C to identify these proteins in human tissues. All three chromogranins were present in human chromaffin vesicles. Their molecular masses differ slightly from those of the bovine proteins. All three chromogranins also are found in the anterior pituitary. In endocrine pancreas only chromogranin A and B could be found whereas in the parathyroid gland only chromogranin A is present. Commenting on this article, Angeletti (1986) pointed out that, despite the rapid increase in understanding of chromogranins in the last few years, the function of these proteins remains an enigma. She suggested that chromogranins may be precursors for new biologically active peptides.

Hagn, Klein, Fischer-Colbrie et al. (1986) compared the chromogranins in bovine chromaffin vesicles and large dense-cored vesicles of bovine splenic nerve. Both types of vesicles contain chromogranins A, B, and C. However, the proteolytic processing of these chromogranins within the vesicles is apparently different. Chromogranin B in chromaffin vesicles is processed by more than 80% whereas in nerve vesicles, only 15% is broken down to smaller proteins.

The presence of chromogranins A, B, and C in bovine endocrine and neural tissues was examined by Lassmann, Hagn, Fischer-Colbrie et al. (1986). The three chromogranins occur together in several endocrine organs (adrenal medulla, anterior pituitary, and endocrine pancreas) and in sympathetic ganglion cells. In the posterior pituitary, only chromogranin C is present; and in the intermediate lobe, only chromogranins A and C are found. The parathyroid gland contains only chromogranin A, and the enterochromaffin cells are immunoreactive for chromogranins A and B. Cells of the thyroid gland and some cells of the anterior pituitary apparently do not contain any chromogranins. It was concluded that the three chromogranins are not always stored together and that they all are not present in all endocrine cells.

Rundle, Somogyi, Fischer-Colbrie et al. (1986) used immunoblotting techniques to localize bovine chromogranins A, B, and C in the ovine pituitary. Chromogranin immunoreactivity was found in gonadotrophs, thyrotrophs, and corticotrophs but not in mammotrophs and somatotrophs. Chromogranin C was the only chromogranin located in the pars nervosa; chromogranin B was rarely found in the pars intermedia.

Schober, Fischer-Colbrie, Schmid et al. (1987) compared chromogranins A, B, and C in human adrenal medulla and pheochromocytoma. With immunohistochemical methods, staining of the tumor cells was positive for all three antigens. By immunoblotting, chromogranins A and B were the major components and were present in about equal amounts. The relative concentrations of all of these antigens in pheochromocytoma were similar to those in adrenal medulla.

Schmid, Fischer-Colbrie, Hagn et al. (1987) studied chromogranins A and B and secretogranin II in medullary thyroid carcinomas by using immunoblotting and immunohistochemical methods. All three antigens were identified by immunoblotting; immunohistochemically, positive staining was obtained for all three antigens throughout the tumor tissue. Using similar techniques, Yoshie, Hagn, Ehrhart et al. (1987) detected the same peptides in bovine pancreatic islets. Chromogranin A was found in all pancreatic endocrine cell types with the exception of most pancreatic polypeptide-producing cells. For chromogranin B, only a faint immunostaining was obtained. These results established that chromogranins A and B are present in the endocrine pancreas but that they exhibit a distinct cellular localization.

Weiler, Fischer-Colbrie, Schmid et al. (1988) investigated various endocrine tumors for the presence of chromogranins A, B, and C. These antigens were identified by one- and two-dimensional immunoblotting and, in some cases, by immunohistochemical methods. An antigen corresponding to adrenal chromogranin A in electrophoretic behavior was present in all types of tumors tested. Chromogranin B had a more limited distribution. The occurrence of chromogranin C was similar to that of chromogranin B. This study establishes that, in most cases, chromogranins in tumors are identical to the adrenal antigens, but these antigens are not always stored together.

Ehrhart, Grube, Bader et al. (1986) also localized chromogranin A in the pancreatic islet. In specimens from the bovine pancreas, they found chromogranin A immunoreactivity in the same cells that reacted with antibodies against insulin, glucagon, and somatostatin. The cells containing pancreatic polypeptide showed faint immunostaining.

Grube, Aunis, Bader et al. (1986) looked for chromogranin A immunoreactivity in the endocrine pancreas of 9 mammalian species (man, tupaia [tree shrew], mole, cat, dog, pig, guinea pig, rabbit, and rat). All pancreatic endocrine cell types were immunoreactive for chromogranin A; however, each species had its own pattern of chromogranin A-immunoreactive cell types. Other findings suggested that chromogranin A may function in the storage mechanisms for peptide hormones.

Using an antiserum raised against a synthetic carboxyl-terminal peptide of porcine pancreastatin, Schmidt, Siegel, Kratzin et al. (1988) and Schmidt, Siegel, Lamberts et al. (1988) detected pancreastatin-like immunoreactivity in human adrenal medulla and other tissue. Immunoreactive peptides were isolated and found to be identical to a portion of human chromogranin A. These and other results indicate that chromogranin A may represent the precursor for pancreastatin-related and possibly other peptides of unknown physiologic function.

Further down the phylogenetic scale, Deftos, Björnsson, Burton et al. (1987) demonstrated the presence of chromogranin A immunohistochemically in the rainbow trout, in the ultimobranchial bodies and corpuscles of *Stannius*. Chromogranin A was also detected in the coho salmon. Their observations demonstrate the presence of chromogranin A in endocrine glands of dissimilar species.

The presence of chromogranin-related peptides in a wide range of species was investigated by Rieker, Fischer-Colbrie, Eiden et al. (1988) using one- and two-dimensional electrophoresis followed by immunoblotting. In all species investigated, including mammals, birds, amphibians, fish, and arthropods, chromogranin A- and B-like proteins could be demonstrated. For all species, there was an immunologic cross-reaction with antisera against bovine chromogranins. The molecular sizes and isoelectric points of the chromogranins were similar in all species. It was concluded that chromogranins A and B have a widespread phylogenetic distribution with a significant conservation of molecular size, isoelectric points, and immunologic epitopes. This is consistent with the concept that these peptides have a specific function.

Siegel, Iacangelo, Park et al. (1988) used *in situ* hybridization in a histochemical method to localize chromogranin A biosynthesis in the bovine adrenal gland. The mRNA was found in the chromaffin cells of the medulla but was absent from the cortex. The distribution of the mRNA in the medulla was uneven; cells located at the periphery were more heavily labeled than those in the center of the medulla. The mRNAs for enkephalin and phenylethanolamine *N*-methyltransferase were present in a narrow band of cells at the periphery of the medulla; however, they were detected in only a small number of cells in the inner region, in contrast to the mRNA for chromogranin A. The difference in the distribution of the enkephalin and phenylethanolamine *N*-methyltransferase mRNAs from that of chromogranin A suggests that the expression of these genes is differentially regulated.

Rausch, Iacangelo and Eiden (1988) examined the regulation of chromogranin mRNA in PC12 cells after treatment with nerve growth factor, dexamethasone, or a combination of these two agents. PC12 cells have low levels of the mRNA, and this does not change after treatment with nerve growth factor. However, treatment with dexamethasone resulted in a 4-fold increase in the amount of mRNA. Nerve growth factor did not enhance this effect. Chromogranin B mRNA levels were unaltered by any of these drug treatments. Treatment with dexamethasone and nerve growth factor seems to be required for full expression of the adrenergic neuronal phenotype in PC12 cells. Measurement of chromogranin A mRNA provided more specific delineation of neural differentiation and how it is influenced by hormone and growth factors.

At the light histochemical level, Volknandt, Schober, Fischer-Colbrie et al. (1987) demonstrated chromogranin A immunoreactivity in cholinergic terminals in the rat diaphragm. No immunoreactivity was found for chromogranin B or C. A function for chromogranin A inside cholinergic nerve terminals is still unknown but does add to the list of peptides that are known to be in cholinergic nerve terminals.

Rindi, Buffa, Sessa et al. (1986) localized chromogranin A immunohistochemically in pancreatic islet cells, gut argentaffin enterochromaffin cells, gastrin cells, thyroid C cells, parathyroid cells, adrenal chromaffin cells, certain pituitary cells, and some axons of visceral nerves. Pancreatic α cells, gut enterochromaffin cells, adrenal chromaffin cells, and pituitary cells as well as some gut nerve fibers showed chromogranin B immunoreactivity. Chromogranin C immunoreactivity was detected in pancreatic A cells, pyloric D cells, intestinal L cells, thyroid C cells, adrenal chromaffin cells, pituitary cells, and some gut nerve fibers. Rindi et al. (1986) suggested that the widespread distribution of chromogranins may give new insights as to the function of such proteins in intravesicular hormone biosynthesis and storage.

Immunoreactive chromogranin A was demonstrated immunocytochemically in the cytoplasm of neuroendocrine cells and neuroepithelial bodies in human, monkey, and pig respiratory mucosa by Lauweryns, vanRanst, Lloyd et al. (1987). By correlating the distribution of chromogranin A immunoreactive cells with other specific stains, it was concluded that chromogranin is a useful histologic marker for APUD (*a*mine *p*recursor *u*ptake and *d*ecarboxylation cells) in the respiratory mucosa of several species.

Localization of chromogranin A and dopamine β-hydroxylase was achieved immunocytochemically at the electron microscopic level by Matsumoto, Tanaka, Yamamoto et al. (1987). Both proteins were localized exclusively on chromaffin vesicles from specimens of the bovine adrenal medulla. At saturation, the maximum number of gold particles bound to the vesicles roughly corresponded to the number of dopamine β-hydroxylase or chromogranin A molecules estimated to be in the vesicles.

The cellular distribution and amount of chromogranin A in bovine endocrine pancreas was examined by Ehrhart, Jörns, Grube et al. (1988). They used a polyclonal antibody against bovine adrenal medullary chromogranin A. Only chromogranin A could be detected, and it was located in the pancreatic β cells.

In addition to the localization of chromogranin A in various tissues and species mentioned above, several laboratories are interested in the localization of this peptide in tumors. Lloyd (1987) reviewed and expanded the assortment of normal and neoplastic endocrine tissues that have been shown to contain chromogranin A. He concluded that chromogranin immunostaining is useful in the characterization and diagnosis of endocrine cells and tumors.

Lloyd, Sisson, Shapiro et al. (1986) used immunohistochemical methods to localize chromogranin-, epinephrine-, norepinephrine-, and catecholamine-synthesizing enzymes in neuroendocrine cells and tumors. There was close agreement between the localization of chromogranin and catecholamines and of their synthesizing enzymes in the tissues examined. These results indicate that the presence of catecholamines and chromogranin in neuroendocrine cells in tumors within the adrenal medulla and in many other sites may be closely related.

Lloyd (1988) reviewed the literature and presented evidence on the distribution of chromogranins, catecholamines, and catecholamine-synthesizing enzymes in neuroendocrine cells and tumors. The value of a combined immunohistochemical approach with various antibodies, including antibodies against synthetic enzymes, was stressed as a model to study both localization and *de novo* production of specific antigens by cells and tumors. Chromogranin A is associated with secretory vesicles in most endocrine cells and tumors. The distributions of chromogranins B and C are generally similar to the distribution of chromogranin A in normal endocrine cells in humans. The presence of catecholamine-synthesizing enzymes can be used as an indirect index of the presence of epinephrine and norepinephrine in endocrine cells. Also, the presence of these enzymes can help to discriminate between uptake of catecholamines in cells where there are no synthesizing enzymes and in cells that are capable of synthesizing catecholamines

The secretion of chromogranin A by peptide-producing endocrine tumors was described by O'Connor and Deftos (1986). Using a radioimmunoassay, they examined the plasma of patients with the following disorders: pheochromocytoma, parathyroid adenoma, primary parathyroid hyperplasia, medullary carcinoma of thyroid, thyroid C cell hyperplasia, carcinoid tumor, oat cell carcinoma of lung, pancreatic islet cell tumor, and aortic body tumor. In all of these patient groups, the plasma chromogranin A concentration was increased. The sensitivity and specificity of increased plasma chromogranin A in the diagnosis of peptide-producing endocrine neoplasms were 81% and 100%, respectively. The increase in plasma chromogranin A concentration in their patients suggests that neoplasms release chromogranin A along with the usual hormone of the tumor, that these neoplasms could be characterized as "chromograninomas," and that measurement of plasma chromogranin A may be a useful diagnostic procedure in patients with endocrine tumors, especially multiple endocrine neoplasia.

Tischler, Dayal, Balogh et al. (1987) studied three adrenal medullary tumors that showed admixtures of elements of pheochromocytoma with ganglioneuroma or ganglion neuroblastoma to determine the distribution of immunoreactive chromogranins, S-100 protein, and vasoactive intestinal peptide

(VIP). In all cases, chromogranin staining was absent or weak in neuronal perikarya and moderate to intense in varicosities of neuronal processes, a finding consistent with the presumed distribution of secretory vesicles in neurons. Chromogranin staining was also intense in chromaffin cells.

Chromogranins A, B, and C were localized in bronchial and intestinal carcinoids by Weiler, Feichtinger, Schmid et al. (1987). Chromogranins A and B appeared to be present in all tumors; chromogranin C also was present but its concentrations were low and variable.

The distribution of chromogranins A and B in normal and neoplastic human tissues was analyzed by Lloyd, Cano, Rosa et al. (1988). These chromogranins were found in normal medulla, in pheochromocytoma, and in certain other tumors and tissues. The results indicated that chromogranins A and B are useful in the characterization of some neuroendocrine cells and neoplasms.

Immunochemical characterization of a novel secretory protein was performed by Krisch, Horvat, Krisch et al. (1988). This protein is defined by monoclonal antibody HISL-19. Extracts of human pheochromocytoma and other tumors were found to contain a protein immunoreactive with this antibody. The protein was found to be similar to chromogranins but distinct from chromogranins A, B, and C.

A clue to the possible function of chromogranins within secretory vesicles was provided by Reiffen and Gratzl (1986a). They demonstrated that chromogranins isolated from the bovine adrenal medulla bind Ca^{2+}. The chromogranins were shown to behave like other known calcium-binding proteins. The calcium-binding function of chromogranins points to the general importance of these proteins in the metabolism of calcium.

In a related study, Reiffen and Gratzl (1986b) examined calcium binding to chromaffin vesicle proteins as well as to ATP, the main nucleotide in these vesicles. A detailed study of the dissociation constants of calcium binding pointed to the possibility that the intravesicular medium determines whether calcium is preferentially bound to the proteins or to ATP. It was found that chromogranin A provides significant amounts of the calcium binding sites within chromaffin vesicles.

Westermann, Stögbauer, Unsicker et al. (1988) demonstrated the calcium dependence of the interaction between chromogranin A and catecholamines. They immobilized chromogranin A to a newly raised monoclonal antibody. They showed that chromogranin A can bind catecholamines in a non-calcium-dependent manner but bound much more catecholamines in the presence of calcium. These results support the concept that chromogranin A may act as a condensing protein within secretory vesicles.

LIPIDS AND RELATED SUBSTANCES

The lipid composition of plasma membranes isolated from bovine adrenal chromaffin cells was determined by Malviya, Gabellec and Rebel (1986). Choline and ethanolamine phosphatides were predominant; the level of lyso compounds was very low. The amount of cholesterol and the cholesterol/phospholipid molar ratio were low compared to those in the other subcellular fractions of chromaffin cells. A complex pattern of neutral glycolipids was observed in contrast to the pattern of gangliosides.

Lishajko (1986) examined the properties of bovine adrenal chromaffin vesicles with regard to the proteophospholipids as a binding substrate for catecholamines. Several lines of evidence led to the suggestion that the dipolar head group of the vesicolipids, particularly the lipid phosphate groups, is involved in electrostatic or complex interactions with catecholamines, either directly or through water molecules within the group.

Margolis, Greene and Margolis (1986) investigated the prevalence of oligosaccharides of the poly(*N*-acetyllactosamine) series in nerve tissue, specifically PC12 cells and rat sympathetic neurons. Treatment of the glycopeptides derived from the trypsinate and membranes of PC12 cells demonstrated the presence of these oligosaccharides. Treatment of PC12 cells with nerve growth factor led to a small but significant decrease in the proportion of these oligosaccharides.

Margolis, Goossen and Margolis (1988) demonstrated the presence of phosphatidylinositol-anchored glycoproteins in PC12 cells. They presented evidence that Thy-1 and three other PC12 cell proteins use this membrane anchoring but otherwise differ from each other in several respects.

Tischler, Mobtaker, Mann et al. (1986) demonstrated that the monoclonal antibody HNK-1 reacts intensely with normal and neoplastic adrenal medullary cells. Ultrastructural immunocytochemical studies and immunoblot analyses reveal that the antibody reacts with an intracellular 75 kDa protein located within chromaffin vesicles. The size of this protein differs from that of chromogranin A and myelin-associated glycoprotein. Tischler et al. (1986) suggested that this antibody may be useful as a marker for specific subsets of secretory vesicles.

Margolis, Ripellino, Goossen et al. (1987) noted that the monoclonal antibody HNK-1 recognizes a glycoprotein in chromaffin vesicle membranes, a component in PC12 cells, and components in other tissues. It was not present in certain proteoglycans in the chromaffin vesicle matrix.

Margolis, Fischer-Colbrie and Margolis (1988) reexamined the structural features of oligosaccharides of chromaffin vesicle membrane glycoproteins. They found that several chromaffin vesicle membrane glycoproteins, but predominantly glycoprotein IV, contained tri- and tetraantennary complex oligosaccharides with poly(*N*-acetyllactosaminyl)disaccharide repeating units. These oligosaccharides were also present on glycoproteins II and III, whereas they are absent from dopamine β-hydroxylase and carboxypeptidase H which are the major glycoproteins of chromaffin vesicle membranes.

Ariga, Yu, Scarsdale et al. (1988) characterized an unusual neutral glycolipid in PC12h cells. Using a number of methods, they were able to identify this glycolipid as Gal(α1-3)Gal(α1-4)Gal(β1-4)Glc(β1-1')Cer.

Neutral glycolipids in PC12 cells were examined by Shimamura, Hayase, Ito et al. (1988). A major neutral glycosphingolipid was found to contain only galactose and glucose and was identified as seramide tetrahexoside. From several lines of evidence, they were able to determine the structure of seramide trihexoside. This is the first report indicating the presence of this glycosphingolipid in PC12 cells.

Ariga, Macala, Saito et al. (1988) studied the lipid composition of PC12 cells cultured in the presence and absence of nerve growth factor. The concentration of neutral glycolipids was about 1.7 μg/mg of protein for both untreated and growth factor-treated cells. The neutral glycolipid fraction contained a major component that accounted for 80% of the total and was characterized as globoside. This and other glycolipids were characterized further.

Ganglioside expression and tetanus toxin binding were studied in PC12 cells by Walton, Sandberg, Rogers et al. (1988). Seven ganglioside species were readily detected; two were identified as tri- and tetrasialogangliosides. The major species was a predominant mammalian brain ganglioside known to support high-affinity tetanus toxin binding. Direct binding of radioiodinated tetanus toxin to PC12 gangliosides revealed selective binding to the tri- and tetrasialogangliosides. Differentiation of PC12 cells caused an increase in the expression of tri- and tetrasialogangliosides and a closely matched increase in tetanus toxin binding to cell membranes. These data provide evidence that complex gangliosides may act as tetanus toxin receptors.

The effect of uranyl acetate on the mesomorphic phase state of lipids in model membranes and in membranes such as those isolated from chromaffin vesicles was studied by Caffrey, Morris and Feigenson (1987). Small concentrations of uranyl acetate were seen to induce a liquid crystal-to-gel phase transformation in the chromaffin vesicle membranes and other membranes. These results support the idea that uranyl acetate used to stain sections for electron microscopy could alter membrane morphology.

Ganglioside expression and tetanus toxin binding were studied in PC12 cells by Walton, Sandberg, Rogers et al. (1988). They established the presence of trisialoganglioside (and other complex gangliosides) as a major ganglioside species on PC12 cells. They also demonstrated that radioiodinated tetanus toxin binds with high affinity to isolated PC12 gangliosides, plasma membranes, and intact cells. Furthermore, they reported that differentiation of PC12 cells causes an increase in production of gangliosides and a concomitant increase in tetanus toxin binding. These findings strongly support the contention that the tetanus toxin receptor is composed of complex gangliosides.

The means by which proteins are attached to plasma membrane or chromaffin vesicle membranes from the bovine adrenal medulla were investigated by Fouchier, Bastiani, Baltz et al. (1988). They incubated membranes at 37°C or used a specific phospholipase C to convert two proteins, with molecular masses of 82 and 68 kDa, respectively, to a hydrophilic form. These proteins were immunoreactive with an antibody known to be revealed after removal of a diacylglycerol anchor. There was further evidence to suggest the presence of a nonacetylated glucosamine residue in the determinant. This is one of the first reports suggesting that a glycosylphosphatidylinositol anchor might exist in membranes other than the plasma membrane.

OTHER CONSTITUENTS

Obata, Kojima, Nishiye et al. (1987) produced monoclonal antibodies against four proteins associated with synaptic vesicles from guinea pig brain. Using immunohistochemical methods, they demonstrated that all of these synaptic vesicle proteins were located in the adrenal medulla and not in the cortex. Their micrographs indicate that the proteins are located within the adrenal chromaffin cells. Perhaps these proteins are involved in common pathways for neurotransmitter accumulation, exocytotic release, and membrane recycling of neurosecretory vesicles.

Bruhn, Engeland, Anthony et al. (1987a) localized bioactive corticotropin-releasing factor in the adrenal glands of dogs, rats, and humans. Radioimmunoassay and immunohistochemical experiments clearly demonstrated that the peptide is confined to the adrenal medulla. Cells containing corticotropin-releasing factor have a characteristic appearance and often are found in close association with blood vessels. It was suggested that the factor is secreted at blood vessels within the vasculature of the adrenal medulla. Corticotropin-releasing factor was also identified in pheochromocytomas. The releasing factor was only a weak secretagogue for catecholamines. It was suggested that corticotropin-releasing factor may affect local blood flow within the adrenal medulla and may modify catecholamine secretory rates via this mechanism.

Minamino, Uehara and Arimura (1988) further characterized the corticotropin-releasing activity in the bovine adrenal medulla. They used two different antisera against corticotropin-releasing factor to identify the immunoreactivity and bioactivity in the adrenal medulla demonstrated the presence of biologically and immunoreactive releasing factor in the bovine adrenal medulla that is chromatographically indistinguishable from bovine hypothalamic-releasing factor.

The output of corticotropin releasing-factor-like immuno-reactivity from the adrenal gland was investigated by Edwards and Jones (1988) by the "adrenal clamp" method in conscious calves. The output of immunoreactivity could be increased greatly by electrical stimulation of the splanchnic nerve, but the output of epinephrine and norepinephrine increased more abruptly. Additional evidence suggested that the corticotropin-releasing factor secreted by the bovine adrenal is structurally similar to its pituitary counterpart.

Doris (1988) presented immunologic evidence that the adrenal gland is a source of an endogenous digitalis-like factor. Experiments had shown that an antibody raised against digoxin in the rabbit is capable of detecting a material in the plasma of rats never treated with glycosides. When rats were adrenalectomized, and plasma levels of this antigen decreased, these findings suggest that the endogenous digitalis-like factor may be derived from the adrenal gland. It was not determined if it was from the adrenal cortex or medulla.

Chang and Ramirez (1988) demonstrated the presence of a potent dopamine-releasing factor in the rat adrenal gland. This novel protease-sensitive factor was present in a partially purified preparation from the rat adrenal gland. It stimulated the release of dopamine from rat striatal tissue. This putative dopamine-releasing factor is probably a glycoprotein, is resistant to boiling, is partially inactivated by trypsin, and is completely inactivated by the protease Pronase E. Because the extract was made from whole adrenal glands, it could not be determined if this dopamine-releasing factor is in the adrenal cortex or medulla.

Pryde and Phillips (1986) solubilized membrane fractions and then separated membrane proteins by temperature-induced phase separation. Using the chromaffin vesicle membrane as a model, they found that many intrinsic membrane glycoproteins are found in an aqueous phase, probably maintained in solution by adherent detergent. Using this technique, they achieved better resolution of membrane separation than obtained in previous studies.

The role of adenine compounds in autonomic neurotransmission was reviewed by White (1988). Although the presence of ATP within chromaffin vesicles is well documented, possible functions for released ATP have not been clearly established. It was suggested that the substantial amounts of ATP released during secretion from the adrenal medulla could exert effects on tissues distal to the adrenal gland, such as the adrenal vein, vena cava, or right atrium. Such effects remain to be demonstrated.

The subcellular distribution of diadenosine tetraphosphate and diadenosine pentaphosphate was studied by Rodriguez del Castillo, Torres, Delicado et al. (1988). In bovine adrenal medullary tissue, these diadenosine polyphosphates were most concentrated within chromaffin vesicles. Enzymatic degradation with phosphodiesterase produced AMP as the final product. The diadenosine polyphosphates were potent inhibitors of adenosine kinase.

Instrumentation and techniques have been developed by Ornberg (1985) for the cryo-ultramicrotomy of isolated bovine adrenal chromaffin cells. Cells were frozen with the liquid helium-cooled copper block method. X-ray microanalysis and electron energy loss spectroscopy were used to measure both light and heavy element contents of the chromaffin vesicle. Values for elements such as phosphorus and nitrogen were in reasonable agreement with the known biochemical composition of chromaffin vesicles.

Leapman and Ornberg (1988) examined cryosectioned adrenal chromaffin cells in order to demonstrate the potential for applying electron energy loss spectroscopy to biologic quantitation. Water content could be assessed from the amount of inelastic scattering as measured by the low-loss spectrum. The nitrogen/phosphorus ratio could be measured to provide information about the relative concentrations of ATP, chromogranins, and catecholamines. Electron energy loss spectroscopy of chromaffin cells was used to measure the distribution of light elements.

Ornberg, Kuijpers and Leapman (1988) used electron probe microanalysis and other techniques to quantitate elements in chromaffin vesicles, both *in situ* and *in vitro*. In units of mmol/kg dry weight, chromaffin vesicles *in situ* contained: phosphorus, 523; potassium, 124; sulfur, 82; chloride, 74; calcium, 13; magnesium, 6; and sodium, 0. When vesicles were isolated in isotonic sucrose buffer, the content of potassium and chloride decreased and the sodium content increased. The water content of various organelles was also quantitated.

Wiedenmann, Franke, Kuhn et al. (1986) raised a monoclonal antibody to synaptophysin. They found by immunohistochemical and immunoblotting methods that an identical or a similar protein is found not only in the adrenal medulla but also in pheochromocytomas and other neuroendocrine tumors. They concluded that synaptophysin is produced independently of other neuronal differentiation markers and proposed that it be used as a differentiation marker in tumor diagnosis.

Obendorf, Schwarzenbrunner, Fischer-Colbrie et al. (1988a) presented an immunologic characterization of a membrane glycoprotein of chromaffin vesicles. This glycoprotein was isolated from detergent-solubilized membranes of bovine chromaffin vesicles by high pressure liquid chromatography. Specific antisera raised against this glycoprotein reacted in one- and two-dimensional immunoblots. The antiserum against bovine glycoprotein II cross-reacted with an analogous component in several species. The antiserum reacted with an analogous antigen in several other endocrine and exocrine glands. The results demonstrated for the first time that secretory vesicles of endocrine and exocrine tissues have at least one common antigen, glycoprotein II. It seems likely that this protein is involved in a basic function common to secretory vesicles.

Obendorf, Schwarzenbrunner, Fischer-Colbrie et al. (1988b) demonstrated that synaptophysin (p38) is present in chromaffin vesicles and in a special vesicle population in the bovine adrenal medulla. Less than half of the amount of synaptophysin present in the adrenal medulla copurified with chromaffin vesicles. The evidence led to the conclusion that, in the adrenal medulla, synaptophysin is present in special membranes, probably in high concentration, but in membranes of chromaffin vesicles it is present either in a low concentration in all or in a higher concentration in a subset.

Schilling and Gratzl (1988) quantitated the amount of synaptophysin in highly purified chromaffin vesicles isolated from the bovine adrenal medulla. Differential sucrose centrifugation showed that synaptophysin had the same distribution as established markers for chromaffin vesicles. Quantification of immunoblots revealed 750 ng of synaptophysin per mg of protein. Thus, the amount of synaptophysin in clear synaptic vesicles is about 100 times greater than in chromaffin vesicles. However, because of the large difference in surface area and protein content, the amount of synaptophysin per single vesicle is about the same in both types of organelles.

The content and distribution of synaptophysin in subcellular fractions from bovine adrenal medulla, PC12 cells, and rat brain were determined by Wiedenmann, Rehm, Knierim et al. (1988) using a sensitive dot immunoassay. Synaptophysin-containing fractions appeared as monodispersed populations similar to synaptic vesicles in density and size distribution. Membranes from synaptic vesicles contain about 100 times more synaptophysin than do those from chromaffin vesicles. Synaptophysin appears to be located in vesicles of brain and PC12 cells but not chromaffin vesicles

Gaardsvoll, Obendorf, Winkler et al. (1988) showed that synaptin and synaptophysin/p38 are identical immunochemically. Discrepancies in certain characteristics of these proteins in previous reports are probably due to variations in glycosylation. Because a protein normally is designated by its original name, Gaardsvoll et al. (1988) suggest that hereafter synaptophysin/p38 (and probably also SVP38) be termed "synaptin."

Romano, Nichols, Greengard et al. (1987) characterized synapsin I in PC12 cells; the molecule was immunochemically similar to that molecule in the brain. Chronic nerve growth factor treatment of the PC12 cells induces a significant increase in the total amount of synapsin I relative to total cell protein.

The presence of two phosphoproteins in the adrenal medulla was reported by Browning, Huang and Greengard (1987). These two phosphoproteins of molecular weights 74 kDa and 55 kDa are known to be present in the mammalian brain. Several organs were investigated, but Browning et al. (1987) found the two proteins only in nervous tissue and the adrenal medulla of the rat.

Sajovic, Moraru, Greene et al. (1986) produced a monoclonal antibody against glycoproteins prepared from PC12 cells. This antibody stains PC12 cells in a rimlike fashion whether they are fixed or living. It appears as though at least some of the binding is to an exposed component of the plasma membrane. They also showed that the monoclonal antibody selectively stains a subset of neurons in brain.

Galanin was measured by radioimmunoassay in extracts of pig, cat, and rat adrenals by Bauer, Adrian, Yanaihara et al. (1986). The galanin immunoreactivity was present in substantial quantities in pig and cat adrenals but not in rat adrenals. Their methods did not allow localization between the adrenal medulla and cortex.

Bauer, Hacker, Terenghi et al. (1986) measured galanin immunoreactivity by radioimmunoassay in tissue extracts of normal human adrenals and pheochromocytomas and also in the plasma of normal subjects and patients with pheochromocytomas. Using a non-COOH-terminus-directed antibody, they were able to detect galanin in the adrenal gland. Pheochromocytomas had a much greater concentration. The concentration of galanin in plasma was below the detection limit of the assay. By immunocytochemical methods, galanin was localized in a few cells in the pheochromocytoma and in only a few chromaffin cells and cortical nerve fibers in normal adrenals.

Rökaeus and Brownstein (1986) constructed a 1.1-million recombinant cDNA library derived from a RNA from pig adrenal medulla rich in galanin-like immunoreactivity, and they isolated and characterized clones encoding preprogalanin. The precursor protein contains 123 amino acids and is composed of a leader sequence, a single galanin sequence, and a 55-amino acid sequence of unknown function.

Rökaeus and Carlquist (1988) described the primary structure of a precursor protein for galanin isolated from the bovine adrenal medulla. This protein had 123 amino acids and contained a nonrepetitive galanin sequence. This protein, named "preprogalanin," was found to be coded for by an apparent single species of mRNA present in a population of chromaffin cells scattered throughout the bovine adrenal medulla.

Helman, Thiele, Linehan et al. (1987) constructed a human pheochromocytoma cDNA library and used differential hybridization to isolate genes that are highly expressed in pheochromocytoma. Two of the cDNA clones were more highly expressed in normal and neoplastic chromaffin tissue than they were in neoblastoma. They are expressed in a remarkable number of other human tumors and normal tissues.

Hacker, Bishop, Terenghi et al. (1988) investigated pheochromocytomas and normal adrenal glands by using a wide range of well-characterized antibodies to active peptides and other derivatives of their precursors as well as general markers of neuroendocrine differentiation. Of the general neuroendocrine markers, neuron-specific enolase was found in all tumors; chromogranin and protein gene product 9.5 was found in most of the specimens. A range of regulatory peptide immunoreactivities could be demonstrated in some of the specimens. In addition, two peptides were found that have not been reported previously in these tumors, peptide histidine methionine and the tryptic fragment of the precursor of vasoactive intestinal peptide. In the normal adrenal medulla, all the peptides previously reported to be present could be demonstrated immunocytochemically. Galanin was present in a subpopulation of cells that was also immunoreactive for enkephalin.

The cellular localization of mRNA for preprotachykinin A in the bovine nervous system was reported by Goedert and Hunt (1987). The highest levels of the mRNA were found in the brain, and lower levels were found in the adrenal medulla, spinal cord, and other structures. Their study showed that *in situ* hybridization can allow for the unambiguous identification of cellular sites of synthesis of preprotachykinin A mRNA.

Zurawski, Benedik, Kamb et al. (1988) identified a cDNA that encodes for mouse preproenkephalin mRNA from an activated mouse T-helper cell line. The abundance of this mRNA was almost as much as has been found in the adrenal medulla, suggesting that this cell line can synthesize almost as much enkephalins as can the classic adrenal medullary site.

Langley and Aunis (1986) used two different immuno-cytochemical methods to demonstrate the neural cell adhesion molecule on cultured bovine adrenal chromaffin cells. With one method, the reaction product was distributed in an apparently continuous layer over the surface. With another method, large areas of membrane were unlabeled. These differences are explained by differences in steric hindrance inherent to each method.

The neural cell adhesion molecule was demonstrated in bovine adrenal medulla, PC12 cells, and anterior pituitary by Langley, Aletsee and Gratzl (1987). A new 49 kDa neural cell adhesion molecule determinant was found in the adrenal medulla and pituitary, although it was absent from PC12 cells and cerebellum. Langley et al. (1987) suggested that this molecule may be functional during the latter stages of adrenal medulla histogenesis, perhaps at the phase at which chromaffin cells aggregate in the adrenal medulla.

Pruss (1987) prepared a number of monoclonal antibodies to

chromaffin cell membranes. One of these antibodies recognized a number of antigenically related proteins that were present in all of the tissues she examined. In the adrenal gland, these proteins are completely excluded from chromaffin vesicles but are present in other subcellular membrane fractions. A second antibody bound to a single protein that segregates specifically into chromaffin vesicles. The protein recognized by the second antibody is cytochrome b_{561}, one of the major components of the chromaffin vesicle membrane. Using this second antibody, Pruss and Shepard (1987) demonstrated that cytochrome b_{561} is present in many neural and endocrine tissues and that its distribution is correlated with the presence of either catecholamines or amidated peptides in the tissue. These tissues include neuronal cell bodies and fibers in the gut, blood vessels, retina, and posterior pituitary; endocrine cells of the gut; anterior and intermediate lobe of the pituitary; heart muscle; and all adrenal chromaffin cells. The discovery of cytochrome b_{561} in many neuropeptide-containing tissues is consistent with a general role for this cytochrome as a secretory vesicle membrane electron carrier.

Clathrin is a protein known to be associated with coated vesicles in secretory cells, including adrenal chromaffin cells. Pearse (1987) reviewed general aspects of this protein.

Purification and characterization of kinesin from bovine adrenal medulla was reported by Murofushi, Ikai, Okuhara et al. (1988). The purified kinesin preparation contained a major polypeptide with a molecular mass of 120 kDa and minor peptides with molecular masses of 71, 68, and 65 kDa. Bovine adrenal kinesin had an ATPase activity that was stimulated severalfold by microtubules. Kinesin molecules absorbed through a glass slide promoted the movement of microtubules on the glass surface. Immunostaining studies demonstrated that some of the kinesin was located on microtubules and the rest was distributed diffusely throughout the cytoplasm.

Toutant, Aunis, Bockaert et al. (1987) demonstrated the presence of interesting proteins within both the plasma membrane of chromaffin cells and chromaffin vesicle membranes. Three of these proteins were shown to be substrates of pertussis toxin, which is known to catalyze ADP-ribosylation of several GTP-binding proteins. One of these appeared to be identical to a GTP-binding protein.

Terashima, Katada, Takayama et al. (1988) used immuno-histochemical methods to localize the GTP-binding regulatory protein G_o in adrenal medulla and other peripheral structures. They used an affinity-purified antibody against the α-subunit of

G_o. The adrenal medulla was stained with this antibody, but the adrenal cortex was not. Their study suggested that G_o is localized in the peripheral autonomic nervous system and plays a role in transmembrane signal transmission in this system.

Nilaver, Beinfeld, Bond et al. (1988) used immunohistochemical methods to localize motilin in the adrenal medulla and PC12 cells. Radioimmunoassay of PC12 cells confirmed these observations. Motilin is a 22-amino acid peptide first isolated from the porcine gut and sequence-analyzed.

Parysek and Goldman (1988) used an antiserum to a 57 kDa intermediate filament protein, that is not vimentin, to detect the protein in various rat tissues. Using immunofluorescence, they showed that chromaffin cells, ganglion cells, and nerve fibers of the adrenal medulla were labeled. They found that this protein is distributed in only certain neuronal elements. They suggested that this protein may be part of a unified but not yet recognized pattern of function in the nervous system.

Satoh, Endo, Nakamura et al. (1988) analyzed the *ras* gene product p21 bound to guanine nucleotides in PC12 cells. They found that conditionally produced exogenous p21 bound mainly GTP and was able to induce morphologic changes, whereas the endogenous p21 mostly bound GDP. This observation provides strong support that p21 causes morphologic changes in PC12 cells because it exists in a GTP-bound confirmation in these cells.

Toutant, Bockaert, Homburger et al. (1987) used the same methods by which they demonstrated GTP-binding proteins in bovine chromaffin vesicles to examine other tissues. In contrast to chromaffin vesicles, no GTP-binding protein could be detected in synaptic vesicles of the torpedo electric organ.

High concentrations of a novel pituitary protein termed "7B2" have been shown to be present in PC12 cells by a radioimmunoassay by Suzuki, Tischler, Christofides et al. (1986). 7B2-like immunoreactivity was released from PC12 cells in response to stimulation. The protein appeared to be identical to a protein in the porcine pituitary gland.

The location of 7B2 and its release from the bovine adrenal medulla was observed by Iguchi, Natori, Nawata et al. (1987). They found that the protein was distributed in the chromaffin vesicle fraction prepared from the bovine adrenal medulla and was released, by nicotine or high potassium concentration, from cultured bovine adrenal chromaffin cells. These and other findings indicate that 7B2 is a secretory protein in the bovine adrenal medulla.

Marcinkiewicz, Fischer-Colbrie, Falgueyret et al. (1988) used two-dimensional immunoblotting analysis and immunocytochemistry to localize secretory polypeptide 7B2 in the bovine adrenal medulla. This 24 kDa pituitary polypeptide was localized in chromaffin vesicles. It was established that 7B2 represents a minor component of the acidic proteins in chromaffin vesicles.

Ong, Lazure, Nguyen et al. (1987) purified acid extracts from bovine chromaffin vesicles and identified both precursor and mature forms of atrial natriuretic factor. The occurrence of both forms indicates the endogenous character of atrial natriuretic factor in the adrenal medulla and suggests the potential usefulness of cultured adrenal chromaffin cells for investigating the synthesis, maturation, and secretion of atrial peptides.

The secretion and biosynthesis of atrial natriuretic factor by bovine adrenal chromaffin cells was characterized by Nguyen, Ong and De Léan (1988). They demonstrated that nicotine and high potassium increased the cosecretion of the precursor and mature forms of atrial natriuretic factor. These atrial peptides in chromaffin cells are increased by phorbol ester, and this is potentiated by forskolin.

Extremely acidic copper-containing proteins were isolated from soluble and membranous fractions of bovine chromaffin vesicles by Mikaelyan, Grigoryan and Nalbandyan (1987). The soluble form of the protein isolated from the vesicles was practically identical to a group of proteins called "neurocupreins." The copper protein isolated from the vesicle membranes was slightly higher in molecular weight and had a somewhat different amino acid composition. It was suggested that the membranous form differs from the soluble one in having a peptide that prolongs the protein chain without changing its antigenic properties.

Mikaelyan, Markossian, Paitian et al. (1988) demonstrated that secretory vesicles from different glands contain a neurocuprein-like protein. The bovine adrenal medulla was shown to contain this extremely acidic copper-containing protein.

Kristensen, Hougaard, Nielsen et al. (1986) stained rat adrenal glands immunocytochemically by using antibodies against plasminogen activators of the tissue type and urokinase type. A subpopulation of cells in the adrenal medulla showed intense immunoreactivity to the tissue-type plasminogen activator; no immunoreactivity was detected for the urokinase type. This substantiated that tissue-type plasminogen activators can be found outside of endothelial cells, and they may have a function in endocrine cells.

A novel human plasma protein called tetranectin was shown to be present in the adrenal medulla by Christensen, Johansen, Jensen et al. (1987). The protein was localized on ganglion cells in the adrenal medulla and in endocrine cells of other organs. It was suggested that this protein may serve as a regulator in the secretion of certain hormones.

Corcoran, Korner, Caughey et al. (1986) carried out studies to characterize the properties of extravesicular and vesicular pools of ATP in bovine chromaffin cells. They examined the effects of metabolic inhibitors, adenosine, and digitonin on ATP utilization and subcellular localization immediately after and 48 hours after labeling with tritiated adenosine and radioactive phosphate. Their studies showed that the ATP contained within chromaffin vesicles is metabolically inert and does not readily exchange with cytosolic ATP, but extravesicular pools of labeled ATP could not be detected. Calculation of ATP disappearance after abrupt cessation of ATP production indicates that the ATP within cultured adrenal chromaffin cells has a turnover time of 12.5 minutes in the extravesicular pool.

Mitsuma, Sun, Nogimori et al. (1987) examined the release of thyrotropin-releasing hormone from the rat adrenal gland. They noted that the amount of immunoreactive thyrotropin-releasing hormone was lowered by the addition of dopamine and enhanced with the addition of pimozide or domperidone to the medium. These and other findings suggest that the dopaminergic system inhibits release of the hormone from the rat adrenal gland, but it is not clear if the hormone is released from the cortex or the medulla.

Benzodiazepine-binding inhibitory activity was detected in aqueous extracts of brain and peripheral tissues of rats, including the adrenal glands, by Wildmann, Möhler, Vetter et al. (1987). It was shown that at least some of this activity was due to the presence of diazepam and N-desmethyldiazepam. Other tests showed that the activity could be due to naturally occurring compounds in food.

Hutton, Peshavaria, Davidson et al. (1987) compared the insulin secretory vesicle to other secretory vesicles, including chromaffin vesicles. Specifically, they examined the distribution of four insulin secretory vesicle antigens. These antigens cross-reacted to a greater or lesser degree with proteins within the bovine chromaffin vesicle.

The formation of GABA from putrescine was demonstrated in the rat adrenal gland by Caron, Cote and Kremzner (1988). GABA in the adrenal gland is almost exclusively derived from putrescine. Diamine oxidase, but not monoamine oxidase, is essential for GABA formation from putrescine. It was suggested that the role of GABA may relate to the role of putrescine as a growth factor and regulator of cell metabolism.

A new facile trinitrophenylated substrate for peptide α-amidation in chromaffin vesicles was described by Katopodis and May (1988). They also reported detection and characterization of peptidyl α-amidating monooxygenase activity in chromaffin vesicles. They suggested that neuropeptides could be modified by this enzyme. The new substrate exhibits high turnover and has the important advantage of allowing quantitative activity determinations by standard spectrophotometric techniques, thus facilitating studies of mechanisms and inhibitor development.

Monoclonal antibodies to insulin secretory vesicle membranes were obtained from a rat insulinoma by Grimaldi, Hutton and Siddle (1987). Using immunologic techniques, they were able to detect immunoreactive antigens in the adrenal medulla and other endocrine organs.

The distribution administered of tritiated deoxyglucose and dopamine in the adrenal medulla and nerve endings of the mouse was examined by Hirano (1988). Nerve endings and adrenal chromaffin cells accumulated dopamine but not deoxyglucose, indicating that their amine uptake system does not require a high glucose consumption rate. Peripheral cells in the adrenal medulla incorporated more dopamine than did those in the center, and this was not correlated with a different rate of glucose utilization.

The immunohistochemical localization of a novel human plasma protein, tetranectin, in human endocrine tissues was reported by Christensen, Johansen, Jensen et al. (1987). Tetranectin is a plasminogen kringle four-binding tetrameric protein with a molecular mass of 80 kDa. Most ganglion cells of the human adrenal medulla showed a moderate, finely granular reaction for tetranectin. The adrenal cortex and adrenal chromaffin cells were negative.

Uvnäs and Åborg (1988) investigated the storage and release behavior of an ion exchanger with amphoteric properties obtained by mixing two synthetic exchanger resins. They found that this combined ion exchanger stored and released catecholamines and ATP very much like chromaffin vesicles did. They thought that this supported their view that the chromaffin vesicle operates like an ion exchanger with amphoteric properties.

A monoclonal antibody (MRK 16), specific to Adriamycin-resistant human myelogenous leukemia cells was used by

Sugawara, Nakahama, Hamada et al. (1988) to examine whether the antigen molecules (P-glycoprotein) recognized by the antibody are present in the adrenal glands. Immunohistochemical analysis revealed that human adrenal specimens were stained positively by the MRK 16 antibody. Immunoprecipitation showed that the 170- to 180 kDa glycoprotein was present in the all of the adult specimens but not in fetal and neonatal adrenal glands. Immunofluorescent studies showed that there was more P-glycoprotein in the cortex than in the medulla, with a tendency to occur in cell clusters in the medulla. It was suggested that this glycoprotein may be related to maturation of the adrenal, in which it may play a physiologic role.

A large polymer with glucose as the sole component monosaccharide was found in PC12 cells by Rasilo and Yamagata (1988a). Except for those capable of degrading glycogen, no exo- or endoglycosidases cleaved the polymer. This is the first report of the occurrence of a glucose polymer in undifferentiated PC12 cells. Rasilo and Yamagata (1988b) characterized this glucose polymer. A similar glucose polymer was isolated from neuronal cells of rat embryos but was not present in mouse neuroblastoma cells.

Mazzanti, Scalori, Mian et al. (1988) investigated the distribution of aclacinomycin in tissues of the rat. This anthracycline antibiotic was preferentially located in the cell nucleus in all tissues examined except the adrenal medulla where a slight fluorescence appeared only in the cytoplasm.

Kim, Sessler and Malvin (1988) measured active renin and inactive trypsin-activatable renin in adrenal homogenates and plasma and found that the adrenal glands contained both of these compounds. It was established that multiple renin forms are present in the adrenal gland, but they were not localized to the medulla.

Deschepper and Ganong (1988) reviewed the role of renin and angiotensin in endocrine glands. It was pointed out that angiotensin II is known to stimulate catecholamine release from the adrenal medulla, and receptors for angiotensin II have been reported in the medulla of bovine and rat adrenals.

A novel peptide similar to parathyroid hormone has recently been purified and cloned from human tumors. Ikeda, Weir, Mangin et al. (1988) surveyed the expression of mRNAs encoding this peptide. Among other organs tested, the peptide was shown to be present in the adrenal medulla.

Chianese, Roy-Choudhury, Murthy et al. (1988) used an antibody to a 34 kDa trophoblast-derived peptide growth factor to localize this factor in several organs. The human adrenal cortex was immunoreactive for this growth factor, but the adrenal medulla was not.

CHAPTER 10

PEPTIDES

Peptides, mainly but not exclusively opioid peptides, are contained within adrenal chromaffin cells, mostly within the chromaffin vesicles. Much of our understanding of the biochemistry and synthesis of these peptides is derived from studies of the adrenal medulla. A review article on the distribution of neuropeptides with special reference to their coexistence with classic transmitters was published by Hökfelt, Johansson, Holets et al. (1987). They emphasized peptides within the central nervous system but also pointed out that the adrenal medulla contained endocrine cells that were among the first in which the coexistence of opioid peptides and catecholamines in the same storage vesicle was demonstrated. Inagaki and Kito (1986) reviewed the distribution of peptides in the peripheral nervous system. Among other things, they pointed out that the peripheral nervous system is more useful than the central nervous system for analyzing the significance of the coexistence of peptides and other neurotransmitters in single neuron systems because the neuronal circuitry of the peripheral system is more fully understood.

In an extensive review of neuropeptides in the autonomic nervous system, Marley and Livett (1985) pointed out that one of the many fertile areas for further research is the need to establish the physiologic conditions under which a given peptide may be released and may exert its effect. A historical overview of progress on the secretion of neuropeptides was presented by Scharrer (1987). She pointed out that work in this area is multidisciplinary and is progressing with increasing rapidity. Techniques for the isolation and identification of opioid peptides were reviewed by Matsuo (1986). Approaches to the study of neuropeptides by using hybridization techniques were reviewed by Schwartz and Costa (1986). The use of gene transfer approaches to study regulation of expression of opioid peptide genes was reviewed by Herbert, Comb, Seasholtz et al. (1986).

Carlsson (1988) made the intriguing suggestion that peptide neurotransmitters are redundant vestiges. Although neurohormones were specifically omitted, a case was made that peptide neurotransmitters are simply left over from an evolutionary process because there is no specific negative survival value associated with them. However, the presence of peptide receptors allows for pharmacologic manipulation of the nervous system.

ENKEPHALINS AND RELATED PEPTIDES

Metters and Rossier (1987) reported a method for purifying synenkephalin-containing peptides. Synenkephalin contains the first 70 amino acids of proenkephalin. Using newly developed affinity columns, Metters and Rossier (1987) purified synenkephalin-containing peptides from bovine adrenal chromaffin vesicles. They also purified the 23.3 kDa putative processing intermediate. This is the largest naturally occurring enkephalin-containing peptide yet found in the bovine adrenal medulla.

Wan, Scanlon, Power et al. (1987) described the synthesis of a specific oligodeoxyribonucleotide probe and its use for studying proenkephalin A gene expression. Northern blot hybridization analysis identified a single band in RNA extracts from bovine, ovine, and porcine adrenal medullas. *In situ* hybridization revealed that proenkephalin A mRNA is localized selectively in cells at the outer margin of the medulla, a region rich in epinephrine-containing cells. This study both confirms and extends previous findings with cloned cDNA probes of the presence of high concentrations of proenkephalin A mRNA in cells at the periphery of the adrenal medulla. In addition, it demonstrates the usefulness of synthetic oligodeoxy-ribonucleotides as specific and sensitive probes for the study of proenkephalin A gene expression.

Birch, Davies and Christie (1986) investigated the possibility that proenkephalin may be associated with chromaffin vesicle membranes. They used the technique of electroimmunoblotting to characterize proenkephalin in both lysate and membranes prepared from bovine adrenal chromaffin vesicles. Although peptides of various sizes were identified in the lysate, the major form in the membranes corresponded to a 27 kDa enkephalin-containing protein. Twenty-two percent of this material was membrane-associated. It was pointed out that the association of hormone and neuropeptide precursors with membrane components may be important for targeting of precursors to secretory vesicles for correct processing.

Northern blot analysis with a single-stranded human proenkephalin A anti-sense probe was used by Monstein and Geijer (1988) to reveal the existence of two different proenkephalin A-like sequences in human pheochromocytoma RNA extracts. Under highly stringent hybridization conditions, the proenkephalin A-like RNA bands still appeared, indicating that the RNA species sequences as either identical or closely related to the sequence of the well-characterized human proenkephalin A mRNA.

The distribution of large enkephalin-containing peptides between soluble and membrane components of bovine chromaffin vesicles was examined by Hook and Liston (1987). Immunoblots showed that 23.3- and 18.2 kDa peptides were present in both soluble and membrane vesicle compartments, but there was a 12.6 kDa peptide that was present only in the soluble fraction. These results suggest that the larger enkephalin-containing peptides may be preferentially associated with the vesicle membrane and may be redistributed to the soluble vesicle compartment upon proteolytic processing.

Parmer and O'Connor (1988) studied the subcellular localization and precursor-product status of enkephalin immunoreactivity in human pheochromocytomas. Enkephalin immunoreactivity was found in the tumors, and the reactivity could be localized to chromaffin vesicles. The immunoreactivity could be released by lysis of the vesicles, indicating that enkephalins are present in the soluble fraction of the chromaffin vesicles, along with catecholamines. Enkephalin immunoreactivity was not contained in purified chromogranin A, either before or after trypsin cleavage. Enkephalin immunoreactivity was augmented by trypsin alone and by trypsin plus carboxypeptidase B, suggesting that the majority of the enkephalins are present in pheochromocytomas as higher molecular weight precursor molecules.

Birch and Christie (1986) generated an antiserum against a synthetic peptide corresponding to a defined fragment of bovine proenkephalin. They developed a sensitive radioimmunoassay and compared the reactivities of the synthetic peptide, its specific cleavage products, and other specific peptides. Their experiments led to the conclusion that antiserum against this synthetic peptide recognizes a wide range of intermediates in the processing of proenkephalin in bovine adrenal medulla and in rat adrenal, brain, and spinal cord, making it a useful tool for further studies of the production and post-translational processing of proenkephalin.

Boarder, Evans, Adams et al. (1987) analyzed extracts of bovine adrenal medulla to investigate the hypothesis that the primary site of cleavage of peptide E in the chromaffin cells is between the amino acids lysine and arginine. Cleavage of this 25-amino acid opioid peptide would generate two opioid fragments. They examined levels of peptide E and its two fragments in cultured chromaffin cells and looked for evidence that the peptides are released under basal and stimulated conditions. Under basal conditions, peptide E and one of the fragments are released; but upon stimulation with nicotine, all three peptides are released.

Katzenstein, Lund, Schultz et al. (1987) studied the tissue binding distribution of adrenal enkephalin-containing peptides F, E, and B. Using radioiodinated forms of the peptides, they found that each has a unique distribution profile. However, all three are found to bind to the pituitary in rats, rabbits, and guinea pigs. A number of lines of evidence point toward the enkephalin-containing peptides studied as having physiologic roles distinct from those of the enkephalins.

The distribution of enkephalin in tissues and cells of the developing and adult rat was examined by Zagon, Rhodes and McLaughlin (1986). Using immunocytochemical methods with antibodies to [Met]- and [Leu]enkephalin, they found evidence of the peptides in the cytoplasm of virtually all developing cells. Not only the cells of the adrenal medulla, but also the cortex, exhibited immunoreactivity. In contrast to cells of the 10-day-old rat, staining was much diminished in the tissues of adults.

Using specific antibodies to [Met]enkephalin, [Met]enkephalin-Arg^6-Phe^7, and [Met]enkephalin-Arg^6-Gly^7-Leu^8, Leboulenger, Cupo, Castanas et al. (1986) studied the distribution of these opioid peptides in the frog adrenal gland. With an indirect immunofluorescence technique, they observed bright staining of all chromaffin cells after application of antisera-containing antibodies to [Met]enkephalin and [Met]enkephalin-Arg^6-Phe^7. [Met]Enkephalin-Arg^6-Gly^7-Leu^8 appeared to be absent from frog chromaffin cells.

Related series of studies characterized the enkephalins of human pheochromocytomas. Yanase, Nawata, Kato et al. (1986a) found that human pheochromocytomas are heterogeneous with regard to storage and secretion of opioid peptides and catecholamines. Yanase, Nawata, Kato et al. (1987a) used immunologic techniques to detect the presence of preproenkephalin B-derived opioid peptides in human pheochromocytomas and normal adrenal medullas. Small amounts of dynorphin and α-neoendorphin were detected in all of the pheochromocytomas tested, whereas leumorphin-like immunoreactivity was detected in only a few tumors. Yanase, Nawata, Haji et al. (1987a) detected proenkephalin A-derived opioid peptides and mRNA for preproenkephalin A in human pheochromocytomas. Remarkably wide distributions in the amounts of immunoreactive peptides were observed. The quantities of preproenkephalin A mRNA in the tumors significantly correlated with the quantities of immunoreactive [Met]enkephalins. These results indicate that the remarkable difference in tissue immunoreactivities for preproenkephalin A-derived opioid peptides is decided at the transcriptional level but not at the post-translational level.

The question of whether the action of reserpine to decrease proenkephalin mRNA is related to an inhibition of the release of enkephalin or to a change in the steady-state levels of catecholamines was explored by Naranjo, Wise, Mellstrom et al. (1988). In cultured bovine adrenal chromaffin cells, the basal release of a [Met]enkephalin-containing heptapeptide into the medium was significantly decreased; when the cultures were pretreated with reserpine. The addition of etorphine also decreased the basal release of the peptide without depleting stores of catecholamines. Neither drug modified the total amount of the peptide. These data establish a correlation between inhibition of the secretion of enkephalin and decreased accumulation of its specific mRNA, suggesting a negative feedback inhibition by low molecular weight enkephalins.

Vindrola, Ase, Finkielman et al. (1988) compared the enkephalin-like material derived from proenkephalin released from perfused cat adrenal glands stimulated with pilocarpine and nicotine. Free and total [Met]enkephalin immunoreactivities were coreleased with catecholamines. Pilocarpine caused a greater release of immunoreactivity than nicotine did. The variations may be related to the heterogeneous distribution of enkephalins in the adrenal medulla. It appears that, under physiologic conditions, different enkephalin-containing peptides can be secreted from the cat adrenal gland.

[Met]Enkephalin concentrations in plasma after endocrine and pharmacologic modifications were examined by Panerai (1988). Plasma concentrations of [Met]enkephalins were greatly increased after adrenalectomy, both in normal and in hypophysectomized rats. It was concluded that the adrenal glands are not a main site of origin for [Met]enkephalin in the plasma.

A sensitive method for the detection, quantification, and purification of peptides was described by Lemaire, Dumont and Nolet (1988). The method is based on pre-column derivatization of peptides with phenyl isothiocyanate to form phenylthiocarbamyl derivatives. Using this method, they were able to measure enkephalins secreted from perfused bovine adrenal glands.

Possible modulatory effects of μ, δ, and κ receptor agonists on the concurrent secretion of catecholamines and [Met]enkephalins from the adrenal medulla evoked by hemorrhage were examined in cats by Gaumann, Yaksh, Tyce et al. (1988a). After 25% and 50% hemorrhage, there were no differences in adrenal vein hormone levels in cats receiving the opioid peptide receptor agonists and in controls. No differences were observed with regard to the proportions of catecholamines and [Met]enkephalin in the adrenal vein during the course of the experiment. It was concluded that opioids are not involved in the regulation of the response of the adrenal medulla to hemorrhage.

In a related study, the effects of hemorrhage and opiate antagonists on the secretion of neuropeptides from the adrenal of cats were described by Gaumann and Yaksh (1988). Levels of [Met]enkephalin, neuropeptide Y, peptide YY, neurotensin, vasoactive intestinal polypeptide, cholecystokinin, and bombesin were measured concurrently in the adrenal lumbar vein, femoral vein, and femoral artery under baseline conditions and during hypotensive hemorrhage. A blood loss of about 40% evoked selective increases in levels of [Met]enkephalin, neuropeptide Y, peptide YY, and neurotensin in the adrenal vein. No differences in regard to hemodynamics and pattern of neuropeptide levels were observed between these animals and animals given naltrexone prior to hemorrhage. Administration of naloxone to bled and control animals at the end of the experiment led to a significant increase in blood pressure in both groups but did not evoke changes in neuropeptide levels. It was concluded that adrenal neuropeptide release during hypotensive hemorrhage is not modulated by actions on opiate receptors in the cat.

<u>NEUROPEPTIDE Y</u>

The anatomic distribution and possible function of neuropeptide Y (NPY) in the mammalian nervous system was reviewed by Gray and Morley (1986). This 36-amino acid peptide is widely distributed within neurons of the central and peripheral nervous systems, including the adrenal medulla. Many anatomic and physiologic studies suggest varied, but important, functions for NPY in the mammalian nervous system.

The human NPY gene was isolated from a human genomic DNA library by Minth, Andrews and Dixon (1986). The transcription unit spans approximately 8 kilobase pairs and is interrupted by three intervening sequences. The gene was characterized and the nucleotide sequence for NPY was reported.

Kuramoto, Kondo and Fujita (1986) used immunohistochemical methods to demonstrate that approximately 50% of the chromaffin cells of the rat adrenal medulla exhibit NPY immunoreactivity. The immunoreactive material was localized in the core of the chromaffin vesicles as well as diffusely in the cytoplasm. Neuropeptide Y appears to be stored with epinephrine. In addition to chromaffin cells, NPY also appeared to be present in nerve fibers. Some of these fibers made synaptic contacts with both chromaffin and cortical cells.

The mRNA for NPY was localized in the rat adrenal gland by *in situ* hybridization studies by Schalling, Seroogy, Hökfelt et al. (1988). Catecholamine-synthesizing enzymes also were localized. Neuropeptide Y-like immunoreactivity was observed in chromaffin cells, ganglion cells within the medulla, and nerve fibers. The chromaffin cells were both epinephrine- and norepinephrine-containing types. *In situ* hybridization revealed a high density of mRNA in ganglion cells, a moderate density in chromaffin cells, and a low density in the adrenal cortex. The intense NPY-like immunoreactivity and low content of mRNA suggested that chromaffin cells have fairly large peptide stores but that the peptide turnover is low. In contrast, the ganglion cell bodies seem to contain low amounts of the peptide but had a high synthesis rate.

Antisera against NPY, [Met]enkephalin, and bombesin were used for characterizing the immunoreactive material in subcellular fractions of bovine adrenal medulla by Fischer-Colbrie, Diez-Guerra, Emson et al. (1986). Neuropeptide Y was identified by high-performance liquid chromatography (HPLC) and by immunoblotting. Subcellular fractionation established that NPY is present in chromaffin vesicles and is released concomitantly with catecholamines during stimulation of the gland. Each chromaffin vesicle contains about 429 copies of NPY. [Met]Enkephalin showed a more variable distribution; and the antiserum against bombesin cross-reacted with chromogranin B, making it difficult to ascertain if this peptide is present in the adrenal medulla.

deQuidt and Emson (1986) applied radioimmunoassay, chromatography, and immunohistochemical methods to determine the occurrence of NPY-like immunoreactivity in the adrenal glands of rat, guinea pig, cat, and cow. The levels of immunoreactivity varied considerably between species; the lowest amounts were found in the rat and highest amounts in the cow. Administration of insulin and reserpine to rats showed that the turnover of adrenal NPY-like immunoreactivity is regulated by splanchnic nerve activity.

Bastiaensen, De Block and De Potter (1988) demonstrated that NPY is co-localized with enkephalins in chromaffin vesicles of the bovine adrenal medulla. During differential centrifugation, NPY co-sedimented with chromaffin vesicle markers. Like [Met]enkephalin immunoreactivity, NPY immunoreactivity was concentrated in fractions containing the lighter chromaffin vesicles. These vesicles were predominantly epinephrine-containing.

The functional implications of the coexistence of NPY with

norepinephrine were presented by Håkanson, Wahlestedt, Ekblad et al. (1986). These two compounds are known to coexist in the peripheral nervous system because NPY is present with dopamine β-hydroxylase. They concluded that, although the coexistence of NPY and norepinephrine in the sympathetic nervous system is apparently of physiologic importance, the nature of their cooperation is only partially understood. The functional role of NPY in the periphery is probably not related exclusively to noradrenergic neurotransmission.

Pruss, Mezey, Forman et al. (1986) found that NPY is localized with enkephalin in bovine adrenal chromaffin cells, specifically within chromaffin vesicles. These are epinephrine- containing cells because they also contain phenylethanolamine N-methyltransferase. Despite their coexistence, production of the two peptides is independently regulated. Enkephalin levels are increased after nicotinic depolarization or after treatment with reserpine, and neither of these treatments has any effect on the levels of NPY.

Lundberg, Hemsén, Fried et al. (1986) measured the co-release of NPY-like immunoreactivity and catecholamines in newborn infants. Their data are compatible with the idea that NPY is released with catecholamines in response to a strong stress such as vaginal delivery.

Hexum, Majane, Russett et al. (1987) investigated the secretion of NPY and the involvement of the cholinergic receptor. They compared this phenomenon with the secretion of enkephalin-like peptides in the perfused bovine adrenal gland system. Using appropriate agonists and antagonists, they demonstrated that NPY can be released from the adrenal through stimulation of the nicotinic cholinergic receptor. The adrenal medulla appears to be one of the sources for circulating NPY.

The effect of reserpine treatment on the NPY content of the rat adrenal gland and other organs was studied with a specific radioimmunoassay by Allen, Schon, Yeats et al. (1986). One hour after reserpine administration, the NPY concentration in the adrenal gland was significantly decreased. Allen, Martin and Heinrich (1987) demonstrated that levels of mRNA for NPY in PC12 cells are regulated by nerve growth factor but not by dexamethasone. They found that the differentiation program initiated by nerve growth factor, which elicits a change to neuronal morphology, enhances neuropeptide Y gene expression in the cells.

Lundberg, Hökfelt, Hemsén et al. (1986) investigated the occurrence of NPY-like immunoreactivity in the adrenal gland of several species as well as in tumor tissue and plasma from pheochromocytoma patients. The peptide appeared to be present in epinephrine-containing cells of the mouse, rat, guinea pig, and cat but not of the pig. Neuropeptide Y also appeared to be present in nerve fibers within both the adrenal medulla and the cortex. High levels of NPY were found in plasma and tumors from patients with pheochromocytoma. After manipulation of the tumors during surgical removal, there was a marked increase in plasma peptide levels in parallel with the increase in catecholamines and blood pressure. They postulated that NPY contributes to the adrenal cardiovascular response and the hypertension seen in patients with pheochromocytoma.

Neuropeptide Y and its flanking peptide were measured in extracts of pheochromocytomas, ganglioneuroblastomas, and other tumors and plasma by Allen, Yeats, Causon et al. (1987). Neuropeptide Y was present in high concentration in most, but not all, pheochromocytomas. There was a good correlation between the concentration of NPY immunoreactivity and that of the flanking peptide. The other tumors had large variations in NPY content.

Osamura, Tsutsumi, Yanaihara et al. (1987) localized NPY, a form of [Met]enkephalin, and somatostatin in human adrenal medulla and pheochromocytomas. The normal adrenal medullas and pheochromocytomas contained NPY and the enkephalin. The distribution of somatostatin was more variable.

High plasma concentrations of NPY were found in a patient with bilateral pheochromocytomas and medullary carcinoma of the thyroid by Connell, Corder, Asbury et al. (1987). Manipulation during operation produced a remarkable increase in plasma NPY concentrations, coinciding with increases in plasma catecholamine levels and arterial pressure.

Age-related changes in NPY regulation were studied in rat adrenal glands by Higuchi, Yang and Costa (1988). Neuropeptide Y immunoreactivity increased with age. Biochemical characterization showed that this increase was due to an increase in NPY and methionine sulfoxide NPY. It was suggested that age-dependent changes in NPY adrenal glands may have physiologic relevance to the regulation of catecholamine release from the adrenal medulla.

OTHER PEPTIDES

Cheung and Holzwarth (1986) examined the distribution and effect of vasoactive intestinal peptide (VIP) on adrenal medullary catecholamine secretion in fetal sheep. Nerve fibers immunoreactive for this peptide were observed in the medulla and cortex. In chromaffin cells isolated from the fetal adrenal medulla, VIP stimulated catecholamine release into the medium in a dose-dependent manner. Cheung and Holzwarth (1986) suggested that the presence of VIP in the fetal adrenal medulla and cortex and the ability of it to stimulate catecholamine release from chromaffin cells suggest that VIP may be an important modulator of adrenal medullary catecholamine secretion during fetal life.

Cheung (1988) investigated the presence of VIP in ovine fetal adrenals to determine changes in adrenal peptide content associated with maturation and explored the factors that regulate fetal adrenal vasoactive peptide release. The content of the peptide in the adrenal was low at 80 days' gestation and increased to a plateau at about 120 days. A significant decrease in adrenal peptide content occurred immediately prior to term. Acetylcholine, high potassium, and angiotensin II stimulated release of VIP from the fetal adrenal. These results suggest that the VIP neuronal system in the ovine fetal adrenal matures between 80 and 110 days of gestation. Furthermore, the release of the peptide from the fetal adrenal gland may be regulated by angiotensin II and cholinergic neurotransmitters.

VIP-like immunoreactivity was found in a few chromaffin cells, ganglion cells, and nerve fibers in the rat adrenal gland by Kondo, Kuramoto and Fujita (1986). Immunoreactive material appeared to be localized in chromaffin vesicles. Immunoreactive nerve fibers were observed in both the medulla and the cortex.

Yoshikawa and Saito (1986) demonstrated the existence, subcellular distribution, and mode of release of VIP-like immunoreactivity in the bovine adrenal medulla by using a radioimmunoassay. The peptide was present in 100-fold and 300-fold greater amounts than neurotensin and somatostatin, respectively, on a molar basis. The peptide was released from the perfused gland after stimulation with carbachol, potassium, or barium. The peptide was found mainly in the mitochondrial fraction; little was found in the highly purified chromaffin vesicle fraction. Yoshikawa and Saito (1986) suggested that VIP is mainly localized in the splanchnic nerve endings in the adrenal gland and may play a role as a neuromodulator in adrenal medullary function.

The regulation of VIP synthesis in chromaffin and neuroblastoma cells was studied by Beinfeld, Brick, Howlett et al. (1988). They demonstrated that many factors modify biosynthesis of the peptide, and these factors act through various second messengers.

They include cyclic nucleotides, phorbol esters, ascorbate, nerve growth factor, dexamethasone, and nicotine. These factors may act independently or together to increase mRNA transcription for VIP.

Three laboratories have independently localized atrial natriuretic polypeptides in the adrenal medulla. Inagaki, Kubota, Kito et al. (1986) found immunostaining for the peptide in most chromaffin cells of the rat adrenal gland. The peptide appeared to be associated with epinephrine-containing cells, primarily in association with chromaffin vesicles. Pruss and Zamir (1987) showed that cultured bovine adrenal chromaffin cells can be stimulated to produce atrial natriuretic peptide-like immunoreactivity. This increase appears to be mediated by second messenger-mediated pathways. Using an antiserum to rat atrial natriuretic peptide, Flügge, Inagami and Fuchs (1987) did not detect immunoreactive cells in the adrenal glands of the tree shrew (*Tupaia belangeri*).

The synthesis, internalization, and localization of atrial natriuretic peptide in the rat adrenal medulla were studied by Morel, Chabot, Garcia-Caballero et al. (1988). *In situ* hybridization studies indicated the presence of mRNA to the peptide in about 15% of the adrenal medullary cells. Systemic administration of radioiodinated atrial natriuretic peptide resulted in a preferential and specific radiolabeling of epinephrine-containing cells over norepinephrine-containing cells. The finding suggests that atrial natriuretic peptide may be synthesized primarily by norepinephrine-containing cells and that epinephrine-containing cells primarily bind and internalize the extracellular (endogenous or exogenous) peptide.

Mukoyama, Nakao, Morii et al. (1988) used radioimmunoassay to measure the level and nature of human α-atrial natriuretic polypeptide in the bovine adrenal gland. A considerable amount of immunoreactivity was detected in the adrenal medulla; no detectable amount was present in the cortex. Further analysis showed that the peptide is composed of two components with high and low molecular weights. These and other findings indicate that a discrete atrial natriuretic peptide system is present in the bovine adrenal medulla and that the peptide is co-secreted with catecholamines, suggesting a possible involvement of atrial natriuretic peptide in the function of the adrenal medulla.

In a developmental study, García-Arrarás, Fauquet, Chanconie et al. (1986) examined the co-expression of somatostatin-like immunoreactivity and catecholamine-synthesizing enzymes in neural crest derivatives. They found evidence of somatostatin and tyrosine hydroxylase coexisting in cells of the sympatho-adrenal system. The development of sympatho-adrenal precursors closely parallels the early stages of adrenergic neuron differentiation in a number of respects.

Somatostatin-like immunoreactivity was detected within the adrenal gland of the cat by using specific monoclonal antibodies by Vincent, McIntosh, Reiner et al. (1987). A large population of chromaffin cells displayed intense immunoreactivity; few nerve fibers were reactive. Somatostatin appeared to account for more than 90% of the observed immunoreactivity. Vincent et al. (1987) suggested that somatostatin may have a paracrine or endocrine role in the feline adrenal medulla.

Wilson, Holscher, Kasselberg et al. (1986) examined somatostatin and [Leu]enkephalin immunoreactivities in canine and equine pheochromocytomas. [Leu]Enkephalin was present in both species but only the dog had immunoreactivity for somatostatin.

Neurotensin was demonstrated in the cat adrenal gland by Ferris, Carraway, Brandt et al. (1986) and was shown to be released during splanchnic nerve stimulation. A subpopulation of norepinephrine-containing cells was shown to contain neurotensin immunoreactive material. Electrical stimulation of the splanchnic nerve releases neurotensin and some of its metabolites into the circulation.

Dobner, Tischler, Lee et al. (1988) demonstrated that lithium dramatically potentiates the expression of the neurotensin-neuromedin N gene in PC12 cells. This effect on a specific neuropeptide gene by lithium suggests that changes in gene expression might be involved in its therapeutic activity.

Pelto-Huikko, Salminen and Hervonen (1987) used immunocytochemical methods to localize neurotensin and enkephalin in the cat adrenal medulla. Neurotensin was found only in norepinephrine-containing cells, of which 60% to 70% were immunoreactive. Enkephalin was mainly in epinephrine-containing cells. Neurotensin and enkephalin were not localized in the same cells.

An antibody that recognizes the COOH-terminal portion of the vasopressin precursor was used by Mezey, Seidah, Chrétien et al. (1986) to visualize this peptide in peripheral and central nerve tissues of the Brattleboro rat. Immunoreactivity was detected in the adrenal medulla and other tissues. These results suggest that Brattleboro rats may make a small amount of normal vasopressin precursor.

The concentrations of immunoreactive oxytocin and arginine

vasopressin and their respective neurophysins were compared in bovine adrenal medulla and cortex by Nussey, Prysor-Jones, Taylor et al. (1987). The concentration of arginine vasopressin was similar in both organs, but the neurophysin of vasopressin was higher in the medulla. The medulla also contained much more oxytocin and its corresponding neurophysin. These authors were able to conclude that the bovine adrenal medulla and cortex contain authentic arginine vasopressin and oxytocin. The peptides are in chromaffin vesicles within the adrenal medulla. The presence of the related neurophysins and high-affinity receptors in the medulla suggests that the peptides are both synthesized and have a site of action within this organ.

A historical overview of substance P by Lembeck (1988) pointed out that there is now ample evidence that afferent neurons containing substance P and related peptides exist in all organs and tissues of the body. They have peripheral functions and convey a great number of messages to the central nervous system by the release of these peptides from their terminals. There is also evidence to implicate substance P in the modulation of catecholamine secretion from the adrenal medulla.

Lemaire, Chouinard, Mercier et al. (1986) used a specific antibody to demonstrate bombesin immunoreactivity in the bovine adrenal medulla. The peptide was detected in the medulla, particularly in a cell preparation enriched in norepinephrine-containing cells. At the subcellular level, immunoreactive bombesin was mainly concentrated in chromaffin vesicles, but a relatively high concentration was also found in the microsomal fraction. Analysis of the bombesin revealed four molecular forms; the major one appeared to resemble synthetic neuromedin-B, and another one corresponded to gastrin-releasing peptide.

Liebisch, Weber, Kosicka et al. (1986) isolated and determined the sequence of a COOH-terminally amidated peptide from bovine brain that was similar to amidorphin which they isolated from the bovine adrenal medulla. The nonopioid peptide present in the brain could not be detected in the adrenal medulla. This indicates differential post-translational processing of proenkephalin in different tissues. In the brain, as opposed to the adrenal medulla, amidorphin is further processed at the typical cleavage signals of two basic amino acid residues.

Eiden and Zamir (1986) found that exposure of cultured bovine adrenal chromaffin cells to reserpine caused a dose- and time-dependent increase in intracellular levels of the amidated enkephalin peptide metorphamide. Metorphamide increases were at least 5-fold greater than those of either [Met] or

[Leu]enkephalin, suggesting that reserpine stimulates both enkephalin processing and amidation in the chromaffin vesicle.

Yanase, Nawata, Kato et al. (1987b) studied the secretion and tissue contents of adrenorphin in human pheochromocytomas. They found a remarkably wide range of immunoreactive adrenorphin levels. They also presented evidence that adrenorphin may play an important role in regulating catecholamine secretion from pheochromocytomas.

Bjartell, Ekman and Sundler (1987) developed a sensitive and specific radioimmunoassay for γ_2-melanotropin and demonstrated that this peptide is present in the adrenal and pituitary glands of pigs. Using immunocytochemical methods, they demonstrated that the peptide is located in a subpopulation of the norepinephrine-containing cells of the adrenal medulla.

Kuramoto, Kondo and Fujita (1987) demonstrated that a small number of chromaffin cells in the rat adrenal medulla exhibit calcitonin gene-related peptide-like immunoreactivity. These appeared to be epinephrine-containing cells and they also exhibited immunoreactivity for NPY. Nerve fibers reactive for calcitonin gene-related peptide could be detected in the medulla and cortex, and they appeared to be extrinsic in origin. Also, all fibers immunoreactive for substance P were reactive for calcitonin gene-related peptide. This suggests that calcitonin gene-related peptide-positive fibers in the adrenal medulla may be derived from the spinal ganglia, as has been demonstrated for substance P-containing fibers.

Peptides derived from both proenkephalin and prodynorphin have been identified in the guinea pig adrenal medulla by Evans, Erdelyi, Hunter et al. (1985). Although many products of both opioid peptide precursors were identified, no metorphamide or dynorphin could be detected. Using immunocytochemical methods, they demonstrated that prodynorphin and proenkephalin products were present in the same cells within the medulla. The results show that two opioid peptide precursors can be localized in the same cells and have some features of their processing in common.

Bostwick, Null, Holmes et al. (1987) looked for opioid peptides and their precursors in 108 tumors of both neuroendocrine and non-neuroendocrine organ. They used a monoclonal antibody which recognizes the sequence responsible for pharmacologic activity in all known opioid peptides. They observed consistent cytoplasmic immunoreactivity in pheochromocytomas and other tumors. A few of the neuroendocrine and all of the non-neuroendocrine tumors demonstrated no immunoreactivity.

Changes in the endogenous opioid peptide system caused by painful stimuli were examined by Kudo, Wei, Inoki et al. (1986). The stimulus was adjuvant-induced arthritis in rats. Twelve days after adjuvant inoculation, they measured an increase in peptide content in the adrenal medulla.

Ekman, Bjartell, Ekblad et al. (1987) demonstrated delta sleep-inducing peptide-immunoreactivity in the porcine adrenal medulla and pituitary by radioimmunoassay and immuno-cytochemically. The peptide-reactive cells in the adrenal medulla were identical with a major population of the norepinephrine-containing cells.

An antiserum raised against the synthetic tyrosinylated COOH-terminal sequence of synenkephalin has been used by Corder, Gaillard and Rossier (1987) to characterize the synenkephalin-like immunoreactivity extracted from pheochromocytomas. Their results indicate that the authentic peptide is present in this adrenal medullary tumor.

Panula, Kivipelto, Nieminen et al. (1987) localized a cardioexcitatory peptide derived from mollusks in the adrenal medulla and other organs of the rat. Although the peptide was present in discrete areas of the brain and in other structures, only a few adrenal medullary cells were immunoreactive.

The post-translational processing of preprotachykinins was examined by Kage, Thim, Creutzfeldt et al. (1988). A polyclonal antiserum to preprotachykinin was used in a radioimmunoassay to identify a single immunoreactive peptide in an extract of human pheochromocytoma. Partial microsequence analysis and determination of the amino acid composition of the peptide indicated identity with a fragment of the preprotachykinin molecule. The data indicate the cleavage site of preprotachykinin.

DeBold, Mufson, Menefee et al. (1988) examined human pheochromocytomas for the presence of proopiomelanocortin-like mRNA. They demonstrated that some proopiomelanocortin-like mRNAs in these tumors contain sequences immediately 5' of exon 1 and thus may arise from transcription initiation upstream of the normal site used in the pituitary gland. This probably also occurs in the normal human adrenal medulla.

REGULATION AND PROCESSING

Höllt (1986) reviewed the literature on opioid peptide processing and receptor selectivity. Black, Adler and La Gamma (1986) gave an overview on how mRNA levels for peptides and neurotransmitters are regulated differentially by impulse activity. They pointed out that regulation in the adrenal medullary chromaffin cell is feasible because the availability of appropriate cDNA probes has helped to identify molecular loci governing plasticity resulting from depolarization. They stressed that membrane depolarization and Na$^+$ influx specifically and differentially alter levels of the mRNA coding for transmitter molecules.

Fischer-Colbrie, Iacangelo and Eiden (1988) presented evidence of separate regulation of enkephalin and NPY and chromogranins A and B by neural and humoral factors. They also demonstrated that this regulation occurs at a pretranslational site. Twenty-four hours after splanchnic nerve activity was increased reflexly, levels of mRNAs encoding for enkephalin and NPY were increased, whereas those of mRNAs for chromogranins were unchanged. Bilateral transsection of the splanchnic nerves completely prevented this increase. Hypophysectomy decreased levels of chromogranin A mRNA with respect to controls, suggesting a dependence on hormones of the pituitary-adrenal axis. This could be restored by treatment with dexamethasone. Chromogranin B mRNA levels were not changed by either treatment.

An excellent overview of enkephalin synthesis and secretion throughout the neuroendocrine system was published by Eiden (1987b). Enkephalinergic cells are found throughout the diffuse neuroendocrine system, including the adrenal medulla. In each of these diverse cell types, the enkephalin phenotype is established during development and is modified by the particular environment of the cell. Enkephalin biosynthesis has been studied in several neuroendocrine cell types. Transcriptional, translational, and post-translational factors can play a role in the regulation of enkephalin production. Cyclic nucleotides, glucocorticoids, and calcium all may act to control the overall level of enkephalin biosynthesis pretranslationally, whereas regulation of post-translational processing of proenkephalin seems to be important in determining the pattern of proenkephalin-derived opiate peptides produced in a given tissue.

Konecka (1987) presented an overview of adrenal opioid peptides. The cellular distribution of enkephalins and other peptides was described. The biosynthesis and secretion of opioid peptides from the adrenal medulla was discussed, along with the physiologic role of these secreted peptides.

Kanamatsu, Unsworth, Diliberto et al. (1986) investigated the effect of reflex splanchnic nerve stimulation on proenkephalin A biosynthesis in the rat adrenal medulla. Two hours of insulin-induced hypoglycemia resulted in a decrease in catecholamine levels in the adrenal medulla, accompanied by a 3-fold increase in proenkephalin A mRNA levels in this organ. By 24 hours, the mRNA levels reached maximum, a 15-fold increase over control values. These and other data indicate that reflex splanchnic nerve discharge stimulates proenkephalin biosynthesis, probably at the level of gene expression.

The effects of barium and potassium on the secretion and biosynthesis of enkephalin in bovine adrenal chromaffin cells were examined by Waschek, Dave, Eskay et al. (1987) to determine whether calcium-dependent secretion and biosynthesis are mediated by the same or by different calcium targets within the cell. In the presence of calcium, barium and potassium stimulated the secretion of enkephalin and later increased levels of proenkephalin mRNA. This effect was inhibited by a calcium channel blocker. When extracellular calcium concentration was decreased, secretion elicited by potassium was blocked whereas secretion elicited by barium was enhanced, indicating that barium wholly substitutes for extracellular calcium in mediating peptide secretion. On the other hand, stimulation of proenkephalin mRNA by both potassium and barium was inhibited when the extracellular calcium concentration was decreased. Waschek et al. (1987) concluded that calcium acts at two different intracellular targets to activate secretion and not the biosynthesis of enkephalin. Parallel results were seen in the pituitary gland.

Waschek and Eiden (1988) examined the calcium requirements for barium stimulation of enkephalin and VIP biosynthesis in adrenal chromaffin cells. Barium chloride stimulated the secretion of [Met]enkephalin in a dose-dependent manner and increased the total amounts of [Met]enkephalin and VIP in cultured bovine cells. A greater than 6-fold increase in proenkephalin mRNA was observed by 24 hours after barium stimulation. The calcium channel blocker D-600 inhibited the barium-stimulated secretion of enkephalin and blocked the stimulation of synthesis of VIP and the mRNA for enkephalin. Decreasing the calcium concentration in the medium enhanced the barium-stimulated release of both peptides but blocked the induction of their biosynthesis. The data indicate that calcium targets involved in secretion can be activated by barium or calcium but calcium targets involved in biosynthesis specifically require calcium. Waschek and Eiden (1988) proposed that pathways leading to peptide secretion and biosynthesis in the adrenal medulla diverge just after secretagogue-stimulated calcium influx.

Pruss and Stauderman (1988) reported that the increased

enkephalin synthesis after potassium depolarization can be blocked by a phorbol ester that increases protein kinase C activity. To explain these results, they measured the effect of the phorbol ester on changes in intracellular free calcium in response to potassium-induced depolarization. The data suggest that phorbol ester activation of protein kinase C inhibited the sustained opening of dihydropyridine-sensitive calcium channels and consequently abolished the prolonged increase in calcium level that occurs after depolarization. The ability of phorbol ester to decrease intracellular calcium rapidly after potassium treatment explains why treatment with the phorbol ester inhibits potassium-stimulated calcium-dependent enkephalin synthesis.

Kley (1988) investigated the role of protein kinase C-mediated processes in the regulation of proenkephalin gene expression in primary cultures of bovine adrenal chromaffin cells. Activators of protein kinase C induced a time-dependent increase in proenkephalin mRNA levels. This increase was potentiated by increasing levels of calcium, suggesting an interaction between protein kinase- and calcium-mediated processes in the regulation of proenkephalin mRNA. At least three receptor-coupled second-messenger systems were implicated in preproenkephalin gene expression.

Using the cholinergic agonist 1,1-dimethyl-4-phenylpiperazinium, Cherdchu and Hexum (1988) characterized an increase in proenkephalin processing in bovine chromaffin cells. This agonist increased the secretion of [Met]enkephalin into the culture medium without a decrease in the cellular content, suggesting the replenishment of peptides. The repletion of cellular [Met]enkephalin immunoreactive material appeared to result from an increase in processing of large peptides containing enkephalins. Using several lines of evidence, they concluded that acute stimulation of nicotinic receptors on the chromaffin cell enhances the processing of proenkephalin precursors to keep pace with the secretion of low molecular weight peptides.

Concentrations of mRNA coding for proenkephalin A were measured in serum-free cultures of bovine adrenal chromaffin cells by Kley, Loeffler, Pittius et al. (1986). Use of a sensitive solution hybridization assay revealed increases in the mRNA levels with potassium, barium, and veratridine. The highest increase (3-fold) was obtained with the sodium channel agonist veratridine, and this effect was inhibited by a sodium channel antagonist. Increases in mRNA were potentiated by a calcium channel agonist and inhibited by calcium channel antagonists. Increases in mRNA were associated with an increase in immunoreactive proenkephalin A-derived peptides, indicating an enhanced production of opioid peptides. These results suggest that membrane depolarization may play an important role in the regulation of proenkephalin A gene expression in bovine adrenal chromaffin cells.

To investigate the possibility that the opioid peptide precursor proenkephalin A was glycosylated, Watkinson, Dockray and Young (1988) utilized an antiserum raised against the carboxyl terminus of a [Met]enkephalin octapeptide to identify and characterize enkephalin-containing peptides from extracts of bovine adrenal medulla. This allowed them to isolate and characterize peptides extended from the amino terminus. They presented evidence that precursor polypeptide proenkephalin A can be glycosylated during translation in the rough endoplasmic reticulum.

Kley, Loeffler, Pittius et al. (1987) analyzed the involvement of sodium and calcium channels in the stimulatory effect of nicotine and cyclic adenosine monophosphate (cAMP) on levels of mRNA for proenkephalin A in cultured bovine adrenal chromaffin cells. They provided evidence suggesting that both stimuli involve the activation of sodium or calcium channels and that the nicotinic effect is more likely due to an activation of calcium channels and changes in calcium flux than to changes in intracellular cAMP.

The effect of histamine on levels of mRNA for proenkephalin A was studied in serum-free cultures of bovine adrenal chromaffin cells by Kley, Loeffler and Höllt (1987). Histamine increased the levels up to a maximum response of 5-fold, and the response could be abolished by an H_1 receptor antagonist but not by an H_2 receptor antagonist. Calcium channel blockers partially decreased the response. Muscarinic receptor stimulation did not alter the mRNA levels. These data show that levels of mRNA for proenkephalin A may be modulated by activation of the H_1 receptor and that this effect is partially dependent on the activation of voltage-dependent calcium channels.

Maret and Faucéhre (1986) detected a heterogeneity in prohormone processing activity among bovine chromaffin vesicles. When chromaffin vesicles were stratified by differential centrifugation, a subpopulation of vesicles was found to be enriched about 10-fold in prohormone processing activity. This enhanced activity is not due to lysosomal contamination. This subpopulation was of relatively low density and probably consisted of immature vesicles.

Bloch, Popovici, Le Guellec et al. (1986) reviewed their experience with *in situ* hybridization histochemical analysis of gene expression in the endocrine and nervous systems. As an example, they reviewed techniques by which they detected cells in the adrenal medulla and brain that expressed the preproenkephalin A gene. Their experience shows that *in situ* hybridization is as sensitive as immunohistochemical methods in detecting peptide-producing cells in the adrenal medulla.

The mRNA coding for preproenkephalin A was detected in the rat adrenal medulla and brain by Bloch, Popovici, Chouham et al. (1986) using *in situ* hybridization. mRNA for preproenkephalin A was present in cells that did not demonstrate immunoreactivity for enkephalin. They concluded that *in situ* hybridization is an efficient technique for detecting the sites of synthesis of preproenkephalin A.

Giraud, Affolter, Hotchkiss et al. (1986) showed that drugs mimicking the effects of cAMP increase enkephalin biosynthesis in cultured bovine adrenal chromaffin cells by increasing the mRNA for enkephalin, but they are ineffective in increasing enkephalin secretion. Nicotine, which causes a rapid release of enkephalin from these cells, also induces increases in enkephalin biosynthesis and in enkephalin mRNA. Reserpine enhances enkephalin precursor processing. The actions of these different pharmacologic treatments show that enkephalin biosynthesis is regulated at different levels in adrenal chromaffin cells.

Cherdchu, Scholar and Hexum (1987) demonstrated that acute stimulation of bovine adrenal chromaffin cells with nicotinic agonists or high potassium concentration gives rise to a significant increase in the release of enkephalin-like material into the medium. The cellular content of enkephalins remained constant after the actual release induced by these secretagogues, suggesting replenishment of the cellular peptides. Both secretion and repletion of cellular enkephalin were calcium-dependent. Cherdchu et al. (1987) concluded that secretion and repletion of enkephalins are tightly coupled and activated by nicotinic receptor stimulation.

Using the rabbit reticulocyte cell-free translation system, Antreassian, Benlot, Thibault et al. (1986) compared the relative proportions of mRNAs of three proteins in human pheochromocytoma: proenkephalin A, tyrosine hydroxylase, and dopamine β-hydroxylase. Tyrosine hydroxylase expression appeared to be rather constant. In contrast, expression of proenkephalin A and dopamine β-hydroxylase was more variable. The ratios between the mRNAs could suggest a coordination in the expression of proenkephalin A and dopamine β-hydroxylase.

Birch, Davies and Christie (1987) used an antiserum that recognizes all the major intermediates of bovine proenkephalin to characterize and to assess the stability of intermediates of proenkephalin in intact chromaffin vesicles as well as soluble and membrane components of the vesicles. Using a combination of immunologic techniques, they showed that processing was slower in intact vesicles than in lysates. Preliminary evidence of the regulation of the processing of a 27 kDa intermediate peptide by soluble factors present in chromaffin vesicles was obtained. It appears that membrane-associated intermediates of proenkephalin are relatively stable.

Lindberg (1986) used antisera directed toward eight different portions of the proenkephalin molecule to examine the processing rates and patterns of proenkephalin-derived peptides in chromaffin cell cultures in the presence and absence of reserpine. Reserpine treatment produced profound effects on the molecular weight profile of nearly all enkephalin-containing peptides. Among other observations, the production of two amidated opioid peptides, amidorphin and metorphamide, was greatly accelerated in the presence of reserpine. The increased levels of low molecular weight enkephalins were not due to decreased basal release. The results indicate that reserpine treatment is able to increase the extent of post-translational processing of proenkephalin within chromaffin cells.

The potential phosphorylation of peptide B, a proenkephalin-derived peptide, was investigated in primary cultures of bovine adrenal chromaffin cells and fresh adrenal medullary tissue by D'Souza and Lindberg (1988). Cultures were labeled with radioactive phosphate and extracts were subjected to immunoprecipitation. These and studies with fresh tissue demonstrated that more than 85% of the peptide B molecules present in the bovine adrenal medulla are phosphorylated. The peptide was phosphorylated at multiple sites in the acidic regions of the peptide.

Wilson (1987a) observed that reserpine increases chromaffin cell enkephalin stores without a concomitant decrease in other proenkephalin-derived peptides. Reserpine increased the levels of enkephalins and of enkephalin-containing peptides as large as 3 kDa without decreasing the levels of larger enkephalin-containing peptides in cultured bovine adrenal chromaffin cells. Similar results were obtained with another catecholamine-depleting drug, tetrabenazine. In contrast, treatment of chromaffin cells with theophylline or forskolin increased the levels of enkephalins and related peptides of all sizes. The results suggest that new synthesis of proenkephalin is required for the effects of reserpine, although proenkephalin processing is also altered by this drug.

Wilson (1987b) investigated the effects of substance P and VIP on the levels of enkephalin-containing peptides in cultured bovine adrenal chromaffin cells. This is of interest because substance P and VIP exist in the splanchnic nerve terminals innervating adrenal chromaffin cells. Cellular stores of enkephalin-containing peptides were increased by both peptides. The results suggest that substance P and VIP may be transsynaptic modulators of enkephalin-containing peptides within chromaffin cells.

Adams and Boarder (1987) reported on investigations of the effects of reserpine and agents that increase cellular cAMP levels on the level and release of [Met]enkephalin-Arg^6-Phe^7 immunoreactivity in cultured bovine adrenal chromaffin cells. Treatments that increase cAMP increase the amount of peptide immunoreactivity in these cells. Treatment with reserpine produces no change in total level of immunoreactive peptide but does result in increased accumulation of the heptapeptide [Met]enkephalin-Arg^6-Phe^7 at the expense of immunoreactivity that appears to be peptide B.

In cultured bovine adrenal chromaffin cells, changes in the dynamic state of enkephalin stores elicited experimentally were studied by Naranjo, Mocchetti, Schwartz et al. (1986) by measuring cellular proenkephalin mRNA as well as enkephalin precursors and authentic enkephalin content of cells and culture media. In parallel, tyrosine hydroxylase mRNA and catecholamine content were also determined. Low concentrations of the steroid dexamethasone increased the cell contents of proenkephalin mRNA and enkephalin-containing peptides. Tyrosine hydroxylase mRNA and catecholamines were increased at higher concentrations of the steroid. These and other experiments demonstrated that dexamethasone exerts a specific permissive action on the stimulation of the proenkephalin gene elicited by depolarization. Although the catecholamines and enkephalins are localized in the same chromaffin vesicles and are co-released, the genes coding for the processes that are rate-limiting in the production of these neuromodulators apparently can be regulated differentially.

La Gamma and Adler (1987) sought to determine whether adrenal opioid peptides are subject to the same hormonal control as catecholamine biosynthesis. Pharmacologic destruction of the adrenal cortex resulted in a decrease in [Leu]enkephalin level *in vivo*. This suggests a tonic regulatory effect of adrenal cortical steroids on enkephalin synthetic pathways. They found that explanted medullas required medium supplemented with dexamethasone or corticosterone to achieve maximal levels of [Leu]enkephalin. The effects of glucocorticoid treatment were blocked by specific antagonists. Based on these and previous results, it appears that adrenal opioid peptides are subject to dual hormonal and transsynaptic regulatory influences in a manner similar to catecholamines.

Yoburn, Franklin, Calvano et al. (1987) showed that the denervation-induced increase in enkephalin-containing peptides in the adrenal medulla can be blocked by hypophysectomy and partially reinstated by corticosterone, dexamethasone, or adrenal corticotropin treatment. In the intact rat, intermediate doses of corticosterone or dexamethasone decreased the denervation-induced increase in peptides; a high dose of dexamethasone restored this response. These results indicate that glucocorticoids exert a permissive effect *in vivo* on the denervation-induced stimulation of enkephalin-containing peptide biosynthesis.

In follow-up studies, Inturrisi, Branch, Robertson et al. (1988) and Inturrisi, Franklin, Shapiro et al. (1988) further established the role of glucocorticoids in regulating enkephalins in the adrenal medulla. The effect of dexamethasone on enkephalin-containing peptide levels and preproenkephalin mRNA levels was determined in adrenal medullary explants from sham and hypophysectomized rats. Culture without dexamethasone resulted in 13- and 4-fold increases in enkephalin-containing peptide levels in sham and hypophysectomized glands, respectively. The addition of dexamethasone produced a large increase in peptides in both explants. This effect could be partially blocked with a glucocorticoid antagonist. Dexamethasone resulted in a 6-fold increase in preproenkephalin mRNA in explants from hypophysectomized animals. These and other results indicated that glucocorticoid treatment is required for maximal proenkephalin gene expression and biosynthesis of enkephalin-containing peptides.

The effects of hypophysectomy and dexamethasone on the levels of [Met]enkephalin precursors and proenkephalin mRNA were examined in the adrenal medulla and superior cervical ganglia by Stachowiak, Lee, Rigual et al. (1988). They found that at 3 weeks after hypophysectomy the proportion of mRNA for proenkephalin to the total RNA content in the adrenal medulla was not significantly changed, whereas the total immunoreactive [Met]enkephalin was decreased by 45%. Dexamethasone increased the relative abundance of mRNA for proenkephalin and the total [Met]enkephalin immunoreactivity in hypophysectomized animals. Their findings indicate that the pituitary-adrenal cortical axis controls the relative proportions of [Met]enkephalin to its precursors and that this control involves a glucocorticoid-independent mechanism in the adrenal medulla. Their studies did not suggest specific control of proenkephalin synthesis by the pituitary adrenal cortical axis.

La Gamma and Adler (1988) looked for the existence of a transsynaptic effect on opioid peptides and catecholamines in the adrenal medulla of rats during development. Adrenal denervation increased [Leu]enkephalin levels, but this was not observed in denervated or explanted medulla from neonatal rats. Prolonged denervation prevented even the normal maturational increase in [Leu]enkephalin. These and other results indicated that, during development, adrenal opioid peptides are not under transsynaptic control but require presynaptic terminals to mature normally. Therefore, like catecholamines, co-localized adrenal opioid peptides require presynaptic regulatory signals to achieve normal development and function.

La Gamma, White, McKelvy et al. (1988) sought to determine whether the depolarization blockade of the increase in [Leu]enkephalin and preproenkephalin mRNAs is mediated by calcium ions. Explanted medullas were depolarized in the presence of a calcium chelator or calcium channel blocker. Inhibition of calcium ion influx prevented the effects of potassium-induced depolarization on the increase in [Leu]enkephalin and preproenkephalin. Increasing intracellular calcium levels, in the absence of depolarizing agents, reproduced the effects of depolarization. There apparently is an absolute requirement for calcium. These and other data suggest that inhibitory effects of transsynaptic activity and membrane depolarization on adrenal enkephalin occur through calcium and perhaps through a protein kinase C-dependent pathway, mechanisms known to augment catecholamine biosynthesis. It appears that the same or similar molecular mechanisms may result in differential regulation of these co-localized transmitter systems.

Inturrisi, La Gamma, Franklin et al. (1988) characterized enkephalins in rat adrenal medullary explants. The increase in enkephalin-containing peptides began after 1 day in culture and reached a peak by 7 days. This apparently is due to an accumulation of proenkephalin. Culturing the explants in the presence of depolarization concentrations of potassium prevented the accumulation of the proenkephalin-like peptide. Thus, both denervated and explanted rat adrenal medullae accumulate proenkephalin which, in turn, is subject to regulation by depolarization stimuli.

Byrd, Naranjo and Lindberg (1987) demonstrated that the differentiation promoter sodium butyrate increases the content of enkephalin-containing immunoreactive peptides in PC12 cells which, unlike mature adrenal chromaffin cells, contain very low levels of opioid peptides. These butyrate-induced enkephalin-immunoreactive peptides, which are specific products of the proenkephalin gene, consist principally of two forms with molecular masses of 20 kDa and 10 kDa. In PC12 cells, the butyrate-stimulated increase in the content of enkephalin-immunoreactive peptides may involve changes in transcription because the peptide increase is preceded by an increase in proenkephalin mRNA. Byrd et al. (1987) suggested that the PC12 cell may be useful for investigating the factors that control the initial production and processing of proenkephalin-derived peptides.

A quantitative assessment of the ontogeny of [Met]enkephalin, epinephrine, and norepinephrine in the human fetal adrenal medulla was presented by Wilburn and Jaffe (1988). By radioimmunoassay, [Met]enkephalin was detectable in human fetal adrenals at 11 weeks gestation. Both epinephrine and norepinephrine were low until 16 weeks gestation and then increased markedly by 21 weeks. Concentrations of epinephrine and norepinephrine were approximately equal throughout gestation. Their data demonstrated that enkephalins are present in the human fetal adrenal medulla by 11 weeks and suggest that the fetal adrenal gland may be capable of secreting enkephalins as well as catecholamines.

Jaffe, Mulchahey, Di Blasio et al. (1988) reviewed the regulation of peptides in the pituitary and its target tissues, including the adrenal medulla, in the primate fetus during growth. They pointed out that relatively little work has been done on the fetal adrenal medulla, whereas the adrenal cortex of the primate fetus has attracted more attention. Among other things, they pointed out that [Met]enkephalin levels in the human fetal adrenal medulla increased to an initial peak at about 15 weeks' gestation, decreased somewhat, and then increased again over the gestational period. The concentration of [Met]enkephalin varied much more during the period of the initial increase, suggesting that adrenal enkephalin production is less tightly regulated during this early period.

Zurawski, Benedik, Kamb et al. (1986) pointed out that enkephalin may be serving as a signal to mediate interactions between the neuroendocrine and immune systems. They showed that activation of mouse T-helper cells induces abundant preproenkephalin mRNA synthesis. The production, by activated T cells, of a peptide neurotransmitter identifies a signal that potentially can permit T cells to modulate the nervous system.

Busby, Quackenbush, Humm et al. (1987) investigated the mechanism of the post-translational conversion of glutamine to pyroglutamic acid on the NH_2 terminus of newly synthesized peptides. They developed an assay to demonstrate that the spontaneous cyclization of the NH$_2$-terminal glutamine occurs only slowly under physiologic conditions. Furthermore, they identified an enzyme that may be responsible for this conversion. In the adrenal medulla, this enzyme appears to be contained in the soluble fraction in the chromaffin vesicle.

Maret, Blanot, Welti et al. (1987) investigated the action of extracts from bovine chromaffin vesicles on synthetic substrates containing lysine and arginine pairs as the central sequence. Two compounds were recognized as substrates for proenkephalin-processing enzymes and were used for the development of a quantitative assay for the enzymes present in chromaffin vesicles.

Tischler, Lee, Costopoulos et al. (1986) examined the cooperative regulation of neurotensin in PC12 cells by nerve growth factor, dexamethasone, and activators of adenylate cyclase. The three factors act cooperatively to increase the content of neurotensin; relatively small increases in neurotensin occur in the presence of nerve growth factor alone or dexamethasone alone but not of forskolin or cholera toxin alone. A combination of the growth factor and either of the other two factors increases neurotensin up to 100-fold. With all three factors, neurotensin content increases up to 600-fold.

In a study to characterize some of the proteases involved in opioid peptide processing, Lindberg (1987) investigated a serine protease that had been purified from bovine adrenal chromaffin vesicles. She specifically wanted to determine if this protease is related to the kallikrein family of enzymes. The data indicate that the adrenal serine protease is involved in the biosynthesis of enkephalins but is not closely related to the kallikreins.

Hook (1988) reviewed the evidence for the role of carboxy-peptidase H in the regulation of neuropeptide biosynthesis. The different mechanisms by which this enzyme can be inhibited and activated support a regulatory role for this peptide-processing enzyme.

An antiserum against carboxypeptidase H, an enzyme involved in the biosynthesis of neuropeptides such as enkephalins, was used to identify the enzyme proteins in bovine chromaffin vesicles by Laslop, Fischer-Colbrie, Hook et al. (1986). Two-dimensional immunoblots and lectin blots revealed that two closely migrating glycoproteins, previously named J and K, represented the enzyme. The same protein doublet was present in the vesicle membrane preparation and in the soluble lysate from chromaffin vesicles; however, the membrane contained significantly more enzyme protein.

Hook and La Gamma (1987) demonstrated that carboxypeptidase H can be inhibited by its peptide products at concentrations similar to those seen within chromaffin vesicles. Inhibition by amidated peptide products (vasopressin, oxytocin, luteinizing hormone-releasing hormone, substance P, and thyrotropin-releasing hormone) shows that the final products of the precursor processing pathway can regulate carboxypeptidase H. The enzyme was inhibited by proenkephalin-derived peptides but at higher concentrations. Hook and La Gamma (1987) suggested that product inhibition of carboxypeptidase H and perhaps other processing enzymes may serve to limit the peptide concentration within chromaffin vesicles.

Fricker, Evans, Esch et al. (1986) cloned and determined the sequence of cDNA for bovine carboxypeptidase E. This peptidase, formerly called enkephalin convertase, was first identified in bovine adrenal chromaffin vesicles. The predicted amino acid sequence of the cDNA clone contains the partially determined sequences of carboxypeptidase E and several pairs of basic amino acids and displays some homology with carboxypeptidases A and B. Restriction analysis suggests that there is only one gene for carboxypeptidase E. This is consistent with a broad role for the enzyme in the biosynthesis of many neuropeptides.

A neutral, divalent cation-sensitive endopeptidase was characterized by Supattopone, Strittmatter, Fricker et al. (1988). This trypsin-like endopeptidase was partially purified from bovine adrenal chromaffin vesicles and other tissues. The purified enzyme had a dimeric structure with subunit molecular masses of 40 and 42 kDa and a native molecular mass of 85 kDa. It was suggested that this endopeptidase may have a role in neuropeptide processing.

Adrenal enkephalin and enkephalin-containing peptides were studied during postnatal development in normotensive and spontaneously hypertensive rats by Vindrola, Aloyz, Ase et al. (1988). Both total and free [Met]enkephalin contents in the adrenal of the spontaneously hypertensive rats were significantly diminished at all ages tested. The analysis showed that spontaneously hypertensive rats displayed decreased levels of high and low molecular weight materials at 11 days and 16 weeks of age; however, levels of intermediate molecular weight compounds were high in the glands of these animals. These and other results presented suggest that the basic alteration of the adrenal proenkephalin system in spontaneously hypertensive rats may be due to a genetic decrease in proenkephalin levels. An alternative explanation is that the free enkephalin decrease could be related to changes in nerve input to the adrenal medulla.

The ability of human plasma kallikrein to hydrolize several
proenkephalin-derived peptides was studied by Metters, Rossier,
Paquin et al. (1988) to establish whether this enzyme also displays
strict selectivity toward cleavage sites within proenkephalin.
Several proenkephalin-derived peptides were purified from
bovine adrenal chromaffin vesicles and used as substrates. All of
the identified cleavages occurred either on the carboxyl terminal
side or between pairs of basic amino acids, with plasma kallikrein
recognizing processing signals between lysine and arginine. In
addition, plasma kallikrein was found to cleave at the carboxyl
terminus of the basic pairs of amino acids preceding enkephalin
sequences, thereby releasing the biologically active form of the
peptide with the free amino terminal tyrosine needed for receptor
recognition.

CHAPTER 11

UPTAKE MECHANISMS

<u>UPTAKE INTO CHROMAFFIN VESICLES</u>

Two excellent reviews cover the recent developments on mechanisms of uptake into secretory vesicles, particularly the uptake of catecholamines into chromaffin vesicles. Njus, Kelley and Harnadek (1986) emphasized the bioenergetics involved in uptake mechanisms. Kanner and Schuldiner (1987) reviewed mechanisms of uptake and storage. A comprehensive review of uptake mechanisms was presented by Johnson (1988). The uptake of catecholamines into adrenal chromaffin vesicles was cited as a model system for the study of hormone and neurotransmitter uptake and storage. It was pointed out that the molecular mechanism for uptake is just beginning to be characterized.

Henry, Scherman, Roisin et al. (1986) examined the molecular pharmacology of the catecholamine transporter of bovine adrenal chromaffin vesicles. Two inhibitors of the monoamine transporter, tetrabenazine and reserpine, bind to sites on the vesicle membrane with different affinities. The two sites are involved in monoamine translocation. Two different substrates are translocated by the same transporter, which confirms the existence of two sites with different properties.

Gasnier, Roisin, Scherman et al. (1986) showed that *m*-iodobenzylguanidine, an adrenal imaging agent used clinically in detection of pheochromocytoma, is a substrate for the monoamine uptake system of adrenal chromaffin vesicles. It is accumulated by chromaffin vesicles in the presence of adenosine triphosphate (ATP), and it can be released by osmotic shock. The energetics and kinetics of uptake are similar to those for norepinephrine. It is apparent that this imaging agent is taken up and stored in the adrenal medulla in a fashion similar to that for monoamines.

Scherman and Weber (1987) showed that tritiated dihydrotetrabenazine, a specific ligand for the monoamine transporter in the chromaffin vesicle membrane, also can be detected in cultured sympathetic neurons from newborn rats. Two other neuronal structures, the electric organ of the torpedo and the ciliary ganglion of the chick embryo, did not contain measurable amounts of binding sites. The monoamine transporter such as that in adrenal chromaffin vesicles is present in some, but not all, neurotransmitter-containing vesicles.

Gasnier, Ellroy and Henry (1987) investigated the possibility that the monoamine transporter has an oligomeric structure by using the radiation inactivation technique, assaying the transporter by its capacity to bind tritiated dihydrotetrabenazine and tritiated reserpine. This technique permits determination of the functional molecular mass of membrane-bound proteins *in situ*. They were able to estimate the functional molecular mass of tetrabenazine binding sites as 68 kDa. The binding site for reserpine was about 37 kDa. They concluded that the monoamine transporter of chromaffin vesicles has an oligomeric structure. Furthermore, using these techniques, they determined that cytochrome b_{561} and dopamine β-hydroxylase have functional molecular masses of 25 kDa and 123 kDa, respectively.

Gasnier, Scherman and Henry (1987) investigated changes in the activity of catecholamine transporter during the preparation of chromaffin vesicle membrane "ghosts." The reserpine binding activity of the transporter decreases during ghost preparation. In contrast, the number of binding sites for dihydrotetrabenazine is constant. Dihydrotetrabenazine apparently binds to an inactive transporter, whereas reserpine binds only to active sites. Inactivation occurs during lysis of the vesicles, possibly because of incomplete resealing. The ratio of reserpine binding to dihydrotetrabenazine binding may be used as an estimate of the proportion of correctly resealed vesicles.

The hydrophobicity of the tetrabenazine-binding site of the chromaffin vesicle monoamine transporter was demonstrated by Scherman, Gasnier, Jaudon et al. (1988). They tested various drugs of different lipophilicities to see how they would displace tritiated dihydrotetrabenazine. The potency of a compound to displace the dihydrotetrabenazine from its binding site was correlated with its lipophilicity. The data indicated that the binding site is hydrophobic and is in equilibrium with the ligand present in the membrane phase.

Deupree and Hitchcock (1988) examined the effect of the ATPase inhibitor N-ethylmaleimide on the binding of reserpine to the transporter in chromaffin vesicle membranes. Specifically, they sought to determine if the inhibition of catecholamine uptake is due to the action of this inhibitor on ATPase or on the catecholamine transporter. They demonstrated that low concentrations of the inhibitor decreased binding of reserpine to the transporter. This inhibition could be blocked by ATP, and the inhibition probably was due to an indirect effect of N-ethylmaleimide on the protein translocater. These results support the concept that reserpine is binding to the catecholamine site on the vesicle membranes and that binding of reserpine requires the presence of a membrane potential.

The *in vivo* storage relationship between catecholamines and ATP in chromaffin vesicles of cultured bovine adrenal chromaffin cells was investigated by Caughey and Kirshner (1987). Three-day treatment with reserpine and tetrabenazine to cause a 90% depletion of catecholamines resulted in a 45% decrease in cellular ATP content. The ATP apparently is lost from the chromaffin vesicle pool. These inhibitors of catecholamine uptake also interfered with adenosine uptake into the cells and vesicles. The observed loss of adenine nucleotides from catecholamine-depleted vesicles *in vivo* is evidence that interactions between ATP and catecholamines are important in the vesicular storage of high concentrations of these compounds.

Parti, Özkan, Harnadek et al. (1987) used reserpine and reserpine derivatives to probe the norepinephrine transporter in bovine adrenal chromaffin vesicles. Reserpine consists of a trimethoxybenzoyl group esterified to an alkaloid ring system. Derivatives of reserpine that include the alkaloid ring system competitively inhibited norepinephrine transport into chromaffin vesicles, although not as efficiently as reserpine. Reserpine derivatives that did not incorporate the ring system did not inhibit either norepinephrine transport or reserpine binding. Therefore, the amine binding site of the catecholamine transporter appears to bind to the alkaloid ring system of reserpine rather than to the trimethoxybenzoyl moiety. The more-potent inhibitors are the more hydrophobic compounds, suggesting that the binding site is hydrophobic.

The effects of anions and membrane potential on the reconstituted proton pump from chromaffin vesicles were investigated by Moriyama and Nelson (1987c). When acetate was present inside the vesicles, ATP-dependent proton uptake was dependent on external chloride. Substantial protein uptake was observed in the absence of external chloride when chloride or sulfate was present inside the vesicles. An inside-negative membrane potential drove ATP-dependent proton uptake regardless of the anion species present. Moriyama and Nelson (1987c) concluded that the internal anion binding site and the membrane potential regulate the proton pumping activity of the ATPase.

King, Ellenberger and Goldin (1988) presented biochemical and immunologic evidence for a calcium pump in chromaffin vesicles. They demonstrated that the electrochemical gradient is not needed for the uptake of radioactive calcium ions into chromaffin vesicles and observed that chromaffin vesicles bind to a monoclonal antibody against a calcium pump from bovine brain. They suggest that chromaffin vesicles contain a calcium-translocating ATPase that directly catalyzes the uptake of calcium.

Weinbach, Costa, Nelson et al. (1986) investigated the effects of imipramine and chlorimipramine on energy-linked reactions in mitochondria and chromaffin vesicles. The tricyclic antidepressant drugs weakly inhibited the ATPase activity of chromaffin vesicle membranes and enhanced the ATPase activity of mitochondria. Weinbach et al. (1986) suggested that these divergent effects may be related to the hydrophobicity of the drugs.

Daniels and Reinhard (1988) reported that MPP$^+$ is actively transported into isolated chromaffin vesicles, apparently through the catecholamine carrier. The uptake is completely blocked by reserpine and is dependent on the potential across the chromaffin vesicle membrane. Furthermore, it was shown that a permanently charged compound is capable of entering the vesicle through the catecholamine carrier.

Darchen, Scherman, Desnos et al. (1988) also demonstrated that MPP$^+$ is a substrate of the monoamine uptake system of chromaffin vesicles. They found that uptake of MPP$^+$ occurs through the monoamine transporter because it is completely inhibited by serotonin, and MPP$^+$ competitively inhibits serotonin uptake. Also, MPP$^+$ efficiently displaces tritiated reserpine and dihydrotetrabenazine from their binding sites on the transporter.

Darchen, Scherman, Laduron et al. (1988) examined the binding of tritiated ketanserin, a specific ligand for serotonin receptors, to purified chromaffin vesicle membranes. Binding was found to occur on one class of sites and was temperature dependent. Tetrabenazine displaced ketanserin binding. Conversely, ketanserin inhibited the binding of tritiated dihydrotetrabenazine. This inhibition was of the competitive type, indicating that both drugs bind to the same site. It was concluded that nonspecific displaceable binding sites for ketanserin, previously described in the striatum, are tetrabenazine binding sites associated with the monoamine transporter.

Born (1988) reviewed some of the similarities in uptake and storage mechanisms between platelets and adrenal chromaffin cells. For example, catecholamines are co-stored with ATP in chromaffin vesicles, and serotonin is co-stored with ATP within platelets.

Fröhlich (1988) reviewed the "tunneling" mode of biologic carrier-mediated transport. Tunneling is used to mean the movement of a substrate along a narrow, channel-like pathway within a biologic carrier protein. During this movement the substrate traverses the protein without undergoing any significant conformational change. Evidence for this mode being involved is more extensive for anion transport than for transport of cations. It would be interesting to determine if the "tunneling" mode is operative within chromaffin vesicles.

UPTAKE INTO CHROMAFFIN CELLS

The uptake and release of catecholamines was investigated in the isolated perfused adrenal gland of the rat by Wakade, Malhotra, Wakade et al. (1986). Large quantities of tritium were found in the adrenal medulla after either intravenous injection of tritiated norepinephrine in the intact animal or perfusion of the isolated gland with tritiated norepinephrine. Agents that enhanced the secretion of catecholamines did not enhance the release of tritium. Other agonists enhanced the secretion of tritium but not catecholamines. This led to the conclusions that chromaffin cells do not possess the norepinephrine uptake mechanism and that the uptake of tritiated norepinephrine occurs mainly in nerve terminals present in the adrenal gland.

Banerjee, Lutz, Levine et al. (1987) investigated the biochemistry and pharmacology of the catecholamine transport system in the chromaffin cell plasma membrane, using knowledge of the chromaffin vesicle transport system. They found evidence of multiple catecholamine transport sites at the cell surface, of which only one proved to be sensitive to cocaine. The cocaine-sensitive site has a high affinity for catecholamines and depends on sodium in the medium. The cocaine-insensitive sites have a low affinity for norepinephrine. Both classes of transport sites in the membrane of the cultured bovine adrenal chromaffin cell appear to be different from the catecholamine transport sites in the chromaffin vesicle membrane. This is based on differences in substrate specificity and sensitivity to inhibitors.

Jaques, Tobes and Sisson (1987) compared the uptake of the scintigraphic imaging agent *m*-iodobenzylguanidine to that of norepinephrine in primary cultures of human pheochromo-cytomas. Two different uptake systems were identified for both the guanidine analog and norepinephrine. One system was sodium-dependent and appeared to be an uptake-1 system. A sodium-independent uptake system was also observed and most likely is a passive diffusion process. Comparative inhibition studies demonstrated that both compounds share a common uptake system.

Miras-Portugal, Torres, Rotlan et al. (1986) investigated adenosine transport in cultured bovine adrenal chromaffin cells. They presented the kinetics for this high-capacity high-affinity uptake system. They showed clearly that adenosine can be taken up from the extracellular medium in order to restore or replenish the nucleotide pool lost by exocytosis after stimulation.

In a follow-up study, Torres, Molina and Miras-Portugal (1986) used an analog of dipyridamole to quantify the adenosine transporters in cultured bovine adrenal chromaffin cells. They found that adenosine transport could be inhibited by this analog with the same effectiveness as described for the nonmodified compound. Binding studies showed that about 630,000 receptors are located on each cell.

Adenosine transporters in freshly isolated and cultured chromaffin cells were quantified by Torres, Delicado and Miras-Portugal (1988). Using the tritiated dipyridamole binding technique, they characterized receptors for adenosine in the plasma membrane of chromaffin cells and found the binding capacities to be similar to those in chromaffin vesicle membranes. Using other radioligands and photoaffinity labels, they found adenosine transporters of 115-, 58-, and 42 kDa in the plasma membrane. Chromaffin vesicle membranes had only 105- and 51 kDa molecular species.

The kinetics of the uptake of norepinephrine into PC12 cells was studied by Friedrich and Bönisch (1986). Extensive studies with various concentrations of Na^+ and Cl^- in the medium and replacement of these ions suggest that sodium and chloride are co-transported with norepinephrine.

Bönisch and Harder (1986) characterized the binding of tritiated desipramine, an uptake inhibitor, to the norepinephrine transporter of PC12 cell plasma membranes and compared the characteristics of binding with those of transport of norepinephrine in PC12 cells. They found and studied a specific binding site for desipramine. The desipramine binding site appears to be the neuronal norepinephrine transporter. This binding site was further characterized by Schmöig and Bönisch (1986). They solubilized the binding sites with the nonionic detergent digitonin. The binding characteristics were essentially unaltered by solubilization, allowing characterization of the binding site.

Characterization of xylamine binding to proteins of PC12 cells was reported by Koide, Cho and Howard (1986). Xylamine, an irreversible inhibitor of norepinephrine uptake that interacts with the norepinephrine transporter, irreversibly inhibited uptake into PC12 cells. Xylamine apparently is transported into the cells and binds covalently to intracellular proteins.

Bitler, Zhang and Howard (1986) isolated variants of PC12 cells that are deficient in transporter-mediated uptake of dopamine.

Most of the variants appear to be similar to PC12 cells, except for the defect in catecholamine uptake. One variant behaves like a multiple deletion mutant in that it fails to exhibit several processes found in PC12 cells. This variant, also lacking several proteins and mRNA species present in PC12 cells, offers the opportunity to identify previously unrecognized proteins involved in neurotransmitter uptake, storage, and release.

Matsuoka, Satake and Kurihara (1986a) demonstrated that choline accumulates in PC12 cells through two uptake systems. Culture of PC12 cells in the presence of glioma-conditioned medium led to a 5-fold increase in the velocity of one of the uptake systems. The norepinephrine uptake decreased. From these and other observations, it was concluded that glioma-conditioned medium contains factors that induce the cholinergic neuronal differentiation of PC12 cells.

The characteristics and regulatory nature of sugar transport in freshly isolated bovine adrenal chromaffin cells were investigated by Bigornia and Bihler (1986). The data show that 3-O-methyl-D-glucose transport in these cells is mediated by a specific facilitated diffusion mechanism that is sensitive to insulin, thus resembling the characteristics of glucose transport in muscle. The regulation of this uptake was investigated in a subsequent study (Bigornia, Wattis and Bihler, 1986). Uptake of this nonmetabolizable glucose analog was stimulated by hyperosmotic medium, and this effect was abolished in the absence of external calcium. Other experiments showed that glucose transport in adrenal chromaffin cells is stimulated by insulin as well as by hyperosmotic conditions in a calcium-dependent manner. Sensitivity to secretory stimuli, a regulatory feature characteristic of chromaffin cells, also was demonstrated. In contrast to muscle, sugar transport was not affected by sodium pump inhibition, metabolic inhibitors, or the sodium ionophore monensin, suggesting that calcium influx by sodium/calcium exchange does not play a significant role in the activation of sugar transport in chromaffin cells.

Isolated chromaffin cells from the bovine adrenal medulla were used by Delicado and Miras-Portugal (1987) to study glucose transport. The kinetics of glucose transport were elucidated. Among other things, they showed that insulin and secretagogues increase glucose transport by increasing the transporter number at the plasma membrane without changing the affinity.

The glucose transporter in chromaffin cells was identified and characterized by Delicado, Torres and Miras-Portugal (1988). Using cytochalasin B binding in subcellular membrane fractions of the adrenal medulla, they found the most binding in plasma membrane fractions, less in microsomal membrane preparations, and the least in chromaffin vesicle membranes. Irreversible photoaffinity labeling of the glucose-protectable binding sites led to the demonstration of three molecular species of 97-, 51.5-, and 30 kDa. The chromaffin vesicle membranes showed only a molecular species of 80 kDa.

The membrane transport of glucose was studied in bovine adrenal chromaffin cells by Bigornia and Bihler (1988). They studied the distribution of the nonmetabolizable glucose analog 3-*0*-methyl-D-glucose. Uptake of this sugar was more rapid in cultures undergoing rapid morphologic changes than in established cultures. The response to insulin was greater in established cultures. Culture of chromaffin cells and the presence of dexamethasone did not inhibit the formation of processes but decreased 3-methylglucose uptake by an apparently competitive effect.

Takahashi, Naoi and Nagatsu (1987) examined the uptake and accumulation of MPP^+ into PC12h cells. The uptake was mediated by saturable, carrier-mediated transport systems with two different kinetic properties: a high-affinity low-capacity system and a low-affinity high-capacity system. The uptake was inhibited by certain inhibitors, including metabolic inhibitors. Subcellular fractionation localized MPP^+ mainly in the cytosol fraction, but a detectable amount was also found in the mitochondrial fraction.

CHAPTER 12

NEURAL REGULATION

INTERMEDIOLATERAL CELL COLUMN AND INNERVATION OF THE ADRENAL MEDULLA

Two studies by Appel, Wessendorf and Elde (1986, 1987) examined the composition of fiber terminals afferent to identified preganglionic sympathetic cell bodies in the intermediolateral cell column of the rat. Preganglionic cell bodies of fibers projecting to the adrenal medulla were identified by the retrograde transport of injected dye. In the earlier study, 35.5% of the cell bodies appeared to be apposed by fibers or terminals in which serotonin and substance P immunoreactivity coexisted; the authors suggested that these fibers arise in the brainstem. The later study expanded these observations to include localization of thyrotropin-releasing hormone (TRH). Populations of preganglionic cell bodies to the cervical sympathetic ganglion and adrenal medulla were localized at T1-3 and T7-9, respectively. In both populations of cells, approximately 95% of the fibers immunoreactive for TRH also stained for serotonin. In another series of sections, approximately three-quarters of the TRH-positive fibers were also immunoreactive for substance P. The authors noted that this close apposition of TRH-containing fibers to preganglionic sympathetic neurons provides an anatomic substrate for the action of TRH in regulating sympathetic outflow.

Appel and Elde (1988) extended their previous investigations to identify separate populations of rat preganglionic sympathetic neurons in the intermediolateral cell column of segments T-1 to T-4 that innervated either the cervical sympathetic trunk or the adrenal medulla. Neurons to the adrenal medulla were located most laterally in the intermediolateral nucleus. The efferent projections of identified neurons were target-specific, with no neurons identified as innervating both the cervical sympathetic chain and the adrenal medulla. Both populations of neurons were apposed by varicosities immunoreactive for either serotonin or somatostatin. In contrast, oxytocin-immunoreactive varicosities were found only in apposition to preganglionic neurons of the cervical sympathetic trunk. These data provide anatomic evidence of the selective control of different components of the sympathetic nervous system by higher centers.

The preganglionic input to the adrenal medulla was eliminated by the administration of β–bungarotoxin to 17-day embryonic rats. Tümmers, Müller, Schmidt et al. (1986) reported that such treatment did not affect the morphologic and biochemical maturation of the adrenal medulla as examined on embryonic day 21. Although adrenal weight was decreased in treated rats, the total and relative amounts of catecholamines were not changed. These data suggest that an intact nerve supply is not required for the normal embryonic development of adrenal chromaffin cells.

Kesse, Parker and Coupland (1988) provided the first evidence, from tracer studies with Fast Blue, of a postganglionic innervation to the rat adrenal medulla. Injections of Fast Blue into the adrenal medulla labeled cells in the ipsilateral intermediolateral cell column between T-1 and L-1; the greatest number of labeled cells was found at T-9, and 95% of the cells were located in the pars principalis of the nucleus. Additionally, postganglionic cell bodies were localized in the sympathetic chain ganglia between T-4 and T-12, with the greatest number at T-9 and T-10 as well as in the suprarenal ganglion. These postganglionic cells accounted for about 11% of the total number of cells labeled by injection of dye into the adrenal medulla.

Bernstein-Goral and Bohn (1988) examined the ontogenetic development of spinal adrenergic projections from the C1 cell group of the medulla and the relationship between these fibers and preganglionic neurons to the adrenal medulla in the rat. A period of adrenergic hyperinnervation of spinal sympathetic nuclei occurred in neonatal rats; the density of adrenergic innervation was decreased in the adult. Adrenal preganglionic neuronal cell bodies were surrounded with phenylethanolamine *N*-methyltransferase-immunoreactive terminals in the neonatal animal, whereas in the adult the adrenergic terminals were more frequently associated with preganglionic dendrites. These data suggest that the synaptic organization between adrenergic fibers and adrenal preganglionic neurons changes during development.

Bacon and Smith (1988) used retrograde tracing and immuno-cytochemical techniques to determine the types of synaptic input to preganglionic sympathoadrenal cells. They identified four distinct types of input to these cells: substance P neurons, GABAergic neurons, and two morphologically distinct types of serotonergic neurons. In addition, they noted that some synapses were made on proximal dendrites of preganglionic neurons that extended across the white matter.

The study described above was expanded by Mohamed, Parker and Coupland (1988) to investigate the sensory innervation of the adrenal medulla in the guinea pig. Injection of the retrograde tracer Fast Blue labeled cell bodies in the dorsal root ganglia between T-3 and L-2; the greatest number of labeled cells (15% of the total) was seen at T-10. Within the adrenal medulla presumed sensory nerve endings were found in association with chromaffin cells and vascular elements.

McIlhinney, Bacon and Smith (1988) reported a method for coupling cholera B-chain to horseradish peroxidase for use in neuronal tracing. The efficacy of this method was tested *in vivo* by injecting the tracer into the adrenal medulla of rats; retrogradely labeled cells were seen in the spinal cord.

Hirano (1986) used autoradiography to determine that surgical denervation of one adrenal gland in mice -- *i.e.*, removal of secretagogues -- significantly enhanced the uptake of [^{3}H]dopamine in subcortical epinephrine-containing cells without affecting uptake in norepinephrine-containing cells or epinephrine-containing cells in the center of the medulla. The author interpreted these results to suggest a different innervation, or sensitivity to suppression of dopamine uptake by secretagogues, in subcortical, rather than central, epinephrinecontaining cells. Such a functional differentiation in a single type of endocrine cell has not been considered previously.

Preganglionic fibers also form the efferent limb of a reflex involving sensory input from the knee joint. Sato, Sato and Schmidt (1986) examined the effects of articular stimulation on adrenal catecholamine secretion and adrenal nerve activity in halothane-anesthetized cats. Movement of the joint within normal ranges produced no change in catecholamine secretion. However, extension of joint movement into the noxious range as well as movement of joints with active inflammation induced by injection of kaolin and carrageenan into the joint cavity elicited increases in both adrenal catecholamine secretion and adrenal nerve activity. Because spinal transection at C-2 abolished such responses, the authors concluded that supraspinal structures contribute to the reflex response of adrenal medullary activity induced by knee joint stimulation.

Edwards and Jones (1987) reported the effects of splanchnic nerve stimulation in conscious calves in the presence and absence of adrenocorticotrophin (ACTH). Splanchnic nerve stimulation markedly increased the adrenal secretion of both epinephrine and norepinephrine. Although the amount of epinephrine was always greater than the amount of norepinephrine, this difference was significant only in the presence of ACTH. ACTH also decreased both the output of [Met]enkephalin immunoreactivity from the denervated adrenal gland prior to stimulation and the increase in secretion of this substance that occurred during splanchnic nerve stimulation.

Bloom, Edwards and Jones (1988) determined the effects of intermittent and continuous high frequency stimulation of the peripheral end of the splanchnic nerve on adrenal medullary secretion in conscious calves. Intermittent high-frequency stimulation increased levels of vasoactive intestinal peptide, free and total [Met^5]enkephalin, norepinephrine, epinephrine, and dopamine. Levels of neuropeptide Y and gastrin-releasing peptide were also increased by burst of high-frequency stimulation, but the adrenal gland was not the primary source of these peptides.

Yoshioka, Togashi, Minami et al. (1986) sought to determine whether there was a correlation between the tachycardia produced by hydralazine, an antihypertensive drug, and the increase in cardiac sympathetic nerve activity in rats with intact or sectioned buffer nerves -- *i.e.*, nerves from the baroreceptors. In both groups, injection of hydralazine elicited a decrease in blood pressure followed by an increase in heart rate and cardiac sympathetic nerve activity. Adrenal nerve activity increased only in the debuffered rats. The authors concluded that the tachycardia produced by hydralazine administration in rats is associated at least in part with its stimulation of the central sympathetic nervous system.

Changes in the activity of the adrenal nerve, a branch from the celiac ganglion to the adrenal gland, may also be related to aging. Ito, Sato, Sato et al. (1986) measured adrenal medullary catecholamine output and nerve activity in 100- to 900-day-old rats anesthestized with a urethan-chloralose mixture. Although the resting secretion of adrenal catecholamines varied among rats, the secretion rates gradually increased after 300 days of age. Secretion levels at 800 to 900 days were 2- to 4-fold higher than the levels at 100 days. Similarly, unitary activity of the resting adrenal nerve increased with age. The authors proposed that these results explain both the increased levels of peripheral catecholamines and the decreased catecholamine receptor activity of various peripheral tissues such as blood vessels and cardiac muscle that are observed with increased age.

Mahata and Ghosh (1986b) evaluated the effect of intact adrenal innervation on the restoration of medullary catecholamine levels after reserpine treatment in the pigeon (*Columbia livia*). Restoration of norepinephrine and epinephrine was severely retarded in the denervated gland, although both neurohormones were resynthesized proportionally, most likely under the influence of corticosteroid. In contrast, catecholamines were replenished more quickly in the innervated adrenal gland. However, norepinephrine was resynthesized faster than epinephrine. The authors suggested that the selective synthesis of this catechol-amine was related to the association between intramedullary adrenergic fibers and norepinephrine-containing cells.

In a companion study, Mahata and Ghosh (1986c) evaluated the effects of age and intact innervation on the response of the adrenal medulla of the pigeon to insulin. Catecholamine depletion and resynthesis after insulin administration was related to both age and intact innervation of the adrenal gland. In contrast, the hypoglycemic effects of insulin were not related to either innervation or age.

Neural stimulation of the adrenal medulla may be mimicked by the administration of reserpine or insulin. Sietzen, Schober, Fischer-Colbrie et al. (1987) combined these techniques with hypophysectomy in rats to determine the effects on the components of the chromaffin vesicle. Insulin and reserpine induced synthesis of enkephalins, dopamine β-hydoxylase, the amine carrier, and chromogranin B. The level of chromogranin A was dependent on corticosteroids. The authors concluded that both the functional components of the vesicle membrane and the soluble content of the chromaffin granule may be altered significantly by various manipulations. Such abilities may permit the chromaffin cell to respond to variations in internal conditions.

Although adrenal medullary secretion of catecholamines has classically been related to the action of acetylcholine released from the preganglionic fibers, peptides also may be involved in this process. Bouloux, Perrett, Sopwith et al. (1986) examined the opioid control of neurally mediated adrenal medullary catecholamine release by means of *in situ* perfusion in the rat. Catecholamine secretion was elicited by electrical stimulation of the descending thoracic sympathetic chain. The administration of morphine, D-Ala^2-MePhe^4-[Met]enkephalin-(O^5)-ol (DAMME; a [Met]enkephalin agonist) or naloxone did not significantly alter the secretion of medullary catecholamines. These observations indicate that, in the model used, there is no evidence of the involvement of μ or δ opioid receptors in neurally-mediated catecholamine release from the rat adrenal medulla.

Vindrola, Ase, Finkielman et al. (1988) determined the effect of muscarinic and nicotinic stimulation on the release of low- and high-molecular weight enkephalin-like material from the feline adrenal gland perfused *in vitro*. Selective activation of muscarinic receptors by by pilocarpine resulted in the release of low-molecular weight material; selective stimulation of nicotinic receptors with nicotine released both small and large peptides. The authors discussed their results in relation to the existence of a heterogeneous population of chromaffin vesicles that contain various pro-enkephalin-derived peptides.

Lima and Sourkes (1986) investigated the interaction between γ-aminobutyric acid (GABA) and cholinergic mechanisms in regulating dopamine β-hydroxylase (DβH) in the rat adrenal medulla. Oxotremorine, a muscarinic agonist, administered together with methylatropine produced a dose-dependent increase in adrenal medullary DβH. Progabide, a GABA$_A$ and GABA$_B$ receptor agonist which crosses the blood-brain barrier, decreased the effects of oxotremorine in a dose-dependent manner; baclofen, a GABA$_B$ receptor agonist, had no effect. A GABA$_A$ receptor blocker, bicuculline, reversed the action of progabide on DβH activity in oxotremorine-treated rats and enhanced the increase in DβH obtained with oxotremorine. Therefore, the authors concluded that GABA$_A$ receptors are involved in the regulation of adrenal medullary DβH activity and that part of this effect of GABA is mediated by interaction with a central cholinergic system.

Ricordi, Shah, Lacy et al. (1988) investigated the time course of extra-adrenal epinephrine secretion in bilaterally adrenal-ectomized rats. For 1 week after adrenalectomy, plasma epinephrine levels were below detection levels. However, plasma epinephrine concentration increased thereafter, with a mean 31 ± 6 pg/mL at 4 weeks after adrenalectomy. These data suggest that extra-adrenal epinephrine secretion is a delayed response and cannot be assumed to be active in the presence of the adrenal medulla.

Sato (1987) reviewed studies of the reflex activation of the adrenal medulla by somatic stimuli in anesthetized animals. Somato-sympathetic reflexes have both a spinal component with segmental organization and a generalized nonsegmental, supraspinal component. Excitatory responses are mediated by unmyelinated C afferents and inhibitory responses are mediated by myelinated A afferents. The role of intracerebroventricularly administered corticotropin-releasing factor in controlling the somato-sympathetic reflex was discussed.

The effects of yohimbine (an α_2-antagonist), and clonidine (an α_2-agonist) on adrenal medullary and neuronal catecholamines during bilateral carotid occlusion in the dog were determined by Yamaguchi and Brassard (1988). Bilateral carotid occlusion in anesthetized, vagotomized dogs caused increases in blood pressure, heart rate, adrenal catecholamines, and renal norepinephrine. Yohimbine decreased the output of adrenal medullary catecholamines and increased the release of neural norepinephrine in the kidney. In contrast, clonidine abolished both adrenal medullary and neural catecholamine release. The authors concluded that, although the blockade of a presynaptic α_2-adrenoreceptor-mediated mechanism enhances norepinephrine release in many tissues, this mechanism does not appear to be involved in the modulation of adrenal catecholamine secretion during bilateral carotid occlusion.

Morton (1987) reported that the extraneuronal accumulation of [³H]isoprenaline by the vas deferens and atria in rats was not affected by adrenal demedullation. The author concluded that neither plasma catecholamines nor adrenergic nerves have a major role in regulating extraneuronal uptake in these tissues of the rat.

EFFECTS OF INSULIN AND
ROLE OF SENSORY AFFERENTS

Khalil, Marley and Livett (1986) identified both neuronal and non-neuronal mechanisms in the biphasic secretion of adrenal medullary catecholamines induced by insulin in the rat. The first phase of the response, lasting about 30 minutes, was neurogenic because it was blocked by denervation or administration of hexamethonium and atropine. This phase was accompanied by a moderate increase in epinephrine output but little or no change in norepinephrine output. The second or non-neurogenic phase occurred from 30 to 60 minutes later and was initiated when the plasma glucose value fell below 75 mg/100 ml. This phase was accompanied by large increases in both epinephrine and norepinephrine secretion and was not blocked by denervation. However, it was abolished by the intravenous administration of glucose, suggesting that the non-neurogenic release of adrenal medullary catecholamines is sensitive to the plasma glucose concentration. Adrenalectomy abolished both phases of the response, indicating that the adrenal medulla was the source of the increased levels of plasma catecholamines.

In contrast, Levin and Sullivan (1987) found evidence of a role of the peripheral sympathetic nerves in the sympathoadrenal response to insulin. Intravenous administration of glucose to rats activated primarily the sympathetic component of the sympatho-adrenal system as measured by increased plasma norepinephrine levels. Insulin-induced hypoglycemia activated primarily the adrenal medulla as reflected by increased plasma epinephrine levels. The increase in plasma norepinephrine level caused by insulin was smaller and this was attributed to activation of the peripheral sympathetic nerves because this response was not blocked by adrenal demedullation. The authors concluded that changes in plasma glucose concentration alone are capable of selectively activating the two components of the sympathoadrenal system.

Lau, Bartolome, Bartolome et al. (1987) reported that peripheral administration of insulin depleted adrenal catecholamines in adult rats but not in 2-day-old rats, although hypoglycemia was produced in both groups. Central administration of insulin was ineffective in inducing hypoglycemia or adrenal catecholamine secretion in either adult or neonatal rats. The authors concluded that insensitivity of the neonatal adrenal to insulin is associated with immaturity of splanchnic neurotransmission and that direct action of insulin on central receptors is not necessary for activation of the adrenal medulla.

The above results were supported by those of Souto, Piezzi and Bianchi (1988) who also examined the response of the rat adrenal medulla to insulin between birth and adulthood. Insulin administration resulted in a depletion of medullary epinephrine in adults (62%), 10-day-old neonates (45%), and 7-day-old neonates (35%). In contrast, insulin did not deplete medullary epinephrine in rats 4 days old or younger. Norepinephrine was not decreased in any group. The authors suggested that the ontogenetic effects of insulin are mediated by the gradual innervation of chromaffin tissue.

Carbonaro, Mitchell, Hall et al. (1988) used chronic hypo-glycemia, induced by administration of long-acting insulin, to investigate alterations in the reactivity of the isolated, perfused adrenal to a standard dose of acetylcholine. Such chronic hypoglycemia caused a selective depletion of medullary epinephrine and a decreased response to acetylcholine; norepinephrine levels or release were unaffected. Electron microscopy showed degranulation and vacuolization of chromaffin cells in this tissue. Both the biochemical and morphologic effects of chronic hypoglycemia recovered gradually over a period of about 5 days after cessation of treatment with insulin. The authors suggested that the chromaffin cells of the adrenal medulla may be a useful model of adrenergic plasticity.

Sidey, Dean and Furman (1988) investigated the effect of infection with *Bordetella pertussis* or treatment with pertussis toxin on adrenal medullary activity after acute ether stress in mice. Such stress resulted in hyperinsulinemia when the release of adrenal catecholamines was prevented or when inhibition of insulin secretion was blocked as with α_2-adrenoreceptor antagonists or pertussis toxin.

Neostigmine, a cholinergic agonist, injected into the third ventricle elicited hyperglycemia in fed rats. Iguchi, Gotoh, Matsunaga et al. (1988) used demedullated rats to determine the contribution of epinephrine to this hyperglycemic effect. Epinephrine was found to be responsible for 29% of the hyperglycemic effect of intraventricular neostigmine; glucagon induced 22% of the effect, and 49% of the hyperglycemic effect was thought to be due to direct neural innervation of the liver, perhaps by an α-adrenergic receptor.

Khalil, Livett and Marley (1986) expanded the observations of their previous study, discussed above, to identify the role of substance P-containing sensory fibers in regulating adrenal medullary secretion. Both insulin administration and cold stress evoked catecholamine release from the adrenal medulla in control rats. Pretreatment with capsaicin, a selective neurotoxin for sensory fibers, resulted in depletion of 70% of the substance P in the splanchnic nerve. Such treatment blocked the neurogenic (early) but not the non-neurogenic (late) phase of catecholamine release in response to insulin administration. The neurogenic mechanism that increased adrenal medullary secretion in response to cold stress also was blocked by capsaicin pretreatment. The authors suggested that substance P within the splanchnic nerve may act as a neuromodulator of acetylcholine to modify the response of the adrenal medulla to stress.

The involvement of capsaicin-sensitive afferents in the adrenal medullary response to glucose is supported by a study by Amann and Lembeck (1986). Adrenal medullary secretion in rats was induced by either hypoglycemia (insulin administration) or intracellular glucopenia in the brain (2-deoxy-D-glucose [2-DG] administration). Pretreatment of rats with capsaicin blocked the increased adrenal medullary secretion seen with insulin but did not affect the increase after 2-DG administration. The authors concluded that insulin-induced hypoglycemia produces increased adrenal medullary secretion through afferent capsaicin-sensitive C fibers in the hepatic portal vein which transmit information from glucose receptors to the brain.

Donnerer (1988) used capsaicin to destroy C-fiber afferents in newborn rats and later tested the activation of the adrenal gland during hypoglycemia, hypotension, and hypovolemia. The reflex mechanisms that activate the adrenal gland were decreased during these conditions. It was concluded that stimulation of the adrenal medulla under these conditions is based on a reflex mechanism initiated by capsaicin-sensitive afferents.

Histamine, like insulin, can induce catecholamine secretion from the adrenal medulla by both neurogenic and non-neurogenic mechanisms. Khalil, Livett and Marley (1987) demonstrated that, in the rat, pretreatment with capsaicin abolished the neurogenic increase in adrenal secretion induced by histamine. Such pretreatment additionally abolished the neurogenic compensatory increase in plasma norepinephrine observed after adrenalectomy or denervation. The authors concluded that these observations support a role for capsaicin-sensitive neurons in regulating neurally controlled secretion from the adrenal medulla in addition to suggesting a similar role for these neurons in modulating neurogenic release of norepinephrine from sympathetic nerves.

Zhang (1986) reported that afferent impulses along both parasympathetic and somatic nerves may influence adrenal medullary secretion in the cat. Stimulation of the cervical vagus nerve induced a vasopressor response in 50% of the cats tested, and this response was accompanied by secretion of epinephrine from the adrenal medulla. Physostigmine, an anticholinesterase, blocked the increase in blood pressure but did not affect the release of medullary catecholamines. Stimulation of the tibial nerve affected adrenal medullary catecholamine secretion in a dose-dependent manner: low frequency stimulation decreased norepinephrine release and high frequency stimulation increased epinephrine release.

Soltanov and Karpovitch (1987) also found support for the influence of peripheral afferents on adrenal medullary function. Stimulation of the central stumps of splenic or superior mesenteric nerves increased the adrenal medullary content of all catecholamines. Additionally, the catecholamine content of the myocardium, brainstem and hypothalamus increased, which the authors interpreted as indicative of tissue uptake of catecholamines that had been released into the circulation.

CNS STIMULATION AND LESION

Various investigators have used electrical stimulation or creation of lesions in regions within the central nervous system as a means of evaluating central nervous system control of adrenal medullary secretion.

Katafuchi, Yoshimatsu, Oomura et al. (1986) stimulated various regions of the hypothalamus in rats anesthetized with ketamine and measured catecholamines in plasma collected from the adrenal vein. Stimulation of individual sites elicited three types of responses: 1) increases in epinephrine alone; 2) decreases in both epinephrine and norepinephrine; and 3) decreases in norepinephrine alone. Subsequent lesions placed through the stimulating electrodes invariably decreased adrenal catecholamine secretion, suggesting that the lateral hypothalamus is predominantly facilitatory for adrenal medullary activity.

Behaviorally identified medial hypothalamic sites were stimulated in pentobarbital-anesthetized cats by Stoddard, Bergdall, Townsend et al. (1986a,b). The sites had previously elicited defensive (31 sites) or escape (29 sites) in the freely moving animal. Catecholamines were measured in bilateral adrenal venous and peripheral plasma. Although various combinations of both preferential increases and decreases in catecholamine secretion were noted, the most common response to stimulation of both types of hypothalamic sites was a bilateral increase $\geq$10 ng/min in the adrenal secretion of both epinephrine and norepinephrine. Additionally, stimulation at both types of hypothalamic sites altered the ratio of catecholamines secreted from the adrenal medulla by increasing the proportion of epinephrine. Hypothalamic defense sites also appeared to activate the peripheral sympathetic nerves as determined by an increase in peripheral norepinephrine from non-adrenal sources. Increases in catecholamines elicited by stimulation of hypothalamic escape sites were correlated with cardiovascular parameters, suggesting concurrent activation of adrenal medullary and cardiovascular systems.

Yoshimatsu, Oomura, Katafuchi et al. (1987) found that the hypothalamus of the rat had effects on adrenal medullary activity somewhat different from those seen in the cat. They recorded activity in the adrenal nerve after stimulation and creation of lesions in the lateral and medial hypothalamus of rats anesthetized with chloralose urethan. Stimulation of the middle portion of the lateral hypothalamus increased adrenal nerve activity; stimulation of the anterior portion decreased nerve activity. Lesions in these areas generally produced effects opposite to those of stimulation. Stimulation of the ventromedial nucleus had no effect on adrenal nerve activity, in contrast to observations in cats, whereas lesions in this region caused notable increases in nerve activity. Cerebroventricular infusion of 2-DG evoked a large increase in adrenal nerve activity which was suppressed by lateral hypothalamic lesions or ventromedial stimulation. The authors concluded that individual hypothalamic regions regulate the activity of the adrenal nerve differently.

In a later study, Katafuchi, Oomura and Kurosawa (1988) reported that injections of L-glutamate into the hypothalamic paraventricular nucleus of anesthetized rats resulted in an increase in ipsilateral adrenal nerve activity and decreases in renal nerve activity and blood pressure. The authors suggested that, although the release of corticotropin-releasing factor from paraventricular neurons could account for the increased activity of the adrenal nerve, this mechanism would not explain either the decreased renal nerve activity or the depressor response. Thus, it is possible that there are distinct groups of neurons within the paraventricular nucleus that subserve different functions.

Specific regions of the subthalamus also may influence the

secretion of adrenal medullary catecholamines differently. Matsui (1987) measured catecholamines in adrenal venous blood after stimulation of the zona incerta in pentobarbital-anesthetized rats. Stimulation of sites in the lateral zona incerta elicited increases in the adrenal secretion of epinephrine; stimulation of sites in other regions decreased both epinephrine and nor-epinephrine secretion. The author suggested that the dopaminergic incerta-hypothalamic system may be involved in mediating both facilitatory and inhibitory responses to stimulation.

Bereiter, Engeland and Gann (1986) compared left adrenal and peripheral venous catecholamine levels after 56 stimulations at approximately 30 sites in the caudal brainstem of chloralose urethan-anesthetized cats. The best correlation between peripheral catecholamine levels and adrenal secretion was seen with epinephrine. Peripheral and adrenal venous levels of this catecholamine were correlated prior to stimulation; stimulation-induced increases and decreases in adrenal epinephrine secretion were reflected in peripheral venous levels. Similarly, baseline dopamine levels in adrenal and peripheral venous plasma were correlated, although changes in adrenal dopamine secretion after stimulation were not reflected in the periphery. Baseline norepinephrine levels in adrenal and peripheral venous plasma were poorly correlated. Although stimulus-induced increases in adrenal norepinephrine secretion were sometimes reflected peripherally, decreases were not. Furthermore, the time of sampling was found to influence the correlation between levels of adrenal and peripheral venous catecholamines, the maximum correlation being at 3 min after stimulation. The authors concluded that measurement of peripheral catecholamines reflects only large increases in adrenal medullary secretion.

In a later study, Bereiter and Gann (1988) microinjected L-glutamate into various laminae of the trigeminal nucleus caudalis in cats anesthetized with α-chloralose. Injections into laminae that receive noiciceptive input (laminae I-II and V-VI)evoked an increase in the secretion of epinephrine from the ipsilateral adrenal gland, accompanied by a smaller, but consistent increase in the secretion of norepinephrine. Increases in epinephrine secretion were not correlated with changes in adrenal venous blood flow, arterial pressure, or heart rate. Injections of L-glutamate into laminae III-IV of the trigeminal nucleus caudalis, which receive primarily non-noxious sensory input, were without effect on adrenal medullary secretion. The authors concluded that efferent fibers from the trigeminal nucleus contribute to the autonomic response resulting from noxious trigeminal stimulation.

The role of the raphe nuclei on the induction of adrenal DβH by reserpine was investigated in rats by Lima, Sourkes and Dubrovsky (1986). Lesions of the dorsal raphe nucleus had no effect on the reserpine-induced synthesis of this enzyme. Lesions of the medial raphe nucleus, however, potentiated the inducing action of reserpine. The authors concluded that serotonergic systems from the medial raphe may be involved in the regulation of adrenal DβH.

Sved (1986) reported that bilateral lesions of the solitary nucleus produced hypertension in unanesthetized, freely moving rats along with at least 10-fold increases in peripheral norepinephrine and epinephrine concentrations as well as increases in vasopressin. In the lesioned animals, an acute decrease in blood pressure was elicited by blockade of either the sympathoadrenal system by chlorisondamine (a ganglionic blocking agent) or of vasopressin by the administration of an antagonist. Such observations suggest that the sympathoadrenal system, interacting with vasopressin and the renin-angiotensin system, is involved in the mediation of hypertension produced by bilateral lesions of the solitary nucleus.

Drolet and Gauthier (1987) reported that stimulation of the locus ceruleus in rats evoked a biphasic pressor response with concomitant increases in plasma norepinephrine and epinephrine. Adrenalectomy 1-2 days prior to brain stimulation abolished the second, post-stimulation phase of the pressor response. Acute adrenalectomy, however, abolished the secondary pressor response only in sympathectomized animals. Thus, the similar post-stimulation pressor responses observed in intact and adrenalectomized animals were mediated by different underlying mechanisms. The authors suggest that intraneuronal epinephrine may mediate the response seen in acutely adrenalectomized animals.

Electrical stimulation of the anterior hypothalamus in the rat has also been shown to increase plasma catecholamines and elicit a biphasic pressor response. The increase in blood pressure at stimulus onset is mediated by the sympathetic nerves; a second increase after stimulus cessation is mediated by epinephrine release from the adrenal medulla. Using specific α- and β-receptor antagonists, Gauthier (1988) determined that the secondary adrenal medullary component of the hypothalamically-stimulated biphasic pressor response is decreased in intact rats by epinephrine-induced vasodilation and reflex bradycardia. In rats with both β_1- and β_2-adrenergic blockade, stimulus intensity and frequency could be used to separate the sympathetic vasomotor and adrenal medullary pressor responses. These data support the hypothesis of separate hypothalamic systems for the control of specific components of the sympathetic nervous system.

Tashiro, Hirata, Maki et al. (1988) investigated the mechanisms underlying the cardiac arrhythmias that occur after electrical stimulation of the anteromedial hypothalamus in anesthetized cats. Such arrhythmias were not observed when the stimulus intensity was decreased gradually rather than being turned off suddenly. Adrenalectomy decreased the frequency of post-stimulus arrhythmias whereas intravenous injection of catecholamines caused cardiac arrhythmias similar to those following hypothalamic stimulation. The authors concluded that post-stimulus arrhythmias result from the sudden change from sympathetic to parasympathetic influences on the heart and that the occurrence of arrhythmias is prolonged by the action of adrenal medullary catecholamines.

Two reports by Iadecola, Lacombe, Underwood et al. (1986, 1987) described the role of adrenal medullary catecholamines in the cerebrovascular vasodilatation elicited by stimulation of the dorsal medullary reticular formation (DMRF) in chloralose-anesthetized rats. Stimulation of the DMRF both increased regional cerebral blood flow throughout the brain and increased plasma catecholamine concentrations. In addition to preventing the epinephrine increase, bilateral adrenalectomy abolished the increases in cerebral blood flow in all brain regions except the cortex, where blood flow was only decreased. DMRF stimulation did not alter the permeability of the blood-brain barrier. The authors concluded that increases in cerebral blood flow evoked by stimulation of the DMRF has both catecholamine-dependent and -independent components. The mechanism by which adrenal catecholamines effect cerebral vasodilatation has not been determined.

Stoddard, Tyce, Carmichael et al. (1988) determined adrenal medullary secretion in response to visceral and somatic stimuli in cats at 0 to 5 days and 15 to 37 days after spinal cord transections at T-3. At the longer interval, both bladder distention and sciatic nerve stimulation caused increases in blood pressure and prominent increases in the release of adrenal medullary catecholamines. These effects were not seen at the shorter interval. These data suggest that there is reorganization of the spinal control of the adrenal medulla that occurs over time after thoracic cord transection. Furthermore, the similarity between this experimental paradigm and the clinical condition of autonomic hyperreflexia suggests that activation of the adrenal medulla may contribute to the symptoms of this condition in the quadriplegic or high paraplegic patient.

MONOAMINERGIC AND CHOLINERGIC SYSTEMS IN THE CENTRAL NERVOUS SYSTEM

Barbeito, Fernández, Silveira et al. (1986) investigated the role of central noradrenergic systems in sympathoadrenal function in rats. The urinary excretion of epinephrine and norepinephrine in animals pretreated with DSP-4, a neurotoxic agent that depletes central norepinephrine, was higher than that in controls. Furthermore, epinephrine in the adrenal gland was significantly increased in rats pretreated with DSP-4. Clonidine, a central α_2-adrenergic receptor agonist, decreased urinary excretion of norepinephrine in control rats but not in the pretreated rats. From these results the authors concluded that lesions of central noradrenergic systems elicit hyperactivity of the sympathoadrenal system in addition to impairing the response to α_2-adrenergic receptor agonists.

Nakamura, Kamata, Inoue et al. (1988b) investigated the mechanism by which intracerebroventricular administration of norepinephrine blocked the immobilization-induced increase in plasma epinephrine in unanesthetized rats. β- and α_2-adrenergic receptor agonists failed to mimic the effect of norepinephrine. Pretreatment with α_1-adrenergic antagonists, but not with $\beta-$ or α_2-adrenergic antagonists, blocked the ability of norepinephrine to inhibit immobilization-induced increases in plasma epinephrine. The authors concluded that norepinephrine mediates its response in this paradigm through action at a central α_1-adrenoceptor.

Central noradrenergic systems also may interact with opioid neurons in the control of the adrenal medulla. Appel and Van Loon (1986) reported that intracisternal administration of β-endorphin increased plasma epinephrine concentrations in conscious, unrestrained rats. Administration of norepinephrine simultaneously with β-endorphin decreased the plasma epinephrine response elicited by β-endorphin alone. In contrast, depletion of brain norepinephrine by pretreatment with 6-hydroxydopamine potentiated the plasma epinephrine response to intracisternal β-endorphin. The authors concluded that central noradrenergic systems inhibit the action of a group of central opioid neurons that are facilitatory for adrenal medullary activity.

Because they found that concentrations of norepinephrine in the peripheral plasma and cerebrospinal fluid were similar in rats, Peskind, Raskind, Wilkinson et al. (1986) undertook a study of whether peripheral norepinephrine contributed to that in cerebrospinal fluid. Rats were chemically sympathectomized at age 1 week by repeated administration of guanethidine. Six weeks later, both adrenal glands were enucleated. Such destruction of the peripheral sympathetic nervous system had no effect on the norepinephrine concentration in the cerebrospinal fluid. Based on these data, the authors concluded that cerebrospinal fluid norepinephrine levels are primarily a reflection of the activity of central noradrenergic systems.

Lau, Ross, Whitmore et al. (1987) investigated the role of central dopaminergic and serotonergic systems in the development of the adrenal medullary response to insulin in the rat. The activation of the adrenal medulla by insulin-induced hypoglycemia was found to be a response that developed over time postnatally. Depletion of epinephrine was seen at 7 days but developed further at later times. Depletion of norepinephrine was not seen at all during the first 3 weeks of life, suggesting that the maturation of the secretory responses of the two types of adrenal medullary cells are under different control. Lesioning of the central noradrenergic systems by neonatal administration of 6-hydroxydopamine resulted in a preferential increase in epinephrine in the developing adrenals as well as a precocious response of both epinephrine- and norepinephrine-containing cells to insulin. In contrast, lesioning of the central serotonergic systems by neonatal administration of 5,7-dihydroxytryptamine decreased norepinephrine content of the adrenals. The authors concluded that central monoaminergic systems may have both facilitatory and inhibitory actions on the maturation of specific types of adrenal medullary chromaffin cells.

Yokotani, Okuma and Osumi (1987) reported the effect of activation of central cholinergic systems on sympathoadrenal function. Intraventricular administration of nicotine to urethan-anesthetized rats resulted in increased plasma catecholamine levels and blocked the increase in gastric acid secretion elicited by vagal stimulation. These actions of nicotine were blocked when the animals were pretreated with bilateral adrenalectomy and peripherally administered 6-hydroxydopamine. The authors concluded that intraventricular administration of nicotine activates the central sympathoadrenal outflow, thus inhibiting one parasympathetic action of vagal stimulation.

Zhang and Yang (1986) also manipulated central cholinergic systems. Administration of both acetylcholine and physostigmine, an anticholinesterase, into the third ventricle of cats increased arterial blood pressure. Acetylcholine also significantly increased plasma levels of epinephrine and norepinephrine, whereas physostigmine only increased levels of norepinephrine. In a companion study, Zhang and Cheng (1986) found that administration of atropine into the third ventricle of rats decreased the vasopressor response but did not affect the epinephrine increase elicited by electrical stimulation of the cervical vagi. Therefore, the authors suggested that central muscarinic receptors are not involved in the release of catecholamines from the adrenal medulla.

PEPTIDERGIC AND OTHER SYSTEMS IN THE CENTRAL NERVOUS SYSTEM

Riphagen, Bauce, Veale et al. (1986) administered arginine vasopressin and substance P into the intrathecal space of rats anesthetized with sodium thiobutabarbital (Inactin) and collected plasma samples from the adrenal vein. Both peptides evoked pressor responses, but only substance P elicited significant increases in the secretion of adrenal epinephrine. Therefore, the authors suggested epinephrine may mediate the pressor response evoked by intrathecal substance P but is not involved in the pressor response that follows intrathecal administration of arginine-vasopressin.

In a related study, Martin, Malkinson, Bauce et al. (1988) reported that intracerebroventricular injection of arginine vasopressin in conscious rabbits increased arterial epinephrine and venous norepinephrine. The increases in these catecholamines were coincident with changes in heart rate and blood pressure. The authors suggested that the central action of arginine vasopressin on cardiovascular function may be mediated through the sympathoadrenal system.

Intrathecal administration of substance P at the T-8 level has been shown to increase the secretion of adrenal medullary catecholamines. Cridland and Henry (1988) reported that such treatment also increased the response time in the tail-flick test in unanesthetized, restrained rats. Intrathecal administration of thyrotropin-releasing hormone or oxytocin, which also increases sympathetic output, did not alter the tail-flick response. The authors suggested that substance P may activate spinal sympathetic neurons at T-8 to cause the release of opioid peptides from the adrenal medulla. The central action of such circulating opioid peptides may be responsible for the depression of the tail-flick response.

Donoso and Barontini (1986) reported that intraventricular administration of histamine in conscious, freely moving rats elicited a significant increase in plasma levels of epinephrine and norepinephrine. The increase in epinephrine was more rapid, reaching a peak in 5 min, and was dose-dependent. In contrast, the norepinephrine increase developed more slowly and its peak was independent of histamine dose. Pretreatment with mepyramine, an H_1 antagonist, decreased the later but not early increase in epinephrine and had no effect on norepinephrine. An H_2 receptor antagonist, ranitidine, potentiated the histamine-induced increase in norepinephrine but had no effect on epinephrine. The authors concluded that central histaminergic systems are involved in the control of adrenal medullary secretion.

Kurosawa, Sato, Swenson et al. (1986) administered corticotropin-releasing factor (CRF) intraventricularly to halothane-anesthetized rats which resulted in a dose-dependent increase in adrenal nerve activity and significant increases in the concentrations of epinephrine and norepinephrine in the adrenal venous blood. Such results indicate that the CRF-induced increase in adrenal secretion of catecholamines is directly mediated through the adrenal nerve.

Kalra, Dube and Kalra (1988) reported that continuous 4-hour infusion of neuropeptide Y into the third ventricle of satiated rats caused continued episodic feeding; the cumulative food intake and the latency to begin feeding were increased and decreased, respectively, in relation to the dose of neuropeptide Y infused. Adrenalectomy caused a significant decrease in the cumulative food intake in infused rats, which was primarily caused by inhibition or marked suppression of feeding during the last 2 hours of neuropeptide Y infusion. The neural or hormonal factors that mediate this effect of adrenalectomy have yet to be elucidated.

The effects of acetylcholine, arginine vasopressin, and oxytocin on both steroid and catecholamine secretion were evaluated by Porter, Whitehouse, Taylor et al. (1988) in the isolated rat adrenal gland perfused *in situ*. Both catecholamine and steroid secretion were increased by acetylcholine and high concentrations of arginine vasopressin, while oxytocin was without effect. These data do not support a role for the modulation of catecholamine secretion by arginine vasopressin and oxytocin at physiological concentrations.

Appel, Kiritsy-Roy and Van Loon (1986) reported that intracisternal administration of β-endorphin to conscious, freely moving rats significantly increased plasma concentrations of all three catecholamines. Similar changes, but of greater magnitude, were obtained after administration of [D-Ala^2,N-MePhe^4,Gly-ol^5]enkephalin (DAGO, a selective μ-receptor agonist), suggesting that the effect of β-endorphin was mediated at μ-receptors. Direct administration of DAGO into the paraventricular nucleus of the hypothalamus or the nucleus of the solitary tract also elicited significant increases in the plasma levels of all three catecholamines and this response was blocked by pretreatment with naloxone. These data suggest that central opioid-induced activation of adrenal medullary secretion is mediated, at least in part, by μ-receptors in discrete and anatomically distinct brain regions.

These results are supported by the work of Gaumann, Yaksh, Tyce et al. (1988b) who reported that intravenous administration of sufentanil, a selective μ-receptor agonist, to halothane-anesthetized cats resulted in significant increases in blood pressure and levels of epinephrine, norepinephrine, dopamine, and [Met]enkephalin in the adrenolumbar vein. Sufentanil was without effect in cats pretreated with naloxone. In cats with acute spinal transections at T-3, peripheral sufentanil administration did not evoke increased release of adrenal medullary catecholamines. Injection of sufentanil into the lateral, but not the fourth, ventricle caused sympathetic activation which could be reversed by naloxone. The authors concluded that the effects of sufentanil to activate the sympatho-adrenal system are mediated through central receptors in the vicinity of the lateral and third ventricles.

In contrast to the results of the above two studies were the results of Nakamura, Kamata, Inoue et al. (1988a) who investigated the mechanism by which endogenous opioid peptides regulate adrenal medullary secretion of epinephrine. Plasma catecholamines were measured in conscious, unrestrained rats after the intracerebroventricular injection of various opioid peptides. The results indicate that the action of central δ-receptors increases basal medullary secretion of epinephrine; central μ- and κ-receptors are without effect.

Amir (1986) injected glucagon into the cerebral ventricular system of mice. Such treatment produced a dose-dependent hyperglycemia that was blocked either by pretreatment with chlorisondamine, a ganglionic blocking agent, or by adrenalectomy combined with chemical sympathectomy with 6-hydroxydopamine. These results suggest that centrally-administered glucagon effects an increase in plasma glucose concentration through the combined action of the adrenal medulla and peripheral sympathetic nerves to stimulate hepatic glucose production to inhibit pancreatic insulin release or both.

CHAPTER 13

GROWTH FACTORS

NERVE GROWTH FACTOR RECEPTORS

Of the growth factors that have been shown to affect adrenal chromaffin cells and their tumor derivatives, particularly PC12 cells, nerve growth factor is by far the best characterized. Greiner, Lloyd and Guroff (1986) demonstrated specific binding of radioiodinated nerve growth factor to neural crest cells maintained in culture. After 5 days in culture, about 28% of the neural crest cells had specific receptors for nerve growth factor.

Hofmann, Ebener and Unsicker (1987) investigated the existence of nerve growth factor receptors on chromaffin cells from newborn and 10-day-old rats and compared the receptor characteristics as revealed by equilibrium binding and dissociation kinetics. Under equilibrium conditions, no differences were found between cells from the two age groups with respect to dissociation constant and receptor number. In dissociation experiments, chromaffin cells from 10-day-old rats exhibited two classes of nerve growth factor receptors, similar to those found in other nerve growth factor-responsive cells. The "slow" receptor class was not found on cells from newborn rats. These and other age-dependent differences seem to indicate regulatory developmental changes in the binding properties of nerve growth factor from rat chromaffin cells.

The internalization and subsequent fate of the two populations of nerve growth factor receptors on PC12 cells were explored by Hosang and Shooter (1987). This was accomplished by identifying the relative amounts and sizes of the receptor after incubation of cells with the radioiodinated factor, by cross-linking with a photoreactive heterobifunctional reagent, or by following the topologic distribution of the cross-linked receptors with time. The ratio of the slow high-affinity receptor to the fast low-affinity receptor decreased over a 5-hour incubation in a process that did not involve proteolytic conversion of the slow to the fast receptor. During this period, the cross-linked slow receptor moved from a trypsin-labile to a trypsin-stable site, suggesting internalization. In contrast, the cross-linked fast receptor remained sensitive to trypsin.

Koike (1987a) examined the effects of extracellular potassium on initial neurite elongation mediated by low concentrations of nerve growth factor in PC12 cells. Neurite outgrowth was stimulated in medium containing high concentrations of potassium. The binding of iodinated nerve growth factor to the cells also was enhanced. Apparently, potassium induces alterations in the binding of nerve growth factor.

The binding and internalization of radioiodinated nerve growth factor by PC12 cells and the effect of extracellular potassium concentration was studied by Koike (1987b). Both surface-bound and internalized fractions of labeled growth factor associated with the cells decreased under depolarizing conditions. This voltage-dependent phenomenon was reversible and also was observed in the presence of veratridine. The effect was abolished in the absence of calcium, indicating that this effect may be mediated by calcium.

Tocco, Contreras, Koizumi et al. (1988) demonstrated that treatment of PC12 cells with dexamethasone leads to a 60% decrease in the binding of radioiodinated nerve growth factor. The decrease in binding was due largely to a decrease in the number of low-affinity receptors. Other nerve growth factor-induced changes, such as the induction of ornithine decarboxylase and the generation of neurites, were only minimally inhibited in dexamethasone-treated cells.

Rosenberg, Hawrot and Breakefield (1986) used a chemically modified derivative of nerve growth factor to characterize the binding and intracellular processing of the factor as well as the structure/function relationships of this polypeptide hormone. They prepared nerve growth factor that was biotinylated on carboxyl groups. This modified peptide was as effective as the native growth factor in binding to receptors. Other experiments showed that the biotinylated nerve growth factor would be useful in a number of potential applications for further study of nerve growth factor.

The association kinetics of radioiodinated β nerve growth factor binding to the PC12 cells were examined by Woodruff and Neet (1986a). When examined by utilizing a reversible second-order integrated rate equation, the data were not consistent with a simple bimolecular process. Two association rates were required to explain the results adequately. Numerous studies were presented that indicated that the interactions of the β subunit of nerve growth factor are more complex than previously described. Woodruff and Neet (1986b) continued their studies by showing that binding of β nerve growth factor to PC12 cells is inhibited by α nerve growth factor and γ nerve growth factor. The findings contrast with those from dorsal root ganglion cells and thus imply differences in the nerve growth factor receptor systems between the two different types of cells.

Several techniques were used by Kasaian and Neet (1988) to distinguish subsets of cell-bound nerve growth factor and to track the route of nerve growth factor through the PC12 cell. The effects of inhibitors of cell trafficking on the distribution of nerve growth factor among the subsets were also investigated. Quantitative and temporal relations among five cellular pools were defined. For example, some experiments suggested that both microfilaments and microtubules are involved in pathways leading to the degradation of nerve growth factor.

Taniuchi, Johnson, Roach et al. (1986) examined the phosphorylation of the nerve growth factor receptor in PC12 cells and cultured sympathetic neurons. They showed that nerve growth factor receptor components of 80 kDa and 210 kDa were phosphorylated when incubated with radiolabeled phosphate. Additional studies led to the conclusion that binding units of the nerve growth factor receptor are phosphorylated constitutively at at least two sites in intact cells.

Green, Rydel, Connolly et al. (1986) selected and cloned four mutant PC12 cell lines that apparently lack all responses to nerve growth factor but are otherwise very similar to the parent PC12 cells. These mutant lines also fail to internalize the growth factor and nearly or completely lack high-affinity nerve growth factor binding sites. Because these mutants have almost exclusively low-affinity nerve growth factor receptors but otherwise are similar to PC12 cells, they are of value in selective studies of the nerve growth factor receptor.

An enzyme-linked immunoadsorbent assay (ELISA) was used by Doherty, Seadon, Flannigan et al. (1988) to study the relative expression of nerve growth factor receptor in PC12 cells. Both nerve growth factor and fibroblast growth factor were found to induce time-dependent increases in receptor expression. Cholera toxin and cordycepin inhibited the induction of receptors by both growth factors, whereas the kinase inhibitor K-252a inhibited receptor induction by nerve growth factor but not by fibroblast growth factor.

Sano, Kato, Totsuka et al. (1988) used a new subline of PC12 cells, PC12D, to develop a convenient bioassay for nerve growth factor. The assay is readily adaptable for the purification of nerve growth factor and for the measurement of nerve growth factor in tissue.

EFFECTS OF NERVE GROWTH FACTOR

The mode of action of nerve growth factor in PC12 cells was

extensively reviewed by Levi, Biocca, Cattaneo et al. (1988). It was concluded that PC12 cells, notwithstanding their neoplastic nature, provide an excellent model for studying the mechanism of action of nerve growth factor.

The effects of pre- and postnatal administration of antibodies to nerve growth factor on the morphologic and biochemical development of the rat adrenal medulla were reinvestigated by Bode, Hofmann, Müller et al. (1986). Specifically, they reexamined the effects of a specific antiserum to nerve growth factor on the morphology, catecholamine and neuropeptide content, and choline acetyltransferase activity of the rat adrenal medulla. Fetuses were injected with the antiserum on day 17 of gestation and daily for 7 days postnatally. The morphology, both at the light and electron microscope levels, was not notably different except that the number of chromaffin vesicles was greater and the diameters of epinephrine-storage vesicles were smaller in antibody-treated animals. Catecholamine and substance P contents, ratios of epinephrine to norepinephrine, and choline acetyltransferase activities were identical in treated and control groups. Antibodies caused a decrease in adrenal [Met]enkephalin of 40%. The authors concluded that administration of antibodies against nerve growth factor to embryonic and early postnatal rats induces only subtle changes in the morphology of chromaffin cells without altering the development of normal catecholamine levels. The small, yet significant, effects of the antibodies on adrenal [Met]enkephalin suggest a role for endogenous growth factor in the regulation of opioid peptide metabolism in developing chromaffin cells.

Hagag, Halegoua and Viola (1986) demonstrated that microinjection of antibody to *ras* p21 into PC12 cells inhibits nerve growth factor-induced differentiation. Whereas nerve growth factor usually induces a number of phenotypic characteristics of sympathetic neurons in PC12 cells, the antibody to *ras* p21 inhibited neurite formation and resulted in temporary regression of partially extended neurites. Neurite formation induced by cyclic adenosine monophosphate (cAMP) was unaffected by injection of anti-p21 antibody. These results indicate that p21 is involved in the initiation phase of nerve growth factor-induced neurite formation in PC12 cells and has a role in hormone- mediated cellular responses distinct from cell proliferation.

Milbrandt (1986) investigated the nerve growth factor-mediated increase in c-*fos* gene expression in PC12 cells. Nerve growth factor treatment results in an increased level of c-*fos* mRNA within 15 minutes. The half-life of this RNA transcript was found to be extremely short. These and other results supported the hypothesis that the *fos* gene product plays a role in signal transduction.

Tiercy and Shooter (1986) studied changes in rates of protein synthesis during treatment with nerve growth factor and dibutyryl cAMP in inducing differentiation of PC12 cells, focusing on the first 8 hours of nerve growth factor treatment. They detected nerve growth factor-induced *de novo* synthesis of a cytoplasmic protein with an apparent molecular mass of 92 kDa. They also presented evidence that most of the early changes in rates of synthesis include a transcription-dependent step and that induction of several of these proteins was not observed in cultures treated with dibutyrl cAMP.

Masiakowski and Shooter (1988) demonstrated that nerve growth factor induced the genes for two proteins related to a family of calcium-binding proteins in PC12 cells. The mRNA species detected by differential hybridization of these cDNA sequences, designated 42A and 42C, reached maximal levels after 24 hours of treatment with nerve growth factor. Epidermal growth factor transiently induced both mRNAs but at much lower levels. The conservation of primary and secondary structure among 42A, 42C, and other proteins suggested a possible role for them in the regulation of cell growth and differentiation.

The catecholamine content and *in vitro* activities of tyrosine hydroxylase and phenylethanolamine N-methyltransferase were measured in cultures of adrenal chromaffin cells from newborn and young postnatal rats by Müller and Unsicker (1986) to study the effects of nerve growth factor and glucocorticoids. Nerve growth factor did not affect the activity of the methylating enzyme but did increase tyrosine hydroxylase and total catecholamine content to about 160% of control values. Dexamethasone increased total catecholamine content to about 200% over the control level without change in tyrosine hydroxylase activity, and it prevented a decrease in phenylethanolamine N-methyltransferase activity.

Kongsamut and Miller (1986) demonstrated that nerve growth factor modulates the drug sensitivity of neurotransmitter release from PC12 cells. Release of labeled norepinephrine from undifferentiated PC12 cells resembles that from chromaffin cells with regard to sensitivity to calcium channel blockers. After nerve growth factor-induced differentiation, release of catecholamines becomes insensitive to such antagonists and agonists; therefore, the cells resemble sympathetic neurons. These results can be explained by the development of new calcium channels with a sensitivity resembling that seen in neurons.

Acheson and Thoenen (1987) compared nerve growth factor-mediated activation of tyrosine hydroxylase (short-term) and tyrosine hydroxylase induction (long-term). They showed that short-term exposure of calf adrenal chromaffin cells to nerve growth factor results in an activation of tyrosine hydroxylase that is potentiated by plating the cells on a laminin substrate. This short-term activation, as well as long-term induction, is blocked by several *S*-adenosylhomocysteine hydrolase inhibitors. Both short- and long-term effects may be related to a common signaling process.

Reed and England (1986) examined the effect of nerve growth factor on the development of sodium channels in PC12 cells. PC12 cells are electrically inexcitable and have a low content of sodium channels. In an assay using saxitoxin to measure sodium receptors, nerve growth factor-treated cells had a receptor density comparable to that of other excitable cells. Levels in untreated cells were below the limit of detection. Their findings showed that nerve growth factor treatment of PC12 cells leads to a substantial increase in the production of neurotoxin-sensitive sodium channels. These channels are pharmacologically similar, if not identical, to those in undifferentiated cells and therefore do not appear to result from the conversion of preexisting channels.

Rudy, Kirschenbaum, Rukenstein et al. (1987) also demonstrated that nerve growth factor increases the number of functional sodium channels in PC12 cells. They also demonstrated that the growth factor induces tetrodotoxin-resistant sodium channels. They used the uptake of radioactive sodium to measure the sodium permeability of PC12 cells before and after long-term nerve growth factor treatment. The treatment does not change the resting sodium permeability, but they did find that nerve growth factor increases the number of functional channels that otherwise behave similarly to those present before nerve growth factor treatment.

The increase in cytosolic calcium content induced by nerve growth factor in PC12 cells and bovine adrenal chromaffin cells was investigated by Pandiella-Alonso, Malgaroli, Vicentini et al. (1986). Using fluorescence techniques, they found that the effect of nerve growth factor on calcium levels has a lag phase of about 40 seconds and a half-life of 40 seconds, increases calcium levels up to 75%, and can persist for more than 10 minutes. The effect apparently is due to calcium influx because it requires extracellular calcium. Perhaps an effect on intracellular calcium plays a messenger role in the action of nerve growth factor.

Streit and Lux (1987) investigated voltage-dependent calcium currents in relation to the nerve growth factor-induced outgrowth of neurites in PC12 cells. Calcium currents were recorded in isolated growth cones of PC12 cells by the whole-cell patch-clamp method. The calcium currents of the cell body were not modified during the early phase of nerve growth factor application. In a later phase, maximal current amplitudes were higher in treated cells than in untreated cells, indicating an increase in current density. Because no evidence was found for a short-latency effect of nerve growth factor on calcium currents, these currents probably do not have an active role in nerve growth factor-induced neurite outgrowth. Long-term effects of the growth factor on calcium currents appeared to be a consequence of cell growth rather than a trigger for it.

Miller, Tischler, Jumblatt et al. (1988) showed that PC12 cells had increased binding of the peripheral-type benzodiazepine Ro 5-4864 in membrane preparations after treatment with nerve growth factor. Forskolin acted synergistically with nerve growth factor to produce further increases in binding but had no effect by itself. The increased binding appeared to reflect increases in receptor number. The physiologic role of benzodiazepine binding sites on PC12 cells is unclear. Treatment of cells with Ro 5-4864 produced no changes in basal or stimulated release of catecholamines or in other variables.

It is known that nerve growth factor regulates the intracellular phosphorylation of several proteins. Hama, Huang and Guroff (1986) investigated the effect of protein kinase C on the nerve growth factor-sensitive phosphorylation of a 100 kDa protein substrate known as Nsp100. Treatment of PC12 cells with either nerve growth factor or a phorbol ester that increases protein kinase C activity caused a decrease in the phosphorylation of Nsp100. The activity of protein kinase C was increased by the nerve growth factor treatment. These and other results suggested that the binding of nerve growth factor to its receptor on PC12 cells causes an increase in the activity of protein kinase C and the phosphorylation of Nsp100 kinase, which in turn lowers the ability of this kinase to phosphorylate Nsp100.

Hashimoto, Iwasaki, Kuzuya et al. (1986) focused on the mechanism of the nerve growth factor-induced decrease in phosphorylation of Nsp100. They showed that the actions of nerve growth factor in this system are mimicked by effectors of cellular calcium metabolism and are prevented by a calcium chelator. Additionally, although the action of nerve growth factor in this system is also mimicked by agents that increase cAMP levels, these agents appear to act through their effects on cellular calcium levels. Mobilization of cellular calcium may be an intimate and early event in the nerve growth factor-induced decrease in the specific phosphorylation of Nsp100.

Matsuda, Nakanishi, Dickens et al. (1986) showed that soluble extracts from nerve growth factor-stimulated PC12 cells show a greater ability to phosphorylate the ribosomal protein S6 than do extracts from control cells. The partially purified nerve growth factor-sensitive S6 kinase has a molecular mass of 45 kDa. Other data suggest that the S6 kinase stimulated by nerve growth factor is neither the cAMP-dependent protein kinase or protein kinase C nor the result of tryptic activation of a proenzyme. Matsuda and Guroff (1987) purified this nerve growth factor-sensitive S6 kinase. They characterized the enzyme and found, among other things, that it is highly specific for S6. Other results indicated that phosphorylation is involved in the mechanism of S6 kinase activation.

Blenis and Erikson (1986) reported on the regulation of S6 kinase and protein kinase A activities in PC12 cells. They also demonstrated the existence of protein kinase C-independent and -dependent mechanisms for regulation of the S6 kinase activity by nerve growth factor. These protein kinase activities may have a role in signal transduction and regulation of cell growth and differentiation.

Mutoh, Rudkin, Koizumi et al. (1988) showed that the treatment of PC12 cells with nerve growth factor led to an increase in the activity of an S6 kinase, whereas treatment of the cells with epidermal growth factor resulted in an increase in the activity of another S6 kinase. They presented data that suggested that the patterns of phosphorylation and, in turn, the functional properties of S6 are different in cells instructed to differentiate in contrast to cells instructed to divide.

Rowland, Müller, Goldstein et al. (1987) developed a cell-free assay to detect and to characterize nerve growth factor-activated protein kinase activity. Cultured PC12 cells were briefly exposed to nerve growth factor and then extracts of these cells were assayed for phosphorylating activity. Nerve growth factor-treated cells showed 2 to 3 times more incorporation of phosphate than controls did. Activation did not occur if the growth factor was added directly to cell extracts. The increase in phosphorylation appeared to be due to regulation of a protein kinase rather than of a phosphoprotein phosphatase. The kinase appears to be a novel enzyme that Rowland et al. (1987) designated "kinase N." A number of characteristics of kinase N were evaluated.

Romano, Nichols and Greengard (1987) showed that nerve growth factor treatment of PC12 cells caused a rapid increase in the state of phosphorylation of synapsin I. Analysis of the synapsin I revealed that the phosphorylation occurred on a particular phosphopeptide, designated peptide N. Phosphoserine was the only phosphoamino acid detected in peptide N. Synapsin I was a substrate for several protein kinases, but none of these kinases phosphorylated synapsin I on peptide N. The results suggested that the nerve growth factor-stimulated phosphorylation of synapsin I may be mediated by a novel protein kinase.

Van Hooff, De Graan, Boonstra et al. (1986) identified a 48 kDa phosphoprotein in PC12 cells as the specific protein kinase substrate B-50. B-50 is present in both undifferentiated and differentiated PC12 cells. Exposure of the cells to nerve growth factor for 2 days results in a 2.5-fold increase in the amount of B-50. This and other observations support the hypothesis that B-50 plays a role in neurite outgrowth and indicate that PC12 cells are a suitable model for study of this hypothesis.

The effect of nerve growth factor on the utilization and fate of ^{14}C-labeled glucose and on the content of several pyridine and purine nucleotides was studied in PC12 cells by Morelli, Grasso and Calissano (1986). After incubation for 72 hours with nerve growth factor, PC12 cells exhibited a 2.7-fold increase in glucose utilization and a 4.7-fold increase in carbon dioxide release. During the same incubation period, the concentrations of all the nucleotides tested increased significantly. The authors suggested that the effect of nerve growth factor on energy metabolism may be an important factor in neuronal differentiation in these tumor cells.

Davis and Kauffman (1987) studied the effects of nerve growth factor and 6-aminonicotinamide on metabolism via the pentose phosphate pathway in PC12 cells. Nerve growth factor appeared to enhance the metabolism of glucose via this pathway. This apparently is a late response to the growth factor because the effect was not noted within the first 6 hours. 6-Aminonicotinamide inhibited metabolism via the pentose pathway but failed to inhibit nerve growth factor-stimulated neurite outgrowth. This suggests that the oxidative enzymes of the pentose pathway are not required for the neurite outgrowth stimulated by nerve growth factor.

Suzuki, Nakanishi and Yamada (1988) demonstrated that nerve growth factor transiently increased tetrahydrobiopterin and total biopterin contents of PC12 cells. The transient increases were maximal at 24 hours and decreased after 3 days.

Torres, Bader, Aunis et al. (1987) determined the effects of nerve growth factor on adenosine transport capacity and affinity in cultured bovine adrenal chromaffin cells. Although they noted a decrease in the affinity of adenosine transporter, transport capacity increased more than 2-fold in nerve growth factor-treated cells.

The effect of nerve growth factor on the hydrolysis of phosphoinositides in PC12 cells was examined by Contreras and Guroff (1987). Addition of the growth factor to cells prelabeled with tritiated inositol resulted in an increase in the formation of inositol trisphosphate, inositol bisphosphate, and inositol monophosphate, indicating that nerve growth factor stimulated hydrolysis of the polyphosphoinositides. The increase in these inositol phosphates was detected as early as 15 seconds after addition of the growth factor. These and other results suggest that nerve growth factor rapidly stimulates the hydrolysis of phosphoinositides in PC12 cells by a calcium-dependent mechanism.

van Calker and Heumann (1987) showed that nerve growth factor potentiates the agonist-stimulated accumulation of inositol phosphates in PC12 cells. This is promoted by agonists under conditions such that the factor by itself does not stimulate the accumulation of inositol phosphates.

The effect of nerve growth factor on the production of neutral glycosphingolipids and gangliosides was examined by Schwarting, Gajewski, Barbero et al. (1986) in PC12 cells grown in spinner culture. They found that PC12 cells contain an unusual group of fucose-containing neutral glycolipids and gangliosides. Nerve growth factor enhances the incorporation of fucose into glycolipids and gangliosides by as much as 80%. All of the gangliosides appeared to increase in concentration when PC12 cells were treated with nerve growth factor. However, only the complex fucose-containing neutral glycolipids increased in concentration during this treatment; the nonfucosylated precursors did not.

Saarma, Toots, Raukas et al. (1986) demonstrated that nerve growth factor induces changes in (2'-5')oligoadenylate synthetase and 2'-phosphodiesterase activities during differentiation of PC12 cells. The synthetase activity increases rapidly and the activity of the phosphodiesterase simultaneously decreases. These changes in the enzyme activities lead to a significant increase in the intracellular concentration of (2'-5')oligoadenylate.

Yavin, Hama, Gil et al. (1986) investigated the possibility that PC12 cells respond to nerve growth factor by releasing glycoproteins. PC12 cells were found to secrete increased amounts of glycoproteins into the medium after the addition of nerve growth factor or brain gangliosides. The release of glycoproteins may relate to the neurotrophic properties that these two entirely different ligands have for PC12 cells.

Noguchi, Muramatsu and Koike (1987) examined the temporal increase of two proteins in PC12 cells induced by nerve growth factor. When cells were stimulated to differentiate with nerve growth factor, a 70 kDa protein increased; later, a 69 kDa protein increased. Then, the amount of both proteins gradually decreased to undetectable levels. These proteins can be used as markers to follow the differentiation of PC12 cells.

Stein, Orit and Anderson (1988) described features of the regulation of a neural-specific gene (SCG10) that is induced by nerve growth factor during differentiation of PC12 cells. The characteristics of SCG10 induction appeared to constitute a reliable molecular index of the transcription-dependent neuronal differentiation induced by nerve growth factor. Glucocorticoids partially blocked SCG10 induction. These and other data suggest that environmental signals such as nerve growth factor may act on specific genes, both positively and negatively, to control the choice of alternate fates by developing neural crest cells including adrenal chromaffin cells.

Differential screening of cDNA libraries was used by Leonard, Ziff and Greene (1987) to detect and prepare probes for mRNAs that are regulated in PC12 cells by treatment with nerve growth factor. One aim of this study was to identify growth factor-regulated mRNAs that may be associated with how the cells attain neuronal properties. Eight nerve growth factor-regulated mRNAs were described. The mRNAs were characterized with respect to the time course of the response to nerve growth factor, regulation by other agents, and tissue distribution within the rat. The findings suggest that the nerve growth factor-regulated mRNAs may play roles in the establishment of the neuronal phenotype.

Kondratiev, Alakhov, Movsesyan et al. (1986) examined the effect of nerve growth factor on the level of endogenous ADP-ribosylation in PC12 cells. The growth factor inhibited ADP-ribosylation in the cellular homogenate, in serum-free cultivation, and in the presence of serum in the culture medium. The ribosylation of several proteins was inhibited. The authors discussed a possible interrelationship between nerve growth factor and the adenylate cyclase system via a receptor-dependent ADP-ribosylation of regulatory components of adenylate cyclase.

Taniguchi, Morisawa, Ogawa et al. (1988) studied the changes in levels of the enzyme molecule and mRNA for poly(ADP-ribose) synthetase during nerve growth factor-promoted neurite outgrowth in PC12 cells. When PC12 cells were cultured in the presence of nerve growth factor, the content of enzyme molecules and the neurite outgrowth decreased to 50% of the original amounts. The mRNA for the enzyme also decreased. These results suggest that the decrease in the enzyme molecule may be due to depression of expression of the gene for the synthetase during nerve growth factor-promoted neurite outgrowth.

Laasberg, Pihlak, Neuman et al. (1988) demonstrated that nerve growth factor causes a rapid increase in cGMP levels and enhances the cGMP phosphodiesterase activity in PC12 cells. No changes in the level of cAMP or the activity of cAMP phosphodiesterase were found. These and other results suggested that the cGMP system may be one of the second messengers of nerve growth factor action in PC12 cells.

In an elegant morphologic study, Jacobs and Stevens (1986) demonstrated changes in the organization of the neurite cytoskeleton during nerve growth factor-activated differentiation of PC12 cells. Using computer-assisted three-dimensional serial electron microscopic reconstruction, they described the progressive cytoskeletal and structural changes of neurites at different stages in differentiation. Developmental changes were characterized by two major transitions. First, microtubules increase in number, leading to a more cylindrical and uniform neurite shape. Second, there are major changes in the relative numbers of other organelle types, which reflect the functional specialization of the neurite. They concluded that many of the observed changes seen in developing PC12 neurites are due simply to the production of a greater number of microtubules in the cell and that many of the other important features that can be measured and that contribute to neurite shape remain constant during development.

Lindenbaum, Carbonetto and Mushynski (1987) showed that the three subunits of neurofilaments in PC12 cells are phosphorylated to various degrees and that nerve growth factor treatment results in increased stability of the more highly phosphorylated forms of the medium (145 kDa) and large (200 kDa) subunits. A highly phosphorylated medium subunit was found in PC12 cells not treated with the growth factor; with nerve growth factor treatment, there was rapid phosphorylation of the medium subunit, suggesting that high levels of neurofilament phosphorylation can be attained in the cell body of PC12 cells. PC12 cells contained detergent-soluble forms of all three neurofilament subunits but their relative proportions were different from those found associated with the detergent-insoluble cytoskeletal fraction. These results indicate that aberrant phosphorylation and retarded assembly may be additional abnormal features of neurofilament metabolism in PC12 cells.

Fernyhough and Ishii (1987) reported that nerve growth factor can increase tubulin transcript levels in PC12 cells in a manner that correlates with its capacity to enhance neurite formation. The increase is due, at least in part, to transcript stabilization. These results, together with those of earlier studies, support the generalization that tubulin transcript levels are specifically increased whenever neurite elongation is initiated by polypeptide neuritogenic factors.

Black and Keyser (1987) showed that α-tubulin is acetylated in PC12 cells, but only after prolonged exposure to nerve growth factor. The evidence for this is that a protein of PC12 cells treated with nerve growth factor comigrates with α-tubulin, is recognized by the monoclonal antibody specific for acetylated α-tubulin, and is prominently labeled in PC12 cells incubated with tritiated acetate. Additional evidence suggests that the stabilization resulting from the acetylation of α-tubulin may be an indirect effect, possibly due to an enhanced interaction between tubulin and microtubule-associated proteins.

The regulation of microtubule protein levels during cellular morphogenesis in nerve growth factor-treated PC12 cells was examined by Drubin, Kobayashi, Kellogg et al. (1988). Nerve growth factor increased the levels of several proteins known to be associated with microtubules in a fashion parallel to the increased neurite outgrowth. The changes were due to changes in protein synthesis rates, as shown by increases in mRNA levels. The turnover of these proteins was increased when differentiated PC12 cells were withdrawn from nerve growth factor and the neurites had retracted.

Paves, Neuman, Metsis et al. (1988) demonstrated that nerve growth factor induces the redistribution of F-actin in PC12 cells. Epidermal growth factor had no effect on microfilament organization. The effect of nerve growth factor did not depend on protein synthesis, but it was blocked by methyltransferase inhibitors.

Brugg and Matus (1988) used several monoclonal antibodies against microtubule-associated proteins to help determine which proteins are expressed during nerve growth factor-induced neurite outgrowth in PC12 cells. One of these proteins, designated microtubule-associated protein 5, underwent the largest change in production and a significant change in its cytoplasmic distribution. The other results suggested that other proteins are produced during maintenance of neuronal form.

The neuron-glia cell adhesion molecule was identified in PC12 cells and brain tissue by Friedlander, Grumet and Edelman

(1986). The molecule in PC12 cells is 230 kDa; a 200 kDa molecule is found in brain. When PC12 cells were treated with nerve growth factor, the amount of the neuron-glia cell adhesion molecule at the cell surface increased about 3-fold whereas the amount of the neural cell adhesion molecule remained unchanged. Other evidence showed that the neuron-glia cell adhesion molecule may be related to a nerve growth factor-inducible large external glycoprotein.

Prentice, Moore, Dickson et al. (1987) determined the effects of nerve growth factor on the production of the neural cell adhesion molecule in PC12 cells. A quantitative immunoassay showed that nerve growth factor induces a 4- to 5-fold increase in relative neural cell adhesion molecule level over a 3-day period. Treatment with nerve growth factor also led to the expression of a species of mRNA not seen in untreated cells. This is the first description of control of neural cell adhesion molecule production by a growth factor.

Shaw and Letourneau (1986) examined the developmental effects of nerve growth factor on embryonic chick adrenal cells in culture. Their findings suggest that a growth factor with physiologic and antigenic similarities to nerve growth factor is produced by cultured embryonic chick adrenal cells but that this growth factor affects different subpopulations of chromaffin cells in different ways. These and other observations led to the suggestion that developing chromaffin cells may represent a more diverse assembly of phenotypes than has been realized previously.

Naoi, Shibahara, Suzuki et al. (1988) examined the effect of nerve growth factor and gangliosides on the specific binding of glycosylated β-galactosidase to PC12h cells. A marked morphologic differentiation was observed in the presence of nerve growth factor, and the cells lost the selectivity of the glycoside binding.

Anderson and Axel (1986) found that the embryonic adrenal medulla and sympathetic ganglia both are initially populated by precursors in which nerve-specific genes are expressed. In primary culture of embryonic rat cells, the medullary precursors have three developmental fates: when treated with nerve growth factor, they continue to mature into neurons and survive; when treated with glucocorticoid, they either lose their neuronal properties and exhibit an endocrine phenotype, or they continue to develop into neurons and then die. These data suggest that, *in vivo*, the adrenal medulla develops through both the glucocorticoid-induced differentiation of bipotential progenitors and the degeneration of committed neuronal precursors that have migrated into the gland.

INTERACTIONS BETWEEN NERVE GROWTH FACTOR AND OTHER FACTORS

Koizumi, Contreras, Matsuda et al. (1988) identified a specific inhibitor of the action of nerve growth factor on PC12 cells. This kinase inhibitor, K-252a, prevented the neurite outgrowth initiated by nerve growth factor but not the outgrowth initiated by fibroblast growth factor or dibutyryl cAMP. Additional evidence was presented that K-252a may be an important new tool for the dissection and study of nerve growth factor-requiring processes.

These findings were supported by data developed by Hashimoto (1988) in PC12h cells. It was shown that pretreatment of PC12h cells with K-252a almost completely blocked the effect of nerve growth factor on the phosphorylation of Nsp100. K-252a also blocked the generation of neurites elicited by nerve growth factor.

Richter-Landsberg and Jastorff (1986) elucidated the role of cAMP in nerve growth factor-induced neurite outgrowth by using the adenylate cyclase activator forskolin, cAMP, and cAMP analogs. Their data show that the establishment of a neurite network, as observed from PC12 cells treated with nerve growth factor alone, could not be induced by forskolin, cAMP, or cAMP analogs alone. The presence of nerve growth factor in combination with these agents potentiated the initiation of neurite outgrowth from PC12 cells. They concluded that the morphologic differentiation of PC12 cells stimulated by nerve growth factor does not require cAMP as a second messenger. The increase of intracellular cAMP caused by either forskolin or cAMP or its analogs in combination with nerve growth factor not only rapidly stimulated early neurite outgrowth but also exerted a maintenance effect on the neuronal network established by nerve growth factor.

The fates of cAMP, dibutyryl cAMP, and an isomer of cAMP were studied in PC12 cell cultures by Braumann, Jastorff and Richter-Landsberg (1986). In the absence of PC12 cells, cAMP was rapidly degraded by nonspecific esterases and a phosphodiesterase, both originating from the serum commonly used in the culture medium. The isomer was completely stable. They found that cAMP metabolites have the capacity to induce an effect on nerve growth factor-induced neurite outgrowth that has been described as cAMP-specific. Using serum-free medium, they estimated the uptake and metabolism of externally applied cyclic nucleotides by PC12 cells. The results indicate that uptake and metabolism are related to the effects of nucleic acid components on nerve growth factor-induced neurite outgrowth.

Tischler, Mobtaker, Kwan et al. (1987) showed that PC12 cells treated with nerve growth factor in combination with high concentrations of forskolin or cholera toxin become more neuron-like in size than cells treated only with nerve growth factor or with activators of adenylate cyclase alone. Cells treated simultaneously with nerve growth factor plus forskolin or cholera toxin paradoxically show less neurite outgrowth than cells treated with the growth factor alone. Addition of forskolin or cholera toxin to cells pretreated with nerve growth factor, however, produces enlarged cells with intact processes that are indistinguishable from cultured neurons. Nerve growth factor may act in concert with agents that increase intracellular cAMP to cause neuronal maturation during embryogenesis, and the proper sequence of exposure to these signals may be necessary for normal development.

Katoh-Semba, Kitajima, Yamazaki et al. (1987) identified a new subline of PC12 cells (PC12D) in which neurites are extended in response to cAMP-enhancing reagents as well as in response to nerve growth factor but not in response to epidermal growth factor or phorbol diester. It is unclear whether the neurite growth from PC12D cells, which is stimulated both by forskolin and nerve growth factor, is mediated by enhanced intracellular levels of cAMP.

Improta, Salvatore, Di Luzio et al. (1988) demonstrated that natural or recombinant murine interferon-γ caused an irreversible arrest of proliferation of PC12 cells. On the other hand, other antimitotics led to mitotic arrest followed by cell death. Interferon-γ-treated PC12 cells responded more rapidly to nerve growth factor. From these and other findings, they proposed the use of a combined treatment of PC12 cells with nerve growth factor and interferon-γ for a more rapid induction of neuronal differentiation.

Katoh-Semba, Skaper and Varon (1986) examined the incorporation of tritiated galactose into endogenous gangliosides, other glycolipids, and glycoproteins as well as incorporation of [^{14}C]acetate into lipids by PC12 cells cultured with nerve growth factor, different combinations of serum, and exogenous serum ganglioside G_{M1}. They reported that nerve growth factor stimulates the synthesis of all four classes of endogenous constituents, regardless of the amount of serum accompanying it. Exogenous G_{M1} stimulates the synthesis of the three lipid classes but not of glycoprotein. For all constituents stimulated by G_{M1}, concurrent treatment with nerve growth factor produces cumulative effects, suggesting independent mechanisms of action by the two molecules.

Seidl, Manthorpe, Varon et al. (1987) examined the effects of nerve growth factor and ciliary neuronotrophic factor on catecholamine content and *in vitro* activities of tyrosine hydroxylase and phenylethanolamine *N*-methyltransferase in adrenal chromaffin cells cultured from 8-day-old rats. Both growth factors enhanced chromaffin cell survival and partially prevented losses of epinephrine. The ciliary neuronotrophic factor was more potent, although cellular levels of epinephrine and norepinephrine were not maintained. Nerve growth factor did not add to the effect of ciliary neuronotrophic factor. Nerve growth factor induced tyrosine hydroxylase but not phenylethanolamine *N*-methyltransferase activity, and ciliary neuronotrophic factor did not change the activity of either enzyme. These data indicate that there are differences between the mechanisms by which nerve growth factor and ciliary neuronotrophic factor affect adrenal chromaffin cells.

Hall, Fernyhough, Ishii et al. (1988) demonstrated that nerve growth factor-directed neurite outgrowth in PC12 cells could be suppressed by sphingosine, an inhibitor of protein kinase C. This was a reversible dose-dependent inhibition. These and other results indicated the presence of a sphingosine-sensitive pathway in neurite outgrowth and indicated that protein kinase C plays a role in mediating the neuritogenic effects of nerve growth factor. Furthermore, the results suggested that protein kinase C acts at a distal segment of the neurite growth pathway.

Kobayashi, Izumi and Meldolesi (1986) demonstrated that, during culture, chromaffin cells become sensitive to α-latrotoxin even before their phenotype is modified. The sensitivity to the toxin was greater in cells treated with nerve growth factor or obtained from young rats. Acquisition of sensitivity to α-latrotoxin appears to be an early marker of chromaffin cell commitment to differentiate toward a neuron.

Because changes in the activity of glycogen phosphorylase may reflect changes in either cytosolic free calcium or cAMP, Davis and Kauffman (1986) examined the effects of nerve growth factor on the activity of this enzyme in PC12 cells. Nerve growth factor increased the amount of the active phosphorylated form of the enzyme within minutes. In contrast, insulin and epidermal growth factor decreased the enzyme activity. Activation of the phosphorylated enzyme by nerve growth factor was not accompanied by increases in cAMP; however, removal of extracellular calcium or incubation of cells with calcium channel blockers inhibited activation of glycogen phosphorylase by nerve growth factor.

Damon, D'Amore and Wagner (1988) evaluated the effect of glycosaminoglycans on the ability of PC12 cells to extend neurites in response to nerve growth factor, basic fibroblast growth factor, and acidic fibroblast growth factor. Sulfated glycosaminoglycans altered the neurite outgrowth response of PC12 cells to all three growth factors. The effects varied with the concentration of glycosaminoglycans and the growth factor. They presented a large amount of data to indicate that not only heparin but also other sulfated glycosaminoglycans found in the extracellular matrix can alter the activity of neuronal growth factors *in vitro* and thus may also influence growth factor-mediated functions *in vivo*.

Volonté (1988) demonstrated that lithium stimulates the binding of GTP to membranes of PC12 cells cultured with nerve growth factor. The binding was increased about 5-fold. It was suggested that this may be related to the action of lithium in its therapeutic effects.

The effects of bradykinin and lithium on the phosphatidylinositol cycle were examined in PC12 cells cultured in the presence or absence of nerve growth factor by Volonté, Parries and Racker (1988). A direct link between bradykinin and nerve growth factor with respect to the phosphatidylinositol cycle of PC12 cells was demonstrated. A small increase in phosphatidylinositol metabolism was demonstrated in the absence of nerve growth factor, and an approximately 4-fold stimulation was found in the presence of nerve growth factor. This was enhanced in the presence of lithium.

OTHER GROWTH FACTORS

The differentiation and phenotypic conversion of adrenal medullary cells were reviewed by Unsicker (1986). He pointed out that chromaffin cell differentiation and phenotype expressions are affected by neuronotrophic, neurite-promoting, hormonal, and neuronal signals. Although much of our understanding comes from *in vitro* studies, *in vivo* validation of the effective agents is essential.

Unsicker, Reichert-Preivsch, Schmidt et al. (1987) demonstrated that astroglial and fibroblast growth factors have neurotrophic effects on cultured peripheral and central nervous system neurons. Their results support the emerging evidence that mitogenic and neuronal growth factors are not strictly separate entities.

Gospodarowicz, Baird, Cheng et al. (1986) isolated and characterized basic fibroblast growth factor from bovine adrenal gland. Homologies were noted between this growth factor and bovine pituitary and brain fibroblast growth factors. The biologic activity of the adrenal fibroblast growth factor was indistinguishable from the activities of other fibroblast growth factors. The adrenal fibroblast growth factor was mitogenic for vascular endothelial cells and stimulated the proliferation of a wide variety of mesoderm- and neuroectoderm-derived cells.

Stemple, Mahanthappa and Anderson (1988) examined the effect of basic fibroblast growth factor on cultured rat adrenal chromaffin cells. Like nerve growth factor, basic fibroblast growth factor induced cell division and neurite outgrowth from the cells. Dexamethasone inhibited neuronal differentiation but not the proliferation induced by basic fibroblast growth factor. Unlike nerve growth factor, basic fibroblast growth factor did not support the survival of chromaffin cell-derived sympathetic neurons. The overlapping but distinct responses to nerve growth factor and basic fibroblast growth factor may underlie a sequence of events in sympathetic differentiation. In the adrenal gland, glucocorticoids may permit basic fibroblast growth factor to amplify the chromaffin cell population while preventing neuronal differentiation.

Neufeld, Gospodarowicz, Dodge et al. (1987) analyzed the effect of acidic fibroblast growth factor on neurite outgrowth from PC12 cells and how it is modified by heparin. Their results indicate that, given sufficient time, heparin can potentiate the biologic effect of acidic fibroblast growth factor, making it as active as basic fibroblast growth factor. In addition, heparin modulates the effects of basic fibroblast growth factor and nerve growth factor. Both acidic and basic fibroblast growth factors can support the long-term neurite outgrowth produced by PC12 cells, and both interact with the same cell surface receptor.

In a search for agents that can mimic some or all of the responses of PC12 cells to nerve growth factor, Rydel and Greene (1987) examined acidic and basic fibroblast growth factors for their ability to promote both early and delayed responses of PC12 cells and compared these responses with those elicited by nerve growth factor. The fibroblast growth factors mimicked all of the nerve growth factor responses tested. They produced stable neurite outgrowth and neuronal differentiation in PC12 cell cultures. These findings support a neurotrophic role for fibroblast growth factors in nervous system development.

Claude, Parada, Gordon et al. (1988) demonstrated that acidic fibroblast growth factor stimulates cultured rat adrenal chromaffin cells to proliferate and to extend neurons, but it is not a long-term survival factor. Acidic fibroblast growth factor was as potent as nerve growth factor in stimulating these cultured cells but failed to support long-term survival. The effects of basic fibroblast growth factor were similar, but not identical, to those of acidic factor. It was demonstrated that the regulatory pathways controlled by acidic fibroblast growth factor, basic fibroblast growth factor, and nerve growth factor are partially distinct.

The distribution of epidermal growth factor receptors was studied in the adrenal gland by Chabot, Walker and Pelletier (1986). They used light-microscopic autoradiography after injection of radioiodinated epidermal growth factor into the adult rat. The labeling was associated with cells of the capsule and the adrenal chromaffin cells. The results clearly indicate that epidermal growth factor receptors are present in the majority of adrenal medulla cells and suggest that this growth factor may have a physiologic role in the regulation of the rat adrenal medulla.

Lazarovici, Dickens, Kuzuya et al. (1987) examined the long-term heterologous down-regulation of the epidermal growth factor receptor in PC12 cells induced by nerve growth factor. Treatment of the cells with nerve growth factor induces a progressive and nearly total decrease in the specific binding of epidermal growth factor, beginning after 12 hours and completed within 4 days. Experiments demonstrated that there is a decreased amount of receptor. These findings suggest that long-term heterologous down-regulation of epidermal growth factor receptors is mediated by an alteration in receptor synthesis. It is further suggested that this down-regulation is part of the mechanism by which differentiating cells become insensitive to mitogens.

Boonstra, Mummery, Feyen et al. (1987) observed an increased number of high-affinity epidermal growth factor binding sites during treatment of PC12 cells with dibutyryl cAMP, despite extensive neurite formation. A number of rapid responses of epidermal growth factor were comparable in the cAMP-treated and untreated PC12 cells, in contrast to nerve growth factor-treated cells.

Wagner (1986) reported the partial purification, from mouse submaxillary glands, of a factor that causes neurite extension in PC12 cells but appears to be distinct from nerve growth factor. This novel neurite-inducing factor can be distinguished from nerve growth factor by immunologic, pharmacologic, and biochemical criteria, suggesting that it is a distinct gene product.

Haselbacher, Irminger, Zapf et al. (1987) demonstrated insulin-like growth factor in human pheochromocytomas. Two forms of the molecule were found with molecular masses of 10- and 7.5 kDa. The tumors contained about 20 times more of the insulin-like growth factor than normal adrenal medulla.

Dahmer and Perlman (1988b) examined bovine adrenal chromaffin cells to determine whether they contain insulin-like growth factor I receptors and whether insulin-like growth factor I affects the differentiated functions of the cells. The radioiodinated factor bound to chromaffin cells at a higher affinity than insulin did. They partially characterized the receptor as a protein with a molecular mass of 125 kDa. Cells cultured with the factor exhibited an almost 2-fold increase in high-potassium-evoked catecholamine secretion, and insulin alone was much less potent in enhancing catecholamine secretion. These data indicate that binding of insulin-like growth factor I to its receptors on chromaffin cells can modulate the function of the cells.

Insulin-like growth factor I was localized immunohistochemically in the adult rat by Hansson, Nilsson, Isgaard et al. (1988). The factor was in the adrenal medulla and several other organs.

Nielsen and Gammeltoft (1988) addressed the issue of a mitogenic effect of insulin-like growth factors I and II on PC12 cells. The proliferation of PC12 cells cultured in a serum-free medium was stimulated 3-fold by both factors, significantly higher potency than that of several growth factors. The mitogenic response of the PC12 cells appeared to be mediated by insulin-like growth factor I. It was suggested that insulin-like growth factors act as mitogens on pluripotent chromaffin cells during the development of the sympathetic nervous system.

The relationships between chromaffin cells bearing neurite-like processes and nonchromaffin cells were investigated in cultures of adult mouse adrenal medulla by Derer, Grynszpan-Winograd and Thibault (1987). After 20 days in culture, 8.5% of the chromaffin cells had extended at least one neurite, and this was not enhanced by the addition of nerve growth factor to the culture medium. Neurites sometimes were ensheathed by nonchromaffin cells. In the absence of adjacent nonchromaffin cells, chromaffin cells retained a glandular phenotype. It is suggested that a definite proportion of adult mouse chromaffin cells in culture are able to extend neurite-like processes if they are in close relationship to nonchromaffin cells. The influence of the nonchromaffin cells could be mediated by an unidentified growth factor.

Tomaselli, Damsky and Reichardt (1987) examined interactions of PC12 cells with laminin, collagen IV, and fibronectin. The PC12 cells were shown to adhere readily to these substrates in a specific fashion. To identify PC12 cell surface proteins that mediate interactions with the substrates, two different antisera to putative receptors were tested. These antisera were effective in limiting adhesion and allowed identification of three prominent cell surface glycoproteins. At least one of these glycoproteins could be a receptor for extracellular matrix constituents, laminin, fibronectin, and some collagens.

Ignatius, Shooter, Pitas et al. (1987) examined the uptake of lipoprotein by growth cones of PC12 cells. Labeled lipid particles and lipoproteins were internalized by the neurites and their growth cones. The lipoprotein uptake appeared to be performed by an apolipoprotein receptor.

An adrenal growth factor was partially purified from a culture of ovine and anterior pituitary cells by Samsoondar and Kudlow (1987). This adrenal growth factor appeared to be distinct from epidermal growth factor.

Demeneix and Grant (1988) found that chromaffin cells from adult bovine adrenal medulla developed neurites when cocultured with pituitary intermediate lobe cells. A soluble factor released by intermediate lobe cells apparently was involved. Moreover, the addition of α-melanocyte-stimulating hormone, one of the peptides secreted by intermediate lobe cells, reproduced the effect in a dose-dependent manner. This provided evidence of a neurotrophic role of α-melanocyte-stimulating hormone.

Sugimoto, Noda, Kitayama et al. (1988) examined the mechanism by which the Harvey sarcoma virus induces some neuron-associated properties in PC12 cells. The extent of the production of neuron-associated properties in transfected cells after the addition of dexamethasone seemed to correlate well with the levels of the gene expression. These cells were able to sustain increased levels of cAMP as well as phosphatidylinositol metabolites, inositol trisphosphate, and diacylglycerol. In cells not transfected with the Harvey sarcoma virus, dexamethasone had no effect. These observations suggested that the role of the gene product in the system may involve simultaneous activation of two signaling pathways, one mediated by cAMP and one mediated by phosphatidylinositol turnover.

Sastry, Chirwa, May et al. (1988) found that samples collected from the rabbit neocortical surface during a tetanic stimulation of the neocortex induced neurite outgrowth in PC12 cells. This was prevented if the samples were preheated and cooled. It was suggested that neurite-inducing factors are released during tetanic stimulation.

CHAPTER 14

CELL CULTURE

Several of the studies reviewed in previous chapters utilized chromaffin cells or PC12 cells in culture. This chapter will deal with studies of cultured cells that do not fit into the categories of previous chapters.

Detailed methods for the isolation and culture of bovine adrenal chromaffin cells were published by Livett, Mitchelhill and Dean (1987). This laboratory has had the longest experience with isolated chromaffin cells, and the step-by-step methods presented would be useful to anyone entering this field.

One of the continuing problems in extracting data from studies of isolated chromaffin cells is the question of the purity of the preparation. The presence of nonchromaffin cells can confuse the results. Wilson (1987c) presented a new method that may help to resolve this problem in which bovine adrenal chromaffin cells are purified on a discontinuous gradient of the radiopaque contrast agent Renografin. Cells prepared in this way contained about 20% more catecholamines than did those prepared by standard techniques. Morphologic examination indicated that the cells prepared on Renografin gradients were approximately 90% chromaffin cells compared with a mean of 75% purity obtained by another method.

Unsicker, Stahnke and Müller (1987) investigated the survival, morphology, and catecholamine storage of chromaffin cells in serum-free culture. Chromaffin cells dissociated from adrenal glands of 8-day-old rats were maintained in a defined medium for up to 6 days. Omission of serum caused a small reduction in the number of surviving cells but did not diminish the survival effects of trophic agents. They also demonstrated that bovine adrenal chromaffin cells release, into their culture medium, trophic substances that support the survival of chromaffin cells.

Various populations of bovine adrenal chromaffin cells were isolated by successive digestions with collagenase followed by sedimentation through a stepwise bovine serum albumin gradient by Lemaire, Dumont, Mercier et al. (1983). Each cell layer had the same total amount of catecholamines, but the epinephrine-norepinephrine ratio varied from 0.6 in the top layer to 3.3 in the bottom layers. Each layer also responded differently to acetylcholine. These results indicate bovine adrenal chromaffin cells can be separated according to their content of epinephrine and norepinephrine and their response to specific stimuli.

Chromaffin cells from the monkey adrenal medulla were maintained *in vitro* by Notter, Gupta and Gash (1986). The cells appeared to maintain the morphologic characteristics of chromaffin cells, and many had neurites with the morphologic features of axons. Notter et al. (1986) pointed out the potential use of adult primate chromaffin cells as a source of neuronal tissue for neural transplantation.

An autoradiographic study of RNA synthesis in adrenal cells of the young rat in primary culture was conducted by Magalhães, Bonito-Vitor and Magalhães (1988). Control and adreno-corticotropin-stimulated cultures were incubated with tritiated uridine. Labeling over nucleolar and extranucleolar areas was always lower in the stimulated specimens. Pulse-chase experiments suggested a slow metabolism of RNA in adrenal cells.

Jaques and Tobes (1986) maintained primary cell cultures from 18 human pheochromocytomas for up to 12 days. The characteristics of the cultured cells included a wide variation in cell size, a wide range in catecholamine content, and a decrease in cell catecholamine content with time. This study presents a large data base on culturing cells from these tumors and describes many of the morphologic and biochemical characteristics of this cell system.

The effects of added soluble glycosaminoglycans on adhesion and neurite formation by cultured PC12 cells on several substrates were tested by Akeson and Warren (1986). With PC12 cells, adhesion and neurite formation can be inhibited by sulfated glycosaminoglycans on some substrates, including fibronectin, but not other substrates, suggesting that these cells have at least two independent molecular adhesion mechanisms.

Turner, Flier and Carbonetto (1987) reported a study of the substratum and medium requirements for attachment and neurite outgrowth of PC12 cells. In the presence of calcium and magnesium, the cells attached to substrata coated with several types of collagen, laminin, wheat germ agglutinin, or poly-L-lysine. In the absence of magnesium, the cells failed to attach to collagen or laminin. Omitting calcium from the medium did not have an obvious effect on cell attachment. These results suggest that a magnesium-dependent adhesion mechanism is operative.

Bethea, Rønnekleiv and Kozak (1987) examined the effect of extracellular matrix on PC12 cells, including the processing of dopamine. PC12 cells were grown on extracellular matrix or plastic and incubated with tritiated tyrosine in the presence and absence of serum or unlabeled tyrosine. PC12 cells grown on extracellular matrix released significantly more labeled dopamine, whereas cells grown on plastic had a significantly higher content of labeled dopamine. Morphometric relationships also were different between the cells grown on the two surfaces. The data suggest that there is an increase in the ratio of cell surface area to cell volume in PC12 cells grown on an extracellular matrix. The change in cell shape does not appear to alter the synthesis of dopamine immediately.

The effect of an extracellular matrix derived from an Englebreth-Holm-Swarm tumor on PC12 cell growth was examined by Bethea and Borg (1988). PC12 cells grown on plastic form a single layer of cells, but when they were grown on the extracellular matrix they formed multicellular aggregates. The plating efficiency of PC12 cells was decreased on the matrix, but the doubling time of cells on the matrix was comparable to that on plastic. Dopamine secretion and cellular content determined with a radioenzymatic assay, as well as dopamine synthesis, were similar on a per cell basis in cultures of PC12 cells on plastic and on the matrix.

Tomaselli, Damsky and Reichardt (1987) studied the interactions of PC12 cells with substrates coated with an extracellular matrix protein. Using quantitative cell attachment assay, PC12 cells were shown to adhere readily to laminin or collagen IV but poorly to fibronectin. The specificity of attachment to these extracellular matrix proteins was demonstrated by using ligand-specific antibodies and synthetic peptides. These authors used two additional antisera that recognize specific extracellular matrix receptors. They presented evidence for both a structural and a functional relationship between the PC12 cell surface glycoproteins recognized by these antisera and extracellular matrix receptors belonging to the integrin family of adhesive protein receptors.

PC12 cells were used by Kittner, Bräutigam and Herken (1987) as a model system to examine drug effects on dopamine synthesis and release. Specifically, they examined the interdependence among potassium-induced release of endogenous dopamine, production of dopamine, and the supply of tetrahydrobiopterin. They also examined the influence of various drugs such as calcium channel blockers and calmodulin inhibitors. They were able to demonstrate the versatility of PC12 cells for studying drug effects on catecholamine release and synthesis.

In a clinically related study, Bitler and Howard (1986) examined dopamine metabolism in variants of PC12 cells that were deficient in hypoxanthine-guanine phosphoribosyltransferase. They found no correlation between the enzyme activity and endogenous dopamine levels, dopamine uptake, dopamine release, or monoamine oxidase. Transformation of the cells with a hypoxanthine-guanine phosphoribosyltransferase retrovirus restored the enzyme activity and did not adversely affect dopamine metabolism. This study has clinical relevance to the Lesch-Nyhan syndrome which is linked to a deficiency in hypoxanthine-guanine phosphoribosyltransferase.

Matsuoka, Satake and Kurihara (1986b) showed that adhesion of PC12 cells to the substratum is promoted by factors contained in glioma-conditioned medium. They were able to separate two active factors from the conditioned medium--proteins of apparent molecular masses 40- and 10 kDa. The larger protein had the abilities to induce neurite outgrowth and to enhance the activity of choline acetyltransferase in addition to the ability to promote the adhesion of the cells. The smaller protein did not induce neurite outgrowth but had the other two properties.

Byrd and Alho (1987) showed that PC12 cells cease dividing and show increased cell-to-cell and cell-to-substratum adhesion in response to treatment with sodium butyrate. These changes are accompanied by the rapid appearance of neuron-specific enolase. However, neurofilament proteins are not induced. These results suggest that sodium butyrate induces differentiation of PC12 cells, perhaps along the chromaffin cell pathway.

Guerrero, Wong, Pellicer et al. (1986) showed that, when activated mouse N-*ras* gene is inserted into PC12 cells, proliferation is suppressed and neuronal differentiation is promoted. They presented findings to suggest that, in certain cells, *ras* genes can play a role in promoting differentiation and suppressing proliferation, in contrast to their established oncogenic neoplasia-promoting activity in other cells.

Cornet, Delpire and Gilles (1987) examined the microfilament network during the volume regulation process of cultured PC12 cells. Hypo-osmotic conditions did not significantly change the general configuration and density of the microtubules. Microfilaments did change in their distribution, becoming more concentrated around the nucleus with radial extensions to the cell periphery.

Enhancement of neurite outgrowth in PC12h cells by a protease inhibitor was described by Saito and Kawashima (1988). Of 14 protease inhibitors examined, only a leupeptin analog was stimulatory for neurite outgrowth in the presence of nerve growth factor. Leupeptin and soybean trypsin inhibitor, which had been reported to induce neurite outgrowth in other systems, had no effect. These results suggest an endogenous protease of a new type is involved in restricting neurite outgrowth in PC12h cells.

Ruzzier, Lee, Dryden et al. (1988) co-cultured adult rat muscle fibers and PC12 cells. Functional synapses were formed between the two cell types. This demonstrated that dissociated adult muscle fibers retain the ability to be reinnervated and can form the function synapses with transformed chromaffin cells.

CHAPTER 15

TRANSPLANTATION

In the eyes of the lay public, undoubtedly the most exciting current information about the adrenal medulla involves the autologous transplantation of tissue from this organ into the brain as a treatment for Parkinson's disease. This chapter will cover the experimental work with nonhumans and then the clinical trials, including the controversies surrounding them.

ANIMAL MODELS

Much of the earlier work on transplantation of the adrenal medulla in animal models was reviewed by Olson, Backlund, Freed et al. (1985). This review stressed experiments directed toward a treatment for Parkinson's disease. Papers from a meeting on cell and tissue transplantation into the adult brain were assembled by Azmitia and Björklund (1987a). They also published a summary of this meeting (Azmitia and Björklund, 1987b). Björklund, Lindvall, Isacson et al. (1987) reviewed the mechanisms of action of intracerebral neural implants, focusing on how implants exert their functional effects. A brief review of studies in lower animals that indicate a future for brain implants was published by Friedman (1986).

Olson, Backlund, Gerhardt et al. (1987) reviewed studies on rodent models and presented preliminary information on transplanted patients. Among other topics, they discussed the advantage of putting the adrenal medulla transplant into the putamen rather than the caudate nucleus because recent studies have implicated the putamen as a more important component of the motor disturbances in Parkinson's disease. This question cannot be studied in rodents because the striatum is not anatomically subdivided into caudate and putamen.

Olson, Strömberg, Bygdeman et al. (1987) reported on experiments in which human fetal tissues were grafted into rodent hosts. Catecholamine-rich tissues such as adrenal chromaffin cells, sympathetic ganglia, central dopaminergic neuroblasts, and neuroblasts from the locus ceruleus all survived grafting into the anterior chamber of the eye of immunocompromised rats and mice. The data suggest that, under their experimental conditions, transplantation of human fetal neural tissues is a unique model system for studies of human brain development, developmental disturbances, connectivity, and the action of drugs. This model also allows evaluation of adrenal medullary implants.

Strömberg, Hultgårdh-Nilsson, Hedin et al. (1988) investigated the ability of adult adrenal chromaffin cells to transform to a neuronal phenotype in the presence of nerve growth factor when injected as a cell suspension into the anterior eye chamber. After 6 days *in oculo*, all cells were immunoreactive for epinephrine; almost none displayed processes, even in the presence of nerve growth factor. One month after weekly intraocular injections of nerve growth factor, many cells were surrounded by nerve fiber networks, and all cells were immunoreactive for dopamine β-hydroxylase. The number of ganglion cells was remarkably increased with time by nerve growth factor; the number of chromaffin cells decreased compared with controls. A single treatment with nerve growth factor at the time of grafting had no marked effects.

Morihisa, Nakamura, Freed et al. (1987) transplanted adrenal medulla and fetal substantia nigra tissue into brains of monkeys that had been treated with 6-hydroxydopamine, an agent that is neurotoxic to dopaminergic neurons. In this animal model of Parkinson's disease, they found that in all five animals with adrenal medulla implants there was at least some catecholamine histofluorescence in the grafts up to 8 months after implantation. In contrast, no catecholamine-containing graft tissue was detected in the two animals that received the fetal substantia nigra implants.

Becker, Adams and Robinson (1988) developed an intraventricular microdialysis method that could be used to measure monoamines from adrenal medullary grafts in the brains of freely-moving rats. Using this technique, Becker and Freed (1988) found that the functional activity of the striatal dopamine system could be enhanced in lesioned rats, although dopamine could not be detected in cerebrospinal fluid. It was determined that adrenal medullary grafts increase striatal dopamine activity without an appreciable release of dopamine into the cerebrospinal fluid.

The effects of intracerebral transplantation of the adrenal medulla in rats with experimental Parkinson's disease was studied by Dymecki, Póltorak, Markiewicz et al. (1986). Histo-fluorescence examination indicated that fragments of adrenal medulla implanted into the striatum survived and were not rejected. However, behavioral tests showed no improvement after transplantation, even though dopamine levels in isolated striatum on the side opposite to the implant were increased. They suggested this is due to the inability of neurons on the transplanted side to utilize dopamine, and dopamine released by the transplant is captured by intact neurons on the opposite side.

Long-term survival and behavioral effects of intrastriatal adrenal medullary grafts in rats was studied by Freed, Cannon-Spoor and Krauthamer (1986). They found that much of the implanted tissue did not survive, although 200 chromaffin cells per recipient rat were found to have survived for at least 6 months. All of the surviving cells developed cytoplasmic extensions, although these processes did not appear to have reinnervated host brain tissue. These and other observations lead to the conclusion that, although the corpus striatum does not appear to provide a particularly favorable environment for the implantation of adrenal medulla grafts, striatal implants of adrenal medulla might become a promising procedure if a means of improving the survival of the tissue was developed.

Jiao, Wang, Cai et al. (1987) found that adrenal medulla grafts to the denervated caudate nucleus in rats survive and lead to reversal of behavioral abnormalities. These grafted adrenal chromaffin cells contained catecholamines, as indicated by histofluorescence. Some cells retained the typical appearance of normal chromaffin cells, and other cells resembled neurons. The concentration of dopamine in the grafts was found to be very high.

Nishino, Ono, Shibata et al. (1988) implanted adrenal medullary cell suspensions, derived from newborn rats, into the head of the caudate nucleus in 35 rats with unilateral lesions in the nigrostriatal pathway. The circling behavior indicative of nigrostriatal pathway lesions decreased significantly in 43% of the rats. That decrease was concurrent with transmutation of the tyrosine hydroxylase immunopositive cells into mature neurons that had abundant elongated neurites with varicosities and synapses on neuronal elements in the host caudate. In the animals that did not show the behavioral recovery, tyrosine hydroxylase-positive cells were very sparse. These and other results suggested that some grafted adrenal medullary cells transformed into dopaminergic neurons, and the release of dopamine from these grafted cells functionally affected behavior improvement.

Pezzoli, Fahn, Dwork et al. (1988) demonstrated that ventricular grafts of either nonchromaffin or adrenal medullary tissue are equally effective in decreasing lesion-induced circling in rats. The intraventricular infusion of nerve growth factor was a more effective treatment than implantation of adrenal medulla without nerve growth factor. In addition, the effects persisted, although at a reduced level, after discontinuation of the nerve growth factor infusion. The results suggest that trophic factors may be crucial to the beneficial effects of intracerebral transplanted tissues.

Kamo, Kim, McGeer et al. (1987) placed cultured human fetal adrenal chromaffin cells into the striatum of young adult rats that previously had had created a unilateral 6-hydroxydopamine lesion of the nigrostriatal dopamine pathway. Correction of the behavioral abnormalities was noted in four of eight rats given such transplants when tested up to 4½ months later; two rats improved slightly and two rats deteriorated. Six rats given sciatic nerve grafts as controls all showed deterioration. Catecholamine fluorescence and immunohistochemical examination demonstrated neurons and neuronal processes positive for catecholamines or tyrosine hydroxylase in the area of the transplant. This transplantation of cultured human fetal cells to an animal model may provide the necessary basic experimental system for assessing the possible utility of human neuronal transplants.

Brundin, Strecker, Widner et al. (1988) transplanted dopaminergic neurons from fetal human mesencephalon into the striatum of rats with a lesion in their nigrostriatal pathway. Grafts within the striatum gave rise to an extensive axonal network throughout the entire poststriatum in rats that had been immunosuppressed. There was no survival or improvement in behavioral effects of human dopamine neurons implanted in rats that did not receive immunosuppression.

An important consideration when studying transplants in the brain is the condition of the blood-brain barrier. Rosenstein (1987a,b) showed that grafts of adrenal medulla into the rat brain produced significant blood-brain barrier dysfunction, whether the graft was in the parenchyma of the brain or in a ventricle. Systemically administered protein rapidly entered either the grafts, the surrounding brain tissue, or the cerebrospinal fluid. The grafted chromaffin cells avidly took up systemically administered radiolabeled amines. Thus, transplantation of neural tissue prevented normal reconstitution of the blood-brain barrier after operation and exposed host brain tissue to blood-borne compounds indefinitely.

In a related study, Krum and Rosenstein (1987) examined patterns of angiogenesis in neural transplant models. They pointed out that functional vascular connections must form rapidly to prevent ischemic damage to grafts of neural tissues. They examined the time course and mechanism of vascular reperfusion in allografts of adrenal medulla or superior cervical ganglia inserted either into the fourth ventricle or directly into the parietal cortex of perinatal rats. They were able to demonstrate that intraventricular graft vessels initially collapsed but sustained minimal ischemic damage and were completely reperfused within 24 hours. Vessels in intraparenchymal grafts sustained more severe damage. This study also demonstrated that mature autonomic tissue stimulates the growth of adjacent host vessels when transplanted to undamaged brain surfaces. This also occurs with intraparenchymal grafts, although the rapidity of reperfusion appears to be predicated on the amount of trauma at the graft site.

Grynszpan-Winograd and Jousselin-Hosaja (1988) and Jousselin-Hosaja (1988a, 1988b) transplanted adrenal medullary pieces into the brains of mice. These grafts survived for periods of at least one month. Cortex-free transplants placed within cerebral tissue developed processes with the characteristics of neurites that extended from the chromaffin cells and formed synapses with other chromaffin cells. Additionally, reinnervation of chromaffin cells in cortex-free grafts was observed. These results suggest that the grafts became integrated into the host brain.

In an elegant series of experiments, a group from the University of Rochester investigated use of both the adrenal medulla and other sources of catecholamine-secreting tissue for transplantation. Hansen, Notter, Okawara et al. (1988) described the heterogenous character of transplanted pieces of adrenal medulla; this tissue contained isolated islands of cortical cells, in addition to Schwann cells, nerve endings, endothelial cells, pericytes, isolated ganglionic neurons, and connective tissue components, such as smooth muscle cells and fibroblasts. To circumvent the potential problems that could arise from transplanting mitotically active tissue, Hanson et al. differentially plated the chromaffin cells prior to transplantation. This procedure yielded relatively pure populations of cells that were viable if processed within two hours after interruption of the blood supply to the adrenal gland. Addition of nerve growth factor to the cultured chromaffin cells caused the cells to exhibit a neuronal phenotype. Bing, Notter, Hansen et al. (1988) and Hansen, Bing, Notter et al. (1988) examined the suitability of adrenal medullary and carotid body glomus cells as implants. Both types of cells survived for at least 30 days in the 6-hydroxydopamine-denervated rat striatum, and effectively reversed amphetamine-induced rotational behavior in this animal model. However, the average diameter of the dense-cored vesicles of the grafted chromaffin cells decreased in size when compared to normal adrenal medullary cells. Grafted glomus cells appeared to dedifferentiate in that they possessed few dense-cored vesicles, although they extended numerous processes surrounded by a thick basal lamina.

Hansen, Kordower, Fiandaca et al. (1988) and Fiandaca, Kordower, Hansen et al. (1988a, 1988b) transplanted adrenal medullary tissue into the striatum of normal and MPTP-treated *Cebus* monkeys. At 30 days posttransplantation, the transplantation site resembled a chronic inflammatory focus. Grafted chromaffin cells were identified ultrastructurally in only a few of the monkeys; these cells showed signs of degeneration and were surrounded by phagocytic macrophages. In spite of the poor survival of the transplanted chromaffin cells, robust sprouting of tyrosine hydroxylase-like immunoreactive fibers was seen up to several millimeters from the graft site. Control stereotactic implants also enhanced immunoreactive fibers. The authors suggested that parenchymal injury during the implantation procedure may induce recovery of the remaining host dopaminergic systems.

In further attempts to identify sources of dopamine-secreting tissue for transplantation to correct Parkinson's disease, some laboratories have transplanted PC12 cells. Freed, Patel-Vaidya and Geller (1986) implanted PC12 cells into the striatum of rats and examined the grafts 1 day to 20 weeks afterward. They found that, in some animals, the number of PC12 cells increased initially but in these animals the graft was ultimately rejected. In other animals, small numbers of PC12 cells survived for the 20 weeks and many of these cells eventually developed processes. Continued uncontrolled tumor growth was not observed. Jaeger (1987), on the other hand, found that PC12 cells continued to undergo mitosis and the growth of such grafts occurred at the expense of the host brain. The transplanted PC12 cells did synthesize transmitters that could serve as dopamine replacement sources. If the growth of the PC12 cells could be limited, they may have some clinical usefulness.

Pappas and Sagen (1986) examined PC12 implants in the rat spinal cord. PC12 cells were injected into the subarachnoid space and the lumbar spinal cord of adult rats. These cells invaded the spinal cord as a metastatic tumor. In spite of the destruction and invasion of the spinal cord parenchyma, no glial reaction could be seen.

Nilsson, Dahlström, Tisell et al. (1986) transplanted tissue pieces from benign human pheochromocytoma to the anterior eye chamber of cyclosporine-treated rats. The grafts were vascularized within 2 days, and no increase in the size of the transplants was noted during the 4 weeks of observation. Tumor transplants grown in nonimmunosuppressed rats began to show signs of rejection 1 week after transplantation.

The effects of neuronotrophic factors on adult adrenal medulla grafts transplanted into the rat caudate nucleus after a lesion was created in the nigrostriatal pathway were investigated by Korfali, Doygun, Ulus et al. (1988). The neuronotrophic factors were extracted by placing a gelatin sponge in the frontal lobe of 10-day-old rats. These sponge pieces then were placed on the adrenal medulla grafts. Two months after transplantation, all of the adrenal medulla grafts were treated with neuronotrophic factors; but only 45% of the untreated grafts survived. The levels of tyrosine hydroxylase activity in the caudate nucleus were not significantly different between groups. The results indicate that treatment with neuronotrophic factors significantly enhances the survival rate of adrenal medullary grafts, although they may not influence catecholamine biosynthesis.

An interesting series of publications explored another purpose for transplantation of adrenal medulla tissue to the central nervous system. Sagen, Pappas and Perlow (1986) showed that transplants of adrenal medulla in the rat spinal cord decreased sensitivity to pain. The rationale was that opioid peptides and catecholamines are known to induce analgesia when injected directly into the spinal cord. Because these and other substances are released from the adrenal medulla, adrenal medullary transplants could induce analgesia similarly. Pieces of rat adrenal medulla were placed in the subarachnoid space of rat spinal cords and stimulated by a low dose of nicotine. This induced a potent analgesia in animals with adrenal medullary transplants but not in animals with control transplants. Furthermore, this analgesia was reversed to pre-nicotine levels by naloxone.

When Sagen, Pappas and Pollard (1986) implanted bovine chromaffin cells into the subarachnoid space in adult rats, they also obtained results suggesting that analgesia is due to the stimulated release of opioid peptides and catecholamines from the implanted chromaffin cells. Sagen and Pappas (1987) and Sagen, Pappas and Perlow (1987a) transplanted rat and bovine adrenal medullary cells into the rat midbrain periaqueductal gray, a region known to play a primary role in the modulation of pain. Potent analgesia was induced when the grafts were stimulated with a low dose of nicotine. The bovine chromaffin cell implants were more effective in inducing this response. The results suggest that the implantation of cells that release neuroactive substances can decrease sensitivity to pain.

Sagen, Pappas and Perlow (1987b) examined the fine structure of adrenal medullary grafts into the rat periaqueductal gray. A striking change found 8 weeks after transplantation was that pronounced myelination had taken place, both in the graft and in the host tissue. The new myelin formation in the graft had the typical appearance of myelination in the peripheral nervous system; in the host, it had the appearance of myelination in the central nervous system. Chromaffin cells were found protruding into the host tissue and sometimes forming synapses, presumably with the host neuronal processes. The behavioral relevance of these synaptic contacts is unclear because similar implants of adrenal medullary tissue into the subarachnoid space also induce potent analgesia but do not contain synapses. Thus, it is more likely that behavioral changes are brought about by diffusion of neuroactive substances from graft chromaffin cells to host receptors.

In continuing studies, Sagen and Pappas (1987, 1988a, 1988b) and Pappas and Sagen (1988a, 1988b, 1988c) transplanted adrenal chromaffin cells into portions of the central nervous system associated with the perception of pain. Two articles emphasized the nature of the vasculature associated with these grafts (Sagen and Pappas, 1988a; Pappas and Sagen, 1988a).

A clinical trial of this concept was attempted by Vaquero, Martinez, Oya et al. (1988) in a terminally ill man in severe pain. He received pieces of the adrenal medulla from a traffic accident victim. There was some amelioration of the pain for 4 days after the operation; then, the pain recurred and persisted until the patient died. This first clinical study suggests that grafting of the adrenal medulla to the spinal cord will not produce sustained analgesia in patients.

Patil, Fillmore, Valentine et al. (1987) transplanted adrenal medulla tissue into the lateral ventricle in adult rats. Half of the animals received human chorionic gonadotropic hormone chronically after the transplantation. Eight weeks later, the brains were removed, a volumetric study was done, and catecholamines were measured. The animals receiving the hormone had a higher graft survival than the other group. The level of norepinephrine was significantly higher in the treated group; the dopamine level was higher in the untreated group. The higher volume of graft tissue in hormone-treated rats suggests that human chorionic gonadotropic hormone may increase survival and growth of the transplanted tissue. The higher levels of norepinephrine seen in the treated group suggest a tendency for such grafts to be more active.

It is controversial whether the beneficial effects of adrenal medulla transplants in Parkinson's disease are due to production of dopamine by the graft or the host tissue. In a significant study that attempted to address this issue, Bohn, Cupit, Marciano et al. (1987) grafted adrenal medullary tissue into the brain of mice that had been treated with a neurotoxin that specifically depletes striatal dopamine, resulting in the virtual disappearance of tyrosine hydroxylase-immunoreactive fibers. After the grafts were implanted, dense tyrosine hydroxylase-immunoreactive fibers were observed in the grafted striatum; only sparse fibers were seen in the contralateral striatum. In all cases, tyrosine hydroxylase-immunoreactive fibers appeared to be from the host rather than from the grafts which survived poorly. These observations suggest that, in mice, adrenal medullary grafts exert a neurotrophic action in the host brain to enhance revival of dopaminergic neurons. This effect may be relevant to the symptomatic recovery in parkinsonian patients who have received adrenal medullary grafts.

CLINICAL TRIALS

The first report of successful amelioration of symptoms in patients with Parkinson's disease was published by Madrazo, Drucker-Colín, Díaz et al. (1987). They described the treatment of two young (35 and 39 years old) patients with intractable incapacitating Parkinson's disease who underwent autologous transplantation of fragments of the adrenal medulla to the right caudate nucleus. Clinical improvement was noted in both patients within 2 weeks postoperatively and continued in both. Rigidity and akinesia had virtually disappeared in the first patient at 10 months postoperatively, and his tremor was greatly decreased. A similar degree of improvement was present in the second patient at 3 months. Madrazo et al. (1987) concluded that autografting of the adrenal medulla was responsible for the marked amelioration in the symptoms of Parkinson's disease. They added that the results are preliminary, and further work is necessary to determine whether this procedure will be successful in the long term. A later report summarized the results of 10 cases (Madrazo, Drucker-Colín, León et al., 1987). Ostrosky-Solís, Quintanar, Madrazo et al. (1988) evaluated the neuropsychological effects of brain autografts on 7 patients with Parkinson's disease. Preoperatively all patients had specific cognitive deficits of varying degrees, including frontal lobe syndrome and visuospatial and memory deficits. Evaluations done 3 months postoperatively revealed significant improvement of the frontal lobe-type symptoms and visuospatial deficits, as well as improvement in some memory tasks. The improvements of the operated Parkinson's patients were clearly distinguishable from those of unoperated Parkinson's patients.

It is a considerable understatement to say that the report by Madrazo et al. (1987) created a sensation. Commentaries relating directly to this report were published by Moore (1987), Factor, Sanchez-Ramos and Weiner (1987), Lieberman, Koslow and Maestrone (1987a), White (1987), Osserman (1987), Backlund (1987b), Rose and Nashef (1987), Sladek and Shoulson (1988), Lieberman (1988), Bartal (1988), Pearce (1988), Merz (1988a), Jankovic (1988), Bakay and Barrow (1988), Calne and McGeer (1988), Riopelle (1988) and Sargent, Aydelott and Doll

(1988). Some of these comments were acknowledged by Drucker-Colín and Madrazo-Navarro (1987). Venna and Sabin (1987) also commented on the article by Madrazo et al. (1987) and further added that they had treated a patient by chronic injections of dopamine into the cerebral ventricles. Mild relief of the parkinsonian symptoms was seen but was complicated by reversible confusional states, myoclonus, and dyskinesia. Snyder (1987) also commented on the article by Madrazo et al. (1987) and added further that neural transplantation has promise for other neurologic diseases as well.

Several of these reports included discussions of the possibility that the adrenal medullary grafts are not providing therapeutic benefit for the patients by secreting dopamine into the striatum but, rather, trophic factors or response to the injury in the brain was causing some therapeutic effects. Commenting on this latter possibility, Motti, Pezzoli, Silani et al. (1988) reported that they had performed biopsies of the head of the caudate nucleus in seven patients with Parkinson's disease. In one patient, autologous adrenal tissue was implanted at the site of biopsy 2 weeks later. Five of the other patients remain improved for a follow-up of about 2 years. The patient receiving the adrenal transplant was reported as definitely improved. They did not conclude that surgical damage to the caudate nucleus *per se* substantially influences the course of Parkinson's disease.

Lewin (1987) compared the success of Madrazo et al. with the results of earlier adrenal medulla transplants in Sweden. It was reported that two patients had been treated similarly in the United States in April 1987 (Merz, 1987a). Backlund (1987a) published a letter pointing out that the earlier patients operated on in Sweden did have improvement, although it had not been permanent. All four Swedish patients, including the first patient who has been followed for more than 5 years, are in an unchanged condition compared with their preoperative status. None is any worse.

Lieberman, Koslow and Maestrone (1987b) reviewed the clinical experience with adrenal medullary autografts as a treatment for Parkinson's disease. They pointed out, among other things, that transplants from human fetuses may be more suitable than adrenal medullary autografts. However, present attention is being directed toward adrenal implants. Understanding the mechanism through which the adrenal transplants survive in the brain and reverse the symptoms of Parkinson's disease will require an ongoing basic research effort.

In a follow-up of her earlier report, Merz (1987b) reported on a symposium entitled "Transplantation into the Mammalian Central

Nervous System." This symposium was compiled in a volume by Gash and Sladek (1988). In addition to a report by Drucker-Colín, representatives from China and the United States presented clinical cases of adrenal medullary autografts as a treatment for Parkinson's disease. Several workers reported on experimental work with animals. A detailed report on a symposium relating to transplantation into the nervous system was published by Kordower, Dean, White et al. (1988). This meeting emphasized basic science studies in this area and also included reports on the clinical trials. Additional reports on the basic science background for neural transplantation were published by Hoffer, Granholm, Stevens et al. (1988) and Sladek, Redmond and Roth (1988). Brief reports on adrenal medullary transplants for parkinsonism in China were published by Zhang, Ding and Cao (1987) and Zhang, Ding and Jiao (1988).

Iacono and Sandyk (1988) reported unpublished data from the Madrazo et al. group. It was noted that several of their patients did not suffer abdominal pain after adrenalectomy. The levels of enkephalins in the cerebrospinal fluid of these patients had increased 40-fold over baseline values. This is an interesting aspect of the autologous transplantation of adrenal chromaffin tissue into the central nervous system.

Joynt and Gash (1987) published a thoughtful editorial asking the question "Are we ready for neural transplants?" They presented a serious argument that we may be ahead of ourselves. Fundamental questions remain.

Transplantation of the adrenal medulla as a treatment for Parkinson's disease was performed at several centers in the United States. As reported by Lewin (1988a, 1988b, 1988c) and Merz (1988), the results have been relatively disappointing. Penn, Goetz, Tanner et al. (1988) presented clinical observations from their first five patients. Postoperatively they noted somnolence, delusions, and lack of significant pain in spite of a large abdominal incision. Following the early postoperative period, the patients showed gradual improvement of on-off times and Schwab-England disabilities scores over a period of 20 weeks. McRae-Degueurce, Klawans, Penn et al. (1988) found that an antibody in the cerebrospinal fluid of these patients disappeared following the operation. An additional procedure was reported by Berman (1988); however insufficient postoperative time had elapsed to evaluate the results of the surgery. Although much of the optimism for adrenal medullary transplantation was diminished by these reports, W.J. Freed (1988) pointed out that further work is required before the successes observed in animal studies can be duplicated clinically.

Two reports offered a possible explanation why the transplantation of autologous adrenal medullary tissue from Parkinson's disease patients may not be effective. Carmichael, Wilson, Brimijoin et al. (1988) and Cervera, Rascol, Ploska et al. (1988) measured catecholamine levels in the adrenal medullae of parkinsonian patients at autopsy and found that the levels were lower compared to a non-parkinsonian population. These data suggest that the adrenal medulla of a parkinsonian patient may be a compromised source of catecholamine-producing tissue.

As mentioned above, the first patients to receive autologous adrenal medulla transplants into the brain were in Sweden. The first two procedures were performed in 1982. Backlund, Olson, Seiger et al. (1987) reported further on these two patients and two additional patients who were operated on in March 1985. The third patient improved during the first 2 months but returned to his preoperative state. At a follow-up 6 months after the operation, the patient considered his general condition to be similar to that before the operation. The fourth patient was similar but at 6 months after the operation the results of all of the motor performance tests were slightly better than they were before the operation. A significant difference between the first two patients and the second two patients was that the adrenal medulla autograft was placed in the caudate nucleus in the first two patients and in the putamen in the latter two.

A more detailed report on the two patients who had adrenal medulla autografts into the putamen was published by Lindvall, Backlund, Farde et al. (1987). The first patient exhibited a transient (2 days) improvement of motor performance in the limbs contralateral to the implantation site. He also had significantly longer episodes of normal function for about 2 months. The second patient reported a minor improvement in balance and gait, again lasting for 2 months. Other tests were consistent with increased catecholaminergic activity in the basal ganglia of both patients. Positron emission tomography showed no postoperative alteration of receptor density in the putamen. By a number of criteria, no significant adverse effects of the transplantation were observed. Lindvall et al. (1987) concluded that catecholamine-rich cellular implants in the basal ganglia have transient beneficial effects in patients with severe Parkinson's disease. They added that further clinical examination performed repeatedly during the 2 years after the transplantation have shown a moderate, progressive worsening of symptoms in both patients. There are no indications that the adrenal medullary grafts have altered the course of the disease.

A brief neuropathologic report of a transplanted patient was given by Frank, Sturiale, Gaist et al. (1988). The patient had died of leukemia 5 months after the transplant. All of the implants had become necrotic and adrenal cortical cells predominated. An implant that had unexpectedly been near the choroid plexus of the lateral ventricle appeared to have proliferated after it was transplanted, although it later died.

Five articles related to the nursing care of the transplant patients. Yi (1987) pointed out that the patient receives two simultaneous surgical procedures, involving both the abdomen and cranial cavity; furthermore, the patient in her care was not young. Possible complications, such as infection of wounds, lung infections, retention of urine, etc., should be considered. The patient under her care experienced an excellent recovery. In another article directed to nurses, Williams (1987) reported on two patients who are making a good recovery from the operation. Berry and Ward-Smith (1988), Delgado and Billo (1988), and Hyman, Rogers, Smith et al. (1988) also discussed the peri-operative considerations and nursing care of patients undergoing adrenal medullary transplantation.

Madrazo, León, Torres et al. (1988) transplanted the "adrenal medulla" and substantia nigra from a 13-week aborted fetus into the striatum of two different patients, and reported some amelioration of parkinsonian symptoms at 8 weeks after surgery. C.R. Freed (1988) commented that substantia nigra dopami-nergic cells from a human fetus will not survive if the fetus is older than 9 weeks, and suggested that the improvement seen by Madrazo et al. (1988) may have been due to the lesion effect of the surgery rather than to the survival of fetal cells. Dwork, Pezzoli, Silani et al. (1988) pointed out that the adrenal medulla is not yet formed at the gestational age of the fetus used by Madrazo et al. (1988). They commented that, at 12-13 weeks gestation, the fetal cortex occupies the major, central, portion of the adrenal gland. Such observations again question the extent to which clinical improvements are dependent on the transplan-tation of catecholamine-producing tissue.

In reviewing the animal models and clinical trials, Wyatt, Morihisa, Nakamura et al. (1986) made several interesting points. For example, if it is assumed that a graft in the human brain would have the same sphere of influence that grafts seem to have in the rat brain (noting that the human striatum is approximately 250 times larger than the rat striatum), then the number of grafts needed for even distribution over the human striatum would be approximately 1,700. Twice that number may be needed for a bilateral procedure. They also discussed other factors that suggest far fewer than this large number of grafts could be effective in providing relief in Parkinson's disease.

In discussing the future role of neurosurgery in medicine, Luyendijk (1986) demonstrated how difficult it is to predict future developments. Among other items, he pointed out that nobody could have predicted 25 years ago that the treatment of Parkinson's disease would involve transplantations of adrenal medullary tissue to the striatum.

OTHER CONSIDERATIONS

Perlow (1987) published a general review of treatment of Parkinson's disease. This included a discussion of pharmacologic as well as surgical treatment. He suggested that the transplantation of dopamine neurons, even across species barriers, is a reasonable consideration for the treatment of Parkinson's disease.

Sanberg and Norman (1988) reported on a single clinical case in which autologous adrenal medullary tissue transplanted into the caudate nucleus as a treatment for Huntington's disease. The rationale for the procedure appears to be an unpublished finding that adrenal medulla transplanted into rat caudate nucleus can protect against exocytotoxin-induced neurodegeneration. The hypothesis appeared to be that adrenal grafts would protect against excytotoxin-induced degeneration in patients with Huntington's disease. Sanberg and Norman (1988) stressed that the use of transplantation techniques in the treatment of Huntington's disease is still in the investigative stage and further extensive animal research was required.

Stevens, Phillips, Freed et al. (1988) transplanted adrenal medullary tissue to the lateral ventricles of rats that are models for epilepsy. The procedure did not appreciably decrease the intensity of the audiogenic seizures in these animals.

General considerations for neural transplantation for various pathologic conditions were discussed by Tulipan (1988), including Parkinson's disease, Huntington's disease, Alzheimer's disease, aging, trauma, hormonal defects, genetic defects, and pain. It was pointed out that there is a wealth of experimental evidence to suggest that transplantation of tissues such as the adrenal medulla to the brain may ameliorate various neurologic and endocrine disorders, but many unanswered questions remain. Answers to these questions will only come with carefully controlled studies.

Transplantation of the adrenal gland in patients was presented in an entirely different context by Matsuda (1987). Autotransplantation of the whole adrenal gland into the mesentery with microvascular anastomoses was attempted in three clinical cases of breast cancer. The clinical cases and surgical techniques were presented. The operation appeared to be successful in all three cases, and clinical results suggested prevention of metastasis.

CHAPTER 16

PSYCHOLOGY

Activation of the adrenal medulla classically has been related to the phenomenon of stress. Both psychologic and physiologic stressors have been used in human subjects. More severe physiologic stressors have been tested for their efficacy in activating the sympathoadrenal system in experimental animals.

STRESS AND EMOTIONAL BEHAVIOR

In human subjects, Arnetz and Fjellner (1986) found that psychosocial and personality factors are relevant to predicting stress-induced neurophysiologic reactions. Stress was produced in a group of men, ages 23 to 44 years, with the Colour Word conflict test and mental arithmetic problems. Psychosocial aspects of personality, in addition to the individual's self-assessment of feelings during testing, were correlated with blood biochemistry values and urinary metabolites. Although one-fourth of the variance in the change in urinary epinephrine after mental stress could be predicted by the degree of extrovertness of the individual, psychosocial variables could not be used to explain variances in the change in urinary norepinephrine excretion.

Measurement of urinary epinephrine and norepinephrine in a group of 21 attack pilots revealed that increased excretion of epinephrine during mission preparation was directly related to the perceived challenge of the mission. Svensson, Thanderz, Sjöberg et al. (1988) reported further both that overall catecholamine secretion increased and the ratio of norepinephrine to epinephrine decreased as a function of condition--*i.e.*, from lesson, to mission planning, to mission flying.

Gharib, Gauquelin, Pequignot et al. (1988) demonstrated that horizontal bed rest is not an appropriate control situation to determine the hormonal effects of head-down tilt. Both these positions were compared with a sitting position in 10 healthy adults. Supine and head-down tilt positions caused significant decreases in diastolic blood pressure and plasma renin, aldosterone, and norepinephrine values. Epinephrine concentration was decreased only in the head-down tilt position. The authors suggested the use of the supine position as a control for evaluating the physiologic effects of head-down tilt has led to the observation that head-down tilt has led to observing few effects with this manipulation.

Halbrügge, Gerhardt, Ludwig et al. (1988) described an assay procedure that permits simultaneous measurement of norepinephrine, epinephrine, dopamine, and dihydroxyphenyl ethylene glycol (DOPEG), a metabolite formed by the oxidative deamination of norepinephrine and epinephrine by monoamine oxidase. By using this method, it was determined that standing increased plasma norepinephrine and DOPEG in a group of 32 volunteers. Pretreatment with desipramine, which blocks presynaptic reuptake of norepinephrine, abolished the DOPEG response to orthostasis, thus indicating that the increases in this compound that occur during standing are presynaptic in origin. Mental stress, evoked by reading a difficult text, was accompanied by substantive increases only in plasma epinephrine.

Shabunina (1987) reported that exposure to industrial noise increased the urinary excretion of norepinephrine, epinephrine and dopamine in a group 21 to 30 years old, whereas exposure to residential noise levels was without effect.

The plasma catecholamine response in rats to short-term (10 min), repeated white noise stress was determined by De Boer, Slangen and Van Der Gugten (1988). During the first exposure to noise, plasma epinephrine and norepinephrine values increased rapidly, peaked at 1 and 5 min, respectively, and quickly returned to baseline at the cessation of the noise. Plasma corticosterone levels also increased, but both the increases and decreases were slower in onset. Successive exposures to noise were associated with decreased concentrations of both catecholamines and corticosteroid, in addition to less intense behavioral reactions, indicative of an adaptive response to this stressor.

Sedl'ak (1987) compared the levels of glutathione and oxidized glutathione in innervated and denervated adrenal glands of rats exposed to immobilization stress. These two compounds are thought to be substrates for the γ-glutamyl cycle which may function *in vivo* to transport amino acids into the intracellular space. Levels of oxidized and unoxidized glutathione did not change in either adrenal medulla or cortex in the denervated adrenal, but levels of both compounds decreased in the contralateral innervated gland during immobilization stress. Thus, intact innervation appears to be necessary for stress-induced changes in adrenal glutathione metabolism.

Catecholamine levels also have been linked to psychologically identified states. Airapetiants, Diakova and Maslova (1986) reported that individuals with various neuroses and neurosis-like states had significant increases in plasma norepinephrine concentration as well as pronounced changes in plasma epinephrine and acetylcholine concentrations. They suggested that the consistent pattern of these neurohumoral changes may permit use of these variables for differential diagnosis.

Mekhedova and Ghadirian (1979) used a model of experimental neurosis to investigate adaptive behavior and plasma catecholamine levels in dogs. Four dogs were trained over 2 years with both ordered partial reinforcement (odd-numbered signals were paired with painful stimuli) and with probabilistic reinforcement (conditioned stimulus was accompanied by painful reinforcement in half the trials, in random sequence). Increases in plasma catecholamine values were noted in only two dogs after ordered partial reinforcement, whereas probabilistic reinforcement resulted in increased plasma catecholamine values in all animals.

Rothschild, Schatzberg, Langlais et al. (1987) compared the effects of administration of dexamethasone, a synthetic corticosteroid, on levels of plasma catecholamines in psychotic and nonpsychotic depressed patients. Although dexamethasone increased plasma dopamine in both depressed groups and in a control group, the increase in the control group was 10 times greater than that in either patient group. Baseline values of dopamine and cortisol levels were significantly higher in psychotic depressed patients than in either nonpsychotic patients or controls. After administration of dexamethasone the depressed psychotic patients had significantly higher dopamine and lower norepinephrine levels than the nonpsychotic patients.

O'Connor (1988) reviewed the effects of lithium on the adrenal medulla and the sympathoadrenal system. Lithium administration increased plasma norepinephrine and epinephrine levels which, at least in the rat, appear to be involved in mediating the behavioral and physiologic effects of lithium.

Dopamine β-hydroxylase (DβH) has also been investigated in relation to depressive disorders. Kjellman, Beck-Friis, Ljunggren et al. (1986) observed that, although DβH values correlated significantly with levels of serum thyroid stimulating hormone in acutely ill patients, no significant differences in serum levels of DβH were found among groups of acutely ill patients, patients in remission, and controls. The authors concluded that determination of DβH level is not of practical use in diagnosis.

In contrast to the above study, Mód, Rihmer, Magyar et al. (1986) found DβH to be an important biochemical marker in distinguishing psychotic from nonpsychotic and unipolar from bipolar forms of primary depression. Both psychotic unipolar and nonpsychotic bipolar patients had serum DβH levels that were significantly lower than those in a control group. Furthermore, the differences between mean DβH values in psychotic and nonpsychotic subgroups were significant in patients with unipolar and with bipolar primary depression.

Sakellariou, Markianos, Tsichlakis et al. (1987) investigated the relationship between DβH and schizophrenia by comparing levels of this enzyme in schizophrenic patients and in their healthy first degree relatives. A control group of subjects had similar levels of enzyme activity. The authors concluded that plasma DβH neither differentiates schizophrenic from healthy family members nor identifies individuals with a family history of this illness.

White, Lishman and Wyke (1986) reported a study of a patient with dementia, hypertension, and insulin dependent-diabetes caused by bilateral pheochromocytoma. One year after removal of the tumors there was improvement in the dementia as documented by an increase of 15 points in the full scale IQ. Such reversibility of dementia, which the authors think was clearly related to the pheochromocytoma, has not been reported previously, although such cognitive improvement may have been aided by better control of hypertension and maintenance of a stable blood glucose level.

Stoddard, Bergdall, Conn et al. (1987) measured peripheral venous catecholamine levels associated with an emotional behavior -- defensive behavior in the cat. This behavior was elicited experimentally either by presentation of a large barking dog or a second cat electrically stimulated in the hypothalamus to attack the experimental cat. Interspecies aggressive behavior was accompanied by adrenal medullary release of both epinephrine and norepinephrine, which decreased over repeated trials. Intraspecies aggressive behavior was initially associated with release of epinephrine from the adrenal medulla; later trials activated the peripheral sympathetic nerves. The authors concluded that, although the somatic and autonomic displays associated with defensive behavior are largely stereotyped, the sympathoadrenal response is more specific to the stimulus.

AGE

Jacobs, Mason, Kosten et al. (1986) correlated bereavement with urinary catecholamine excretion in a population of 59 middle-aged and elderly people. The 24-hour output of norepinephrine and epinephrine in both acutely bereaved individuals and those threatened with the loss of a spouse was significantly greater than in a normal population but did not differ between these two groups. Urinary catecholamine values were not correlated with depression scores. The observation that, within the group of acutely bereaved subjects, increased age was associated with increased urinary catecholamine levels suggested that adaptation of the sympathoadrenal system is slower in older persons.

Alterations in sympathoadrenal activity with age have also been documented in animals. McCarty (1986a) summarized studies that evaluated the activity of the sympathoadrenal system in response to stressors of various intensities and durations as a function of age in rats. Brief stress (foot-shock) elicited greater increases in plasma catecholamines in 4-month-old rats than in 12- or 24-month-old rats. An intermediate stress (cold water swim) evoked similar sympathoadrenal responses in 6- and 22-month old rats, although the sympathetic nervous system remained in a state of heightened activity for a longer period in the older animals. Prolonged exposure to a severe stress (glucoprivation) resulted in an exaggerated adrenal medullary response and greater incidence of death in the older animals. The author concluded that the homeostatic regulation of the sympathoadrenal system is impaired in aged rats.

MEMORY

Epinephrine has been shown to enhance memory storage for several types of learning. Gold (1986, 1987) presented evidence from a range of experiments that suggests that this effect is mediated by glucose released from hepatic stores by epinephrine. In one situation, rats were pretrained to drink in a passive avoidance apparatus, after which one period of foot-shock was administered. Immediately after the foot-shock the rats were injected with saline, epinephrine, or various amounts of glucose and tested for retention at 24 hours. Both glucose and epinephrine enhanced memory storage for inhibitory avoidance training. This effect, also shown to be time dependent, was an inverted U function with the greatest effect on retention at a dose of 100 mg of glucose or 0.01 mg of epinephrine per kg of body weight.

Post-training ingestion of glucose also has been shown to improve retention of various learned responses. White and Messier (1988) tested the hypothesis that this action of glucose was mediated through the glucose-stimulated release of adrenal medullary epinephrine. Post-training injections of glucose improved retention of a conditioned emotional response in both demedullated and control rats, thus indicating that the adrenal medulla is not involved in this phenomenon.

A second adrenal medullary neurohormone, [Met]enkephalin, has been shown to be involved in memory. Zhang and McGaugh

(1987) and Zhang, McGaugh, Juler et al. (1987) trained mice in both inhibitory avoidance and Y-maze discrimination tasks. Post-training administration of naloxone enhanced memory retention whereas post-training administration of [Met]enkephalin impaired retention for both tasks. Furthermore, naloxone and [Met]enkephalin each antagonized the action of the other when administered simultaneously. Because neither compound affected memory retention in adrenal-denervated mice, the authors suggested that the observed effects were due to competition between naloxone and [Met]enkephalin at opioid receptor sites in the adrenal medulla, and that endogenous adrenal medullary [Met]enkephalin may be involved in memory consolidation.

Black, Adler, Dreyfus et al. (1987) reported on the use of the adrenal medulla as a model for investigating the molecular mechanisms of memory in the sympathetic nervous system. Denervated rat adrenal medulla transplanted to culture showed a 50-fold increase in [Leu]enkephalin and a concomitant increase in preproenkephalin mRNA which encodes for the enkephalin precursor. Depolarizing influences blocked both of these increases through the mechanism of inhibiting transcription of the preproenkephalin gene. That expression of neurohumoral genes is differentially regulated by depolarizing stimuli at the transcriptional level was shown by the fact that such stimuli did not alter transcription of the tyrosine hydroxylase gene.

Neurohormones of both the adrenal cortex and medulla were shown to be involved in the retention of a behavioral response after stress. Jefferys and Funder (1987) reported that rats subjected to the Porsolt model swimming test become progressively more immobile during the test and retain an extended period of immobility when tested again at 24 hours, whereas adrenalectomized rats do not retain such immobility on retesting. Administration of glucocorticoids to adrenalectomized rats or of MR 2266, an opioid κ-receptor antagonist, to hypophysectomized rats resulted in decreased periods of immobility when the rats were retested which were not significantly different from those in adrenalectomized rats. Although either cortical or medullary mechanisms were sufficient for the retention of this behavior, the data suggest that these components may subserve a complementary action in the intact animal in response to stress.

Martinez, Schulteis, Janak et al. (1988) evaluated the relationship between age and memory after short and long training-testing periods in rats. Both young and old animals retained behavioral tasks at short training-testing intervals but old animals had impaired retention at long testing-training intervals. The authors' review of that literature indicates that the activity of the adrenal medulla in foot-shock stress is impaired in aged rodents, and they suggest that a similar medullary functional impairment may be involved in the rapid forgetting seen in senescent animals.

STRESS INDUCED ANALGESIA, CHRONIC STRESS AND INDICES OF SYMPATHOADRENAL FUNCTION

Amit and Galina (1986) reviewed the causes and mechanisms of adaptive pain suppression. Prolonged foot-shock, intermittent cold water swim, and immobilization all produced an opiate-mediated form of stress-induced analgesia because administration of naloxone blocked pain suppression induced by these treatments. Both pituitary and adrenal medullary opioids have been implicated in producing pain suppression under such conditions. Evidence for the role of adrenal medullary opioids derives from the observation that both demedullation and denervation decreased the analgesia evoked by prolonged or short intense foot-shock.

Three distinct forms of stress-induced analgesia elicited by varying the intensity and duration of foot-shock were discussed in related papers by Terman, Lewis and Liebeskind (1986) and Terman and Liebeskind (1986). Both prolonged intermittent and low-severity continuous foot-shock elicited opiate-mediated analgesia; this form of analgesia was naloxone-sensitive and demonstrated tolerance with repeated stress and cross-tolerance with morphine. High-severity continuous foot-shock, however, produced analgesia through a nonopioid mechanism. In contrast to the results of Amit and Galina (1986), the analgesia produced only by prolonged intermittent stress was shown to be mediated by adrenal medullary opioids because it was blocked by adrenalectomy, adrenal demedullation, or adrenal denervation. This form of stress-induced analgesia was dependent on the animal learning that the shock was inescapable.

Rochford and Henry (1988) reported evidence of an opioid-mediated analgesia in rats evoked by swimming in cold water; this was not dependent on adrenal medullary function. Analgesia assessed by the tail flick response was elicited by either intermittent or continuous cold water swimming; such analgesia was not not affected by bilateral adrenal denervation.

Thompson, Miczek, Noda et al. (1988) investigated the source of endogenous opioids that are responsible for the analgesic response of mice subjected to defeat in a social context paradigm. Manipulations both to enhance and to block the release of either pituitary β-endorphin or adrenal medullary enkephalins had no effect on the defeat-induced analgesic response. The authors concluded that analgesia associated with defeat is mediated primarily by opioids released and acting within the central nervous system.

Wallin, Cassuto, Högström et al. (1988) reported that a single intraperitoneal injection of a local anesthetic, bupivacaine, did not decrease the pain of or the sympathoadrenal response to upper abdominal surgery. Pain scores and plasma catecholamines levels were similar in groups of cholecystectomy patients who received injections of either bupivicaine or saline.

The intensity of stress also affects concomitants of sympatho-adrenal activation other than pain suppression. Natelson, Creighton, McCarty et al. (1987) compared hormonal and behavioral indices of stress in rats subjected to different intensities of foot-shock. Absolute levels of plasma cate-cholamines were found to be a reliable index of the the intensity of foot-shock administered to individual rats, although there was overlap in levels among animals at each intensity. Additionally, plasma epinephrine and norepinephrine levels were significantly correlated with movement during the period of foot-shock, indicating that this behavioral measure was as sensitive and reliable an index of stress intensity as hormonal variables. The authors suggested that further investigation may identify behavioral measurements that can be used effectively as clinical "barometers" of stress.

Roske, Rathsack, Nieber et al. (1987) compared the effects of chronic immobilization and adrenal demedullation on a series of adaptive and regulatory processes in rats. Both procedures induced the same direction of alteration in several physiologic variables: a decrease in the number of animals surviving the procedure, an increase in blood pressure, a decrease in the level of substance P in the pituitary, and an increase in the level of substance P in the hypothalamus. In some cases the effects of adrenal demedullation were wholly or partially antagonized by chronic immobilization stress. In this group of animals, survival rate increased, pituitary substance P increased, and hypothalamic substance P decreased. The authors interpreted these results as suggesting the involvement of similar regulatory mechanisms in adrenal demedullation and chronic stress.

Demedullated and sham-operated August rats, which have a high stress sensitivity, were used by Roske, Ûmatov, Pevcova et al. (1988) to investigate the functional participation of the adrenal glands in the adaptation process. The authors described the interactions between the catecholamine and the regulatory peptidergic systems of the adrenal medulla. Demedullation produced changes that were associated with the removal of the catecholamine system. Additional stress revealed further alterations connected with the loss of the regulatory system.

Vaupel, Jarry, Schlömer et al. (1988) determined the effect of various stressors on adrenal medullary output by measuring catecholamine and substance P in dialysate fractions collected from the adrenal gland of the rat. An exogenous stressor, electric foot-shock, increased release of both adrenal catecholamine and substance P. In contrast, an endogenous stressor, insulin-induced hypoglycemia, increased adrenal catecholamines to a greater degree than did foot-shock but did not affect the release of substance P.

Eremina and Belyakova (1987) described two distinct periods in the initial stress reaction in dogs. The first phase, dissociation of sympathoadrenal secretory synthetic activity, was characterized by increases in tissue epinephrine and dopamine contents, decrease in tissue norepinephrine, decrease in plasma epinephrine, and increase in plasma norepinephrine. The second phase of the stress reaction developed within 1 min. This phase of synchronous system activation was characterized by simultaneous increases in epinephrine, norepinephrine, and dopamine as well as increases in the catabolism of these compounds.

Vogel and Jensh (1988) determined the response of rats to chronic (3 weeks) exposure to immobilization, light, and noise stressors by measuring plasma norepinephrine, epinephrine, and corticosterone concentrations. The norepinephrine and epinephrine responses remained constant during this period; the corticosterone response increased markedly after 2 weeks. Following a 3-week recovery period, the animals were again exposed to the various stressors. Norepinephrine and corticos-terone responses showed some adaptation to the stressor but the epinephrine response was unchanged. Marked differences in response were seen among individual animals.

Mitchell and Vulliet (1987) used the isolated perfused rat adrenal gland to evaluate the effect of chronic stressors -- immobilization, dexamethasone administration, or insulin administration -- on the responsiveness of the adrenal to a fixed concentration of acetylcholine. The secretory response of the perfused rat adrenal to a standard dose of acetylcholine was altered by various pretreatments of the animal. Immobilization increased both the epinephrine content of the gland and the amount released in response to acetylcholine perfusion. In contrast, dexamethasone and insulin decreased both the adrenal content of epinephrine and its stimulated release.

Mitchell, Hattingh and Ganhao (1988) assessed, in cattle, the neuroendocrine response to graded degrees of stress -- handling, transport, and slaughter. Adrenal cortical and pituitary activities were increased with handling; plasma catecholamine levels were increased by both transport and slaughter. The authors suggested that the stress response has two phases: a hypothalamic-adrenal cortical phase associated with environmental changes, such as noise, and a sympathoadrenal phase associated with neurogenic stimuli such as transport and stunning.

McCarty, Horwatt and Konarska (1988) reviewed the literature on the sympathoadrenal response to chronic intermittent stress in rats. Chronic stress elicits several sympathoadrenal adaptations including increased synthesis and storage of catecholamines, increased baseline levels of circulating catecholamines, and decreased release of catecholamines after exposure to the same stressor. Exposure of such chronically stressed animals to a unique stressor elicits an exaggerated sympathoadrenal response. This neuroendocrine adaptation to chronic intermittent stress is compared with the processes of habituation and sensitization.

INTERACTION WITH PITUITARY HORMONES AND PEPTIDES

Part of a comprehensive review by Tilders and Berkenbosh (1986) discussed the role of the "adrenomedullary-pituitary axis" in regulating stress-induced secretion of peptides from the anterior pituitary gland. Immobilization-induced increases in α-melanocyte stimulating hormone (MSH), β-endorphin, and prolactin were attributed to the action of adrenal medullary epinephrine because these increases were significantly less in demedullated rats. Furthermore, peripheral injections of epinephrine in anesthetized rats enhanced levels of these three peptides. It is likely that epinephrine acted on the anterior pituitary through β-adrenergic receptors because the effects of peripheral epinephrine were blocked by propranolol and mimicked by isoproterenol.

A later study by Kvetnansky, Tilders, vanZoest et al. (1987) investigated sympathoadrenal control of peptide secretion from the intermediate lobe of the pituitary during immobilization stress in rats. The authors concluded that adrenal medullary epinephrine was involved in the control of secretory activity of the intermediate lobe during stress because demedullation abolished the stress-induced increases in both plasma epinephrine and in α-MSH and β-endorphin from the intermediate lobe. The acute, stress-induced release of corticotropin (ACTH) was not altered by either demedullation or demedullation plus guanethidine sympathectomy. However, demedullation with or without sympathectomy notably attenuated the decrease in ACTH and, to a lesser degree, in β-endorphin that accompanied prolonged stress; this observation suggests that, under physiologic conditions, catecholamines are important in decreasing circulating levels of these compounds.

Castagne, Corder, Gaillard et al. (1987) compared the release of neuropeptide Y (NPY) and catecholamines during various types of stress in rats. Because the stress-induced increases in NPY were similar in control and adrenalectomized rats, the adrenal medulla was not considered to be a significant source of circulating NPY. Although NPY levels were more closely correlated with changes in norepinephrine than in epinephrine level, NPY was not always concomitantly released with norepinephrine, suggesting that the nature of the stimulus is important in triggering the release of NPY.

Zukowska-Grojec, Konarska and McCarty (1988) measured plasma levels of neuropeptide Y, norepinephrine, and epinephrine in response to two acute stress paradigms in rats. Intermittent foot-shock elicited immediate, intensity-dependent increases in plasma epinephrine and norepinephrine and a delayed increase in plasma neuropeptide Y-like immuno-reactivity. In contrast, prolonged immobilization resulted in greater increases in plasma epinephrine and norepinephrine without changes in neuropeptide Y. Because neuropeptide Y responses correlated with norepinephrine responses but not epinephrine responses, the authors suggested that neuropeptide Y is released from sympathetic nerves under these stress conditions.

Khaisman, Arefolov and Malikova (1988) used fluorescence microscopy and spectrofluorometry to evaluate the activity of the adrenergic nerves to the dura and the levels of catecholamines in the adrenal medulla of rats after immobilization stress. Intraperitoneal administration of dalargin, a synthetic analogue of [Leu]enkephalin, exerted a notable antistress action.

Yntema and Korf (1987) reported that activation of the adrenal medulla by the stress of handling, cold exposure, or immobilization suppressed haloperidol-induced catalepsy in rats. Catalepsy did not decrease in adrenalectomized stressed rats, whereas intracardial infusion of epinephrine had an anticataleptic effect. They suggested that, because neuroleptic-induced catalepsy in animals and antipsychotic-induced parkinsonism in humans are similar, these observations may be related to the phenomenon of kinesia paradoxica, the ability of parkinsonian patients to overcome their akinesia in situations of severe stress.

CHAPTER 17

PHYSIOLOGY

Catecholamines secreted from the adrenal medulla are involved in the mediation of most physiologic processes. Recent advances in the understanding of the autonomic nervous system were reviewed by Brooks (1987). Warren (1986) discussed the evidence for a protective role of epinephrine in the lung because this catecholamine is the only endogenous stimulator of β_2-adrenergic receptors. Ghosh (1986) reviewed the cellular physiology of the avian adrenal gland.

PERINATAL PHYSIOLOGY

Costa, De Filippis, Voglino et al. (1988) measured catecholamines and adrenocorticotropic hormone (ACTH) in plasma of mothers, fetuses (umbilical artery), and newborns to determine the effect of vaginal delivery on these values. In the mother, catecholamine values, especially epinephrine, were increased during labor and delivery, compared with nonpregnant controls. In contrast, in the fetus and newborn norepinephrine shows the predominant increase in response to parturition. During labor, in the mother ACTH concentration is 10 times greater than control, and twice as great as in the fetus; these levels increase further during delivery. The authors suggested that the predominance of epinephrine and the higher ACTH/epinephrine ratio in the mother than in the fetus may be due to the action of the pituitary-adrenal axis to regulate the stress response to parturition in the mother. In the fetus, this stress response may be predominantly controlled by the sympathetic nervous system.

It has been established that the stress of the birth process is a potent activator of adrenal medullary secretion. Lagercrantz and Slotkin (1986b), reviewing this topic, reported that levels of catecholamines in umbilical samples of newborn infants were increased approximately 20-fold over resting levels in the adult. This catecholamine surge occurs during normal vaginal delivery. It is caused, at least in part, by hypoxemia and has been shown to alter blood flow to protect the heart and brain against potential asphyxia and to increase lung compliance to promote normal breathing after delivery.

The contribution of postganglionic sympathetic fibers to the catecholamine surge at birth was investigated by Agata, Padbury, Ludlow et al. (1986). Chemical sympathectomy of fetal lambs by the administration of 6-hydroxydopamine resulted in significantly decreased baseline levels of norepinephrine and slightly increased baseline levels of epinephrine. Cutting the umbilical cord of sympathectomized animals caused a 2-fold increase in norepinephrine level whereas a 4-fold increase was seen in control animals; epinephrine level significantly increased in both groups. The authors concluded that postganglionic sympathetic nerves are the primary source of circulating norepinephrine at birth. The importance of epinephrine at birth was demonstrated by the comparable hemodynamic and metabolic homeostasis in both sympathectomized and control animals.

Padbury, Agata, Ludlow et al. (1987) extended their previous observations by adrenalectomizing near-term ovine fetuses. At birth, plasma epinephrine increased rapidly in control animals but was not detectable in adrenalectomized animals; plasma norepinephrine levels were not different between the two groups. Furthermore, various cardiovascular and metabolic functions, in addition to lung compliance, were impaired in adrenalectomized animals. The authors concluded that epinephrine is required for many of the adaptive events at birth.

Because the release of adrenal medullary catecholamines in response to hypoxia is known to be a protective mechanism for the fetus, Slotkin, Orband-Miller and Queen (1987) investigated the effect of α-adrenergic stimulation in the rat on the recovery of brain and heart from hypoxia-induced damage. Damage was assessed by measuring the activity of ornithine decarboxylase (ODC), an enzyme that reflects disruption of cellular development. Hypoxia at 1 or 8 days of age resulted in increased ODC activity in both the brain and the heart. Pretreatment of the animals with phenoxybenzamine, an irreversible α-adrenergic antagonist, attentuated the long-term ODC response to hypoxia at 8 days. The authors concluded that catecholamines acting at α-adrenergic receptors are important in the adjustment of brain and heart tissue to neonatal hypoxia.

Cohen, Piasecki, Cohn et al. (1987) investigated the sympatho-adrenal response to hypoxia in the primate fetus by measuring catecholamines in fetal carotid artery and maternal femoral artery plasma. Hypoxemia was induced in the fetus by having the mother inspire air with various concentrations of oxygen. Fetal plasma levels of norepinephrine and epinephrine were increased at all levels of hypoxia with norepinephrine levels exceeding epinephrine levels. Maternal catecholamines also increased during hypoxia, but to a lesser extent than observed in the fetus. A negative correlation was found between fetal catecholamine levels and fetal arterial PO_2 similar to that demonstrated in sheep, suggesting the fetal sheep is a suitable model for obtaining information about fetal sympathoadrenal activity in the primate.

A study by Schuijers, Walker, Browne et al. (1986) supports the involvement of the sympathetic function in the neonatal response to hypoxia. Fetal lambs were sympathectomized by treatment with antibodies to nerve growth factor at 80 days of gestation. At 120 days of gestation, hypoxemia for 1 hr elicited increases in heart rate, blood pressure, and plasma norepinephrine levels in control, but not sympathectomized, animals. These results indicate that the peripheral sympathetic nervous system is involved in mediating the cardiovascular changes that accompany neonatal hypoxia.

Jones, Roebuck, Walker et al. (1987) investigated the role of the adrenal medulla in fetal cardiovascular control. Fetal sheep were sympathectomized by intravascular injection of guanethidine, were demedullated by the intra-adrenal injection of acidic formalin, or were subjected to both treatments. Demedullation resulted in the removal of epinephrine from the fetal plasma but did not affect norepinephrine levels. Plasma norepinephrine levels were decreased to half after both demedullation and sympathectomy, suggesting that paraganglia in the sheep provide some contribution to the resting levels of catecholamines. Although baseline heart rate and blood pressure were not affected in these animals, periods of cardiovascular instability developed after several days in the sympathectomized fetuses. The authors concluded that fine control of the cardiovascular system in the fetus requires an intact sympathetic nervous system because the endocrine control by the adrenal medulla is too slow to modulate vascular reflexes.

Jones, Roebuck, Walker et al. (1988) expanded their previous work to investigate the role of the adrenal medulla and peripheral sympathetic nerves in the physiologic responses to hypoxia in fetal sheep. Hypoxia depressed fetal PO_2 by 30% in either adrenal demedullated or sympathectomized fetuses, whereas PO_2 was depressed by 50% in demedullated and sympathectomized animals. The adrenal medulla was found to be responsible for most of the increases in plasma catecholamines that occurred during hypoxia. Furthermore, demedullation enhanced the increases in vasopressin, cortisol, and ACTH, depressed the insulin response, and abolished the increase in blood pressure associated with hypoxia. The authors concluded that fetal adrenal catecholamines are important in sustaining the responses to hypoxia.

Davidson (1987) investigated the role of angiotensin II in the cardiovascular changes and catecholamine surge that accompany birth. Measurements were made in chronically instrumented lambs before, during and after birth by cesarean section. Twenty minutes prior to birth, the lambs were infused with saline (control), saralasin (an angiotensin II receptor antagonist), or captopril (an inhibitor of angiotensin II formation). The greatest increases after birth in plasma renin, angiotensin II and total catecholamines, as well as more rapid adaptation of the cardiovascular system, were seen in control lambs. These results suggest that angiotensin II has a role in postnatal increases in blood pressure and in PO_2 and that circulating catecholamines may support these postnatal cardiovascular changes.

Seidler and Slotkin (1986) reported that innervation of the adrenal medulla is responsible for suppressing the neonatal ability of the tissue to respond to non-neurogenic stimuli. The adrenals of neonatal rats were denervated at 3 days, prior to the formation of functional neuronal connections. When tested at 14 days, the adrenals of the denervated rats did not respond to insulin-induced hypoglycemia, a neurogenic stimulus, but a non-neurogenic stimulus, hypoxia, evoked depletion of catecholamines. Prolonged denervation (21 days) in adulthood also led to the re-emergence of non-neurogenic capabilities; these animals showed no adrenal response to insulin but did respond to hypoxia. The authors concluded that neural stimulation is responsible for the ontogenetic loss of the immature non-neurogenic response of the adrenal medulla and that this response can reappear in the adult after extended loss of neural input.

In a later report, Slotkin and Seidler (1988) discussed the importance of the non-neurogenic adrenal medullary response to stress in the fetus and neonate. Factors that accelerated the ontogenetic development of neural competence resulted in the loss of the non-neurogenic response and a consequent increase in vulnerability of the organism. The fetus and newborn also have high proportions of adrenergic receptor subtypes in many tissues which may permit a greater range of reactivity during stress and periods of tissue differentiation. The authors concluded that the adrenal medulla provides both trophic and physiologic signals to the fetus and neonate.

Graham, Longo and Cheung (1986) investigated the action of several hormones on the release of catecholamines in cultured fetal, newborn and maternal ovine adrenal medullary cells. Prolactin, angiotensin II, and cortisol all increased catecholamine secretion in fetal cells to a greater degree than was observed in newborn or maternal cells, indicating a greater sensitivity of the fetal cells to hormones. The concentrations of cortisol used *in vitro* were within the physiologic range, suggesting that this hormone may function *in vivo* in regulating fetal adrenal medullary catecholamine secretion.

Norepinephrine, of which more than half is contributed by the paraganglia, has been shown to be the predominant catecholamine in the rabbit at birth. Hypoxia on day 1 caused a decrease in norepinephrine production from the paraganglia but not from the adrenal medulla. Fried, Wilkström and Lagercrantz (1988) reported that epinephrine became the predominant catecholamine after birth, representing 67% of the total pool at day 6 and 95% of the total in the adult.

Lau, Seidler, Cameron et al. (1988) found that, although central nervous system transmitter systems generally are protected from the effects of nutritional deprivation, such treatment adversely affects the maturation of chromaffin cells. In rats nutritional deprivation during the neonatal period resulted in decreased catecholamine stores and tyrosine hydroxylase and phenyl-ethanolamine *N*-methyltransferase levels. In contrast, nutritional deprivation had only mild and temporary effects on the development of the centrally derived preganglionic innervation to the chromaffin cells.

EXERCISE

Increased sympathoadrenal activity is the primary autonomic response that occurs during exercise. Kjaer, Secher and Galbo (1987) reviewed the literature on this topic, with emphasis on the effects of exercise on adrenal medullary catecholamine secretion and metabolism.

Sato (1987) reviewed studies of the reflex activation of the adrenal medulla by somatic stimuli in anesthetized animals. Somato-sympathetic reflexes have both a spinal component with segmental organization, and a generalized, non-segmental, supraspinal component. Excitatory responses are mediated by unmyelinated C afferents while inhibitory responses are mediated by myelinated A afferents. The role of intracerebroventricularly administered corticotropin-releasing factor in controlling the somato-sympathetic reflex was discussed.

The influence of the sympathoadrenal system on energy metabolism and fluid and electrolyte balance during exercise in humans was reviewed by Bunt (1986). Mobilization of glucose and free fatty acids during acute exercise is caused primarily by secretion of catecholamines. These acute alterations in metabolism stimulate other endocrine glands to secrete secondary hormones which potentiate metabolic adaptation to extended exercise.

Mercuro, Gessa, Rivano et al. (1988) studied the involvement of the presynaptic D_2-dopamine receptor in mediating catecholamine release during exercise in untrained healthy males. The selective D_2 antagonist domperidone increased plasma catecholamines, heart rate, and blood pressure during exercise. The authors inferred that, in untreated subjects, stimulation of the D_2 receptor acts to inhibit peripheral norepinephrine release and offered this observation as the first evidence of involvement of dopamine in modulating the sympathetic response to exercise. The results further suggest that D_2 receptors in the adrenal medulla normally exert a tonic inhibitory control on the release of epinephrine.

Lehmann, Berg and Keul (1986) related the catecholamine response during treadmill exercise to sex in a group of 15 healthy adults. At identical work loads, free plasma norepinephrine and epinephrine levels, in addition to glucose level, were higher in women than in men. The authors suggested that the higher sympathetic activity in the women may have been due to their relatively smaller skeletal muscle mass in relation to the workload.

Kjaer, Farrell, Christensen et al. (1986) compared the adrenal medullary response to exercise in trained athletes to that in a control group. Plasma epinephrine concentration was higher in the trained athletes both at rest and at identical relative workloads. Identical increases in plasma norepinephrine and heart rate indicated identical levels of sympathetic nerve activity in the two groups. The authors concluded that the adrenal medulla of trained athletes exhibits increased responsiveness during exercise which may contribute to higher glucose production and less accurate glucoregulation in these subjects.

Kjaer and Galbo (1988) determined that physical training increased the ability of the adrenal medulla to secrete epinephrine by comparing the responses of 8 endurance trained athletes and 7 sedentary men. In the athletes, plasma epinephrine concentration and heart rate increased more in response to glucagon administration, hypercapnia, and hypoxia than they did in the sedentary men. Plasma norepinephrine did not increase in either group under any test condition.

Bramnert (1988) determined the effect of intravenously administered naloxone on the physiologic features of submaximal exercise in a group of 8 healthy men. Naloxone did not affect the exercise-induced changes in plasma norepinephrine, blood pressure, heart rate, aldosterone, or renin activity. In contrast, it markedly increased plasma epinephrine level. The author concluded that opioid receptors are involved in mediating the epinephrine response to submaximal exercise.

Rowell, Brengelmann and Freund (1987) measured plasma catecholamines during exercise with or without heat stress in six normal young men. The relationship between heart rate and the logarithm of the concentration of norepinephrine was maintained during exercise with added heat stress even when both variables were increased. No change in plasma epinephrine concentration was observed in relation to heat stress, suggesting that heat stress during exercise raises sympathetic neural, but not adrenal medullary, activity.

Hoelzer, Dalsky, Clutter et al. (1986) investigated the role of redundant glucoregulatory systems during exercise in young adults. The subjects were tested under control, islet clamp (insulin and glucagon held constant), adrenergic blockade, and islet block plus adrenergic blockade conditions. Exercise was associated with the greatest decrease in glucose production and concomitant hypoglycemia in the adrenergic blockade plus islet clamp condition. The authors concluded that redundant glucoregulatory systems are involved in preventing hypoglycemia during exercise. Although sympathoadrenal activation plays a primary role in glucoregulation, changes in pancreatic islet hormone secretion are critical when catecholamine action is deficient.

Barnett, Maslowski, Livesey et al. (1986) compared changes in blood pressure and heart rate during exercise and postural changes in diabetics with autonomic neuropathy and in patients with adrenomedullary deficiency. There were no differences in heart rate or blood pressure between the groups during exercise, and both groups showed significant increases in plasma norepinephrine concentration. The authors concluded that, because the responses of the subjects with adrenomedullary deficiency were not different from those of the diabetic subjects, circulating epinephrine is not required for a normal hemodynamic response during exercise.

Marker, Arnall, Conlee et al. (1986) evaluated the role of epinephrine in glycogenolysis during high-intensity exercise in treadmill trained rats. Liver glycogenolysis was similar in demedullated and sham-operated animals. However, glycogen depletion in the soleus muscle was less and exercise-induced hyperglycemia was not present in the demedullated animals. From these observations the authors concluded that epinephrine is essential for stimulating muscle glycogenolysis and inducing hyperglycemia during intense exercise.

In a second study, Arnall, Marker, Conlee et al. (1986) investigated whether a threshold concentration of epinephrine was responsible for stimulating liver glycogenolysis in treadmill-trained exercising male rats. Infusion of epinephrine at various concentrations into demedullated and sham-operated animals did not alter the rate of decrease in liver glycogen during exercise -- during exercise, no concentration of epinephrine was more effective in stimulating liver glycogenolysis than was physiologic saline.

Because female rats have higher hepatic β-adrenergic receptor activity than male rats, Winder, Loy, Burke et al. (1986) expanded the previous study to evaluate the role of epinephrine in liver glycogenolysis in both male and female exercising rats. Demedullation did not affect the rate of liver glycogenolysis during exercise in either male or female rats, suggesting that epinephrine is not essential for liver glycogenolysis under these conditions.

Yang, Hammer, Sellers et al. (1988) tested the endurance of three groups of rats (sham-operated, demedullated plus saline, and demedullated plus epinephrine) for running on a treadmill (80 minutes or until exhaustion). Demedullated rats given saline had significantly decreased endurance compared with other groups. In contrast, there was no difference among groups in the extent of the decrease in liver fructose 2,6-biphosphate. The authors concluded that the adrenal medulla is not required for normal endurance exercise as long as liver glycogen is available. After depletion of the liver glycogen, epinephrine is necessary to prevent hypoglycemia and to permit exercise to continue.

Winder and Yang (1987) described a technique for collection of blood from the abdominal aorta of rats that permitted samples to be collected within an average of 43 seconds after treadmill exercise and subsequent pentobarbital anesthesia They also determined that there was no significant decrease in epinephrine level in blood samples kept on ice for 1 hour; significant decreases in norepinephrine level were observed after 15 and 60 minutes.

Winder, Yang, Jaussi et al. (1987) investigated the function of the increase in circulating epinephrine that occurs during exercise in fasted rats. Fasted demedullated rats showed decreased endurance, hypoglycemia, increased plasma insulin concentration and decreased muscle glycogenolysis during exercise. All of these conditions were reversed by epinephrine infusion. Thus, epinephrine appears to be essential for preventing hypoglycemia and permitting endurance in the fasted exercising rat.

In a following study, Yang, Carlson and Winder (1987) investigated whether the decreased muscle glycogenolysis seen during exercise in fasted, demedullated rats was due to lack of epinephrine or to increased plasma insulin concentration. When fasted demedullated rats were infused with saline, epinephrine, or epinephrine plus insulin during treadmill exercise, there was no difference in the decrease in glycogen in the soleus muscle between animals infused with epinephrine or with epinephrine plus insulin. Both groups had significantly higher rates of muscle glycogenolysis than animals infused with saline. The authors concluded that increased plasma insulin level does not impair epinephrine-stimulated muscle glycogenolysis during exercise in fasted demedullated rats.

Sudo (1987) tested adrenalectomized and guanethidine-sympathectomized rats to determine the source of epinephrine found in peripheral tissue of control animals after swimming stress. After bilateral adrenalectomy, epinephrine in heart, spleen, and submaxillary gland disappeared and was not increased after swimming stress. Guanethidine treatment greatly decreased both resting level and stress-induced increase in epinephrine in peripheral tissue. These results suggest that epinephrine in peripheral tissue is found in sympathetic nerve terminals and is primarily derived from the adrenal medulla.

Fosha-Dolezal, Avery, Wagner et al. (1988) reported that neither the adrenal medulla nor the cortex influenced serum potassium dynamics during acute exercise in Hereford steer calves. Serum potassium concentrations increased an average of 2.4 mEq/L during 3.5 minutes of exercise at maximal speed capability in both adrenalectomized and intact animals, and returned to control levels by 10 minutes after exercise.

Klimovskaya and Kokoreva (1986) reported the effect of exposure to a magnetic field on the ability of rats to perform a swimming task. During the first 15 days of exposure the physical work capacity of the rats increased together with the reactivity of the sympathoadrenal system. During the second 15 days of exposure, the work capacity returned to normal and the reactivity of the sympathoadrenal system decreased, although there was no change in the catecholamine content of the adrenal glands. The results suggest that the sympathoadrenal system is involved in mediating changes in work capacity that result from exposure to a magnetic field.

TEMPERATURE REGULATION

Stachowiak, Fluharty, Stricker et al. (1986) investigated the molecular genetic mechanisms underlying the increase in adrenal medullary tyrosine hydroxylase (TH) observed after cold stress or sympathectomy induced by 6-hydroxydopamine. Cold stress evoked a 4-fold increase in adrenal medullary TH mRNA and a

90% increase in brainstem TH mRNA. These increases in mRNA were specific for TH because they were not accompanied by increases in total RNA. The increases in TH mRNA after sympathectomy were smaller than those related to cold exposure. Changes in TH mRNA from either stimulus were eliminated by denervation of the adrenal, indicating that an intact sympathetic input to the adrenal is necessary to induce production of TH mediated through specific increases in TH mRNA.

Häggendal, Jönsson, Johansson et al. (1988) investigated catecholamine release in pigs susceptible to developing malignant hyperthermia in response to stress. Stress was induced by two muscle relaxants: succinylcholine (depolarizing) and pancuronium (nondepolarizing). In normal pigs, plasma norepinephrine during stress and the degree of cardiac necrosis were about the same after administration of either muscle relaxant. In contrast, succinylcholine produced very high plasma norepinephrine and epinephrine levels and severe cardiac necrosis in stress-susceptible pigs; pancuronium elicited much lower catecholamines levels and less cardiac necrosis. Pretreatment of stress-susceptible pigs with dantrolene, a drug that efficiently prevents malignant hyperthermia, attenuated the catecholamine increases and prevented myocardial cell necrosis after succinylcholine-induced stress. The authors suggested that the abnormal reaction of the peripheral sympathetic nerves in pigs susceptible to malignant hyperthermia is caused by defects in calcium turnover.

Dhingra, Bhatia, Chhina et al. (1987) reported that demedullation 6 days prior to testing decreased the incidence of freezing and tissue loss after exposure of the hind limbs of rats to a mixture of ice, ethylene glycol, and alcohol. Demedullation 1 hr prior to exposure was ineffective. Norepinephrine administered just before exposure increased the incidence of freezing in both control and demedullated animals but did not affect tissue loss. Administration of epinephrine affected neither variable. The authors suggested that the loss of epinephrine accompanying demedullation results in a decrease in the reactivity of blood vessels to some vasoactive agent.

Janocko and Mycek (1986) determined that the adrenal medulla is not involved in the prolonged hypothermia that accompanies administration of a hypnotic dose of phenobarbital to adrenalectomized rats. This response is seen in adrenalec-tomized, but not demedullated, animals, and is reversed by the administration of glucocorticoids.

The nonshivering thermogenesis that develops with prolonged cold exposure is stimulated in a nonlinear fashion by the infusion of norepinephrine. Pácha, Mejsnar and Kvetnansky (1986) suggested that this observation could be due to a nonlinear distribution of norepinephrine to the cells and tested this hypothesis by measuring the lymphatic transport of norepinephrine in the thoracic duct of cold-acclimatized rats. Lymphatic transport of norepinephrine was related quadratically to the infusion rate, indicating a similar relationship between infusion rate and influx of norepinephrine into the extravascular space. Furthermore, the correspondence between the metabolic response and the infusion rate suggested that the flux of norepinephrine into the extravascular space may be the primary factor in the adrenergic control of thermogenesis.

Brown, Allen and Fisher (1987) reported that the intracerebroventricular administration of bombesin in rats impaired the cold-induced increase in dopamine in intercapsular brown fat but not in other tissues. In contrast, bombesin plus cold exposure increased the level of epinephrine in intercapsular brown fat over that seen with cold exposure alone; this response was blocked by bilateral adrenalectomy. Levels of plasma epinephrine were increased and levels of norepinephrine were suppressed by bombesin administration plus cold exposure. The authors concluded that bombesin acts within the central nervous system to modify response of the sympathetic nervous system to cold exposure in a viscerotopically specific manner.

Laury, Beuavallet, Zizine et al. (1986) reported that the weight of brown adipose tissue and adrenals after cold exposure in the rat is not affected by hypophysectomy. Hypophysectomy did result in a decrease in the epinephrine content of the adrenals without a change in the norepinephrine content. The authors suggested that activation of the sympathoadrenal system induced by cold exposure is little affected by hypophysectomy.

DIGESTIVE PHYSIOLOGY, FEEDING, AND DRINKING

It is well established that catecholamines released from the adrenal medulla play an important role in the regulation of blood glucose concentration by stimulating hepatic glycogenolysis. Niijima (1985, 1986) reviewed the literature on the interaction of visceral afferents, glucose-responsive neurons in the central nervous system, and the autonomic system in the regulation of blood glucose level.

Vander Tuig, Crist and Romsos (1987) evaluated sympatho-adrenal activity in adults rats made obese by using knife cuts to separate the medial and lateral hypothalamus. Overall sympathetic nervous system activity was increased in obese rats, as indicated by a 2-fold increase in the urinary excretion of norepinephrine, but the turnover of norepinephrine in intercapsular brown and abdominal white adipose tissue was suppressed. Adrenal medullary activity, measured by the urinary excretion of epinephrine, was decreased in obese animals. The authors concluded that obesity induced by hypothalamic knife cuts results in an activation of the sympathetic nervous system, selectively excluding brown and white adipose tissue, and reciprocal inhibition of the adrenal medulla.

Yokotani, Okuma and Osumi (1986) investigated the peripheral effects of nicotine on gastric function in urethan-anesthetized rats. Intravenous injection of nicotine inhibited the gastric acid output elicited by stimulation of the vagus nerve and initially inhibited gastric mucosal blood flow; these effects were blocked by phentolamine (an α–adrenergic antagonist) or by adrenalectomy plus reserpine. Nicotine secondarily increased gastric mucosal blood flow, an effect abolished by propranolol (a β–adrenergic antagonist) or adrenalectomy. Thus, nicotine appears to act on both the adrenal medulla and the sympathetic neurons in the gastric wall. Action at α–adrenoreceptors decreased gastric acid output and mucosal blood flow whereas action at β-adrenoreceptors increased mucosal blood flow.

Gastric acid secretion also may be modulated by central mechanisms. Okuma, Yokotani and Osumi (1987) found that the administration of bombesin into the cerebral ventricular system inhibited the gastric acid secretion induced by vagal stimulation or intravenous injections of 2-deoxy-D-glucose but not that induced by bethanechol, a cholinergic agonist. This inhibitory effect of bombesin was greatly diminished by bilateral section of the splanchnic nerves but was not altered by adrenalectomy or chemical sympathectomy alone. However, combined adrenalectomy and chemical sympathectomy abolished the bombesin inhibition of gastric acid secretion. The authors concluded that bombesin acts centrally to excite the sympathoadrenal system and to inhibit gastric acid secretion; a dysfunction in one division of the sympathoadrenal system may be compensated for by activity in the other division.

Yokotani, Yokotani, Okuma et al. (1988) reported that the intracerebroventricular administration of prostaglandin inhibited secretion of gastric acid evoked by electrical stimulation of the vagus in urethane-anesthetized rats. The inhibitory action of prostaglandin was abolished by splanchnectomy or by adrenalectomy combined with 6-hydroxydopamine treatment. The authors suggest centrally administered prostaglandins excite the sympathoadrenal system; the resultant activation of gastric α-adrenoreceptors inhibits electrically elicited gastric acid output.

Lenz (1988) reported that injection of canine gastrin-releasing peptide into the third ventricle of conscious dogs acts centrally to activate the sympathoadrenal system. This peptide inhibits gastric emptying of a liquid meal via a vagally-mediated mechanism and increases left gastric artery blood flow by releasing adrenal medullary epinephrine. The pathways by which gastrin-releasing peptide mediates these effects remain unknown.

El-Refai and Chan (1986) reported that adrenalectomy results in prominent changes in the mechanism of adrenergic activation of hepatic glycogenolysis in rats. After adrenalectomy, the population of α_1-adrenoreceptors labeled with [^{3}H]epinephrine significantly decreased while β-adrenergic binding sites increased. These observations support an increased role of the β-adrenoreceptor in adrenalectomized animals associated with a decrease in phosphorylase activation mediated by α-adrenoreceptors.

Watanabe, Kawada and Iwai (1987) investigated the mechanism by which capsaicin enhances energy metabolism -- *i.e.*, increases serum glucose concentration -- in rats. Neither ganglionic blockade plus cholinergic receptor antagonism nor chemical sympathectomy plus presynaptic catecholamine depletion with reserpine affected the increase in serum glucose that followed capsaicin administration. However, demedullation completely blocked this response, and the effects of demedullation were reversed by administration of epinephrine. The authors concluded that catecholamines from the adrenal medulla mediate the increase in energy metabolism induced by capsaicin.

In a subsequent report, Watanabe, Kawada, Yamamoto et al. (1987) established that capsaicin directly increases the output of catecholamines from the adrenal medulla. Catecholamines measured in the adrenal vein of pentobarbital-anesthetized rats increased in a dose-dependent fashion after the administration of capsaicin. The evoked increase in epinephrine was greater than that in norepinephrine.

Hwang, Wu and Severs (1986) used an angiotensin II antagonist to evaluate angiotensin II receptor binding in rats after 5 days of water deprivation. Autoradiography revealed decreased binding in the adrenal medulla and increased binding in the subfornical organ of these animals. The authors suggested that the decreased receptor binding in the adrenal medulla may result from down-regulation of the angiotensin II receptors, resulting in decreased sympathetic function. Such a mechanism would explain why dehydration increases the plasma level of angiotensin II but does not affect blood pressure.

Clonidine, an α_2-adrenergic agonist, was used by Tung and Yin (1987) to assess the role of the central sympathetic outflow in the schedule-induced polydipsia that occurs when food-deprived rats are subjected to an intermittent food reinforcement schedule. Injection of clonidine into the cerebral ventricles suppressed all aspects of schedule-induced polydipsia, including drinking, licking, lever pressing and pellet intake; this effect was reversed by pretreatment with yohimbine, an α_2-adrenergic antagonist. Additionally, schedule-induced polydipsia was accompanied by an increase in epinephrine content and a decrease in dopamine content in the adrenal medulla. The authors suggested that these results indicate a relationship between catecholaminergic function and establishment of a chronic behavior.

Morita, Nakanishi, Murakumo et al. (1988) used catecholamine secretion and calcium uptake in cultured bovine chromaffin cells to determine the effect of hypoglycemic sulfonylureas on the function of the plasma membrane of secretory cells. The sulfonylureas appeared to act directly on the plasma membrane to inhibit catecholamine secretion by blocking calcium channels.

<u>RENAL PHYSIOLOGY</u>

Yoshikawa and Nakada (1986) investigated the effects of renal function and hypertension on catecholamine levels in adrenal glands collected from patients with renal tumor or breast cancer or at autopsy. These glands were subdivided into groups as coming from normotensive or hypertensive persons, both with and without impaired renal function. Total adrenal catecholamine values were highest in the group of hypertensive patients with impaired renal function. The authors suggested that, in man, the increased levels of catecholamines associated with hypertension may result in diminished renal function.

Sundaresan, Guarnaccia and Izzo (1987) reported that, under normal physiologic conditions the adrenal medulla modulates the number of adrenergic receptors in the renal cortex. The specific binding of [^{3}H]prazosin, [^{3}H]rauwolscine, and [^{3}H]iodocyano-pindolol was used to quantitate levels of α_1-, α_2-, and β-adrenergic receptors, respectively in renal cortical particulate fractions. Each type of receptor was significantly increased in rats adrenal-ectomized 6 wk prior to measurement. Because plasma norepinephrine levels were unchanged in the adrenalectomized animals but plasma epinephrine levels were greatly decreased, the authors suggested that the adrenal medulla normally may modulate renal cortical receptors through the action of epinephrine.

Yamazaki, Yoshimura, Kanbara et al. (1986a) determined the involvement of adrenal medullary catecholamines in the synthesis and turnover of renal catecholamines in the rat. Both adrenalectomy and demedullation increased the renal content and turnover of norepinephrine. Additional changes, including decreased blood pressure, were noted in the adrenalectomized rats; these were attributed to adrenocortical insufficiency. The authors suggested that the increased synthesis of norepinephrine in the kidney was a compensatory mechanism to enhance blood pressure maintenance and electrolyte handling in the kidney in animals deprived of adrenal medullary catecholamines.

MacLean and Ungar (1986) investigated the role of the renin-angiotensin system in the control of catecholamine secretion from innervated and denervated adrenal glands in pentobarbital-anesthetized dogs. Captopril, which inhibits the conversion of angiotensin, decreased the baseline secretion of catecholamines measured in the adrenal vein and inhibited catecholamine release after either a decrease in carotid sinus pressure or stimulation of the cut splanchnic nerve. Administration of angiotensin II restored both responses. These results indicate that the response of the adrenal medulla to sympathetic stimulation requires a minimum level of circulating angiotensin II and is severely inhibited by impairment of the renin-angiotensin system.

Blackburn, Borkowski, Friend et al. (1986) induced diuresis in saline-loaded rats by the subcutaneous administration of 3 κ-receptor opioid agonists, tifluadom, U50488, and ethylketocyclazocine, which could be blocked by naloxone. These agonists were ineffective in inducing diuresis at 1 week after demedullation. Although the adrenal medulla was involved in κ-receptor induced diuresis, catecholamines did not appear to be required since pretreatment with α- and β-adrenergic receptor antagonists did not block the diuretic response in intact animals. The authors raise the possibility that κ-opioid agonists may induce diuresis through an as yet unidentified adrenal medullary factor.

The effects of short- and long-term sodium loading on the sympathoadrenal system in rats were evaluated by Shigetomi, Ueno, Tosaki et al. (1986) Although short-term sodium loading appeared to depress the sympathoadrenal system, long-term loading increased sympathetic activity as measured by levels of total (free plus conjugated) norepinephrine and epinephrine in urine. Such treatment also significantly decreased the maximal binding capacity of the renal dopamine receptor. The increased catecholamines and decreased renal dopamine may contribute to sodium-dependent hypertension.

Debinski, Kuchel, Buu et al. (1986) reported that decreased natriuresis and dopamine excretion in demedullated rats were corrected by infusion of atrial natriuretic factor. Additionally, atrial natriuretic factor suppressed the compensatory increase in norepinephrine secretion observed after adrenalectomy. The mechanisms by which atrial natriuretic factor abolishes the changes in urinary constituents after demedullation remain to be elucidated.

Westerink and Koolstra (1986) reported that, in rats, the urinary excretion of free epinephrine (per millimoles of creatinine) displayed a pronounced circadian variation which was correlated with the overall locomotor activity of the animals. Excretion of free norepinephrine and dopamine (relative to creatinine) did not display circadian variation. The action of phentolamine, an α-adrenergic antagonist, to increase notably the excretion of norepinephrine suggested that at least a portion of the urinary norepinephrine originated from peripheral noradrenergic nerves.

<u>INTERACTIONS WITH OTHER
ENDOCRINE ORGANS</u>

<u>Thyroid Gland</u>

Neonatal hyperthyroidism, induced by injection of triiodothyronine, was found to accelerate synaptic development in the sympathoadrenal axis of the rat. Lau, Franklin, McCarthy et al. (1988) reported that such treatment initially increased both synaptic density and the activity of choline acetyltransferase, a marker for preganglionic sympathetic nerve terminals, in rat pups. These effects were diminished by 25 days of age, and the hyperthyroidism eventually led to deficits in chromaffin cell number and in biosynthesis and storage of catecholamines.

Gripois, Valens and Diarra (1986) reported that hypothyroidism retards the normal development of the adrenal medullary response to insulin-induced hypoglycemia in neonatal rats; hyperthyroidism facilitates development of the response. Recovery of catecholamine levels after depletion in control animals was accompanied by increases in both tyrosine hydroxylase and dopamine β-hydroxylase (DβH). Both conditions of thyroid imbalance impaired induction of tyrosine hydroxylase, while hypothyroidism additionally suppressed DβH. In the developing rat, different aspects of catecholamine synthesis appear to be controlled differentially by thyroid hormones.

Gripois, Valens, Richter et al. (1987) expanded their previous observations to include an analysis of the postnatal development of adrenal medullary dopamine production. As observed with epinephrine and norepinephrine, hypothyroidism impaired both the magnitude and duration of the increase in dopamine elicited by insulin-induced hypoglycemia compared with the response in control animals.

Uchida and Nomoto (1988) reported that injection of thyrotropin-releasing hormone into the lateral ventricle of urethane-anesthetized rats increased the blood flow to the intercapsular brown fat. This effect was decreased by adrenal demedullation, suggesting that the effect of thyrotropin-releasing hormone in this paradigm may be mediated in part by circulating epinephrine from the adrenal medulla.

Pineal Gland

Kachi (1987) reviewed the literature on the various actions of the pineal gland on the autonomic system, including a discussion of the effects of pinealectomy and melatonin administration on various sympathoadrenal functions.

Mahata, Mandal and Ghosh (1988) examined the influence of melatonin injection on adrenal medullary catecholamine content in the pigeon (*Columbia livia*), the crow (*Corvus splendens*), and the parakeet (*Psittacula krameri*). They found that melatonin modulated catecholamine content in the adrenal medulla of all avian species studied. The regulation of norepinephrine synthesis was mediated by the splanchnic nerve; epinephrine synthesis was not dependent on innervation to the adrenal gland.

Banerji and Quay (1986) reported that pinealectomy decreased adrenal medullary DβH activity; these effects were reversed by melatonin administration in a dose-dependent fashion. The authors concluded that the pineal gland has an inhibitory action on adrenal medullary DβH and suggested that the pineal gland may influence catecholaminergic function through action on this enzyme.

In companion studies, Kachi, Banerji and Quay (1988a,b) used quantitative light and electron microscopy to determined the effects of pinealectomy on the epinephrine- and norepinephrine-containing cells of the adrenal medulla in the rat. Pinealectomy affected primarily epinephrine-containing cells, causing hypertrophy of them, especially in the dark, by increasing the cytoplasm and nucleolar pars granulosa. The authors concluded that removal of the pineal gland disturbs the balance between the epinephrine- and norepinephrine-containing cell systems of the adrenal medulla.

Joshi, Troiani, Milin et al. (1986) investigated the influence of the adrenal gland on the pineal gland by measuring levels of melatonin and *N*-acetyltransferase, the rate-limiting enzyme in melatonin synthesis, in the rat. Stress evoked by injection of saline into the hind limb resulted in significant depression of levels of both melatonin and *N*-acetyltransferase in the pineal gland. These changes were prevented by adrenalectomy, suggesting that a factor from the adrenal, either catecholaminergic or steroid, was involved in mediating the response of the pineal gland to a stress.

Gonzalez-Brito, Reiter, Tannenbaum et al. (1988) examined β–adrenergic receptor density in pinealocyte membranes after adrenalectomy and gonadectomy by measuring the binding of [^{125}I]iodopindolol. Only adrenalectomy affected β–adrenergic receptor density, as indicated by an increase in the morning levels of binding. The authors suggested that these changes may be caused by a decreased input of catecholamines to the pinealocytes because adrenalectomy has been shown to decrease the nocturnal synthesis of melatonin.

Animal Models of Diabetes

Tessari, Travagli and Prosdocimi (1988) reported that the catecholamine content of various organs is altered in mutant diabetic mice (C57BL/Ks *db/db* strain) compared with nondiabetic controls (*db/m*). In 6 month old diabetic mice, norepinephrine was increased in the adrenal glands but markedly decreased in the bladder. Concentrations of epinephrine and dopamine in the adrenal glands did not differ between groups. However, dopamine was significantly decreased in the vas deferens and the kidney

Bitar, Koulu, Rapoport et al. (1987) reported abnormalities of catecholaminergic function in the adrenal medulla in (streptozotocin induced) diabetic rats. In these animals, adrenal medullary concentrations of dopamine, norepinephrine, and epinephrine were increased by 52%, 46%, and 33%, respectively. Activities of medullary enzymes were also increased: tyrosine hydroxylase, 52%; DβH, 28%; and phenylethanolamine *N*-methyltransferase, 39%. In contrast, the density of adrenergic receptors in the myocardium was decreased. The authors suggested that such abnormalities of catecholamine metabolism may contribute to the physiologic defects described in this animal model of diabetes.

An increase in plasma DβH in streptozotocin-diabetic rats was reported by Muñoz, García-Estañ, Canteras et al. (1986). The development of cataracts in this form of chemically induced diabetes correlated with the severity of the disease. Rats about to develop cataracts had higher levels of Dβ–H than animals that were to remain without this complication. The authors suggested that DβH, a glycoprotein, may serve as an index for similar glycoproteins that may be involved in the development of cataracts in severely affected animals.

Karimoto and Bondarenko (1986) reported that the concentration of epinephrine in adrenal slices incubated with L-tyrosine was lower in tissue taken from rats with alloxan diabetes than in control tissue. Immobilization stress increased the level of epinephrine in tissue from control animals but had no effect on tissue from diabetic animals.

Pituitary-Adrenal Axis
Yanase, Nawata, Kato et al. (1986b) measured the 24 hour urinary excretion of epinephrine after glucocorticoid treatment in four patients with lupus erythematosus, one with rheumatoid arthritis, and one with pheochromocytoma. Patients with a deficiency of ACTH alone had a significantly lower urinary output of epinephrine. Although the output of epinephrine was normal in patients with lupus or rheumatoid arthritis, it was significantly depressed by administration of prednisolone. These results suggest that epinephrine synthesis in the human adrenal medulla may be inhibited by the pituitary-adrenal axis. The observation that administration of dexamethasone to the patient with pheochromocytoma greatly increased the already high urinary output of epinephrine may have been caused by a direct action of corticosteroid on the tumor.

Song, Crowley and Grosvenor (1988) investigated the role of adrenal medullary catecholamines in the adrenergic regulation of oxytocin release during lactation. Adrenal demedullation of rats in midlactation did not change basal levels of oxytocin but increased the suckling-induced release of oxytocin; this suggested an inhibitory action of adrenal catecholamines. Such an inhibitory response was mimicked by the administration of the β-adrenergic blocker propranolol. These data support the hypothesis that activation of β-adrenergic receptors inhibits oxytocin secretion and that this effect may be due in part to action by adrenal medullary catecholamines acting centrally or directly on the neurohypophysis.

Lau (1988) evaluated the role of the adrenal medulla and peripheral sympathetic nerves in the response of primiparous rats to suckling at days 8-9 and 13-14. At both times the rats have the same milk supply, but less milk is released at the earlier time. Neither adrenalectomy with corticosterone replacement nor ganglionic blockade altered the amount of milk normally ingested by pups at these two times. The author concluded the smaller amount of milk released at 8-9 days does not result from direct suppression of milk ejection by the sympatho-adrenal system.

Nicholson, DeCherney, Jackson et al. (1987) assayed normal and neoplastic human adrenal medullary tissue to determine whether proopiomelanocortin-derived and hypothalamic peptides coexist within the same tissue. Immunoreactive ACTH, corticotropin-releasing hormone, somatostatin, and growth hormone-releasing hormone were identified in both types of tissue. The immuno-reactive ACTH and corticotropin-releasing hormone both were biologically active. Although the physiologic function of these peptides has yet to be determined, it is of interest that hypothalamic peptides known to regulate proopiomelanocortin secretion by the anterior pituitary coexist with this peptide in the adrenal medulla.

Fernández-Ruiz, Bukhari, Martínez-Arrieta et al. (1988) described the effect of sex steroids on adrenal catecholamines in normal cycling female rats . Epinephrine and phenylethanolamine *N*-methyltransferase were increased during diestrous, while norepinephrine did not change throughout the estrous cycle. Monoamine oxidase was significantly increased during proestrous and catecholamine *O*-methyltransferase was increased during estrous. Ovariectomy or injections of estradiol, 2-hydroxy-estradiol, or progesterone also altered the adrenal catecholamine and enzyme levels. The authors concluded that sex steroids may modify the ability of the adrenal medulla to synthesize and store catecholamines.

Snell dwarf mice, which genetically lack growth hormone, prolactin, and thyrotropin, were used to investigate the effects of these pituitary hormones on the catecholamine content of the adrenal glands. Lewinski, Bartke, Esquifino et al. (1986) found that the adrenal contents of epinephrine and dopamine, as well as the activity of phenylethanolamine *N*-methyltransferase, were significantly decreased in these dwarf mice. Although the dwarf mice had adrenal glands several times smaller than those in control animals, the norepinephrine content was not decreased, suggesting the presence of a much higher concentration of this catecholamine in the glands of the dwarf mice. Treatment with growth hormone abolished all differences in adrenal medullary catecholamine content between the control and dwarf mice; the mechanism of this action remains unclear.

Fernández-Ruiz, Cebeira, Agrasal et al. (1987) investigated the action of prolactin on the synthesis and release of adrenal medullary catecholamines in female rats. Plasma levels of prolactin were increased by grafting the anterior pituitary from a littermate beneath the capsule of the kidney. Such treatment increased plasma prolactin, norepinephrine, and epinephrine without affecting levels of dopamine. Norepinephrine and epinephrine content of the adrenal glands was decreased and dopamine content was increased in grafted animals. The authors concluded that increased circulating levels of prolactin are able to increase both the synthesis and the release of adrenal medullary catecholamines in the female rat.

Fernández-Ruiz, Martinez-Arrieta, Hernandez et al. (1988) determined that the ability of prolactin to alter the activity of the adrenal medulla is dependent on the previous plasma levels of this hormone. The synthesis and release of catecholamines was measured *in vitro* in the adrenal medulla from control animals and those made hypo- or hyperprolactinemic by ovariectomy or pituitary grafts, respectively. The effects of prolactin added to the incubation medium depended on the source of the tissue. Prolactin partly reversed a delayed decrease in catecholamine release in tissue from ovariectomized animals, significantly decreased epinephrine release in control tissue, and increased both epinephrine and norepinephrine release in tissue from hyperprolactinemic animals.

Raza-Bukhari, Fernández-Ruiz, Pais et al. (1988) induced chronic hyperprolactinemia in male rats by implanting ectopic pituitary glands. The adrenal medullary contents of epinephrine and norepinephrine were similar in both control and grafted animals. However, in hyperprolactinemic rats, levels of monoamine oxidase, catechol *O*-methyltransferase, and phenylethanolamine *N*-methyltransferase were significantly decreased concomitantly with an increase in adrenal medullary weight. The authors concluded that hyperprolactinemia causes adrenal medullary hyperplasia accompanied by a decrease in catecholamine-synthesizing enzymes.

Adrenal Cortex and Glucocorticoids
Mpoy and Kolanowski (1986) evaluated the excretion of catecholamines in eight patients with adrenocortical insufficiency secondary to panhypopituitarism. Although the mean epinephrine excretion was only slightly reduced below normal levels in these patients, the amount of epinephrine excreted was significantly correlated with both plasma and urinary cortisol levels. The authors concluded that secondary adrenocortical deficiency results in a decrease in epinephrine excretion proportional to the decrease in adrenocortical activity.

The observations by Valenta, Elias and Eisenberg (1986) support a role for the adrenal cortex in modulating adrenal medullary catecholamine secretion in humans. Administration of ACTH to six patients significantly increased the levels of epinephrine and norepinephrine measured in the adrenal vein. The authors suggested two mechanisms for the observed effect. ACTH may increase catecholamine secretion by enhancing blood flow through the adrenal gland and by inducing tyrosine hydroxylase and dopamine β-hydroxylase.

The action of ACTH to increase blood flow through the adrenal gland is supported by observations in the *in situ*, isolated, perfused rat adrenal system. Using this model, Hinson, Vinson, Whitehouse et al. (1986) found that bolus injections of ACTH increased perfusate flow and steroid secretion in a dose-dependent manner. Hinson, Vinson and Whitehouse (1986) expanded this observation with a report that, in the same perfused rat adrenal system, epinephrine decreased the rate of both perfusate flow and steroid secretion.

Edwards, Hansell and Jones (1986) investigated the effect of ACTH administration on adrenal medullary secretion evoked by splanchnic nerve stimulation in conscious calves. ACTH had little effect on the amount of epinephrine and norepinephrine released after splanchnic nerve stimulation. However, ACTH administration completely blocked the decrease in adrenal vascular resistance that was a normal concomitant of splanchnic nerve stimulation. Additionally, ACTH decreased the baseline and stimulus-induced secretion of free [Met]enkephalin but did not alter the secretion of total [Met]enkephalin-containing peptides. The authors suggested that ACTH or an adrenal steroid may acutely alter the processing of proenkephalin in the adrenal medulla or may influence the population of chromaffin cells activated by splanchnic nerve stimulation.

Brown and Fisher (1986) evaluated the effects of glucocorticoids on sympathoadrenal function in rats. Glucocorticoid excess, induced in intact rats by dexamethasone administration, suppressed the increases in plasma epinephrine and norepinephrine induced by inhalation of ether vapor. Glucocorticoid deficiency, effected by adrenalectomy, increased baseline levels of norepinephrine and accentuated the increases evoked by ether vapor; dexamethasone administration antagonized these effects of adrenalectomy. However, dexamethasone did not cause a nonspecific suppression of the sympathoadrenal system because it did not alter the increase in plasma epinephrine and norepinephrine produced by intraventricular administration of corticotropin-releasing factor. The authors suggested that glucocorticoids may inhibit ether vapor-induced increases in plasma catecholamines by inhibiting the release of corticotropin-releasing factor in central regions that mediate sympathoadrenal activity.

Chronic glucocorticoid treatment in Wistar-Kyoto rats depressed basal adrenal medullary activity without affecting the reflex activation of the sympathoadrenal system in response to stress.

Szemeredi, Bagdy, Stull et al. (1988) reported that 7-day administration of cortisol markedly decreased plasma epinephrine levels, while plasma norepinephrine was depressed to a lesser extent. The responses of plasma catecholamines to nitroprusside-induced hypotension were not affected by chronic cortisol administration.

Chatterjee and Pal (1987) reported that two adrenocortical blockers, metopirone and vitamin C, have antagonistic effects on adrenal catecholamine levels in castrated pigeons. Metopirone tended to increase the epinephrine content, whereas vitamin C increased the norepinephrine content. When both compounds were administered together, the effect of metopirone was dominant.

Bethea (1987) investigated the effect of glucocorticoids on the production of dopamine in PC12 cells, a clonal cell line from rat pheochromocytoma, cultured on plastic and on extracellular matrix, conditions that induce cells of different shapes. Cells on both culture substrates had increased dopamine release after glucocorticoid administration in a dose-dependent manner. However, glucocorticoid-treated cells cultured on extracellular matrix showed a larger increase in dopamine content, and glucocorticoid-treated cells cultured on plastic showed a greater increase in dopamine release. The author suggested that glucocorticoids augment dopamine production in both types of cells but that storage is facilitated in cells cultured on extracellular matrix.

Adrenal Cortex and Aldosterone
Jungmann, Germann, Austin et al. (1986) identified the source of dopamine that inhibited aldosterone secretion from the zona glomerulosa of the adrenal cortex in rats. Haloperidol, a dopamine antagonist, increased aldosterone levels in control and unilaterally adrenalectomized animals but was ineffective in bilaterally demedullated animals. Therefore, the authors concluded dopaminergic inhibition of aldosterone secretion is caused by dopamine originating in the adrenal medulla.

An adrenal medullary source for dopamine in the zona glomerulosa is supported by the anatomic observations by Gallo-Payet, Pothier and Isler (1987). Light and electron microscopy revealed rays of medullary tissue extending across the cortex in rat adrenal glands. These tissue rays contained chromaffin cells, collagen, and nerve fibers and could form the anatomic basis for a paracrine function of medullary catecholamines.

Jungmann (1986) reported that the inhibitory effect of dopamine on aldosterone secretion was closely related to sodium homeostasis in a patient population. Dopamine, acting as an inhibitory transmitter, mediated the effects of increased levels of sodium in decreasing aldosterone secretion. In contrast, both dopamine antagonists and sodium depletion stimulated the release of aldosterone.

Inglis, Kenyon, Hannah, et al. (1987) further investigated the role of dopamine in aldosterone secretion in rats fed low-, normal-, and high sodium diets. Dopamine was found in the zona glomerulosa of normal rats at concentrations similar to those in tissue with dopaminergic innervation. Although plasma aldosterone levels were inversely related to the amount of sodium in the diet, dopamine in the zona glomerulosa did not decrease when dietary sodium was decreased. Furthermore, dopamine levels in the zona glomerulosa were not found to be inversely related to aldosterone levels. From these observations, the authors suggest that dopamine may inhibit aldosterone secretion through an extra-adrenal mechanism involving an increase in the clearance of angiotensin II, the major aldosterone trophic factor.

CARDIOVASCULAR SYSTEM

Effects of Drugs and Neurotransmitters on Blood Pressure
The action of neuropeptide Y on the cardiovascular system was reviewed by Waeber, Aubert, Corder et al. (1988). Neuropeptide Y is released with catecholamines during sympathoadrenal stimulation and has direct vasoconstrictor properties.

Evéquoz, Waeber, Aubert et al. (1988) demonstrated that, in doses that did not evoke a pressor response when administered alone, neuropeptide Y was able to block the hypotension induced by endotoxemia in conscious rats. Neuropeptide Y was equally as effective as epinephrine in preventing endotoxemic hypotension. Endotoxin did not produce hypotension in rats with intact adrenal glands.

Mason, Medbak and Rees (1987) investigated the relationship between circulating [Met]enkephalin and catecholamines after drug-induced hypotension in methohexitone-anesthetized greyhounds. Injection of sodium nitroprusside induced hypotension accompanied by significant increases in plasma levels of [Met]enkephalin and catecholamines. In contrast, although hexamethonium also elicited hypotension, the response was accompanied by a decrease in plasma norepinephrine, an increase in plasma epinephrine, and no change in [Met]enkephalin. This observation suggests that neural mechanisms are required for the release of [Met]enkephalin. The authors concluded that [Met]enkephalin may modulate the catecholamine release induced by hypotension as part of the stress response.

Evéquoz, Waeber, Matsueda et al. (1988) examined the effect of adrenal medullary activation in conscious rats on the hypotension caused by two vasodilating drugs -- sodium nitroprusside and atriopeptin III, a synthetic atrial natriuretic peptide. The hypotensive effect of sodium nitroprusside was enhanced by adrenal demedullation; the effect of atriopeptin III was not altered by this manipulation. The authors concluded that, for a given blood pressure reduction, these compounds activate the sympathetic nervous system reflexly to different degrees; the mechanism of this differential action remains unknown.

Valet, Tran, Damase-Michel et al. (1988) reported that rilmenidine, a new α-adrenoreceptor agonist, decreased sympathetic tone primarily through action on the adrenal medulla. In anesthetized normotensive dogs, this drug decreased heart rate, blood pressure, and catecholamine release from the adrenal medulla. In conscious, sino-aortic denervated dogs, an experimental model of hypertension, the drug significantly decreased cardiovascular parameters and plasma epinephrine and norepinephrine levels.

The effects of activation of a dopamine receptor on adrenal medullary chromaffin cells was investigated by Montiel, Artalejo, Bermejo et al. (1986). Secretion of catecholamines was stimulated in isolated cat adrenal glands by nicotine. Low concentrations of droperidol, a dopamine antagonist, increased the nicotine-induced secretion of catecholamines whereas low concentrations of apomorphine, a dopamine agonist, decreased catecholamine secretion. These observations point to the existence of an inhibitory dopamine system in the adrenal medulla which acts under physiologic conditions to limit catecholamine output. The authors suggested that the action of droperidol to increase arterial blood pressure in patients with pheochromocytoma may be due to an inactivation of such a dopaminergic inhibitory system in the adrenal medulla.

Nagahama, Chen, Lindheimer et al. (1986) evaluated the mechanism of action of LY171555, a specific dopamine D_2 receptor agonist, in eliciting dose-dependent increases in blood pressure in conscious rats. LY171555 also increased plasma levels of norepinephrine, epinephrine, and arginine vasopressin. Pretreatment with metoclopramide, a central D_2 receptor antagonist, abolished the pressor effect of LY171555 but pre-treatment with domperidone, a peripheral D_2 antagonist was without effect. Administration of LY171555 to demedullated rats evoked an initial depressor response followed by a pressor response that was smaller than that seen in intact rats. The authors concluded that the pressor action of LY171555 in conscious rats is dependent on activation of a central D_2 receptor that mediates sympathetic outflow and arginine vasopressin release.

Bagdy, Szemeredi, Zukowska-Grojec et al. (1987) reported that the administration of m-chorophenylpiperazine (m-CPP), a serotonergic agonist, produced dose-dependent increases in blood pressure in pithed rats. This effect was not affected by α_1- and α_2-adrenoreceptor blockade or adrenal demedullation but was blocked by administration of the serotonergic antagonists metergoline and ritanserin. These results suggest that the pressor effect of m-CPP is due to its action at serotonergic receptors within blood vessels.

Tsujimoto, Minegishi, Ishizaki et al. (1987) characterized the receptors involved in the biphasic pressor response to sympathetic stimulation in pithed rats. The direct neurogenic component was mediated by postganglionic sympathetic fibers acting at α_1-adrenoreceptors because it was abolished by prazosin, an α_1-adrenoreceptor antagonist, in adrenalectomized rats. The second response, slowly developing, was mediated by adrenal catecholamines activating α_1- and α_2-adrenoreceptors because it was blocked only by prazosin and yohimbine, an α_2-adrenergic antagonist, together in sympathectomized rats.

The action of γ-aminobutyric acid (GABA) on cardiovascular variables in barbitone- or urethan-anesthetized guinea pigs was investigated by Giuliani, Maggi and Meli (1986). In both groups of animals, GABA produced a transient decrease in blood pressure, heart rate, and cardiac contractility; in barbitone-anesthetized animals this was followed by a transient excitatory phase. Adrenalectomized animals or animals treated with phentolamine and propranolol, α- and β-adrenergic antagonists, respectively, responded to GABA administration with only a depression of cardiovascular variables. These observations suggest that GABA may modulate peripheral cardiovascular function in part through the activity of the sympathetic nerves and catecholamine release from the adrenal medulla.

Kuchel, Buu, Racz et al. (1986) reported that sulfoconjugated catecholamines are inactive at vascular receptor sites and thus have no direct effect on blood pressure. However, sulfation of catecholamines appears to be important in overall blood pressure regulation because sulfoconjugated catecholamines, especially dopamine sulfate, have a longer plasma half-life and so are better markers of catecholamine release than are free catecholamines. Additionally, sulfation represents an important storage mechanism by which circulating catecholamines may be rapidly inactivated, thus decreasing their cardiovascular effects. There is evidence of a genetic component in controlling the process of sulfoconjugation and this may be involved in modulating the cardiovascular actions of circulating catecholamines.

Majewski, Alade and Rand (1986) reported that stress induced in rats by immobilizing the hind limbs and isolating the animals for 12 days resulted in significant increases in blood pressure and cardiac epinephrine levels over the levels in control animals. Demedullation blocked the stress-induced increases in blood pressure and decreased the cardiac level of epinephrine. Administration of both desipramine, which blocks the neuronal uptake of epinephrine, and propranolol, a β-adrenergic antagonist, prevented the stress-induced increase in blood pressure. The authors concluded that both the neuronal accumulation of epinephrine and the action of epinephrine at presynaptic β-adrenoreceptors are primary components in mediating stress-induced increase in blood pressure.

Tran, De Saint-Blanquat, Valet et al. (1988) examined the effects of nicardipine, a calcium channel blocker, on adrenal medullary secretion in anesthetized dogs. Nicardipine administration evoked increases in heart rate and catecholamine secretion accompanied by a decrease in blood pressure. In contrast, this drug did not alter catecholamine secretion from the denervated adrenal medulla, evoked by splanchnic stimulation,indicating that a nicardipine-sensitive functional calcium channel is not required for calcium entry into dog chromaffin cells. The authors concluded that the adrenal medulla is not involved in the antihypertensive action of nicardipine.

Tomori (1986) investigated the role of the adrenal medulla in modulating blood pressure in relation to age in the rat. Blood pressure in both immature and adult rats increased progressively over 5 to 6 weeks after demedullation. Tyramine, a sympathomimetic agent, markedly increased blood pressure in immature demedullated rats but had less effect in adult demedullated and control rats. One day after administration of reserpine, blood pressure levels were maintained in immature demedullated rats and tyramine evoked an increase in blood pressure. In contrast, reserpine decreased resting blood pressure in immature control rats, and tyramine was without effect. The author suggested that the adrenal medulla is involved from prepubertal age on in the development of systems that regulate peripheral norepinephrine release.

Izumi (1988) reported that pretreatment of rats with [D-Ala^2, Met^5]enkephalin enhanced the pressor response evoked by administration of L-epinephrine; this response was blocked by

pretreatment with naloxone. In contrast, pretreatment with
enkephalin nearly completely suppressed the pressor effects of
epinephrine in rats that were demedullated at 3 weeks of
age; administration of naloxone followed by enkephalin and
epinephrine caused an increase in blood pressure. The author
concluded that the lack of adrenal medullary enkephalin and
epinephrine at an early age can affect the cardiovascular response
to these substances in adulthood.

Cardiac Arrest and Cardiac Ischemia
Foley, Tacker, Wortsman et al. (1987) examined the plasma
catecholamine and serum cortisol responses to cardiac arrest via
ventricular fibrillation, cardiopulmonary resuscitation, and
ventricular defibrillation in intact and bilaterally adrenalectomized dogs. There were massive increases in the plasma
norepinephrine and epinephrine levels after cardiac arrest in the
intact animals. In the adrenalectomized dog, there was no
increase in plasma epinephrine and the increase in plasma
norepinephrine was decreased by about 70%. These results
indicate that the adrenal medulla is the primary source of the
circulating catecholamines seen in association with cardiac arrest.
In a companion study, Foley, Tacker, Voorhees et al. (1987)
determined the effect of naloxone on the sympathoadrenal
response associated with cardiac arrest and resuscitation in
pentobarbital-anesthetized dogs. After resuscitation via
ventricular defibrillation, circulating norepinephrine
levels remained slightly increased, whereas epinephrine level
decreased to baseline. Administration of naloxone at 6.5 minutes
into cardiopulmonary resuscitation had no effect on the increased
levels of plasma catecholamines seen in relation to cardiac arrest.

Parvez, Ichihara, Parvez et al. (1986) studied the effects of
regional ischemia of the myocardium induced by ligation of a
branch of the left anterior descending coronary artery on adrenal
medullary catecholamine synthesis in the dog. Ischemia for 24
hours caused a decrease in epinephrine and an increase in
dopamine in the adrenal medulla, accompanied by increases in
phenylethanolamine N-methyltransferase, monoamine oxidase,
and catechol O-methyltransferase. These results indicate that
coronary ligation accelerates both the synthesis and the
catabolism of adrenal medullary catecholamines.

Addicks, Hirche, McDonald et al. (1987) investigated the effects
of morphine administration on blood pressure and catecholamine
levels after myocardial ischemia in rats. Morphine administration
increased plasma level of norepinephrine and the norepinephrine
content of adrenergic nerves in the rat heart. The authors
suggested that the ability of morphine to increase norepinephrine
in the heart could be mediated either through increased uptake

of norepinephrine secondary to increased plasma concentration or through direct inhibition of norepinephrine release, perhaps via presynaptic opiate receptors.

Anikin, Kargina-Terentyeva and Afonskaya (1988) used quantitative histofluorescence to examine the effect of experimental cardiac infarction on adrenergic innervation of the heart and adrenal medullary catecholamines. The luminescence intensity of catecholamines was increased in the adrenal medulla on days 1, 3, and 14 after infarction. The authors questioned the efficacy of their method for assessing changes in adrenergic innervation.

Daugherty, Frayn, Redfern et al. (1986) used coronary artery ligation to elicit ischemia-induced ventricular arrhythmias in pentobarbitone-anesthetized rats. Both acute adrenalectomy and chronic demedullation prevented the increases in plasma catecholamines caused by coronary artery ligation, but neither procedure prevented the occurrence of ventricular arrhythmias. The authors concluded that circulating catecholamines do not mediate the early phase of ventricular arrhythmias in the rat.

Lathers, Flax and Lipka (1986) investigated the mechanism by which thioridazine induces arrhythmia and death in the cat. Neither bilateral adrenal vein ligation nor spinal cord section at C1 significantly altered the interval to arrhythmia or to death after thioridazine administration. These results suggest that neither the adrenal medulla nor the central sympathetic system is significantly involved in acute thioridazine-induced arrhythmia; rather the cardiotoxicity of this compound appears to result from direct action on the myocardium.

Shvalev, Vikhert, Stropus et al. (1986) studied the catecholamine content of the adrenal glands and myocardium in 124 cases of sudden cardiac death. In death from acute myocardial infarction, catecholamine levels were depleted in both the adrenal glands and the adrenergic plexuses of the heart. In contrast, in cases of sudden death, both with scarring from previous myocardial infarction and with alcoholic cardiomyopathy, the catecholamine contents of the adrenals and myocardium were increased. The authors concluded that distinctive patterns of catecholamine response are found in cases of sudden cardiac death.

Ghergut (1987) briefly reviewed the involvement of the adrenal medulla in sudden cardiac death. The production of catecholamines in conditions of repeated stress may affect the incidence of this syndrome.

Ma, Zang, Zhu et al. (1988) determined that increased activity of the adrenal medulla was responsible for changes in blood viscosity, hematocrit value, and plasma fibrinogen concentration that occurred in the early phase of myocardial ischemia in dogs. All changes were abolished by bilateral section of the splanchnic nerves but were observed after epinephrine infusion.

Spontaneously Hypertensive Rats

Hano, Kuchii, Umemoto et al. (1986) studied the acetylcholine-induced release of epinephrine from isolated perfused adrenal glands of spontaneously hypertensive rats (SHR) and normotensive Wistar-Kyoto (WKY) rats. There were no differences in the epinephrine content or basal secretion between glands from SHR and WKY rats. The epinephrine secretion induced by acetylcholine in the presence of calcium was greater in SHR glands. These results suggest that secretion of epinephrine may be increased by a local mechanism in the adrenal glands of SHR rats. Such hyperreactivity may be involved in the development of hypertension in these animals.

Bucher and Stoclet (1987) reported that electrical stimulation of the entire spinal cord increased blood pressure and heart rate to a greater degree in SHR than in WKY rats. These responses were enhanced only in SHR rats by pretreatment with propranolol, a β-adrenergic antagonist. Acute adrenalectomy decreased the stimulation-induced increases in blood pressure to identical levels in both groups of rats. These results suggest that the enhanced blood pressure response induced by stimulation of the entire sympathetic outflow is caused primarily by the adrenal medulla and includes a hypotensive component related to β-adrenergic receptor activity.

Fournier, Chiueh, Kopin et al. (1986) described the effect of diet on blood pressure in SHR and WKY rats. Both groups had significant increases in blood pressure and urinary excretion of norepinephrine and dopamine when the carbohydrate source in the diet was refined rather than natural ingredients. Although urinary catecholamine secretion was similar in both groups, the increases in blood pressure were greater in the SHR rats. These results suggest that increased blood pressure after high carbohydrate ingestion is mediated via a neurogenic mechanism because it involves increased catecholamine production.

Racz, Kuchel and Buu (1986) reported that the chronic administration of bromocriptine, a dopamine agonist, decreased blood pressure in SHR rats without affecting urinary epinephrine excretion, adrenal medullary catecholamine synthesis, or tissue catecholamine content. The authors concluded that the anti-hypertensive action of bromocriptine is not mediated through changes in adrenal medullary catecholamines.

Nieber and Oehme (1986) determined the effect of naloxone on the basal outflow of tritiated norepinephrine from adrenal gland slices from SHR and from WKY rats. The basal output of tritiated norepinephrine was significantly higher from SHR tissue than from WKY rat tissue. Naloxone elicited a concentration-dependent decrease in the output of tritiated norepinephrine in both groups of rats. The authors interpreted these data to suggest a functional interaction between opioid peptides and basal catecholamine secretion during hypertension.

In a companion study, Hilse and Oehme (1987) investigated the effect of bilateral demedullation on blood pressure and heart rate in conscious SHR and WKY rats. Blood pressure increases were noted throughout 12 weeks after demedullation in SHR but not in WKY rats; decreases in heart rate occurred in both groups. The authors suggested that the increase in blood pressure that occurred only in the SHR group could be mediated by the loss of opioid peptides or substance P whereas the bradycardia seen in both groups might be related to decreased catecholamine levels in the plasma.

Ming and Renaud (1985) reported that α–methyldopa, an antihypertensive agent, decreased the systolic blood pressure of SHR rats without affecting the blood pressure of normotensive rats. The authors suggested that this effect was due to a decrease in norepinephrine and epinephrine in the periphery because α–methyldopa had the additional effect of inhibiting DβH activity in the adrenal medulla.

O'Connor, Naylor, Cox et al. (1988) found that daily, intraperitoneal injections of lithium chloride significantly lowered systolic blood pressure in SHR rats but were without effect in WKY rats. In contrast, adrenal medullary epinephrine and norepinephrine values were significantly increased in lithium chloride-treated rats of both strains. The authors concluded that the antihypertensive effects of lithium chloride are independent of its action on adrenal medullary catecholamines.

Sato, Sato, Shimamura et al. (1986) recorded the spontaneous activity of a single adrenal sympathetic nerve fiber in stroke-prone SHR and WKY rats anesthetized with halothane. Both the spontaneous activity of the nerve and the secretion of norepinephrine and epinephrine from the adrenal gland were higher in the stroke-prone SHR rats than in the normotensive animals. The mechanism of the increased sympathetic nerve activity in the stroke-prone SHR rats was not established.

The effect of aerobic training on blood pressure was determined in stroke-prone SHR, borderline hypertensive rats (BHR; the first generation offspring of a cross of SHR and WKY rats), and WKY rats by Lütgemeier, Luft, Unger et al. (1987). Swim-trained SHR and BHR had significantly lower blood pressures than controls. Additionally, increases in norepinephrine, epinephrine, and corticosterone associated with swimming were less in trained rats. The authors concluded that exercise training ameliorates hypertension in rats.

Hendley, Cierpial and McCarty (1988) measured plasma catecholamine levels in response to the acute stress of foot-shock in 4 strains of rats: hypertensive and behaviorally hyperactive (SHR), a new strain that was hyperactive only (WK-HA), a second new strain that was hypertensive only (WK-HT), and a strain that was neither hypertensive nor hyperactive (WKY). The sympathoadrenal hyperreactivity that is characteristic of SHR rats was also seen in animals that were only hyperactive. The authors concluded that exaggerated sympathoadrenal activity in response to stress is associated with the hyperactive trait but not with the hypertensive trait of the SHR rats.

McCarty (1986b) investigated the development of adrenal medullary and cardiac responses to reflex sympathetic stimulation in SHR and WKY rat pups. Injection of insulin at 2, 4, and 8 days of age elicited a greater induction of ornithine decarboxylase activity in the heart of SHR rat pups, indicating functional sympathetic neurotransmission to this organ. Both groups of pups had similar adrenal medullary responses to insulin. These results suggest that alterations in the pattern of development of sympathetic target tissues in early life may contribute to the hypertension that occurs in SHR rats.

Szemeredi, Bagdy, Stull et al. (1988) found conscious, freely moving juvenile SHR rats to have significantly higher plasma levels of norepinephrine, epinephrine, and dihydroxyphenylglycol, the intraneuronal norepinephrine metabolite, than did juvenile WKY rats. Administration of yohimbine, an α_2-adrenergic antagonist, dramatically increased the levels of all three neurochemicals in the juvenile SHR rats. The data support the theory that sympathoadrenal medullary activity is greatly increased in juvenile SHR rats and that in these animals there is abnormal dependence on central α_2-adrenoreceptors for decreasing sympathetic activity.

Sautel, Sacquet, Vincent et al. (1988) described catecholamine turnover in seven organs of 5- and 22-week-old genetically hypertensive, normotensive, and hypotensive rats of the Lyon strain. In 5-week-old hypertensive rats the turnover of dopamine in the adrenal medulla was associated with an increased content of epinephrine. The authors suggested that such changes in catecholamine metabolism in the adrenal medulla, rather than changes in sympathetic cardiac innervation, may be involved in the development of hypertension in this animal model.

Borderline, Salt-Induced and Neurogenic Hypertension
Esler, Jennings, Korner et al. (1988) reviewed the literature on the activity of the human sympathetic nervous system in relation to hypertension. Sympathetic activation was measured by norepinephrine turnover; various aspects of these techniques are discussed.

McCarty, Cierpial, Kirby et al. (1987) extended previous observations to consider the development of adrenal medullary and cardiac responses in BHR. BHR and normotensive rat pups had similar levels of ornithine decarboxylase activity in the heart both under basal conditions and after insulin administration. In contrast, BHR pups had increased basal medullary epinephrine levels and a greater depletion of medullary epinephrine after insulin injection. The authors suggested that the greater capacity of the adrenal medulla in the BHR rat pups for catecholamine synthesis and release may contribute to their susceptibility to hypertension as adults.

A companion study by Murphy, Cierpial, Borom et al. (1987) examined the functional development of the adrenal medulla and cardiac sympathetic response in Dahl salt-sensitive and salt-resistant rat pups. Both groups of pups showed similar insulin-induced changes in ornithine decarboxylase activity of the heart and depletion of adrenal epinephrine after insulin administration. In both strains, the functional innervation of the heart was seen as early as 2 days of age but functional activity of the adrenal medulla did not occur until 8 days. These data indicate that the Dahl salt-sensitive rats do not have accelerated development or hyperresponsiveness of the sympathoadrenal system prior to weaning. Similar times of functional sympathetic neuro-transmission in normotensive Long-Evans and Sprague-Dawley rats were reported by Kirby and McCarty (1987).

Knardahl, Sanders and Johnson (1988) used chronically stressed, adrenal demedullated rats derived from a cross between SHR and WKY rats to determine whether a reduction in plasma epinephrine would attenuate acute stress-induced hypertension. The acute stressor, foot-shock, evoked increases in mean arterial blood pressure of 22.3% in demedullated animals compared with 4.2% in sham-operated controls. The authors suggested that the large pressor response in the demedullated animals was caused by diminished β-adrenoreceptor-mediated dilatation of the vasculature in the skeletal muscles. Any beneficial effect of decreased plasma epinephrine level may have been offset by less

β-adrenergic vasodilatation.

Fujita, Ando, Noda et al. (1987) investigated the effect of taurine, a sulfonic amino acid, on blood pressure and plasma catecholamines in a population of 20 to 25-year-old borderline hypertensive patients. Oral administration of taurine significantly decreased both systolic and diastolic blood pressures and plasma level of epinephrine in the borderline hypertensive patients; the decreases in these values were significantly correlated. Administration of glucagon significantly increased plasma epinephrine concentration in borderline hypertensives over the level in normotensive patients. Taurine diminished the increased response to glucagon in the borderline hypertensive patients but had no effect on normotensive patients. The authors concluded that administration of taurine to borderline hypertensive patients decreases sympathetic tone, causing a decrease in blood pressure.

Bouvier and deChamplain (1986) evaluated sympathoadrenal activity in halothane-anesthetized deoxycorticosterone (DOCA)-salt hypertensive and normotensive rats. Baseline plasma norepinephrine levels were significantly higher in the DOCA-salt hypertensive rats and were significantly correlated with mean arterial pressure. The increases in plasma epinephrine concentration elicited by carotid occlusion were markedly greater in hypertensive than in control rats. Administration of yohimbine, an α_2-adrenergic antagonist, potentiated the epinephrine release induced by carotid occlusion in normotensive rats, but was without effect in the hypertensive rats. These results suggest that the hypertension seen in DOCA-salt hypertensive animals may be related to dysfunction in the α_2-adrenergic mechanism that modulates adrenal medullary activity.

deChamplain, Bouvier and Drolet (1987) extended their previous study to investigate the role of the β–adrenergic receptor in DOCA-salt hypertensive rats. The observation that treatment with sotalol, a β–adrenergic blocker, decreased blood pressure and circulating norepinephrine levels in hypertensive but not in normotensive rats suggested that the DOCA-salt hypertensive rats have a more sensitive sympathetic presynaptic facilitatory mechanism mediated through a β–adrenoreceptor. The authors concluded that the susceptibility of the DOCA-salt hypertensive rats to the development of hypertension may be caused by various abnormalities in the adrenergic receptor-mediated control of the sympathoadrenal system.

Yamazaki, Yoshimura, Kanbara et al. (1986b) reported that demedullated rats fed a high sodium diet had decreased blood pressure and heart rate and increased urinary excretion of norepinephrine compared with sham-operated rats. Both the urinary excretion of norepinephrine and the norepinephrine content of the kidney increased after demedullation in rats fed a diet of normal sodium content. In contrast, demedullation did not affect either of these variables in rats fed a high sodium diet. The authors concluded that the adrenal medulla and sympathetic nerves act in concert to regulate sodium handling and that the adrenal medulla is involved in the hypertension induced by salt loading.

Neurogenic hypertension may be induced in rats by complete denervation of the sino-aortic baroreceptors. Using this procedure, Dominiak, Kees and Grobecker (1986) found that blood pressure, plasma norepinephrine and epinephrine levels, and adrenal medullary epinephrine content were significantly higher in denervated than in sham-operated rats. The authors concluded that this form of neurogenic hypertension is supported by increased plasma levels of catecholamines and increased synthesis in and secretion of catecholamines by the adrenal medulla.

Using the same model of neurogenic hypertension in rats, Saavedra and Krieger (1987) found that levels of all adrenal medullary catecholamines, as well as the synthesizing enzymes, tyrosine hydroxylase and DβH, were increased within 3 days after sino-aortic denervation in rats. These differences were no longer present at 21 days after denervation. Such observations suggest that increased activity of the adrenal medulla is important in initiating neurogenic hypertension.

Lorrain, Dipaola, Lhoste et al. (1987) examined the effect of pergolide, a D$_2$ dopamine receptor agonist, on the sympatho-adrenal responses of pentobarbital-anesthetized dogs with sino-aortic baroreceptor denervation. Denervation of the adrenal glands produced an increase in mean arterial blood pressure that was notably less in magnitude and duration than that seen in animals with intact adrenal innervation. Additionally, the increase in plasma epinephrine associated with neurogenic hypertension was only 12% in animals with denervated adrenal glands. Pergolide completely blocked the sympathoadrenal activation after sino-aortic denervation in adrenal denervated dogs and decreased the responses in sham-operated dogs to about ½. The authors concluded that the adrenal medulla is necessary for the sustained hypertension that accompanies sino-aortic baroreceptor denervation and that pergolide, acting through a D$_2$ dopamine receptor, can inhibit the the nervous and humoral characteristics of the hypertensive response.

Rauch and Campbell (1988) measured the *in vitro* activity of tyrosine hydroxylase in rabbits with Goldblatt hypertension.

Enzymatic activity increased 85% in the one-kidney, one clip group compared with one-kidney controls and increased 49% in the two-kidney one clip group compared with two-kidney controls. The authors concluded that the adrenal medulla contributes to the development of hypertension in both of these experimental models.

Borkowski and Kelly (1986) found that demedullation significantly decreased plasma epinephrine levels, adrenal catecholamine content, and resting heart rate and blood pressure in conscious rats. However, demedullation did not affect increases in blood pressure associated with the alerting response. The authors concluded that, although the adrenal medulla is involved in the maintenance of baseline cardiovascular activity, it is not required for the cardiovascular responses that occur in response to environmental stimuli.

Tan and Tsou (1986) reported that the intrathecal injection of vasopressin in rats elicited a dose-dependent increase in blood pressure and heart rate at a much lower dosage than was required with intracerebroventricular injection. The hypertensive action of vasopressin was blocked by pretreatment with a ganglionic blocker (Ecolid), an α–adrenergic receptor antagonist (phenoxybenzamine), or a monoamine-depleting agent (reserpine), suggesting that vasopressin was exerting its actions through the sympathetic nervous system. The authors suggest that vasopressin may act within the intermediolateral cell column of the spinal cord to mediate descending hypothalamic control of sympathetic activity.

Effects of Hemorrhage

Hemorrhage, which may lead to hypovolemic shock, is a potent stimulator of the adrenal medulla. The interaction of adrenal medullary neurohormones with the adrenal cortical hormones and vasopressin was reviewed by Woolf (1986).

Lilly, Engeland and Gann (1986) examined the effect of two small (about 12%) or large (about 19%) hemorrhages, repeated at 24 hr, on catecholamine levels in the adrenal vein in awake, trained dogs. Blood was reinfused after the first hemorrhage. Dogs with small hemorrhages showed no differences in catecholamine secretion on the two days. In contrast, epinephrine and norepinephrine secretions were greater after the second large hemorrhage. This potentiation of the adrenal medullary response to hemorrhage was similar to that observed in pentobarbital-anesthetized dogs, but a larger hemorrhage was required to evoke potentiation in the conscious animals.

In a later study, Engeland, Bereiter and Gann (1986) investigated the secretion of enkephalins during 10% to 20% hemorrhage in awake, trained dogs. Secretion of [Met]- and [Leu]enkephalin into the adrenal vein increased in proportion to the amount of hemorrhage, with the ratio between the two forms being approximately 4:1. Correlation between the secretory rates of [Met]enkephalin and epinephrine suggested that these substances were cosecreted. Chronic denervation of the adrenal gland abolished the increases in both [Met]enkephalin and epinephrine that occurred in response to hemorrhage. The authors suggested that the secretion of opiate peptides by the adrenal medulla is a component of the sympathoadrenal response to trauma.

Gaumann, Yaksh, Dousa et al. (1987) described the effects of hemorrhage and naloxone on the sympathoadrenal response in halothane-anesthetized dogs. Normotensive hypovolemia, induced by a 24% decreased in blood volume, resulted in a 15 to 20 fold increase in catecholamines and a 5-fold increase in [Met]enkephalin in the adrenal vein. Hypotensive hypovolemia, induced by a 49% decreased in blood volume, further increased the secretion of both catecholamines and [Met]enkephalin from the adrenal medulla. Naloxone administration after the induction of hypovolemia caused a further 2-6 fold increase in both catecholamines and [Met]enkephalin. The authors concluded that graded hemorrhage evokes a graded release of catecholamines and [Met]enkephalin from the adrenal, which can be potentiated by naloxone. The observation that the increase in [Met]enkephalin was much less prominent than the increase in catecholamines, in spite of the fact that it is stored and released with the catecholamines, suggests that there may be a differential regulation of the release of monoamines and enkephalin. Additionally, enkephalins may modulate the release of catecholamines from the adrenal medulla.

Inoue, Kimura, Matsui et al. (1987) also reported that [Met]enkephalin-like substances and epinephrine, in addition to vasopressin, increased in the arterial plasma after hemorrhage in pentobarbital-anesthetized dogs. Adrenalectomy severely attenuated all these hemorrhage-induced responses. The difference in the levels of [Met]enkephalin-like substances in the superior vena cava and thoracic aorta did not change during hemorrhage in intact dogs, suggesting that the source of these compounds was from the adrenal medulla rather than the brain.

Schadt and Gaddis (1988) examined the role of adrenal medullary enkephalins in hemorrhagic hypotension in conscious rabbits. In adrenal denervated and adrenalectomized rabbits, plasma norepinephrine decreased at the onset of hemorrhagic hypotension; in intact animals, plasma norepinephrine reached a plateau, suggesting increased secretion from the adrenal medulla.

The pressor response to naloxone was attenuated in both adrenal denervated and adrenalectomized animals. The authors concluded that, although adrenal medullary neurochemicals are not responsible for the acute hemodynamic changes during hemorrhage, some medullary substance, perhaps epinephrine, has a role in the pressor effect of naloxone in this paradigm.

Khalid, Morat and Merican (1987) investigated the interaction between steroids and naloxone in reversing hypotensive shock in adrenalectomized and control rats. In control rats treated with dexamethasone, deoxycorticosterone, or 17-hydroxypro-gesterone,the hypotension and shock caused by hemorrhage of 1% and 2% body weight were reversed by naloxone in a dose-dependent fashion. In contrast, steroid treatment was ineffective in reversing the effects of hemorrhage in adrenalectomized animals. Because the effects of the steroids, which act on the pituitary gland, were essentially similar, the authors concluded that the action of naloxone was on opioid peptide released from the adrenal medulla. They suggested that hypovolemic shock evokes a concomitant release of pressor catecholamines and depressor enkephalins from the adrenal medulla.

A study by Rutter, Potocnik and Ludbrook (1987) suggested that adrenal enkephalins are not responsible for the pressor effect of naloxone after hemorrhage. These authors used a factorial analysis to determine the contribution of the sympathoadrenal components to the pressor response to naloxone after hemorrhage. Rabbits were allocated to six groups: sham-operated, adrenalectomized with adrenocorticoid replacement, guanethidine-sympathectomized, adrenalectomized and guan-ethidine-sympathectomized, and demedullated animals. Because the pressor responses after total adrenalectomy and adrenal demedullation were similar, the authors concluded that adrenal medullary enkephalins are not required for the action of naloxone.

Muldoon, McKenzie and Collins (1988) examined the effects of nalbuphine, an analgesic with opiate receptor agonist and antagonist properties, on the response to hemorrhage in conscious rats. In control animals, nalbuphine returned blood pressure rate to prehemorrhage levels. This pressor effect was not seen in either β-adrenergic blocked or adrenal demedullated animals and was blocked by the administration of naloxone. Because nalbuphine did not increase plasma catecholamine levels over normal, the authors suggested that its pressor action did not result from activation of the sympathoadrenal system. Rather, they proposed that the beneficial effects of nalbuphine in hemorrhagic shock are mediated through β-adrenergic receptors in the heart and require an intact sympathoadrenal system.

Klein, Yabuno, Peeler et al. (1986) determined the adrenal content of catecholamines and enkephalin after a "closed space" subarachnoid hemorrhage induced in cats by rupture of the middle cerebral artery. Epinephrine, norepinephrine, and [Met]- and [Leu]enkephalin in the adrenal were greatly depleted at 4 hr and 3 days after the hemorrhage. By 10 days, catecholamine levels had returned to baseline, but enkephalin levels were somewhat high. The similarity in the molar ratios of enkephalins to epinephrine associated with depletion suggested that the enkephalins are released from epinephrine-containing cells in the adrenal medulla. The observation that the depletion of epinephrine immediately after hemorrhage was much greater than the depletion of enkephalins suggested that the induction of precursor peptide processing compensated to maintain enkephalin content to a greater degree than new synthesis acted to maintain catecholamine levels.

Several compounds other than the enkephalins have been linked to the adrenal medullary response to hemorrhage. Bruhn, Engeland, Anthony et al. (1987b) identified significant levels of corticotropin-releasing factor (CRF)-like immunoreactivity in the adrenal medulla, but not cortex, of dogs. CRF-like immunoreactive cells were concentrated in the vicinity of blood vessels and at the border between medulla and cortex. Additionally, CRF-like immunoreactivity was found in adrenal venous, but not peripheral, plasma samples. The secretion of CRF-like immunoreactive substance increased 3-fold in response to 20% hemorrhage in awake, trained dogs and was associated with increased catecholamine secretion. A physiologic role for this peptide in hemorrhage has yet to be determined.

Morris, Kapoor and Chalmers (1987) reported that plasma levels of neuropeptide Y, in addition to norepinephrine and epi-nephrine, increased after hemorrhage in conscious rats. Sympathectomy with 6-hydroxydopamine abolished the increase in neuropeptide Y seen in association with hemorrhage; adrenalectomy enhanced the response in neuropeptide Y. The authors concluded that the sympathetic nerves supply the major contribution of neuropeptide Y evoked from the sympatho-adrenal system by hemorrhagic stress.

Koivunen, Seaton, Scaer et al. (1987) reported that the renin-angiotensin system is responsible for the maintenance of reflex adrenal medullary section of catecholamines after hemorrhage in pentobarbital-anesthetized dogs. Bilateral nephrectomy suppressed the increase in adrenal medullary epinephrine and norepinephrine secretion after hemorrhage; these responses were restored in a dose-dependent manner by infusion of renin.

Baker, Wilmoth, Sutton at el. (1988) used videomicroscopy to evaluate the reactivity of the cremasteric microvasculature in pentobarbital-anesthetized Wistar rats after hypotensive and normotensive hemorrhage. There was little difference between intact and adrenal demedullated rats in the responses after normotensive hemorrhage, indicating that adrenal medullary catecholamines are not essential in this response. However, the response of the demedullated animals to hypotensive hemorrhage was more severe than that of intact rats, suggesting that medullary catecholamines are required to compensate for blood loss under these conditions.

Tabsh, Rudelstorfer, Nuwayhid et al. (1986) studied the involvement of the sympathetic nervous system in the circulatory responses to hemorrhage in pregnant and nonpregnant sheep. Ablation of the sympathetic nervous system by the administration of 6-hydroxydopamine did not significantly alter the overall circulatory responses to progressive hypovolemia, shock, blood reinfusion, and recovery. The authors concluded that, in addition to the increased sensitivity of the postsynaptic receptors in the peripheral vascular beds, the increased catecholamine output from the adrenal medulla sufficiently compensated for the lack of sympathetic nerve function.

<u>Blood Flow to the Adrenal Gland</u>
Sparrow and Coupland (1987) used radioactive microspheres to measure blood flow to the adrenal gland of pentobarbital-anesthetized rats under baseline conditions and after hemorrhagic hypotension. Under baseline conditions, approximately 7% of the total adrenal blood flow passed directly to the medulla. Hemorrhage sufficient to cause a 32% decrease in systolic blood pressure resulted in a 52% increase in the total blood flow to the adrenal gland; the proportional distribution of blood between the cortex and medulla remained the same as that seen prior to hemorrhage.

The observations by Breslow, Mennen, Koehler et al. (1986) suggest that there is a species difference in the response of the adrenal cortex to hemorrhagic hypotension. Radiolabeled microspheres were used to determine adrenal blood flow in pentobarbital-anesthetized dogs with mean arterial blood pressures of 100, 80, 60, or 40 mm Hg. Adrenal medullary blood flow increased 2 to 4-fold at all levels of hemorrhage. In contrast, cortical blood flow decreased approximately 50% at mean arterial pressures of 80 mm Hg or less but returned to control level within 25 min at all levels of hemorrhage except 40 mm Hg. Total blood flow to the adrenal gland initially decreased, as would be anticipated because the cortex is the larger portion of the adrenal gland. The authors suggested that there is differential control of the blood flow to the medulla and cortex in the dog.

In a subsequent study, Breslow, Jordan, Thellman et al. (1987) determined whether changes in blood flow to the adrenal gland related to hemorrhage were mediated neurally by comparing blood flow between intact glands and contralateral glands denervated by section of the splanchnic nerve. Hemorrhage that decreased blood pressure to 60 mm Hg effected a 50% decrease in blood flow to the adrenal cortex in both innervated and denervated glands. In contrast, blood flow to the adrenal medulla increased 4-fold in the innervated glands and was decreased 50% in the denervated glands. In a separate group of dogs, splanchnic nerve stimulation increased adrenal medullary blood flow 5 to 10-fold in the ipsilateral gland but had no effect on blood flow to the contralateral gland; there was no effect on cortical blood flow to either gland. The authors concluded that splanchnic nerve activity profoundly affects adrenal medullary blood flow and is involved in mediating the changes in blood flow induced by hemorrhage.

The report by Hamaji, Nakamura, Izukura et al. (1986) supports the observation of differential control of adrenal blood flow in the dog. Total blood flow as measured with nonradioactive microspheres was not significantly decreased in the adrenal gland until hemorrhage decreased blood pressure to below 30 mm Hg. Although cortical blood flow decreased below 50 mm Hg, medullary blood flow increased in contrast below 70 mm Hg; at 30 mm Hg medullary blood flow returned to that in normotensive controls.

The importance of autoregulatory mechanisms in the adrenal medullary vasculature during hemorrhage was emphasized by the report of Hamaji, Nakaba, Nakamura et al. (1987). During hemorrhagic hypotension, nonradioactive microspheres were used to measure the blood flow in an intact adrenal gland and in the contralateral adrenal which was equipped with a transperitoneal adrenofemoral shunt. Blood flow in the intact gland was significantly decreased in the adrenal cortex and unchanged in the medulla. In contrast, both cortical and medullary blood flow in the cannulated contralateral adrenal gland were significantly decreased due to the inhibition of adrenal vascular autoregulation by the shunt.

Bergdall, Levin, Townsend et al. (1986) measured the rate of adrenolumbar venous blood flow related to hypothalamic stimulation in pentobarbital-anesthetized cats. Stimulation of 15% of medial hypothalamic sites, which elicited aggressive behavior in the awake animal, increased ipsilateral adrenolumbar venous blood flow more than 20% over baseline. Such increases in blood flow were accompanied by significant increases in blood pressure, heart rate, and secretion of epinephrine from the ipsilateral adrenal gland. The authors suggested that hypothalamic sites that mediate increased blood flow from the adrenal gland may overlap with sites that selectively control the secretion of epinephrine.

NONHYPOVOLEMIC SHOCK AND HYPOXIA

Kovarik, Jones and Romano (1987) reported that arterial plasma epinephrine levels were increased 5 hr after cecal ligation and puncture in rats. Plasma epinephrine levels at later times and the epinephrine content of the adrenal medulla at 20 hr after cecal ligation and puncture were not increased. In contrast, plasma norepinephrine levels were increased at all times after the induction of septic shock. The authors concluded that increases in catecholamines associated with septic peritonitis result primarily from local nerve-stimulated release of norepinephrine.

Oyama (1985) reported that cAMP and histamine levels increased in plasma and lung tissue during anaphylactic shock in rats. Administration of epinephrine or isoproterenol, a β–adrenergic agonist, enhanced the increase in cAMP and prevented mortality induced by anaphylaxis in all the rats tested. The author suggested that the adrenal medulla may be involved, through β–receptors, in the mechanism that regulates the release of histamine and serotonin from the lung during anaphylactic shock.

The effect of dexamethasone on hypoxic stress was investigated by Johnson, Rock, Young et al. (1988) because dexamethasone is known to prevent the acute mountain sickness that occurs with altitude exposure. In humans dexamethasone blunted the pulse rate increases that occurred with altitude exposure. Furthermore, urinary epinephrine excretion was decreased without effect on norepinephrine excretion. The authors concluded that acute mountain sickness was associated with activation of the adrenal medulla; dexamethasone was effective in combating this syndrome by blocking the adrenal medullary response.

The effects of hypoxia (8% O_2 in N_2) on the turnover of nor-epinephrine and epinephrine in rats was studied by Lee, Miwa, Fujiwara et al. (1987). Four hours of hypoxia increased the turnover of both norepinephrine and epinephrine in the adrenal gland; this was blocked either by pretreatment with hexamethonium, a ganglionic blocker, or by spinal cord transection at C5-6. Somewhat different patterns of catecholamine turnover were seen in the heart, submaxillary gland, and stomach.

The authors concluded that changes in adrenal medullary function in relation to hypoxia are mediated by centers in the brain.

Srinivasan, Yamamoto and Lagercrantz (1988) reported that administration of naloxone caused a sustained increase in respiratory rate during hypoxia in rabbit pups. This response is in contrast to the normal respiratory response to hypoxia which consists of an initial increase in rate followed by a decrease. The authors suggested that the effects of naloxone are mediated through the sympathoadrenal system.

Novíc, Mijatov-Ukropina, Mihíc et al. (1986) reported that exposure of animals to decreased atmospheric pressure resulted in depletion of both norepinephrine- and epinephrine-containing cells in the adrenal medulla.

ETHANOL

Adams and Hirst (1986) reported that intragastric administration of ethanol to rats in severely intoxicating doses for 4 days resulted in increases of 20% in wet and dry proportional heart weights over control animals. The development of cardiac hypertrophy was less apparent in demedullated rats. Control animals given ethanol had decreased levels of epinephrine in their adrenal gland and increased urinary excretion of epinephrine. Urinary norepinephrine was increased in both control and demedullated animals given ethanol. The authors concluded that adrenal medullary catecholamines are involved in the development of the cardiac hypertrophy that accompanies severe, continuing intoxication in the rat.

LS and SS strains of mice are selectively bred for differences in the duration of loss of the righting response after administration of a hypnotic dose of ethanol. Zgombick, Erwin and Cornell (1986) reported that intraperitoneal administration of ethanol elicited dose-dependent increases in plasma norepinephrine and epinephrine only in the LS strain of mice. The ethanol-induced increase in plasma catecholamines was abolished by the administration of chlorisondamine, a ganglionic blocker, suggesting that the effects of ethanol are mediated thought the central nervous system. Elimination of ethanol-induced hypothermia by increasing the ambient temperature did not decrease the hyperglycemic response to ethanol, indicating that hypothermia is not the stimulus for catecholamine secretion and concomitant hyperglycemia associated with ethanol. The authors suggested that the differences in responses between the two strains of mice are due to differential CNS sensitivities to ethanol.

AGING

Changes in plasma catecholamine levels and catecholamine metabolism in the elderly were reviewed by Linares and Halter (1987). The observation that clearance of norepinephrine from the plasma appears to be diminished in the elderly, whereas epinephrine clearance is enhanced, implies a differential effect of aging on catecholamine removal mechanisms.

Sato, Sato and Suzuki (1987) reviewed studies of sympatho-adrenal activity in aged rats. They concluded that the increased catecholamine levels in the peripheral plasma of aged rats were the result of increased secretion from the adrenal medulla, probably mediated by increased sympathetic nerve activity. A later study by Kurosawa, Sato, Sato et al. (1987) provided evidence that reflex adrenal sympathetic nerve activity is not altered by age. Baroreceptor stimulation and cutaneous brushing depressed adrenal nerve activity in both young adult (4 months old) and aged (26 months old) rats. Similarly, adrenal nerve activity reflexly increased in response to cutaneous pinching in both groups of animals. The authors concluded that the maintenance of reflex sympathetic responses does not explain the age-related changes in baseline adrenal sympathetic nerve activity.

Serov and Sokolova (1987) examined the sympathetic innervation of the cardiac ventricles and the catecholamine content of the adrenal medulla as a function of age in inbred white rats. A 20% decrease in the catecholamine content of the adrenal glands, in addition to a decrease in the adrenergic terminals of the heart was seen in old as compared with mature animals.

Odink, Thissen, Zeegers et al. (1986) reported that the 24-hr excretion of catecholamines and catecholamine metabolites increased with age in a population of 2- to 17-year-old subjects. Excretion of epinephrine was more than 6-fold greater and norepinephrine more than 2-fold greater during the day than during the night.

Mahata and Ghosh (1986a) related age to adrenal medullary catecholamines in 25 species of newly hatched and adult birds. In newly hatched animals, norepinephrine was the primary catecholamine in passerine birds and epinephrine was predominant in nonpasserine birds. In adult passerine birds the primary catecholamine changed to epinephrine and in adult nonpasserine birds either norepinephrine was predominant or both catecholamines were found in approximately equal proportions.

Mahapatra, Mahata and Maito (1987) determined that age is without effect on the diurnal rhythms of epinephrine, norepinephrine, and corticosterone secretion in the soft-shelled turtle *Lissemys punctata punctata*. In both juvenile and adult animals, peaks levels of norepinephrine and corticosterone occurred at 24:00 hours; peak epinephrine level occurred a 06:00 hours. The lowest levels of all hormones were observed at 12:00 hours in both juvenile and adult animals.

OTHER PHYSIOLOGIC EFFECTS

Results of a study by Kondo (1988) support the theory that vibration syndrome is a systemic dysfunction. Plasma catecholamines were measured in patients with vibration syndrome and in a control group, both under baseline conditions and after cold stress. Patients with vibration syndrome were found to have hyperfunction of the sympathetic nervous system, as indicated by increased plasma norepinephrine level, and also hypofunction of the adrenal medullary system, as indicated by decreased plasma epinephrine level.

Gross, Bertrand, Ribes et al. (1988) reported that epinephrine potentiated the response of adenosine to stimulate glucagon secretion in the isolated perfused rat pancreas. This effect was mediated by α–adrenergic receptors and appeared to be effected by the adrenal medulla because norepinephrine was ineffective. The authors suggested that epinephrine and adenosine may act synergistically to increase glucagon secretion during stressful energy-deficient situations.

Bohme, Belay, Dettmer et al. (1987) reviewed the interactions between adrenal and pancreatic hormones in controlling the differentiation of the hepatic enzymes tyrosine aminotransferase, serine dehydratase, and phosphofructokinase 2 and renal β-glucosidase in suckling rats. Adrenal medullary catecholamines and glucagon have antagonistic actions on the induction of these enzymes, which are mediated by α_1-adrenergic receptors.

Jones, Westfall and Sayeed (1988) reported that the administration of live *Escherichia coli* to rats caused significant tachycardia and a 7-fold increase in plasma norepinephrine and epinephrine. A second group of rats made tolerant to endotoxin also were tolerant to *E. coli* and had significantly lower plasma catecholamine levels compared with nontolerant bacteremic rats. The authors suggested that bacteremia produces a striking sympathoadrenal activation; this response is attenuated in endotoxin-tolerant rats, perhaps by reduction of the afferent stimuli activated during bacteremia.

Sakata and Iriuchijima (1988) reported that when a rat is moved from its home cage, blood flow to a hind limb increases without a concomitant increase in arterial pressure. During this transposition response plasma epinephrine and norepinephrine values increased 6- and 2-fold, respectively; these increases were blocked by demedullation. Because infusion of epinephrine to concentrations similar to those during the transposition response also increased blood flow to the hind limb, the authors concluded that epinephrine is responsible for this effect.

CHAPTER 18

CLINICAL CONSIDERATIONS

<u>IMAGING</u>

Magnetic resonance imaging (MRI) has revolutionized radiology. Visualization of the adrenal glands by this new technology has become very popular and important. A number of laboratories have reported the features of the normal adrenal gland and of its pathologic states by MRI. (Glazer, Woolsey, Borrello et al. 1986; Reinig, Doppman, Dwyer et al. 1986a; Reinig and Doppman 1986; Reinig, Doppman, Dwyer et al. 1986b; Belenkov, Arabidze, Belichenko et al. 1986; Mezrich, Banner and Pollack 1986; Graif, Leung, Steiner et al. 1986; Chang, Glazer, Lee et al. 1987; Krahe, Steudel, Dewes et al. 1987; McGahan 1988; Glazer and Lee 1988; Glazer, Francis and Quint 1988; and Glazer 1988.) Although in general these studies were preliminary, they all revealed the value of this technology in diagnosing pheochromocytomas and other tumors of the adrenal gland. It was found in most of these investigations that T2-weighted images permit differentiation between pheochromocytomas and metastatic lesions. All of these reports concluded that MRI has great promise for imaging the adrenal glands.

Falke, te Strake, Shaff et al. (1986) evaluated the role of MRI in adrenal disease by correlative imaging with computed tomography (CT). They demonstrated that MRI is capable of identifying most adrenal abnormalities previously detected by CT. The results further suggest that MRI has greater specificity for mass lesions and might be useful for differentiating pheochromocytomas from various other lesions. The ability to perform multiplanar imaging and the superior contrast compared with CT are useful for the assessment of origin and extension of large lesions and the detection of pheochromocytomas in complex cases.

Paling, Brookeman and Mugler (1987) evaluated the potential of phase-contrast imaging, also called "proton chemical shift imaging" or "proton spectroscopy," in detecting and displaying tumors located outside of the liver. Specifically, they assessed the ability of this technique to render such tumors more conspicuous, both visually and quantitatively, than may be possible with standard MRI. In all cases, tumors were most conspicuous with the phase-contrast technique and even tumors that could not be visualized by standard MRI could be visualized with this technique. Among other tumors, they demonstrated that pheochromocytoma could be well visualized. They recommended phase-contrast imaging as an adjunct to standard MRI sequences when evaluating for either the presence or extent of a tumor.

Two cases of nonmalignant adrenal masses with prolonged T2 relaxation time and increased adrenal/liver signal ratios were reported by Baker, Spritzer, Blinder et al. (1987). These two cases emphasized that increased signal intensity in an adrenal mass on T2-weighted images is not always due to a pheochromocytoma or a malignancy.

Levine, deVries and Wetzel (1987) reported on a patient with postoperative intracaval recurrence of an extra-adrenal pheochromocytoma that was discovered by MRI. A patient developed clinical features of tumor recurrence 3 years after resection of an extra-adrenal pheochromocytoma, and MRI showed a tumor in the inferior vena cava. The tumor was resected successfully.

Mathieu, Despres, Delepine et al. (1987) assessed various imaging techniques for the diagnosis of pheochromocytomas in patients with Sipple's disease: MRI, CT, and nuclear imaging. MRI detected all adrenal and ectopic lesions. Nuclear scans gave some false-negative results. CT missed an ectopic lesion and gave a false-positive result. The authors concluded that MRI may replace both CT and nuclear scans in the work-up of Sipple's disease.

Radioiodinated *m*-iodobenzylguanidine has proved to be a useful agent for use in nuclear scanning to diagnose adrenal medullary disease. The uses of *m*-iodobenzylguanidine for such purposes have been reviewed by Beierwaltes (1987), Fischer and Vetter (1986), Nakajo (1986), Chan, Warshawski, Scott et al. (1987), and Sakai, Tadokoro, Makino et al. (1987). It was pointed out that *m*-iodobenzylguanidine is similar to norepinephrine and, therefore, is concentrated, stored, and released from chromaffin vesicles in a manner similar to that of norepinephrine. Use of it not only gives structural information about the size and location of the adrenal medulla or its tumors but also gives important functional information. The high sensitivity and specificity of this imaging agent was emphasized in these reviews.

Troncone, Rufini, Campioni et al. (1985) evaluated *m*-iodobenzylguanidine in experimental studies in animals. The tissue distribution of the imaging agent was studied in rats sacrificed at various time intervals, and a striking affinity of the agent for the adrenal glands was found. These and other studies in dogs demonstrated that *m*-iodobenzylguanidine is a suitable diagnostic agent for adrenal medullary disease.

Carreño Hernández, Estrada, Peréz Maestu et al. (1986) reported on two cases of pheochromocytoma with bone and liver metastasis in which *m*-iodobenzylguanidine was used in the diagnosis. They demonstrated that this agent can be used in a safe and specific way for the localization of extra-adrenal pheochromocytoma and metastasis. High doses of radiolabeled *m*-iodobenzylguanidine can be used as therapy.

Shulkin, Shen, Sisson et al. (1987) showed that *m*-iodobenzylguanidine scintigraphy may be used to determine the presence or absence of metastatic lesions in the appendicular skeleton in cases of malignant pheochromocytoma and neuroblastoma. Discrete concentrations of radioactivity in bone are often seen in such patients.

Bomanji, Flatman, Horne et al. (1987) quantitated radioiodinated *m*-iodobenzylguanidine uptake by the normal adrenal medulla in hypertensive patients. None of these patients had any evidence of pheochromocytoma. This study presents an appropriate control group range for comparison with patients who have proven epinephrine- and norepinephrine-secreting tumors.

A case study in which a hyperfunctioning adrenal medulla and a medullary thyroid carcinoma were imaged by using *m*-iodobenzylguanidine was reported by Ansari, Siegel, DeQuattro et al. (1986). Scintigraphy visualized a pheochromocytoma of the right adrenal gland, adrenal medullary hyperplasia of the left adrenal gland, and a primary medullary thyroid carcinoma.

Jansson, Tisell, Fjälling et al. (1988) presented a case for the early diagnosis and treatment of patients with thyroid and adrenal medullary neoplasia referred to as multiple endocrine neoplasia type 2 syndrome. They used recently improved methods, including CT and radioiodinated *m*-iodobenzylguanidine scintigraphy to provide an early diagnosis of adrenal medullary disease. If this can be accomplished in patients suspected to be carrying the gene for multiple endocrine neoplasia type 2 syndrome, a more conservative strategy for adrenal surgery could be justified for these patients with unilateral pheochromocytomas -- *i.e.*, bilateral adrenalectomy would not always be required.

Mangner, Tobes, Wieland et al. (1986) examined the metabolism of radioiodinated *m*-iodobenzylguanidine in patients with metastatic pheochromocytoma. After diagnostic doses of the agent, about half of the administered radioactivity appeared in the urine within 24 hours; about 80% was recovered within 4 days. Most of the agent was recovered unaltered. The composition of the metabolites varied widely among the patients, and there was no obvious correlation between the presence of metabolites and the location of the tumors or the plasma or urinary catecholamine levels. These studies suggest that *m*-iodobenzylguanidine is a rapidly excreted, relatively stable radiopharmaceutical agent.

Ertl, Deckart, Blottner et al. (1987) collected quantitative biokinetic data from patients who had been given radioiodinated *m*-iodobenzylguanidine and attempted to provide improved estimates of absorbed doses to various organs and tissues. Animal studies also were done. The absorbed doses were high in the thyroid and adrenal glands and relatively low in testes, lungs, and other organs.

A series of 59 patients who underwent imaging with radioiodinated *m*-iodobenzylguanidine was reviewed by Lindberg, Fjälling, Jacobsson et al. (1988). Among other things, they found that imaging the adrenal medulla several days after the administration of the nucleide (late images) produced a low rate of false-negative scans; the background activity diminishes, and even small pheochromocytomas can be detected. They also reported measurements of percentage uptake and calculations of half-time values in the adrenal glands.

An analog of *m*-iodobenzylguanidine, 4-amino-3-[^{123}I]-iodobenzylguanidine, was examined as a new sympathoadrenal imaging agent by Shulkin, Shapiro, Tobes et al. (1986). This analog is easier to prepare. In three patients with metastatic pheochromocytoma, the new agent revealed the same metastatic deposits shown by *m*-iodobenzylguanidine. The uptake of the new agent was greater than that of *m*-iodobenzylguanidine in lung, gut, and spleen, which may pose diagnostic problems in some cases.

Letiec, Guilloteau, Huguet et al. (1986) compared another adrenal medullary imaging agent, iodocarboxyamidino-1 phenyl-4 piperazine, with *m*-iodobenzylguanidine. There was early and preferential uptake of both agents by both adrenal glands and heart and storage in the adrenal gland. However, iodocarboxamidino-1- phenyl-4-piperazine was less stable *in vivo* than *m*-iodobenzylguanidine. Reserpine depletion studies in the rat indicated that the uptakes of the two imaging agents are not similar.

Koizumi, Endo, Sakahara et al. (1986) compared *m*-iodobenzylguanidine scintigraphy and CT in the diagnosis of pheochromocytoma. In a prospective study, they found that CT was superior for locating tumors in the adrenal glands but scintigraphy was more useful in the detection of extra-adrenal pheochromocytoma.

An apparent false-positive localization of a pheochromocytoma by using radioiodinated *m*-iodobenzylguanidine was reported by Krubsack, Arnaout, Hagen et al. (1988). They suggested that this finding could be attributed to the presence of three adenomas in the zona fasciculata of the adrenal cortex.

Mendoza, Lueg, Puyau et al. (1986) reported on a case study in which a presumptive diagnosis of pheochromocytoma was made, and the tumor was localized with CT. A selective angiogram clearly demonstrated the adrenal tumor. They pointed out that angiography plays a major role in localization when CT does not clearly identify an adrenal tumor.

Radin, Ralls, Boswell et al. (1986) reported on detecting pheochromocytoma with unenhanced CT. They recommended unenhanced CT as the initial localizing procedure in patients with suspected pheochromocytoma. This avoids the small but finite risk of hypertensive crisis associated with more invasive techniques.

The accuracy of prospective diagnosis of functional adrenal disorders by CT was evaluated by Kenney, Streeten and Anderson (1986) in patients strongly suspected of having such disorders. This study emphasizes the need to correlate CT findings with biochemical evaluation and the usefulness of venous sampling in selected cases to avoid inappropriate surgery. When CT is interpreted in correlation with complete biochemical analysis, the correct diagnosis usually can be made.

CT was used to evaluate patients with various adrenal lesions by Abe, Koganemaru, Yasuda et al. (1986). They analyzed the CT findings in terms of size, border, structure, and other features and described the typical and atypical appearances of adrenal tumors.

King, Kopecky, Baker et al. (1987) described a patient with agnogenic myeloid metaplasia who was discovered by CT to have bilateral asymmetric adrenal enlargement. CT-guided needle biopsy demonstrated hematopoietic cells in the adrenal gland. By using CT and cytologic findings, they diagnosed extramedullary hematopoiesis.

Coexistent adrenal and extra-adrenal pheochromocytomas were accurately localized with noninvasive investigations by Mathieson, Sandler, Joplin et al. (1986). In this case, the tumor was demonstrated preoperatively by CT and *m*-iodobenzylguanidine scintigraphy. The authors thought that the combination of investigations used in this case was particularly useful for the successful identification of an unexpected combination of tumors.

Schroeder, Wells and Sty (1987) used comparative imaging to diagnose an extra-adrenal pheochromocytoma in a young patient. CT and angiography demonstrated a tumor medial to the lower wall of the left kidney.

Günther (1986) compared ultrasonography with CT for examining the adrenal gland. It was concluded that adrenal ultrasonography is suitable as a screening procedure but is inferior to CT in spatial resolution and overall accuracy. With adrenal lesions exceeding 2 cm in diameter, the sensitivity of ultrasonography equals that of CT.

Yamakita, Yasuda, Goshima et al. (1986) also compared ultrasonography and CT for the diagnosis of adrenal disorders. They found that ultrasonography is better than CT for demonstrating characteristics within the tumor, the relationship between the tumor and surrounding organs, and the organs from which large tumors arise such pheochromocytoma. On the other hand, CT is better able to detect small adrenal tumors and adrenal hyperplasia. Of two ultrasonographic instruments tested, the sector scanner was more useful in the delineation of certain tumors.

The shape of the adrenal glands as visualized by ultrasonography was examined by Winkler, Abel and Helmke (1988). They found that the shape of the glands varied according to the angle and position of the transducer, but the most common appearance was that of a triangle, helmet, or cap. In many cases, the adrenal gland could not be distinguished from the upper pole of the kidney. They described several features to distinguish adrenal gland from kidney in order to provide assistance in measuring the adrenal glands and making the detection of minor abnormalities possible.

Kenney and Stanley (1987) reviewed 106 cases of adrenal masses of all types in all age groups. Thirty-three contained calcium visible on radiographs, ultrasonography, or CT. Two of the cases were pheochromocytomas. CT was found to be the best for showing the presence and pattern of calcium in an adrenal mass.

PHEOCHROMOCYTOMAS

Pheochromocytoma is by far the most common tumor of adrenal chromaffin cells. General reviews of this disease entity were published by Abeatici, Durando, Palestini et al. (1986), Manger (1986), Hull (1986), Albert and Howerton (1986), Moulonguet (1987), Balázs, Ihász and Illyés (1987a), and Samaan and Hickey (1987). An epidemiologic study of 439 cases of pheochromocytoma diagnosed in Sweden between 1958 and 1981 was presented by Stenström and Svärdsudd (1986). Andersen, Lund, Toftdahl et al. (1986) reported on the 47 cases of pheochromocytoma that were diagnosed in Denmark from 1977 through 1981. Welbourn (1987) reviewed the early history of surgical treatment of pheochromocytoma. Among other things, he described the first case of pheochromocytoma seen at autopsy, in 1886, and the first surgical removals of the tumor performed in 1926 in two independent medical centers by César Roux and Charles Mayo.

In addition to the imaging techniques discussed in the preceding section, pheochromocytoma may be diagnosed by analyzing catecholamine metabolism, particularly in hypertensive patients. Therefore, it is important to understand catecholamine metabolism in this pathologic condition. Collste, Brismar, Alveryd et al. (1986) concluded from a retrospective analysis of catecholamines that, owing to the variability of venous drainage from the right adrenal gland, the technique of adrenal venous blood sampling may be unreliable and even misleading in attempts to locate a pheochromocytoma.

Proye, Fossati, Fontaine et al. (1986) described 3 patients with pheochromocytoma who exhibited exclusively dopamine secretion and 12 patients who exhibited dopamine secretion with other catecholamines. All 3 patients with exclusive dopamine-secreting tumors were normotensive. They suggested that hypertension in patients with pheochromocytoma might depend on the ratio of dopamine to the combination of epinephrine and norepinephrine.

Tippett, McEwan and Ackery (1986) reevaluated the use of dopamine as a prognostic factor in pheochromocytoma. Patients with malignant pheochromocytoma who had increased dopamine excretion had a poor prognosis. Further observations suggested that dopamine excretion may be a useful prognostic factor when combined with imaging to determine the extent of the disease. Tippett, West, McEwan et al. (1987) reported on a larger number of patients with malignant pheochromocytoma. They measured the urinary excretion of dopamine and its metabolite, homovanillic acid. Patients with increased dopamine and homovanillic acid excretion had disseminated malignancy and the poorest prognosis. In this study, dopamine excretion appeared to be a more discriminating biochemical index of malignancy, prognosis, and disease progression than homovanillic acid excretion.

Rubio and Barontini (1985) evaluated the levels and characteristics of the catecholamine synthetic enzymes in pheochromocytoma and their correlation with catecholamine secretion. The activities of tyrosine hydroxylase, aromatic amino acid decarboxylase, and dopamine β-hydroxylase were significantly higher in tumor than in normal human adrenal medulla. Phenylethanolamine N-methyltransferase activity in tumors secreting epinephrine was similar to that in normal adrenal medulla and was undetectable in tumors secreting only norepinephrine. The activities of catabolic enzymes such as monoamine oxidase were significantly decreased, allowing for increased liberation of catecholamines from the pheochromocytoma.

The free and sulfoconjugated norepinephrine, epinephrine, and dopamine levels in the plasma were examined in normal patients and in patients with various clinical disorders by Ratge, Knoll and Wisser (1986). As suspected, levels of free catecholamines were higher in patients with pheochromocytoma. About one-third of the total norepinephrine and epinephrine was found in the free form, whereas only 1.5% of the dopamine was in this form, indicating that most of the dopamine in the serum is sulfoconjugated.

The metabolism and storage of catecholamines in rats with implanted pheochromocytoma were investigated by Buu, Kuchel, Manger et al. (1986). The implants caused the rats to become hypertensive, and they showed high plasma concentrations of dopamine and norepinephrine. However, the norepinephrine content of several peripheral tissues of rats with implants did not differ from that in controls. There also was no increase in conjugated dopamine in the plasma or tissue of rats with implants. On the other hand, plasma and urine of tumor-bearing rats had abnormally high concentrations of homovanillic acid. In contrast to control rats, in tumor-bearing rats intravenous infusion of free dopamine had no effect on plasma free dopamine levels but increased homovanillic acid levels considerably. These and other data show the important role of monoamine oxidase in the removal of excessive quantitites of the catecholamines released by pheochromocytoma.

The effect of long-term administration of reserpine on the adrenal medulla and pheochromocytomas was presented by Diener (1988). It was concluded that pheochromocytomas and other tumors of the adrenal medulla induced by reserpine have no carcinogenic relevance to man.

Tischler, DeLellis, Nunnemacher et al. (1988) used reserpine-treated rats to develop a model for adrenal medullary hyperplasia and neoplasia. Treatment of rats with reserpine for 5 days resulted in an 8-fold increase in mitotic figures in chromaffin cells in otherwise histologically normal adrenal glands. The fact that reserpine directly depletes catecholamines and reflexively increases catecholamine synthesis in chromaffin cells suggested that these activities also may regulate proliferation of mature chromaffin cells. Perhaps prolongation of these signals may lead to pathologic proliferative states. Tischler et al. (1988) suggested that this reserpine model may be a useful system for elucidating normal and pathologic mechanisms of signal transduction.

The relationship between tumor and plasma concentrations of neuropeptide Y in patients with pheochromocytoma was examined by Corder, Shapiro, Lowry et al. (1986). The basal plasma levels of neuropeptide Y could be divided into two groups, normal and high. Tumor concentrations similarly could be divided into low and high. Increased plasma levels were observed only in patients with high tumor concentrations. Thus, a subtype of pheochromocytoma has been identified which synthesizes and secretes neuropeptide Y.

Capella, Riva, Cornaggia et al. (1988) reviewed the histopathology, cytology, and cytochemistry of pheochromo-cytomas and paragangliomas including chemodectomas. They presented a functional indentification of the cell types composing many such tumors. They also compared these data with clinicopathologic findings that outline the advantages and limitations of cytologic studies for understanding these tumors and improving diagnostic and prognostic criteria.

Beaser, Guay, Lee et al. (1986) described a patient with Cushing's syndrome due to a pheochromocytoma that was producing ACTH. The preoperative diagnosis was suggested by bilateral adrenal cortical hyperplasia plus a separate unilateral adrenal medullary mass and was confirmed by laboratory studies. The patient was successfully treated with a unilateral adrenalectomy.

Report of a case of pheochromocytoma of the organ of Zuckerkandel was presented by Fischbein, Rio, Wenger et al. (1986). This extra-adrenal tumor had many similarities to pheochromocytoma of the adrenal medulla.

Balázs, Ihász and Illyés (1987b) described a rare adrenal medullary tumor in a 37-year-old patient--the joint occurrence of a pheochromocytoma and ganglioneuroma. Only three similar cases were found in the literature. They presented a detailed light and electron microscopic study that demonstrated a high degree of differentiation of both tumors. The correct diagnosis was established with the aid of arteriography. After removal of the adrenal gland, the patient's condition has improved for a period of 2 years.

A case study by Balázs (1988) found a mixed pheochromocytoma and ganglioneuroma of the adrenal medulla. The patient had a 5-year history of paroxysmal attacks of hypertension, headache, and palpatation but did not have an increased catecholamine level. After removal of the tumor, the patient recovered fully. The morphology of the tumor and the light and electron microscopic features were described.

Aiba, Hirayama, Ito et al. (1988) reported a case of a combined pheochromocytoma and ganglioneuroma plus a cortical adenoma in the ipsilateral adrenal gland. They reported the catecholamine and peptide levels in the compound adrenal medullary tumor and also the steroid levels. They suggested that the tumors in the adrenal glands of this patient may be functionally related.

Miettinen and Saari (1988) reported a case in which a pheochromocytoma was combined with a malignant schwannoma. They suggested that a neoplastic proliferation of cells from the pheochromocytoma may account for the origin of the schwannoma.

Four laboratories assessed the value of a clonidine-suppression test in the diagnosis of pheochromocytoma. Karlberg, Hedman, Lennquist et al. (1986) concluded that the test is a safe and simple diagnostic procedure for verification or exclusion of pheochromocytomas, especially in patients with normal baseline catecholamine levels. Misleading results of a clonidine-suppression test were reported by Taylor, Mayes and Anton (1986); examples of a false-negative and false-positive result were presented. Mulinari, Zanella, Guerra et al. (1987) demonstrated that urinary metanephrine levels, when monitored by the clonidine-suppression test, are useful for the diagnosis of pheochromocytoma. This combination permits the differentiation of patients with tumor from those with essential hypertension. Gross, Shapiro, Sisson et al. (1987) found that use of the clonidine-suppression test to separate normal from abnormal adrenal medulla function was only possible in those patients with abnormal adrenal anatomy and in whom adrenal medulla dysfunction could be documented by other diagnostic means. This suggests that the clonidine test alone may not be useful to identify the earliest stages of adrenal medullary dysfunction.

Needle biopsy is another diagnostic procedure that has been used for pheochromocytoma and other tumors. Nguyen (1987) reviewed percutaneous fine needle aspiration biopsy and cytologic study of the adrenal glands and kidney. Selvaggi, Pace, Zambonin Zallone et al. (1984) obtained samples of human adrenal pheochromocytoma by using open needle biopsy and examined the fine structure in these specimens. The good tissue preservation and absence of mechanical damage allowed confirmation of already-described features of this differentiated tumor. The results of fine needle aspiration cytologic study, immunohistochemical staining, and electron microscopy of a combined pheochromocytoma-ganglioneuroma were presented by Layfield, Glasgow, Du Puis et al. (1987). The neoplastic cells showed positive staining for neuron-specific enolase and vasoactive intestinal polypeptide. This appears to be the first description of the fine needle aspiration cytologic features of this combined tumor. Wernecke and Galanski (1986) reviewed the difficulties of percutaneous fine needle biopsy of the adrenal glands. They pointed out that clinical suspicion of pheochromocytoma is an absolute contraindication for fine needle biopsy. The problems caused by encountering an unsuspected pheochromocytoma during percutaneous adrenal biopsy were presented by Casola, Nicolet, vanSonnenberg et al. (1986). The risk of blood pressure alterations during the procedure was stressed.

The use of formaldehyde-induced fluorescence for cytologic diagnosis of pheochromocytoma was presented by Kimura, Sasano and Ishioka (1986). They described a simple technique that they claim may contribute greatly to the confirmation of the diagnosis of pheochromocytoma.

Hosaka, Rainwater, Grant et al. (1986) performed DNA ploidy studies on paraffin-embedded archival tumor specimens by using flow cytometry. Patients with a normal DNA histogram had a benign clinical course. Thirty-one percent of the patients with polyploidy and 39% of patients with a DNA aneuploid peak had evidence of malignancy. Flow cytometric DNA ploidy measurements of isolated nuclei seems to provide useful prognostic information for patients with pheochromocytoma.

Nuclear DNA content of paraffin-embedded tissue from pheochromocytomas and normal adrenal glands was analyzed by flow cytometry by Amberson, Vaughan, Gray et al. (1987). All control adrenal glands and one-third of the pheochromocytomas were diploid. Aneuploid DNA content was found to be a frequent occurrence in benign adrenal pheochromocytomas. They concluded that aneuploidy *per se* cannot be considered a specific marker of malignancy in adrenal pheochromocytoma.

The difficulties of diagnosing pheochromocytoma in children was presented in two independent publications by Turner, DeQuattro, Falk et al. (1986) and Trofimov and Gubin (1986). Turner et al. (1986) found that surgical findings conflicted with preoperative information obtained by CT and *m*-iodobenzylguanidine scintigraphy. Vena caval sampling for catecholamines confirmed all adrenal tumors but suggested additional tumors not verified at operation in some patients. They concluded that more pediatric experience with diagnostic procedures is needed. Trofimov and Gubin (1986) reported a case in which a child had to undergo reoperation for recurrence of a pheochromocytoma. The second operation located two tumors, and the treatment was considered a success.

Cooper, Goodman, Frauman et al. (1986) suggested that pheochromocytoma in the elderly is a poorly recognized clinical entity. Although less than 3% of reported cases occur in patients over 70 years of age, Cooper et al. (1986) described four elderly patients with pheochromocytoma seen over a 1-year period; two had malignant tumors. This is the first antemortem description of malignant pheochromocytoma in patients over age 70 years. Their experience with these four patients suggests that pheochromocytoma is a potentially remediable condition in the elderly.

Krane (1986) reported cases of clinically unsuspected pheo-chromocytoma in which the typical symptoms were not present. In a review of 32 histologically confirmed tumors, 53% of these tumors were undiagnosed before operation or autopsy in cases before 1962. Since 1962, only 18% of pheochromocytomas have remained clinically unsuspected. Krane (1986) concluded that, by maintaining greater diagnostic acuity and using newer biochemical and imaging techniques, the incidence of clinically unsuspected pheochromocytomas should be reduced.

A histochemical, immunohistochemical, and ultrastructural study of eosinophilic globules in pheochromocytoma from the adrenal medulla was performed by Kawai, Senba and Tsuchiyama (1988). The globules were observed in 7 of 11 cases. These globules were not related to chromaffin vesicles. They suggested that the eosinic globules observed in this study may be a complex protein.

Bronstein, Smirnova and Yuryeva (1988) described the morphology of pheochromocytoma and other catecholamine-producing tumors of chromaffin tissue. They found an inverse correlation between the number of chromaffin vesicles and of other organelles. They also described certain ultrastructural peculiarities that may serve to identify the malignant nature of the tumor.

Tan, Huang, Lai et al. (1988) reported on their experience treating three patients with pheochromocytoma over a 15-year period. Two patients had adrenal pheochromocytomas, and one patient had an extra-adrenal pheochromocytoma.

Roth, Wilson, Eberwine et al. (1986) described a patient with pheochromocytoma, acromegaly, congestive heart failure, and diabetes mellitus. The pheochromocytoma and acromegaly were linked because postmortem examination of the tumor showed that growth hormone-releasing hormone was present in the tumor tissue. Hyperplasia of the somatotrophs in the pituitary was also found. The acromegaly apparently was secondary to a pheochromocytoma that secreted not only catecholamines but also growth hormone-releasing hormone.

Madias, Goorno and Herson (1987) describe a case in which severe lactic acidosis was a presenting feature of pheochromocytoma. The secretion of catecholamines resulted in this condition that featured prominently in the clinical presentation. Pheochromocytoma should be listed among the clinical entities associated with or predisposing to lactic acidosis.

Smith, McPherson and Lynn (1987) described a patient with inferior vena cava involvement by a pheochromocytoma and described a novel approach to such tumors that extend above the diaphragm. A number of diagnostic procedures were used, including ultrasonography and venography. The tumor was removed through a partial sternotomy in combination with a bilateral subcostal incision. The liver was rotated laterally to gain access to the tumor.

Komatsu, Shirasu, Takei et al. (1987) also presented a case of a right adrenal pheochromocytoma with prominent displacement of the inferior vena cava. Their surgical approach was an anterior subcostal incision, after which the liver was easily mobilized.

Dicke, Henry and Minton (1987) reported a case in which the pheochromocytoma extended into the inferior vena cava and simulated pulmonary embolism. The biochemical diagnosis of pheochromocytoma was confirmed by sharply increased catecholamine, metanephrine, and vanillylmandelic acid urinary levels; these returned to normal after resection of the tumor.

A case in which pheochromocytoma was a cause of pulmonary edema was reported by Blom, Karsdorp, Birnie et al. (1987). They suggested that predominantly epinephrine-secreting tumors may predispose patients to this complication. An adrenergic receptor blocker was useful in the initial treatment in this case.

An adrenal composite tumor of pheochromocytoma and malignant peripheral nerve sheath tumor in a 39-year-old woman was described by Min, Clemens, Bell et al. (1988). The peripheral nerve sheath tumor appeared to have undergone further malignant degeneration, resulting in a highly anaplastic sarcoma that led to the death of the patient 8 months after radical resection of the involved adrenal gland. At autopsy, the other adrenal gland was found to be normal.

Jansson, Tisell and Hansson (1988) examined the coincidence between pheochromocytoma and other neuroectodermal abnormalities. Medullary thyroid carcinoma was the most commonly associated neuroectodermal abnormality. This was followed by von Recklinghausen's neurofibromatosis, intracranial tumors, parathyroid hyperplasia, and carcinoid tumors of the midgut. All 7 patients with multiple pheochromocytomas and all 13 patients with hyperplasia of the adrenal medulla outside of the tumor had other neuroectodermal abnormalities. It was pointed out that it is important to detect these abnormalities because they can be fatal. Patients with associated neuroectodermal abnormalities often have hereditary syndromes.

Kamiya, Yasui, Suzuki et al. (1986) reported an autopsy case in which von Recklinghausen's disease was associated with pheochromocytoma. Neurofibroma was confirmed for tumors in the skin, and the histopathologic findings were highly suggestive of pheochromocytoma.

Catecholamine metabolism in pheochromocytoma and normal adrenal medullas was studied by Nakada, Furuta and Katayama (1988). Pheochromocytomas were found to contain significantly larger amounts of norepinephrine, dopamine, dihydroxy-phenylalanine, tyrosine hydroxylase activity, metanephrine, normetanephrine, and vanillylmandelic acid than did normal adrenal medullas. The ratio of epinephrine to norepinephrine was significantly higher in the normal medulla than in pheochromocytoma. These and other results indicate the possible presence of a negative feedback on catecholamine synthesis via tyrosine hydroxylase in normal adrenal medullas but not in pheochromocytoma.

Two independent groups reported on the use of calcium channel blockers in the treatment of pheochromocytoma. Tomaru, Oh, Miura et al. (1986) successfully treated the fluctuation blood pressure in a patient by intravenous administration of diltiazem. There was no evidence that diltiazem affects catecholamine secretion levels, as reported by others for nifedipine. Favre, Forster, Fathi et al. (1986) reported on three patients with pheochromocytoma who were treated with nifedipine. They found that nifedipine appears to control hypertension in pheochromocytoma without altering catecholamine levels. This suggests that calcium-channel blockade interferes with catecholamine vascular action but not with the release of catecholamines from the tumor. They concluded that calcium antagonists may be helpful in preventing the pressor complication of provocative tests used in the screening of pheochromocytoma.

Bao (1987) described six cases in which pheochromocytoma was removed by enucleation. The technique is described and suggested as a procedure of choice for avoiding the risk of injuring the great vessels during surgery.

A new pharmacologic approach to anesthesia during surgery for pheochromocytoma was presented by Braude, Leiman and Moyes (1986). Etomidate was infused in conjunction with other drugs. Cardiovascular stability was well-maintained in the two cases presented.

Three cases of pheochromocytoma that were successfully treated with adenosine triphosphate (ATP) during the operation were reported by Aso, Tajima, Suzuki et al. (1986). The mechanism of the hypotensive effect of ATP has been proved to be due to the antagonizing action of it on the vasoconstriction induced by norepinephrine. In the three cases reported, ATP safely and effectively controlled blood pressure. ATP could be used as the drug of choice to keep the circulatory state stable during surgery for pheochromocytoma.

Zabransky (1988) discussed the general aspects of surgery of the adrenal glands from a pediatric viewpoint. The incidence, diagnosis, treatment, and prognosis of pheochromocytoma and neuroblastoma in children were discussed.

Assyag, Klein, du Cailar et al. (1986) reported a case in which residual hypertension was present after excision of an extra-adrenal pheochromocytoma. The patient was diagnosed as having essential arterial hypertension, although a malignant pheochromocytoma is possible.

Bomanji, Bouloux, Levison et al. (1987) explored the pathophysiologic effects of pheochromocytomas and paragangliomas on the uninvolved adrenal medulla as reflected by its uptake of radioiodinated *m*-iodobenzylguanidine. They showed that, in the presence of increased plasma catecholamine levels originating from a tumor, the unaffected adrenal medullary tissue retains the capacity to take up *m*-iodobenzylguanidine. Catecholamine release from the normal adrenal medulla was within the normal range.

Hoffman (1987) presented a model for studying pheochromocytoma. Specifically, when this tumor transplanted into New England Deaconess Hospital rats, markedly increased concentrations of norepinephrine developed and were associated with severe hypertension. Desensitization of α- and β-adrenergically mediated responses and catecholamine induced cardiomyopathy.

Lairmore, Knight and DeMartini (1987) described a pheochromocytoma and other neoplasms in a goat. They pointed out that the three primary neoplasms in this goat have been reported separately in other species; their simultaneous occurrence in one animal is uncommon.

OTHER CONSIDERATIONS

Weiss (1986) reviewed the autopsy reports in 32 cases in which the patients had died with acquired immune deficiency syndrome (AIDS). She found 15 cases with reported adrenal disease. Adrenal disease was restricted to the medulla in eight cases, but there were no cases of disease restricted to the cortex. It was suggested that the apparent selective destruction of the adrenal medulla is a result of infection by the retrovirus.

Bricaire, Marche, Zoubi, et al. (1987) confirmed a high incidence of adrenal lesions in a systematic study of 100 autopsies in cases of AIDS. In the cases in which the adrenals were studied, the glands were normal in 19 and abnormal in 64. Four additional cases were uninterpretable because of necrosis, and an additional case had a nonspecific depletion. The most common sites of lesions were the adrenal medulla, the cortex, or at the cortico-medullary junction.

Immunoglobulin G antibodies from patients with Lambert-Eaton syndrome were found, by Kim and Neher (1988), to block voltage-dependent calcium channels in bovine adrenal chromaffin cells. Electrophysiologic measurements showed that the IgG from these patients decreased the potassium-stimulated increase in free intracellular calcium concentration. This demonstrated that in patients suffering from the Lambert-Eaton syndrome IgG reacts with voltage-dependent calcium channels and blocks their function. This phenomenon can account for the presynaptic impairment characteristic of this disease. This study is sure to open interesting avenues for investigation.

A case of Cushing's syndrome associated with adrenal medullary hyperplasia was reported by Kazama, Noguchi, Kawabe et al. (1986). The patient showed the typical clinical features and laboratory data of Cushing's syndrome. Norepinephrine excretion was increased; after the adrenal gland was removed, medullary cells were seen to be distinctly hyperplastic. This may be the first case of these two diseases present in the same patient.

Deus Fombellida (1987) reported a case of unilateral primary adrenal medullary hyperplasia. The value of adrenal planimetric analysis of the specimen was discussed. It was suggested that primary adrenal medullary hyperplasia may be a precursor of pheochromocytoma.

Borrero, Katz, Lipper et al. (1987) reported a case of adrenal medullary hyperplasia and adrenal cortical adenoma in a single adrenal gland. Microscopic examination of the surgical specimen confirmed this diagnosis. This is the first reported case of these two conditions in a single gland.

Chalaprawat, Wannakrairot, Vajarapongse et al. (1988) described a patient with bilateral adrenal medullary hyperplasia and presentation similar to that with pheochromocytoma. Subsequent removal of the left adrenal gland resulted in complete amelioration of the symptoms, and the diagnosis was based on the histopathologic findings.

Margulies and Sheps (1988) reviewed Carney's triad--gastric leiomyosarcoma, pulmonary chondroma, and extra-adrenal paraganglioma and presented a protocol for the follow-up of patients with this syndrome.

Hale, Suarez, Williams et al. (1987) reported an insulinoma and an adrenal medullary ganglioneuroma in a 26-year-old woman. The diagnosis was made postoperatively after partial pancreatectomy and removal of a retroperitoneal mass. It was suggested that this case may simply be a coincidence of two rare tumors; however, the possibility of a true association should be considered, particularly in view of the common embryologic derivation of the tissues from which these tumors were derived.

Brown, Kamalesh, Sesh Adri et al. (1988) reported the presence of complement-fixing anti-adrenal medullary antibodies in patients at high risk of developing insulin-dependent diabetes mellitus and in patients with diabetes. The functional significance of the presence of these antibodies needs to be evaluated.

Schurter and Hedinger (1987) reported a case in which *Cryptococcus neoformans* was detected in the left adrenal gland at autopsy. They presumed that this cryptococcoma was the source of the hematogenous dissemination of a septicemia that had contributed to the death of the patient.

The adrenal glands and the central nervous system from five human cases of rabies were studied by De Oliveria Almedia, De Paula Antunes Teixeira, De Oliveria et al. (1986). The adrenal medulla showed diffuse and intense mononuclear exudate associated with changes in the chromaffin cells in three of the patients. Eosinophilic bodies were found in the cytoplasm of chromaffin cells and in the interstitial space. De Oliveria Almedia et al. (1986) suggested that the involvement of the adrenal medulla in the pathology of rabies may be related to the embryologic and metabolic relationships of the chromaffin system and the nervous system.

Adrenal medullary function in patients with atopic dermatitis was assessed by Archer, Dalton, Turner et al. (1987). They measured plasma levels of catecholamines and cAMP in response to the stimuli of standing after lying supine and of 5-minute infusion of histamine in the standing position. Resting plasma levels of catecholamines and cAMP were not statistically different in atopic patients and normal controls. Plasma catecholamines and cAMP in response to the stimuli and the rate of clearance of exogenous epinephrine from the plasma were not significantly different in patients with atopic dermatitis and in normal subjects.

Amsterdam, Marinelli, Arger et al. (1987) assessed adrenal gland volume by CT in psychiatrically depressed patients and in healthy volunteers. They found that, generally, the depressed patients had demonstrably larger adrenal glands. These observations suggest that there may be adrenal hypertrophy during depressive illness.

Oppedal, Brandtzaeg and Kemshead (1987) evaluated a panel of monoclonal neuroblastoma antibodies histochemically on normal structures from the sympathetic nervous system that are known to be sites of origin of neuroblastomas and related tumors. Comparisons were made with corresponding tumors. Antigen distribution varied among normal structures such as ganglion cells, satellite cells, various nerve fibers, and different cells in the adrenal gland. The neoplasms seem to reflect the normal maturation in the structures derived from the neuroectoderm.

A case of familial dysautonomia in a patient with severe postural hypotension was reported by Le Floch, Thomsen, Joly et al. (1987). Study of adrenal medullary secretion showed that it was poorly adapted to the decrease in blood pressure, with no increase in dopamine secretion. Insulin-induced hypoglycemia had no effect on adrenal medullary secretion. Other tests showed that peripheral catecholamine receptors were present in this patient.

Thirty-seven patients with medullary thyroid carcinoma were investigated by Miyauchi, Matsuzaka, Kuma et al. (1987) to determine the status of the adrenal medulla by CT and *m*-iodobenzylguanidine scintigraphy as well as by measurement of urinary catecholamine excretion. Patients were followed for up to 8 years. CT demonstrated adrenal enlargement in all six patients with increased urinary epinephrine excretion. Several other abnormalities of the adrenal medulla were found in many of these patients. None of the patients with sporadic medullary thyroid carcinoma had adrenal abnormalities.

In a comprehensive examination of 16 patients with Rossolimo-Curschmann-Steinert-Batten myotonic dystrophy, Dzhabarov (1986) determined catecholamine levels in the blood and urine. Catecholamine levels generally were increased in these patients. Catecholamines also were increased in the adrenergic structures of skeletal muscles, which suggests an intensified diffusion of catecholamines from adrenergic structures into the adjacent muscles. Myotonic dystrophy was associated with disturbances of both central and peripheral mechanisms of vegetative regulation of muscular activity.

Glazer, Morgenstern, Jeste et al. (1987) performed a case control study to test the effect of serum dopamine β-hydroxylase activity on tardive dyskinesia among schizophrenic outpatients treated with neuroleptics. In contrast to the results of several previous studies, they found no significant association between serum dopamine β-hydroxylase activity and the occurrence or severity of tardive dyskinesia when other predictors were controlled.

Melnikova and Brusov (1987) reported that transcutaneous electrostimulation appears to be an effective method of treating a sympathoadrenal crisis. A pronounced and stabilizing clinical effect was produced by the action of the electrical stimulants. Among the effects of the treatment was a return of the catecholamine content in peripheral blood to normal.

Ito, Fujita and Yamashita (1987) estimated the sympathoadrenal medullary activity in young patients with borderline hypertension and compared it with the activity in age-matched normotensive patients. They subjected the two patient groups to isometric stress and glucagon stimulation. They presented evidence that suggested that the responses of the sympathetic nervous system and the adrenal medulla to stress are increased in young patients with borderline hypertension. Moreover, the augmented response of the sympathoadrenal medullary system to stress may be involved in the development of essential hypertension.

Helman, Thiele, Linehan et al. (1987) constructed a human pheochromocytoma cDNA library and used differential hybridization to human pheochromocytoma and human neuroblastoma cDNA probes to isolate genes that are highly expressed in pheochromocytoma (an adrenal medullary neuroendocrine tumor) but not in neuroblastoma (a more immature embryonic tumor). Two cDNA clones were expressed more highly in normal and neoplastic chromaffin tissue than in neuroblastoma. Furthermore, they were expressed in a remarkably limited number of other human tumors or normal tissues. One clone was highly expressed in medullary thyroid carcinoma, and the other clone was highly expressed in the adrenal cortex. The expression of both clones could be induced in human neuroblastoma cells with dexamethasone, suggesting a mechanism by which glucocorticoids may influence development of a neuroendocrine phenotype.

Prokocimer, Maze and Hoffman (1987) investigated the role of the sympathetic nervous system in the maintenance of hypertension in a strain of rats that can accept a transplantable pheochromocytoma without rejection. They utilized a variety of drugs that affect the sympathetic nervous system, including clonidine, chlorisondamine, naloxone, yohimbine, and prazosin. Their data supported the hypothesis that the sympathetic nervous system is involved in the maintenance of hypertension due to pheochromocytoma.

Fisher, MacPhee, Davies et al. (1987) published a case history of a patient with a 10 year history of watery diarrhea and an acute quadriparesis. The patient had a well-differentiated pheochromocytoma; following surgical resection, the patient recovered fully. The tumor contained vasoactive intestinal polypeptide and calcitonin, which apparently contributed to the patient's symptoms.

BIBLIOGRAPHY

Abajo, F.J., M.A.S. Castro, B. Garijo, and P. Sánchez-García. Catecholamine release evoked by lithium from the perfused adrenal gland of the cat. *Br J Pharmacol 91*:539-546, 1987.

Abate, C., J.A. Smith, and T.H. Joh. Characterization of the catalytic domain of bovine adrenal tyrosine hydroxylase. *Biochem Biophys Res Commun 151*:1446-1453, 1988.

Abe, T., M. Koganemaru, Y. Yasuda, Y. Moriguchi, T. Tanaka, M. Uchida, H. Nishimura, S. Kikuchi, and H. Ohtake. [CT findings in adrenal tumors.] *Rinsho Hoshasen 31*:389-395, 1986.

Abeatici, S., R. Durando, N. Palestini, and M.G. Maquignaz. [Surgery of functioning tumours of the adrenal gland.] *Ann Urol (Paris) 20*:369-372, 1986.

Abou-Donia, M.M., S.P. Wilson, T.P. Zimmerman, C.A. Nichol, and O.H. Viveros. Regulation of guanosine triphosphate cyclohydrolase and tetrahydrobiopterin levels and the role of the cofactor in tyrosine hydroxylation in primary cultures of adrenomedullary chromaffin cells. *J Neurochem 46*:1190-1199, 1986.

Accordi, F. The chromaffin cells of *Siren lacertina* (Amphibia, Urodela): cytological characteristics and evidence of exocytosis. *J Anat 156*:169-176, 1988.

Accordi, F. and F. Corbellini. Morphogenesis of the adrenal gland of *Triturus cristatus carnifex* during larval development and metamorphosis. *Arch Anat Microsc Morphol Exp 75*:241-251, 1986-1987.

Acheson, A., D. Edgar, R. Timpl, and H. Thoenen. Laminin increases both levels and activity of tyrosine hydroxylase in calf adrenal chromaffin cells. J Cell Biol 102:151-159, 1986.

Acheson, A. and H. Thoenen. Both short- and long-term effects of nerve growth factor on tyrosine hydroxylase in calf adrenal chromaffin cells are blocked by *S*-adenosylhomocysteine hydrolase inhibitors. *J Neurochem 48*:1416-1424, 1987.

Adams, M. and M.R. Boarder. Secretion of [Met]enkephalyl-arg[6]-phe[7]-related peptides and catecholamines from bovine adrenal chromaffin cells: modification by changes in cyclic AMP and by treatment with reserpine. *J Neurochem 49*:208-215, 1987.

Adams, M.A. and M. Hirst. The influence of adrenal medullectomy on the development of ethanol-induced cardiac hypertrophy. *Can J Physiol Pharmacol 64*:592-596, 1986.

Adam-Vizi, V., D.E. Knight, and A. Hall. The *ras* protein is not associated with exocytosis [letter]. *Nature 328*:581, 1987.

Adam-Vizi, V., S. Rosener, K. Aktories, and D.E. Knight. Botulinum toxin-induced ADP-ribosylation and inhibition of exocytosis are unrelated events. *FEBS Lett 238*:277-280, 1988.

Addicks, K., H. Hirche, F.M. McDonald, and W. Polwin. Effects of morphine on catecholamine release and arrhythmias evoked by myocardial ischaemia in rats. *Br J Pharmacol* 90:247-254, 1987.

Agata, Y., J.F. Padbury, J.K. Ludlow, D.H. Polk, and J.A. Humme. The effect of chemical sympathectomy on catecholamine release at birth. *Pediatr Res* 20:1338-1344, 1986.

Ahn, N.G. and J.P. Klinman. Activation of dopamine β-monooxygenase by external and internal electron donors in resealed chromaffin granule ghosts. *J Biol Chem* 262:1485-1492, 1987.

Ahn, T.G., D.V. Cohn, S.U. Gorr, D.L. Ornstein, M.A. Kashdan, and M.A. Levine. Primary structure of bovine pituitary secretory protein I (chromagranin A) deduced from the cDNA sequence. *Proc Natl Acad Sci USA* 84:5043-5047, 1987.

Ahnert-Hilger, G., M. Bräutigam, and M. Gratzl. Ca^{2+}-stimulated catecholamine release from α-toxin-permeabilized PC12 cells: Biochemical evidence for exocytosis and its modulation by protein kinase C and G proteins. *Biochemistry* 26:7842-7848, 1987.

Ahnert-Hilger, G. and M. Gratzl. Further characterization of dopamine release by permeabilized PC12 cells. *J Neurochem* 49:764-770, 1987.

Ahnert-Hilger, G. and M. Gratzl. Controlled manipulation of the cell interior by pore-forming proteins. *Trends Pharmacol Sci* 9:195-197, 1988.

Aiba, M., A. Hirayama, Y. Ito, Y. Fujimoto, Y. Nakagami, H. Demura, and K. Shizume. A compound adrenal medullary tumor (Pheochromocytoma and ganglioneuroma) and a cortical adenoma in the ipsilateral adrenal gland. *Am J Surg Pathol* 12:559-566, 1988.

Airapetiants M.G., S.D. Diakova, and A.F. Maslova. [Clinical and experimental studies of the role of neuromediator systems in the pathogenesis of neurosis and neurosis-like states.] *Zh Nevropatol Psikhiatr* 86:1698-1703, 1986.

Akeson, R. and S.L. Warren. PC12 adhesion and neurite formation on selected substrates are inhibited by some glycosaminoglycans and a fibronectin-derived tetrapeptide. *Exp Cell Res* 162:347-362, 1986.

Al-Awqati, Q. Proton-translocating ATPases. *Annu Rev Cell Biol* 2:179-199, 1986.

Albert, T.W. and D.W. Howerton. Pheochromocytoma: case report and review of diagnosis and treatment. *J Oral Maxillofac Surg* 44:657-659, 1986.

Albert, V.R., J.M. Allen, and T.H. Joh. A single gene codes for aromatic L-amino acid decarboxylase in both neuronal and non-neuronal tissues. *J Biol Chem* 262:9404-9411, 1987.

Alho, H., M. Fujimoto, A. Guidotti, I. Hanbauer, Y. Kataoka, and E. Costa. γ-Aminobutyric acid (GABA) in the adrenal medulla: Location, pharmacology, and function. *Neurol Neurobiol 16*:453-464, 1986.

Alho, H., H. Tähti, J. Koistinaho, and A. Hervonen. The effect of toluene inhalation exposure on catecholamine contents in rat sympathetic neurons. *Med Biol 64*:285-288, 1986.

Allen, J.M., J.B. Martin, and G. Heinrich. Neuropeptide Y gene expression in PC12 cells and its regulation by nerve growth factor: a model for developmental regulation. *Brain Res 3*:39-43, 1987.

Allen, J.M., F. Schon, J.C. Yeats, J.S. Kelly, and S.R. Bloom. Effect of reserpine, phenoxybenzamine and cold stress on the neuropeptide Y content of the rat peripheral nervous system. *Neuroscience 19*:1251-1254, 1986.

Allen, J.M., J.C. Yeats, R. Causon, M.J. Brown, and S.R. Bloom. Neuropeptide Y and its flanking peptide in human endocrine tumors and plasma. *J Clin Endocrinol Metab 64*:1199-1204, 1987.

Amann, R. and F. Lembeck. Capsaicin sensitive afferent neurons from peripheral glucose receptors mediate the insulin-induced increase in adrenaline secretion. *Naunyn Schmiedebergs Arch Pharmacol 334*:71-76, 1986.

Amberson, J.B., E.D. Vaughan Jr., G.F. Gray, and G.J. Naus. Flow cytometric determination of nuclear DNA content in benign adrenal pheochromocytomas. *Urology 30*:102-104, 1987.

Ambrosio, S., R. Blesa, G.M. Mintenig, L. Palacios-Araus, N. Mahy, and A. Gual. Acute effects of 1-methyl-4-phenyl-1,2,3,6-tetrahydropyridine (MPTP) on catecholamines in heart, adrenal gland, retina and caudate nucleus of the cat. *Toxicol Lett 44*:1-6, 1988.

Amenta, F., W.L. Collier, S.L. Erdö, S. Giuliani, C.A. Maggi, and A. Meli. GABA$_A$ receptor sites modulating catecholamine secretion in the rat adrenal gland: Evidence from ^{3}H-muscimol autoradiography and in vivo functional studies. *Pharmacology 37*:394-402, 1988.

Amir, S. Central glucagon-induced hyperglycemia is mediated by combined activation of the adrenal medulla and sympathetic nerve endings. *Physiol Behav 37*:563-566, 1986.

Amit, Z. and Z.H. Galina. Stress-induced analgesia: Adaptive pain suppression. *Physiol Rev 66*:1091-1120, 1986.

Amsterdam, J.D., D.L. Marinelli, P. Arger, and A. Winokur. Assessment of adrenal gland volume by computed tomography in depressed patients and healthy volunteers: A pilot study. *Psychiatry Res 21*:189-197, 1987.

Andersen, G.S., J.O. Lund, D. Toftdahl, S. Strandgaard, and P.E. Nielsen. Pheochromocytoma and Conn's syndrome in Denmark 1977-1981. *Acta Med Scand [Suppl] 714*:11-14, 1986.

Anderson, D.J. and R. Axel. A bipotential neuroendocrine

precursor whose choice of cell fate is determined by NGF and glucocorticoids. *Cell 47*:1079-1090, 1986.

Andersson, K.K., D.D. Cox, L. Que Jr., T. Flatmark, and J. Haavik. Resonance Raman studies on the blue-green-colored bovine adrenal tyrosine 3-monooxygenase (tyrosine hydroxylase). Evidence that the feedback inhibitors adrenaline and noradrenaline are coordinated to iron. *J Biol Chem 263*:18621-18626, 1988.

Angeletti, R.H. Chromogranins and neuroendocrine secretion [editorial]. *Lab Invest 55*:387-390, 1986.

Anglade, F., L. Dang Tran, G. DeSaint Blanquat, G. Gaillard, C. Michel-Damase, J.-L. Montastruc, P. Montastruc, M. Rostin, and M.-A. Tran. A study of the action of clonidine on secretion from the adrenal medulla in dogs. *Br J Pharmacol 91*:481-486, 1987.

Anikin, A.Yu., R.A. Kargina-Terentyeva, and N.I. Afonskaya. [Adrenergic innervation of intact zone of the heart and adrenal glands of rabbits in experimental myocardial infarction.] *Arkh Patol 50*:34-40, 1988.

Ansari, A.N., M.E. Siegel, V. DeQuattro, and L.H. Gazarian. Imaging of medullary thyroid carcinoma and hyperfunctioning adrenal medulla using iodine-131 metaiodobenzylguanidine. *J Nucl Med 27*:1858-1860, 1986.

Antreassian, J., C. Benlot, J. Thibault, F. Gros, H. Gozlan, F. Cesselin, J.P. Henry, and J.C. Legrand. Simultaneous evaluation of mRNAs of dopamine β–hydroxylase, tyrosine hydroxylase and proenkephalin A from three human pheochromocytomas. *Neurochem Int 8*:93-101, 1986.

Appel, N.M. and R.P. Elde. The intermediolateral cell column of the thoracic spinal cord is comprised of target-specific subnuclei: Evidence from retrograde transport studies and immunohistochemistry. *J Neurosci 8*:1767-1775, 1988.

Appel, N.M., J.A. Kiritsy-Roy, and G.R. Van Loon. Mu receptors at discrete hypothalamic and brainstem sites mediate opioid peptide-induced increases in central sympathetic outflow. *Brain Res 378*:8-20, 1986.

Appel, N.M. and G.R. Van Loon. β-Endorphin-induced stimulation of central sympathetic outflow: inhibitory modulation by central noradrenergic neurons. *J Pharmacol Exp Ther 237*:695-701, 1986.

Appel, N.M., M.W. Wessendorf, and R.P. Elde. Coexistence of serotonin- and substance P-like immunoreactivity in nerve fibers apposing identified sympathoadrenal preganglionic neurons in rat intermediolateral cell column. *Neurosci Lett 65*:241-246, 1986.

Appel, N.M., M.W. Wessendorf, and R.P. Elde. Thyrotropin-releasing hormone in spinal cord: coexistence with serotonin and with substance P in fibers and terminals apposing identified preganglionic sympathetic neurons. *Brain Res*

415:137-143, 1987.

Archer, C.B., N. Dalton, C. Turner, and D.M. MacDonald. Investigation of adrenomedullary function in atopic dermatitis. *Br J Dermatol 116*:793-800, 1987.

Arefolov, V.A., L.A. Malikova, and A.V. Valdman. [Ultrastructural morphometry of adrenaline- and noradrenaline-accumulating rat adrenal cells in the time-course of immobilization stress.] *Biull Eksp Biol Med 103*:743-746, 1987.

Ariga, T., L.J. Macala, M. Saito, R.K. Margolis, L.A. Greene, R.U. Margolis, and R.K. Yu. Lipid composition of PC12 pheochromocytoma cells: Characterization of globoside as a major neutral glycolipid. *Biochemistry 27*:52-58, 1988.

Ariga, T., R.K. Yu, J.N. Scarsdale, M. Suzuki, Y. Kuroda, H. Kitagawa, and T. Miyatake. Accumulation of a globo-series glycolipid having galα_1-3gal in PC12h pheochromocytoma cells. *Biochemistry 27*:5335-5340, 1988.

Arita, M., A. Wada, H. Takara, and F. Izumi. Inhibition of ^{22}Na influx by tricyclic and tetracyclic antidepressants and binding of [^{3}H]imipramine in bovine adrenal medullary cells. *J Pharmacol Exp Ther 243*:342-348, 1987.

Arnall, D.A., J.C. Marker, R.K. Conlee, and W.W. Winder. Effect of infusing epinephrine on liver and muscle glycogenolysis during exercise in rats. *Am J Physiol 250*:E641-E649, 1986.

Arnetz, B.B. and B. Fjellner. Psychological predictors of neuroendocrine responses to mental stress. *J Psychosom Res 30*:297-305, 1986.

Artalejo, C.R., M.-F. Bader, D. Aunis, and A.G. García. Inactivation of the early calcium uptake and noradrenaline release evoked by potassium in cultured chromaffin cells. *Biochem Biophys Res Commun 134*:1-7, 1986.

Artalejo, C.R. and A.G. García. Effects of BAY K 8644 on cat adrenal catecholamine secretory responses to A23187 or ouabain. *Br J Pharmacol 88*:757-765, 1986.

Artalejo, C.R., A.G. García, and D. Aunis. Chromaffin cell calcium channel kinetics measured isotopically through fast calcium, strontium, and barium fluxes. *J Biol Chem 262*:915-926, 1987.

Artalejo, C.R., M.G. López, M.A. Moro, C.F. Castillo, R. de Pascual, and A.G. García. Voltage-dependence of nitrendipine provides direct evidence for dihydropyridine receptor coupling to calcium channels in intact cat adrenals. *Biochem Biophys Res Commun 153*:912-918, 1988.

Artsruni, G.G., A.V. Zil'fian, N.R. Azgaldian, and R.A. Dovlatian. [Effect of an external electrostatic field on catecholamine secretion by rat adrenals.] *Kosm Biol Aviakosm Med 21*:67-70, 1987.

Asamoah, A., A.F. Wilson, R.C. Elston, E. Dalferes Jr., and G.S. Berenson. Segregation and linkage analyses of dopamine-β–

hydroxylase activity in a six-generation pedigree. *Am J Med Genet* 27:613-621, 1987.

Ashino, N., K. Sobue, Y. Seino, and H. Yabuuchi. Purification of an 80 kDa Ca^{2+}-dependent actin-modulating protein, which severs actin filaments, from bovine adrenal medulla. *J Biochem (Tokyo) 101*:609-617, 1987.

Aso, Y., A. Tajima, K. Suzuki, Y. Ohmi, T. Kanbayashi, T. Mitsuhashi, and K. Ikeda. Intraoperative blood pressure control by ATP in pheochromocytoma. *Urology* 27:512-520, 1986.

Assyag, P., J.M. Klein, G. du Cailar, P.F. Plouin, and C. Brechenmacher. [Residual arterial hypertension after excision of an extra-adrenal pheochromocytoma.] *Ann Cardiol Angeiol (Paris) 35*:487-490, 1986.

Aunis, D. and M.-F. Bader. The cytoskeleton as a barrier to exocytosis in secretory cells. *J Exp Biol 139*:253-266, 1988.

Azmitia, E.C. and A. Björklund, (eds). Cell tissue transplantation into the adult brain. *Ann NY Acad Sci 495*:1-813, 1987a.

Azmitia, E.C. and A. Björklund. Cell and tissue transplantation into the adult brain. *Neurobiol Aging 8*:77-82, 1987b.

Backlund, E.-O. Adrenal-to-brain transplants and Parkinson's disease [letter]. *JAMA 258*:1891, 1987a.

Backlund, E.-O. Transplantation to the brain--a new therapeutic principle or useless venture? *Acta Neurochir [Suppl] (Wien) 41*:46-50, 1987b.

Backlund, E.-O., L. Olson, A. Seiger, and O. Lindvall. Toward a transplantation therapy in Parkinson's disease: A progress report from continuing clinical experiments. *Ann NY Acad Sci 495*:658-673, 1987.

Bacon, S.J. and A.D. Smith. Preganglionic sympathetic neurones innervating the rat adrenal medulla: immunocytochemical evidence of synaptic input from nerve terminals containing substance P, GABA or 5-hydroxytryptamine. *J Autonom Nerv Syst 24*:97-122, 1988.

Bader, M.-F., D. Thiersé, D. Aunis, G. Ahnert-Hilger, and M. Gratzl. Characterization of hormone and protein release from α-toxin-permeabilized chromaffin cells in primary culture. *J Biol Chem 261*:5777-5783, 1986.

Bader, M.-F., J.-M. Trifaró, O.K. Langley, D. Thiersé, and D. Aunis. Secretory cell actin-binding proteins: Identification of a gelsolin-like protein in chromaffin cells. *J Cell Biol 102*:636-646, 1986.

Baer, A. Sugars and adrenomedullary proliferative lesions: The effects of lactose and various polyalcohols. *J Am Coll Toxicol 7*:71-81, 1988.

Baetge, E.E., R.R. Behringer, A. Messing, R.L. Brinster, and R.D. Palmiter. Transgenic mice express the human phenylethanolamine *N*-methyltransferase gene in adrenal medulla and retina. *Proc Natl Acad Sci USA 85*:3648-3652, 1988.

Baetge, E.E., Y.H. Suh, and T.H. Joh. Complete nucleotide and deduced amino acid sequence of bovine phenylethanolamine *N*-methyltransferase: Partial amino acid homology with rat tyrosine hydroxylase. *Proc Natl Acad Sci USA 83*:5454-5458, 1986.

Bagdy, G., K. Szemeredi, Z. Zukowska-Grojec, J. Hill, and D.L. Murphy. *M*-chlorophenylpiperazine increases blood pressue and heart rate in pithed and conscious rats. *Life Sci 41*:775-782, 1987.

Bakay, R.A. and D.L. Barrow. Neural transplantation for Parkinson's disease [letter]. *J Neurosurg 69*:807-810, 1988.

Baker, C.H., F.R. Wilmoth, E.T. Sutton, and J.M. Price. Microvascular responses of intact and adrenal medullectomized rats to hemorrhagic shock. *Circ Shock 26*:203-218, 1988.

Baker, M.E., C. Spritzer, R. Blinder, R.J. Herfkens, G.S. Leight, and N.R. Dunnick. Benign adrenal lesions mimicking malignancy on MR imaging: Report of two cases. *Radiology 163*:669-671, 1987.

Baker, P.F. Calcium channels and other devices for effecting stimulus-exocytosis coupling. pp. 247-265. In: **Structure and Physiology of the Slow Inward Calcium Channel**, Vol 9. (J.C Venter and D. Triggle, eds). Alan R. Liss Inc., New York, 1987.

Baker, P.F. and D.E. Knight. Experimental control of the internal environment of chromaffin cells. pp. 223-233. In: **In Vitro Methods for Studying Secretion**, Vol 3. (A.M. Poisner and J.-M. Trifaró, eds). Elsevier Science Publishers, Amsterdam, 1987a.

Baker, P.F. and D.E. Knight. Theme and variation in the control of exocytosis. pp. 1-20. In: **Cell Calcium and the Control of Membrane Transport**, Vol. 42. (L.J. Mandel and D.C. Eaton, eds). Rockefeller University Press, New York, 1987b.

Baker, P.F., D.E. Knight, and J.A. Umbach. Calcium clamp of the intracellular environment. *Cell Calcium 6*:5-14, 1985.

Baksi, S.N., M.J. Hughes, and H.K. Strahlendorf. Adrenal catecholamine concentration after chronic treatment with bromocriptine and haloperidol. *J Pharm Pharmacol 38*:774-776, 1986.

Balázs, M. Mixed pheochromocytoma and ganglioneuroma of the adrenal medulla: A case report with electron microscopic examination. *Hum Pathol 19*:1352-1354, 1988.

Balázs, M., M. Ihász, and G. Illyés. [Pheochromocytoma: clinico-pathologic analysis of 13 cases.] *Orv Hetil 128*:1397-1402, 1987a.

Balázs, M., M. Ihász, and G. Illyés. [Pheochromocytoma-ganglioneuroma of the adrenal gland. Light and electron microscopy studies.] *Morphol Igazsagugyi Orv Sz 27*:167-175, 1987b.

Banerjee, D.K., R.A. Lutz, M.A. Levine, D. Rodbard, and H.B.

Pollard. Uptake of norepinephrine and related catecholamines by cultured chromaffin cells: Characterization of cocaine-sensitive and -insensitive plasma membrane transport sites. *Proc Natl Acad Sci USA 84*:1749-1753, 1987.

Banerji, T.K., G. Callas, W.J. Meyer, and A. Rassoli. ACTH increases adrenal medullary PNMT activity in neonatal rats. *Life Sci 38*:343-349, 1986.

Banerji, T.K. and W.B. Quay. Effects of melatonin on adrenomedullary dopamine-β-hydroxylase activity in golden hamsters: Evidence for pineal and dose dependencies. *J Pineal Res 3*:397-404, 1986.

Bao, Z.-M. Removal of pheochromocytoma by enucleation. *Urology 29*:211-212, 1987.

Barbeito, L., C. Fernández, R. Silveira, and F. Dajas. Evidences of a sympatho-adrenal dysfunction after lesion of the central noradrenergic pathways in rats. *J Neural Transm 67*:205-214, 1986.

Bardakhch'yan, E.A. and Y.G. Kirichenko. [Changes in the ultrastructure of the adrenal medulla and cortex during endotoxic shock.] *Biull Eksp Biol Med 102*:97-100, 1986.

Bardakhch'yan, E.A. and Y.G. Kirichenko. Ultrastructural changes in the adrenal medulla and cortex in endotoxin shock. *Bull Exp Biol Med [Engl Tr] 102*:988-991, 1987.

Barnett, A., A. Maslowski, J. Livesey, G. Nicholls, H. Ikram, and E. Espiner. Haemodynamic and hormonal responses to exercise: Studies in patients with diabetes mellitus and adrenomedullary deficiency. *Eur J Clin Invest 16*:5-10, 1986.

Barron, B.A. and T.D. Hexum. Muscarinic receptor modulation of release of [Met⁵]enkephalin immunoreactive material and catecholamines from the bovine adrenal gland. *Br J Pharmacol 89*:713-718, 1986a.

Barron, B.A. and T.D. Hexum. Modulation of bovine adrenal gland secretion by etorphine and diprenorphine. *Life Sci 38*:935-940, 1986b.

Barron, B.A., L.C. Murrin, and T.D. Hexum. Muscarine binding sites in bovine adrenal medulla. *Eur J Pharmacol 122*:269-273, 1986.

Bartal, A. [Adrenomedullary-brain transplant: a breakthrough in the management of Parkinson's disease.] *Harefuah 114*:570-572, 1988.

Bartolf, M. and R.C. Franson. Characterization and localization of neutral sphingomyelinase in bovine adrenal medulla. *J Lipid Res 27*:57-63, 1986.

Bartolf, M. and R.C. Franson. Modulation by cytosol and commercial proteins of acid-active phospholipase A₂ from adrenal medulla. *Biochim Biophys Acta 917*:308-317, 1987.

Bastiaensen, E., J. De Block, and W.P. De Potter. Neuropeptide Y is localized together with enkephalins in adrenergic granules of bovine adrenal medulla. *Neuroscience 25*:679-686, 1988.

Batter, D.K., S.R. D'Mello, L.M. Turzai, H.B. Hughes III, A.E. Gioio, and B.B. Kaplan. The complete nucleotide sequence and structure of the gene encoding bovine phenylethanolamine *N*-methyltransferase. *J Neurosci Res* *19*:367-376, 1988.

Bauer, F.E., T.E. Adrian, N. Yanaihara, J.M. Polak, and S.R. Bloom. Chromatographic evidence for high-molecular-mass galanin immunoreactivity in pig and cat adrenal glands. *FEBS Lett 201*:327-331, 1986.

Bauer, F.E., G.W. Hacker, G. Terenghi, T.E. Adrian, J.M. Polak, and S.R. Bloom. Localization and molecular forms of galanin in human adrenals: Elevated levels in pheochromocytomas. *J Clin Endocrinol Metab 63*:1372-1378, 1986.

Bauersfeld, W., D. Ratge, E. Knoll, and H. Wisser. Determination of catecholamines in plasma by HPLC and amperometric detection. Comparison with a radioenzymatic method. *J Clin Chem Clin Biochem 24*:185-188, 1986.

Beaser, R.S., A.T. Guay, A.K. Lee, M.L. Silverman, and L.D. Flint. An adrenocorticotropic hormone-producing pheochromo-cytoma: Diagnostic and immunohistochemical studies. *J Urol 135*:10-13, 1986.

Becker, J.B., F. Adams, and T.E. Robinson. Intraventricular microdialysis: A new method for determining monoamine metabolite concentrations in the cerebrospinal fluid of freely moving rats. *J Neurosci Methods 24*:259-269, 1988.

Becker, J.B. and W.J. Freed. Adrenal medulla grafts enhance functional activity of the striatal dopamine system following substantia nigra lesions. *Brain Res 462*:401-406, 1988.

Beers, M.F., R.G. Johnson, and A. Scarpa. Evidence for an ascorbate shuttle for the transfer of reducing equivalents across chromaffin granule membranes. *J Biol Chem 261*:2529-2535, 1986.

Beierwaltes, W.H. Update on basic research and clinical experience with metaiodobenzylguanidine. *Med Pediatr Oncol 15*:163-169, 1987.

Beinfeld, M.C., P.L. Brick, A.C. Howlett, I.L. Holt, R.M. Pruss, J.R. Moskal, and L.E. Eiden. The regulation of vasoactive intestinal peptide synthesis in neuroblastoma and chromaffin cells. *Ann NY Acad Sci 527*:68-76, 1988.

Belenkov, Y.N., G.G. Arabidze, O.I. Belichenko, and T.S. Pustovitova. [Diagnostic potentialities of MR-tomography of the adrenals in patients with arterial hypertension.] *Ter Arkh 58*:11-14, 1986.

Benedum, U.M., P.A. Baeuerle, D.S. Konecki, R. Frank, J. Powell, J. Mallet, and W.B. Huttner. The primary structure of bovine chromogranin A: A representative of a class of acidic secretory proteins common to a variety of peptidergic cells. *EMBO J 5*:1495-1502, 1986.

Benedum, U.M., A. Lamouroux, D.S. Konecki, P. Rosa, A. Hille,

P.A. Baeuerle, R. Frank, F. Lottspeich, J. Mallet, and W.B. Huttner. The primary structure of human secretogranin I (chromogranin B): Comparison with chromogranin A reveals homologous terminal domains and a large intervening variable region. *EMBO J* 6:1203-1211, 1987.

Bereiter, D.A., W.C. Engeland, and D.S. Gann. Peripheral venous catecholamines versus adrenal secretory rates after brain stem stimulation in cats. *Am J Physiol 251*:E14-E20, 1986.

Bereiter, D.A. and D.S. Gann. Adrenal secretion of catecholamines evoked by chemical stimulation of trigeminal nucleus caudalis in the cat. *Neuroscience 25*:697-704, 1988.

Bergdall, V.K., B.E. Levin, D.W. Townsend, and S.L. Stoddard. Increases in blood flow from the adrenal gland following medial hypothalamic stimulation in the cat. *J Auton Nerv Syst 15*:263-268, 1986.

Berman, S. U-M Medical Center performs state's first brain graft. *Mich Med 87*:224-226, 1988.

Bernstein-Goral, H. and M.C. Bohn. Ontogeny of adrenergic fibers in rat spinal cord in relationship to adrenal preganglionic neurons. *J Neurosci Res 21*:333-351, 1988.

Berry, P. and P.A. Ward-Smith. Adrenal medullary transplant as a treatment for Parkinson's disease: Perioperative consider-ations. *J Neurosci Nurs 20*:356-361, 1988.

Bertossi, M., D. Ribatti, and L. Roncali. Ultrastructural features of the vasculogenetic processes in the thyroid and suprarenal glands of the chick embryo. *J Submicrosc Cytol 19*:119-128, 1987.

Bethea, C.L. Glucocorticoid stimulation of dopamine production in PC12 cells on extracellular matrix and plastic. *Molec Cell Endocrinol 50*:211-222, 1987.

Bethea, C.L. and T.K. Borg. PC12 cell aggregation and dopamine production on EHS-derived extracellular matrix. *Molec Cell Endocrinol 58*:113-128, 1988.

Bethea, C.L., O.K. Rønnekleiv, and S.L. Kozak. Extracellular matrix changes PC12 cell shape and processing of newly synthesized dopamine. *Molec Cell Endocrinol 54*:63-79, 1987.

Bezrukov, V.V. and V.M. Shaposhnikov. [Morphofunctional peculiarities of the adrenals reactivity in rats at various ages.] *Arkh Anat Gistol Embriol 93*:76-81, 1987.

Bianchi, C., J. Gutkowska, C. Charbonneau, M. Ballak, M.B. Anand-Srivastava, A. DeLéan, J. Genest, and M. Cantin. Internalization and lysosomal association of [^{125}I] angiotensin II in norepinephrine-containing cells of the rat adrenal medulla. *Endocrinology 119*:1873-1875, 1986.

Bigham, E.C., G.K. Smith, J.F. Reinhard Jr., W.R. Mallory, C.A. Nichol, and R.W. Morrison Jr. Synthetic analogues of tetrahydrobiopterin with cofactor activity for aromatic amino acid hydroxylases. *J Med Chem 30*:40-45, 1987.

Bigornia, L. and I. Bihler. 3-*O*-Methyl-D-glucose uptake in isolated bovine adrenal chromaffin cells. *Biochim Biophys*

Acta 885:335-344, 1986.
Bigornia, L. and I. Bihler. 3-*O*-Methyl-D-glucose uptake in cultured bovine adrenal chromaffin cells. *Can J Physiol Pharmacol 66*:402-407, 1988.
Bigornia, L., M. Suozzo, K.A. Ryan, D. Napp, and A.S. Schneider. Dopamine receptors on adrenal chromaffin cells modulate calcium uptake and catecholamine release. *J Neurochem 51*:999-1006, 1988.
Bigornia, L., M. Wattis, and I. Bihler. Regulation of 3-*O*-methyl-D-glucose uptake in isolated bovine adrenal chromaffin cells. *Biochim Biophys Acta 886*:177-186, 1986.
Bing, G., M.F.D. Notter, J.T. Hansen, and D.M. Gash. Comparison of adrenal medullary, carotid body and PC12 cell grafts in 6-OHDA lesioned rats. *Brain Res Bull 20*:399-406, 1988.
Birch, N.P. and D.L. Christie. Characterization of the molecular forms of proenkephalin in bovine adrenal medulla and rat adrenal, brain, and spinal cord with a site-directed antiserum. *J Biol Chem 261*:12213-12221, 1986.
Birch, N.P., A.D. Davies, and D.L. Christie. Identification of a 27-kDa enkephalin-containing protein associated with bovine adrenal medullary chromaffin granule membranes by immunoblotting. *FEBS Lett 197*:173-178, 1986.
Birch, N.P., A.D. Davies, and D.L. Christie. An investigation of the molecular properties and stability of intermediates of proenkephalin in isolated bovine adrenal medullary chromaffin granules. *J Biol Chem 262*:3382-3387, 1987.
Bitar, M.S., M. Koulu, S.I. Rapoport, and M. Linnoila. Adrenal catecholamine metabolism and myocardial adrenergic receptors in streptozotocin diabetic rats. *Biochem Pharmacol 36*:1011-1016, 1987.
Bitler, C.M. and B.D. Howard. Dopamine metabolism in hypoxanthine-guanine phosphoribosyltransferase-deficient variants of PC12 cells. *J Neurochem 47*:107-112, 1986.
Bitler, C.M., M.-b. Zhang, and B.D. Howard. PC12 variants deficient in catecholamine transport. *J Neurochem 47*:1286-1293, 1986.
Bittner, M.A. and R.W. Holz. Effects of tetanus toxin on catecholamine release from intact and digitonin-permeabilized chromaffin cells. *J Neurochem 51*:451-456, 1988.
Bittner, M.A., R.W. Holz, and R.R. Neubig. Guanine nucleotide effects on catecholamine secretion from digitonin-permeabilized adrenal chromaffin cells. *J Biol Chem 261*:10182-10188, 1986.
Bjartell, A., R. Ekman, and F. Sundler. γ_2-MSH-like immunoreactivity in porcine pituitary and adrenal medulla. An immunochemical and immunocytochemical study. *Regul Pept 19*:291-306, 1987.
Björklund, A., O. Lindvall, O. Isacson, P. Brundin, K. Wictorin,

R.E. Strecker, D.J. Clarke, and S.B. Dunnett. Mechanisms of action of intracerebral neural implants: Studies on nigral and striatal grafts to the lesioned striatum. *Trends Neurosci 10*:509-516, 1987.

Black, I.B., J.E. Adler, C.F. Dreyfus, W.F. Friedman, E.F. La Gamma, and A.H. Roach. Biochemistry of information storage in the nervous system. *Science 236*:1263-1268, 1987.

Black, I.B., J.E. Adler, and E.F. La Gamma. Impulse activity differentially regulates co-localized transmitters by altering messenger RNA levels. *Prog Brain Res 68*:121-127, 1986.

Black, M.M. and P. Keyser. Acetylation of α-tubulin in cultured neurons and the induction of α-tubulin acetylation in PC12 cells by treatment with nerve growth factor. *J Neurosci 7*:1833-1842, 1987.

Blackburn, N.J., M. Concannon, S.K. Shahiyan, F.E. Mabbs, and D. Collison. Active site of dopamine β-hydroxylase. Comparison of enzyme derivatives containing four and eight copper atoms per tetramer using potentiometry and EPR spectroscopy. *Biochemistry 27*:6001-6008, 1988.

Blackburn, T.P., K.R. Borkowski, J. Friend, and M.J. Rance. On the mechanisms of κ-opioid-induced diuresis. *Br J Pharmacol 89*:593-598, 1986.

Blenis, J. and R.L. Erikson. Regulation of protein kinase activities in PC12 pheochromocytoma cells. *EMBO J 5*:3441-3447, 1986.

Bloch, B., T. Popovici, S. Chouham, and C. Kowalski. Detection of the mRNA coding for enkephalin precursor in the rat brain and adrenal by using an *in situ* hybridization procedure. *Neurosci Lett 64*:29-34, 1986.

Bloch, B., T. Popovici, D. LeGuellec, E. Normand, S. Chouham, A.F. Guitteny, and P. Bohlen. *In situ* hybridization histochemistry for the analysis of gene expression in the endocrine and central nervous system tissues: A 3-yr experience. *J Neurosci Res 16*:183-200, 1986.

Blom, H.J., V. Karsdorp, R. Birnie, and G. Davies. Phaeochromocytoma as a cause of pulmonary oedema. *Anaesthesia 42*:646-650, 1987.

Bloom, S.R., A.V. Edwards, and C.T. Jones. The adrenal contribution to the neuroendocrine responses to splanchnic nerve stimulation in conscious calves. *J Physiol (Lond) 397*:513-526, 1988.

Boarder, M.R., C.J. Evans, M. Adams, E. Erdelyi, and J.D. Barchas. Peptide E and its products, BAM 18 and leu-enkephalin, in bovine adrenal medulla and cultured chromaffin cells: Release in response to stimulation. *J Neurochem 49*:1824-1832, 1987.

Boarder, M.R., D. Marriott, and M. Adams. Stimulus secretion coupling in cultured chromaffin cells: Dependency on external sodium and on dihydropyridine-sensitive calcium channels. *Biochem Pharmacol 36*:163-167, 1987.

Boarder, M.R. and W. McArdle. Release of catecholamines from superfused bovine adrenal chromaffin cells cultured on microcarrier beads. *J Neurochem 46*:1473-1477, 1986.

Boarder, M.R., R. Plevin, and D.B. Marriott. Angiotensin II potentiates prostaglandin stimulation of cyclic AMP levels in intact bovine adrenal medulla cells but not adenylate cyclase in permeabilized cells. *J Biol Chem 263*:15319-15324, 1988.

Bode, K., H.-D. Hofmann, T.H. Müller, U. Otten, R. Schmidt, and K. Unsicker. Effects of pre- and postnatal administration of antibodies to nerve growth factor on the morphological and biochemical development of the rat adrenal medulla: A reinvestigation. *Dev Brain Res 27*:139-150, 1986.

Boelsterli, U.A. and G. Zbinden. Hypothetical mechanism of adrenal medullary hyperplasia in xylitol fed rats. *Arch Toxicol 59*:194, 1986.

Bohme, H.J., D. Belay, D. Dettmer, W. Goltzsch, E. Hofmann, R. Lange, C. Schubert, E. Schulze, G. Sparmann, and E. Weiss. Interaction of adrenal and pancreatic hormones in the control of hepatic enzymes during development. *Adv Enzyme Regul 26*:31-61, 1987.

Bohn, M.C. Expression and development of phenylethanolamine *N*-methyltransferase (PNMT): Role of glucocorticoids. *Neurol Neurobiol 16*:245-271, 1986.

Bohn, M.C., L. Cupit, F. Marciano, and D.M. Gash. Adrenal medulla grafts enhance recovery of striatal dopaminergic fibers. *Science 237*:913-916, 1987.

Bohn, M.C., M. Goldstein, and I.B. Black. Expression and development of phenylethanolamine *N*-methyltransferase (PNMT) in rat brain stem: Studies with glucocorticoids. *Dev Biol 114*:180-193, 1986.

Boksa, P. Studies on the uptake and release of propranolol and the effects of propranolol on catecholamines in cultures of bovine adrenal chromaffin cells. *Biochem Pharmacol 35*:805-815, 1986a.

Boksa, P. Effects of substance P on the long-term regulation of tyrosine hydroxylase activity and catecholamine levels in cultured adrenal chromaffin cells. *Can J Physiol Pharmacol 64*:1548-1555, 1986b.

Bomanji, J., P.-M.G. Bouloux, D.A. Levison, W.D. Flatman, T. Horne, K.E. Britton, G. Ross, and G.M. Besser. Observations on the function of normal adrenomedullary tissue in patients with phaeochromocytomas and other paragangliomas. *Eur J Nucl Med 13*:86-89, 1987.

Bomanji, J., W.D. Flatman, T. Horne, J. Fettich, K.E. Britton, G. Ross, and G.M. Besser. Quantitation of iodine-123 MIBG uptake by normal adrenal medulla in hypertensive patients. *J Nucl Med 28*:319-324, 1987.

Bommer, M., D. Liebisch, N. Kley, A. Herz, and E.P. Noble. Histamine affects release and biosynthesis of opioid peptides primarily via H_1-receptors in bovine chromaffin cells. *J*

Neurochem 49:1688-1696, 1987.

Bönisch, H. and R. Harder. Binding of ^{3}H-desipramine to the neuronal noradrenaline carrier of rat phaeochromocytoma cells (PC-12 cells). *Naunyn Schmiedebergs Arch Pharmacol* 334:403-411, 1986.

Boonstra, J., C.L. Mummery, A. Feyen, W.J. DeHoog, P.T. van der Saag, and S.W. de Laat. Epidermal growth factor receptor expression during morphological differentiation of pheochromocytoma cells, induced by nerve growth factor or dibutyryl cyclic AMP. *J Cell Physiol* 131:409-417, 1987.

Borges, R., J.J. Ballesta, and A.G. García. M_2 muscarinoceptor-associated ionophore at the cat adrenal medulla. *Biochem Biophys Res Commun* 144:965-972, 1987.

Borges, R., F. Sala, and A.G. García. Continuous monitoring of catecholamine release from perfused cat adrenals. *J Neurosci Methods* 16:289-300, 1986.

Borkowski, K.R. and E. Kelly. The effect of adrenal demedullation on cardiovascular responses to environmental stimulation in conscious rats. *Br J Pharmacol* 88:943-945, 1986.

Born, G.V.R. Platelets and neurobiological research. *Experientia* 44:113-115, 1988.

Borowitz, J.L., G.S. Born, and G.E. Isom. Potentiation of evoked adrenal catecholamine release by cyanide: Possible role of calcium. *Toxicology* 50:37-45, 1988.

Borrero, E., P. Katz, S. Lipper, and J.B. Chang. Adrenal cortical adenoma and adrenal medullary hyperplasia of the right adrenal gland--a case report. *Angiology* 38:271-274, 1987.

Bossard, M.J. and J.P. Klinman. Mechanism-based inhibition of dopamine β-monooxygenase by aldehydes and amides. *J Biol Chem* 261:16421-16427, 1986.

Bostwick, D.G., W.E. Null, D. Holmes, E. Weber, J.D. Barchas, and K.G. Bensch. Expression of opioid peptides in tumors. *N Engl J Med* 317:1439-1443, 1987.

Bouloux, P.-M.G., D. Perrett, M. Sopwith, and G.M. Besser. An *in situ* isolated rat adrenal perfusion system for study of neurally mediated catecholamine secretion: Effects of morphine, a Met-enkephalin analogue, and naloxone on catecholamine secretion. *J Endocrinol* 111:7-15, 1986.

Bourke, J.E., S.J. Bunn, P.D. Marley, and B.G. Livett. The effects of neosurugatoxin on evoked catecholamine secretion from bovine adrenal chromaffin cells. *Br J Pharmacol* 93:275-280, 1988.

Bouvier, M. and J. deChamplain. Increased sympatho-adrenal tone and adrenal medulla reactivity in DOCA-salt hypertensive rats. *J Hypertens* 4:157-163, 1986.

Bowsher, R.R. and D.P. Henry. Aromatic L-amino acid decarboxylase: Biochemistry and functional significance. pp. 33-78. In: **Neuromethods. Neurotransmitter Enzymes**, Vol. 5. (A.A. Boulton, G.B. Baker, and P.H. Yu, eds). Humana

Press, Clifton, NJ, 1986.

Boyd, N.D. Two distinct kinetic phases of desensitization of acetylcholine receptors of clonal rat PC12 cells. *J Physiol (Lond) 389*:45-67, 1987.

Boyd, N.D. and S.E. Leeman. Multiple actions of substance P that regulate the functional properties of acetylcholine receptors of clonal rat PC12 cells. *J Physiol (Lond) 389*:69-97, 1987.

Bramnert, M. The effect of naloxone on blood pressure, heart rate, plasma catecholamines, renin activity and aldosterone following exercise in healthy males. *Regul Pept 22*:295-301, 1988.

Braude, B.M., B.C. Leiman, and D.G. Moyes. Etomidate infusion for resection of phaeochromocytoma: A report of 2 cases. *S Afr Med J 69*:60-62, 1986.

Braumann, T., B. Jastorff, and C. Richter-Landsberg. Fate of cyclic nucleotides in PC12 cell cultures: uptake, metabolism, and effects of metabolites on nerve growth factor-induced neurite outgrowth. *J Neurochem 47*:912-919, 1986.

Breslow, M.J., D.A. Jordan, S.T. Thellman, and R.J. Traystman. Neural control of adrenal medullary and cortical blood flow during hemorrhage. *Am J Physiol 252*:H521-H528, 1987.

Breslow, M.J., A. Mennen, R.C. Koehler, and R.J. Traystman. Adrenal medullary and cortical blood flow during hemorrhage. *Am J Physiol 250*:H954-H960, 1986.

Bricaire, F., Cl. Marche, D. Zoubi, Ch. Perronne, S. Matheron, E. Rouveix, and D. Vittecoq. [Adrenal lesions in AIDS;an anatomopathological study.] *Ann Med Interne (Paris) 138*:607-609, 1987.

Brocklehurst, K.W., G. Lee, and H.B. Pollard. Rapid purification and properties of protein kinase C from bovine adrenal medulla. *Biosci Rep 6*:749-757, 1986.

Brocklehurst, K.W. and H.B. Pollard. Synergistic actions of Ca^{2+} and the phorbol ester TPA on K^+-induced catecholamine release from bovine adrenal chromaffin cells. *Biochem Biophys Res Commun 140*:990-998, 1986.

Brocklehurst, K.W. and H.B. Pollard. Pertussis toxin stimulates delayed-onset, Ca^{2+}-dependent catecholamine release and the ADP-ribosylation of a 40 kDa protein in bovine adrenal chromaffin cells. *FEBS Lett 234*:439-445, 1988.

Bronstein, M.E., Ye.A. Smirnova, and N.P. Yuryeva. Catecholamine-producing tumors of chromaffin tissue. Morphologic and ultrastructural characteristics. *Vopr Onkol 34*:826-832, 1988.

Brooks, C.M. The autonomic nervous system and recently acquired understanding of its function and organization. *Neurol Neurobiol 31*:459-463, 1987.

Brooks, J.C. and M.H. Brooks. Thiophosphorylation and phosphorylation of saponin-permeabilized cultured chromaffin cells. *Neurochem Int 11*:31-38, 1987a.

Brooks, J.C. and M.H. Brooks. Drug modification of protein thiophosphorylation in permeabilized chromaffin cells. *Neurochem Int 11*:407-415, 1987b.

Brooks, J.C., M.H. Brooks, and S.W. Carmichael. Effect of chronic and acute 1-methyl-4-phenyl-1,2,3,6-tetrahydro-pyridine (MPTP) administration on the function and structure of cultured chromaffin cells. *Neurochem Int 10*:311-321, 1987.

Brooks, J.C. and S.W. Carmichael. Ultrastructural demonstration of exocytosis in intact and saponin-permeabilized cultured bovine chromaffin cells. *Am J Anat 178*:85-89, 1987.

Brown, E.R., G.T. Coker III, and K.L. O'Malley. Organization and evolution of the rat tyrosine hydroxylase gene. *Biochemistry 26*:5208-5212, 1987.

Brown, F.M., M. Kamalesh, M.N. Sesh Adri, and S.L. Rabinowe. Anti-adrenal medullary antibodies in IDDM subjects and subjects at high risk of developing IDDM. *Diabetes Care 11*:30-33, 1988.

Brown, M.R., R. Allen, and L.A. Fisher. Bombesin alters the sympathetic nervous system response to cold exposure. *Brain Res 400*:35-39, 1987.

Brown, M.R. and L.A. Fisher. Glucocorticoid suppression of the sympathetic nervous system and adrenal medulla. *Life Sci 39*:1003-1012, 1986.

Browning, M.D., C.-K. Huang, and P. Greengard. Similarities between protein IIIa and protein IIIb, two prominent synaptic vesicle-associated phosphoproteins. *J Neurosci 7*:847-853, 1987.

Brugg, B. and A. Matus. PC12 cells express juvenile microtubule-associated proteins during nerve growth factor-induced neurite outgrowth. *J Cell Biol 107*:643-650, 1988.

Bruhn, T.O., W.C. Engeland, E.L.P. Anthony, D.S. Gann, and I.M.D. Jackson. Corticotropin-releasing factor in the adrenal medulla. *Ann NY Acad Sci 512*:115-128, 1987a.

Bruhn, T.O., W.C. Engeland, E.L.P. Anthony, D.S. Gann, and I.M.D. Jackson. Corticotropin-releasing factor in the dog adrenal medulla is secreted in response to hemorrhage. *Endocrinology 120*:25-33, 1987b.

Brundin, P., R.E. Strecker, H. Widner, D.J. Clarke, O.G. Nilsson, B. Astedt, O. Lindvall, and A. Björklund. Human fetal dopamine neurons grafted in a rat model of Parkinson's disease: Immunological aspects, spontaneous and drug-induced behaviour, and dopamine release. *Exp Brain Res 70*:192-208, 1988.

Brust, I. and M.P. Printz. Functional angiotensin receptor on cultured bovine chromaffin cells: Secretory response to angiotensin II and angiotensin III. *Clin Exp Hypertens [A] 10*:1321-1323, 1988.

Bucher, B. and J.-C. Stoclet. Adrenomedullary and β-adrenergic participation in enhanced sympathetic pressor responses of spontaneously hypertensive rats. *Eur J Pharmacol 138*:233-238, 1987.

Bunn, S.J., P.D. Marley, and B.G. Livett. The distribution of opioid binding subtypes in the bovine adrenal medulla. *Neuroscience 27*:1081-1094, 1988a.

Bunn, S.J., P.D. Marley, and B.G. Livett. Effects of opioid compounds on basal and muscarinic induced accumulation of inositol phosphates in cultured bovine chromaffin cells. *Biochem Pharmacol 37*:395-399, 1988b.

Bunt, J.C. Hormonal alterations due to exercise. *Sports Med 3*:331-345, 1986.

Burgoyne, R.D. G proteins: Control of exocytosis. *Nature 328*:112-113, 1987.

Burgoyne, R.D. Calpactin in exocytosis? *Nature 331*:20, 1988.

Burgoyne, R.D. and A.J. Baines. Synapsin or protein 4.1 in chromaffin cells [letter]. *Nature 330*:115-116, 1987.

Burgoyne, R.D. and T.R. Cheek. Reorganisation of peripheral actin filaments as a prelude to exocytosis. *Biosci Rep 7*:281-288, 1987a.

Burgoyne, R.D. and T.R. Cheek. Role of fodrin in secretion. *Nature 326*:448, 1987b.

Burgoyne, R.D., T.R. Cheek, and K.-M. Norman. Identification of a secretory granule-binding protein as caldesmon. *Nature 319*:68-70, 1986.

Burgoyne, R.D., T.R. Cheek, A.J. O'Sullivan, and R.C. Richards. Control of the cytoskeleton during secretion. pp. 612-627. In: **Molecular Mechanisms in Secretion**, Alfred Benzon Symposium. (N.A. Thorn, M.Treiman, O.H. Petersen, eds). Munksgaard, Copenhagen, 1988.

Burgoyne, R.D., A. Morgan, and A.J. O'Sullivan. A major role for protein kinase C in calcium-activated exocytosis in permeabilised adrenal chromaffin cells. *FEBS Lett 238*:151-155, 1988.

Burke, W.J. and T.H. Joh. Effect of dopamine on tyrosine hydroxylase in cultured rat adrenal medulla. *Biochem Pharmacol 37*:1391-1397, 1988.

Busby, W.H. Jr., G.E. Quackenbush, J. Humm, W.W. Youngblood, and J.S. Kizer. An enzyme(s) that converts glutaminyl-peptides into pyroglutamyl-peptides: Presence in pituitary, brain, adrenal medulla, and lymphocytes. *J Biol Chem 262*:8532-8536, 1987.

Bussolino, F., F. Tessari, F. Turrini, P. Braquet, G. Camussi, M. Prosdocimi, and A. Bosia. Platelet activating factor induces dopamine release in PC12 cell line. *Am J Physiol 255*:C559-C565, 1988.

Buu, N.T., O. Kuchel, W.M. Manger, M.C. Hulse, and R.N. Chute.Metabolism and storage of catecholamines in rats with pheochromocytoma implants. *Hypertension 8*:940-946, 1986.

Byrd, J.C. and H. Alho. Differentiation of PC12 pheochromocytoma cells by sodium butyrate. *Dev Brain Res 31*:151-155, 1987.

Byrd, J.C., M. Hadjiconstantinou, and D. Cavalla. Epinephrine synthesis in the PC12 pheochromocytoma cell line. *Eur J Pharmacol 127*:139-142, 1986.

Byrd, J.C. and U. Lichti. Two types of transglutaminase in the PC12 pheochromocytoma cell line: Stimulation by sodium butyrate. *J Biol Chem 262*:11699-11705, 1987.

Byrd, J.C., J.R. Naranjo, and I. Lindberg. Proenkephalin gene expression in the PC12 pheochromocytoma cell line: Stimulation by sodium butyrate. *Endocrinology 121*:1299-1305, 1987.

Caffrey, M., S.J. Morris, and G.W. Feigenson. Uranyl acetate induces gel phase formation in model lipid and biological membranes. *Biophys J 52*:501-505, 1987.

Callachan, H., G.A. Cottrell, N.Y. Hather, J.J. Lambert, J.M. Nooney, and J.A. Peters. Modulation of the GABA$_A$ receptor by progesterone metabolites. *Proc R Soc Lond [Biol] 231*:359-369, 1987.

Calne, D.B. and P.L. McGeer. Tissue transplantation for Parkinson's disease [editorial]. *Can J Neurol Sci 15*:364-365, 1988.

Campbell, D.G., D.G. Hardie, and P.R. Vulliet. Identification of four phosphorylation sites in the *N*-terminal region of tyrosine hydroxylase. *J Biol Chem 261*:10489-10492, 1986.

Cantau, P., N. Bourhim, P. Giraud, C. Oliver, and E. Castanas. Photoaffinity crosslinking of etorphine with opioid binding sites in the bovine adrenal medulla. *Biotechnol Appl Biochem 9*:114-122, 1987.

Capella, C., C. Riva, M. Cornaggia, A.M. Chiaravalli, B. Frigerio, and E. Solcia. Histopathology, cytology and cytochemistry of pheochromocytomas and paragangliomas including chemodectomas. *Path Res Pract 183*:176-187, 1988.

Capponi, A.M., P.D. Lew, W. Schlegel, and T. Pozzan. Use of intracellular calcium and membrane potential fluorescent indicators in neuroendocrine cells. *Methods Enzymol 124*:116-135, 1986.

Carbonaro, D.A., J.P. Mitchell, F.L. Hall, and P.R. Vulliet. Altered reactivity of the rat adrenal medulla. *Brain Res Bull 21*:451-458, 1988.

Cárdenas, A.M., C. Montiel, A.R. Artalejo, P. Sánchez-García, and A.G. García. Sodium-dependent inhibition by PN200-110 enantiomers of nicotinic adrenal catecholamine release. *Br J Pharmacol 95*:9-14, 1988.

Cárdenas, A.M., C. Montiel, C. Esteban, R. Borges, and A.G. García. Secretion from adrenaline- and noradrenaline-storing adrenomedullary cells is regulated by a common dihydropyridine-sensitive calcium channel. *Brain Res 456*:364-366, 1988.

Carlsson, A. Peptide neurotransmitters--Redundant vestiges? *Pharmacol Toxicol 62*:241-242, 1988.

Carmichael, S.W., D. Banerjee, A. Mandal, and A. Ghosh. Electronmicroscopical studies of the adrenal medulla of a migratory snipe, *Capella gallinago*, Charadriformes. *Mikroskopie 42*:287-291, 1985.

Carmichael, S.W., D. Banerjee, A. Mandal, and A. Ghosh. Effect of insulin on adrenal medulla in two avian species. *Z Mikrosk Anat Forsch 101*:561-566, 1987.

Carmichael, S.W. and G.L. Pfeiffer. Monoamine oxidase types A and B in the human adrenal gland and pheochromocytoma. *Neurochem Int 10*:49-53, 1987.

Carmichael, S.W., D.B. Spagnoli, R.G. Frederickson, W.J. Krause, and J.L. Culberson. Opossum adrenal medulla: I. Postnatal development and normal anatomy. *Am J Anat 179*:211-219, 1987.

Carmichael, S.W., R.J. Wilson, W.S. Brimijoin, L.J. Melton III, H. Okazaki, T.L. Yaksh, J.E. Ahlskog, S.L. Stoddard, and G.M. Tyce. Decreased catecholamines in the adrenal medulla of patients with parkinsonism [letter]. *N Engl J Med 318*:254, 1988.

Caron, P.C., L.J. Cote, and L.T. Kremzner. Putrescine, a source of γ-aminobutyric acid in the adrenal gland of the rat. *Biochem J 251*:559-562, 1988.

Carreño Hernández, M.C., J. Estrada, J.C.R. Peréz Maestu, C. Masa Vázquez, and J.M. deLetona. [Use of 131-metaiodine-benzylguanidine for the diagnosis and treatment of malignant pheochromocytomas.] *Med Clin (Barc) 87*:330-333, 1986.

Carydakis, C., N. Bourhim, P. Giraud, P. Cantau, C. Oliver, and E. Castanas. [Direct interaction of tricyclic antidepressants with opioid binding sites in the bovine adrenal medulla.] *C R Acad Aci [III] 302*:419-422, 1986.

Casola, G., V. Nicolet, E. vanSonnenberg, C. Withers, M. Bretagnolle, R.M. Saba, and P.M. Bret. Unsuspected pheochromocytoma: risk of blood-pressure alterations during percutaneous adrenal biopsy. *Radiology 159*:733-735, 1986.

Castagne, V., R. Corder, R.C. Gaillard, and P. Morméde. Stress-induced changes of circulating neuropeptide Y in the rat: comparison with catecholamines. *Regul Pept 19*:55-63, 1987.

Castro, E., M.J. Oset-Gasque, S. Cañadas, G. Gimenez, and M.P. González. GABA$_A$ and GABA$_B$ sites in bovine adrenal medulla membranes. *J Neurosci Res 20*:241-245, 1988.

Cattaneo, A. and A. Grasso. A functional domain on the α-latrotoxin molecule, distinct from the binding site, involved in catecholamine secretion from PC12 cells: Identification with monoclonal antibodies. *Biochemistry 25*:2730-2736, 1986.

Caughey, B. and N. Kirshner. Effects of reserpine and tetrabenazine on catecholamine and ATP storage in cultured bovine adrenal medullary chromaffin cells. *J Neurochem*

49:563-573, 1987.

Cervera, P., O. Rascol, A. Ploska, G. Gaillard, R. Raisman, C. Duyckaerts, J.J. Hauw, D. Scherman, J.L. Montastruc, F. Javoy-Agid, and Y. Agid. Noradrenaline, adrenaline and tyrosine hydroxylase in adrenal medulla from Parkinsonian patients. *J Neurol Neurosurg Psychiatry 51*:1104-1105, 1988.

Chabot, J.-G., P. Walker, and G. Pelletier. Distribution of epidermal growth factor binding sites in the adult rat adrenal gland by light microscope autoradiography. *Acta Endocrinol (Copenh) 113*:391-395, 1986.

Chai, S.Y., P.M. Sexton, A.M. Allen, R. Figdor, and F.A.O. Mendelsohn. *In vitro* autoradiographic localization of ANP receptors in rat kidney and adrenal gland. *Am J Physiol 250*:F753-F757, 1986.

Chalaprawat, M., P. Wannakrairot, R. Vajarapongse, A. Chittinandana, and S. Chandraprasert. Bilateral adrenal medullary hyperplasia. *J Med Assoc Thai 70*:686-688, 1987.

Chan, R., R.S. Warshawski, J. Scott, and R.W. Arnold. Experience with I-131-metaiodobenzylguanidine (MIBG): A retrospective study. *J Can Assoc Radiol 38*:35-39, 1987.

Chang, A., H.S. Glazer, J.K.T. Lee, D. Ling, and J.P. Heiken. Adrenal gland: MR imaging. *Radiology 163*:123-128, 1987.

Chang, A., E. Toloza, and J.C. Bulinski. Changes in the expression of β and γ actins during differentiation of PC12 cells. *J Neurochem 47*:1885-1892, 1986.

Chang, G.D. and V.D. Ramirez. A potent dopamine-releasing factor is present in high concentrations in the rat adrenal gland. *Brain Res 463*:385-389, 1988.

Chatterjee, S. and D. Pal. Effect of adrenocortical blockers on adrenomedullary catecholamines in pigeons. *Ind J Physiol Allied Sci 41*:75-79, 1987.

Chatterjee, S., D. Pal, and A. Ghosh. Effect of vitamin C on adrenomedullary hormones in gonadectomized juvenile pigeons. *Proc Indian Acad Sci 95*:223-226, 1986.

Cheek, T.R. and R.D. Burgoyne. Nicotine-evoked disassembly of cortical actin filaments in adrenal chromaffin cells. *FEBS Lett 207*:110-114, 1986.

Cheek, T.R. and R.D. Burgoyne. Cyclic AMP inhibits both nicotine-induced actin disassembly and catecholamine secretion from bovine adrenal chromaffin cells. *J Biol Chem 262*:11663-11666, 1987.

Cheek, T.R., J.E. Hesketh, R.C. Richards, and R.D. Burgoyne. Assembly and characterisation of a multi-component cytoskeletal gel from adrenal medulla. *Biochim Biophys Acta 887*:164-172, 1986.

Cherdchu, C. and T.D. Hexum. Characterization of 1,1-dimethyl-4-phenylpiperazinium induced increased proenkephalin processing in bovine chromaffin cells. *Life Sci 43*:1069-1077, 1988.

Cherdchu, C., E.M. Scholar, and T.D. Hexum. Nicotinic receptor stimulation enhances enkephalin-like peptides processing in chromaffin cells. *Life Sci 41*:2563-2572, 1987.

Chern, Y.-J., M. Herrera, L.-S. Kao, and E.W. Westhead. Inhibition of catecholamine secretion from bovine chromaffin cells by adenine nucleotides and adenosine. *J Neurochem 48*:1573-1576, 1987.

Chern, Y.-J., K.-T. Kim, L.L. Slakey, and E.W. Westhead. Adenosine receptors activate adenylate cyclase and enhance secretion from bovine adrenal chromaffin cells in the presence of forskolin. *J Neurochem 50*:1484-1493, 1988.

Cheung, C.Y. Ontogeny of adrenal VIP content and release from adrenocortical cells in the ovine fetus. *Peptides 9*:107-111, 1988.

Cheung, C.Y. and M.A. Holzwarth. Fetal adrenal VIP: Distribution and effect on medullary catecholamine secretion. *Peptides 7*:413-418, 1986.

Chianese, D., S. Roy-Choudhury, U. Murthy, A. Sen-Majumdar, C. Ernst, and M. Das. Cell- and tissue-specific expression of a 34,000-molecular-weight peptide growth factor in humans. *Hum Pathol 19*:190-194, 1988.

Christensen, L., N. Johansen, B.A. Jensen, and I. Clemmensen. Immunohistochemical localization of a novel, human plasma protein, tetranectin, in human endocrine tissues. *Histochemistry 87*:195-199, 1987.

Cidon, S. and N. Nelson. Purification of N-ethylmaleimide-sensitive ATPase from chromaffin granule membranes. *J Biol Chem 261*:9222-9227, 1986.

Citovsky, V., Y. Laster, S. Schuldiner, and A. Loyter. Osmotic swelling allows fusion of Sendai virions with membranes of desialized erythrocytes and chromaffin granules. *Biochemistry 26*:3856-3864, 1987.

Clark, S.A., W.E. Stumpf, C.W. Bishop, H.F. DeLuca, D.H. Park, and T.H. Joh. The adrenal: a new target organ of the calciotropic hormone 1,25-dihydroxyvitamin D_3. *Cell Tissue Res 243*:299-302, 1986.

Claude, P., I.M. Parada, K.A. Gordon, P.A. D'Amore, and J.A. Wagner. Acidic fibroblast growth factor stimulates adrenal chromaffin cells to proliferate and to extend neurites, but is not a long-term survival factor. *Neuron 1*:783-790, 1988.

Cobbold, P.H., T.R. Cheek, K.S.R. Cuthbertson, and R.D. Burgoyne. Calcium transients in single adrenal chromaffin cells detected with aequorin. *FEBS Lett 211*:44-48, 1987.

Cohen, W.R., G.J. Piasecki, H.E. Cohn, and B.T. Jackson. Plasma catecholamines in the hypoxaemic fetal rhesus monkey. *J Dev Physiol 9*:507-515, 1987.

Coker, G.T., L. Vinnedge, and K.L. O'Malley. Characterization of rat and human tyrosine hydroxylase genes: Functional expression of both promoters in neuronal and non-neuronal cell types. *Biochem Biophys Res Commun 157*:1341-1347, 1988.

Collste, P., B. Brismar, A. Alveryd, I. Björkhem, C. Hardstedt, L. Svensson, and J. Öestman. The catecholamine concentration in central veins of hypertensive patients--An aid not without problems in locating phaeochromocytoma. *Acta Chir Scand [Suppl] 530*:67-71, 1986.

Colombo, G., N.J. Papadopoulos, D.E. Ash, and J.J. Villafranca. Characterization of highly purified dopamine β-hydroxylase. *Arch Biochem Biophys 252*:71-80, 1987.

Connell, J.M.C., R. Corder, J. Asbury, S. MacPherson, G.C. Inglis, P.J. Lowry, A.D. Burt, and P.F. Semple. Neuropeptide Y in multiple endocrine neoplasia: Release during surgery for phaeochromocytoma. *Clin Endocrinol (Oxf) 26*:75-84, 1987.

Contreras, M.L. and G. Guroff. Calcium-dependent nerve growth factor-stimulated hydrolysis of phosphoinositides in PC12 cells. *J Neurochem 48*:1466-1472, 1987.

Cooper, M.E., D. Goodman, A. Frauman, G. Jerums, and W.J. Louis. Phaeochromocytoma in the elderly: A poorly recognised entity? *Br Med J [Clin Res] 293*:1474-1475, 1986.

Corcoran, J.J., M. Korner, B. Caughey, and N. Kirshner. Metabolic pools of ATP in cultured bovine adrenal medullary chromaffin cells. *J Neurochem 47*:945-952, 1986.

Corcoran, J.J., S.P. Wilson, and N. Kirshner. Turnover and storage of newly synthesized adenine nucleotides in bovine adrenal medullary cell cultures. *J Neurochem 46*:151-160, 1986.

Corder, R., R.C. Gaillard, and J. Rossier. Identification of human synenkephalin-like immunoreactivity in phaeochromocytoma tissue using a novel carboxy-terminal radioimmunoassay. *Neurosci Lett 82*:308-314, 1987.

Corder, R., B. Shapiro, P.J. Lowry, P.-M.G. Bouloux, G.M. Besser, J.M.C. Connell, P.F. Semple, A.C. Nieuwenhuijzen-Kruseman, A.F. Muller, and R.C. Gaillard. Relationship between tumour and plasma concentrations of neuropeptide Y in patients with adrenal medullary phaeochromocytoma. *J Hypertens 4 [Suppl 6]*:S193-S195, 1986.

Cornet, M., E. Delpire, and R. Gilles. Study of microfilaments network during volume regulation process of cultured PC12 cells [letter]. *Pflügers Arch 410*:223-225, 1987.

Costa, A., V. De Filippis, M. Voglino, G. Giraudi, M. Massobrio, C. Benedetto, L. Marozio, M. Gallo, G. Molina, C. Fabris, E. Bertino, and D. Licata. Adrenocorticotropin hormone and catecholamines in maternal, umbilical and neonatal plasma in relation to vaginal delivery. *J Endocrinol Invest 11*:703-709, 1988.

Côté, A., J.-P. Doucet, and J.-M. Trifaró. Adrenal medullary tropomyosins: Purification and biochemical characterization. *J Neurochem 46*:1771-1782, 1986.

Côté, A., J.-P. Doucet, and J.-M. Trifaró. Phosphorylation and dephosphorylation of chromaffin cell proteins in response to stimulation. *Neuroscience* 19:629-645, 1986.

Cottrell, G.A., J.J. Lambert, and J.A. Peters. Modulation of GABA$_A$ receptor activity by alphaxalone. *Br J Pharmacol* 90:491-500, 1987.

Coulter, C.L., I.C. McMillen, and C.A. Browne. The catecholamine content of the perinatal rat adrenal gland. *Gen Pharmacol* 19:825-828, 1988.

Coupland, R.E. Histochemistry and the chromaffin cell: An historic perspective on the identity of adrenaline and noradrenaline storing cells. *J Anat* 155:221-222, 1987.

Craig, S.P., V.J. Buckle, A. Lamouroux, J. Mallet, and I.W. Craig. Localization of the human tyrosine hydroxylase gene to 11p15: Gene duplication and evolution of metabolic pathways. *Cytogenet Cell Genet* 42:29-32, 1986.

Craig, S.P., V.J. Buckle, A. Lamouroux, J. Mallet, and I.W. Craig. Localization of the human dopamine beta hydroxylase (DBH) gene to chromosome 9q34. *Cytogenet Cell Genet* 48:48-50, 1088.

Creutz, C.E., S.L. Snyder, L.D. Husted, L.K. Beggerly, and J.W. Fox. Pattern of repeating aromatic residues in synexin: Similarity to the cytoplasmic domain of synaptophysin. *Biochem Biophys Res Commun* 152:1298-1303, 1988.

Creutz, C.E., W.J. Zaks, H.C. Hamman, and W.H. Martin. The roles of Ca^{2+}-dependent, membrane-binding proteins in the regulation and mechanism of exocytosis. In: **Cell Fusion**, (A. Sowers, ed), Plenum Publishing Corp., New York, 1987a.

Creutz, C.E., W.J. Zaks, H.C. Hamman, S. Crane, W.H. Martin, K.L. Gould, K.M. Oddie, and S.J. Parsons. Identification of chromaffin granule-binding proteins: Relationship of the chromobindins to calelectrin, synhibin, and the tyrosine kinase substrates p35 and p36. *J Biol Chem* 262:1860-1868, 1987b.

Cridland, R.A. and J.L. Henry. An adrenal-mediated, naloxone-reversible increase in reaction time in the tail-flick test following intrathecal administration of substance P at the lower thoracic spinal level in the rat. *Neuroscience* 26:243-251, 1988.

Critchley, J.A.J.H., P. Ellis, C.G. Henderson, A. Ungar, and C.P. West. Muscarinic and nicotinic mechanisms in the responses of the adrenal medulla of the dog and cat to reflex stimuli and to cholinomimetic drugs. *Br J Pharmacol* 89:831-835, 1986.

Critchley, J.A.J.H., M.R. MacLean, and A. Ungar. Inhibitory regulation by co-released peptides of catecholamine secretion by the canine adrenal medulla. *Br J Pharmacol* 93:383-386, 1988.

Cull-Candy, S.G., A. Mathie, and D.A. Powis. Acetylcholine receptor channels and their block by clonidine in cultured bovine chromaffin cells. *J Physiol (Lond)* 402:255-278, 1988.

Cuppoletti, J., J.E. Strasser, and G.E. Dean. An immunoreactive 8-

azido ATP-labeled protein common to the lysosomal and chromaffin granule membrane. *Biochim Biophys Acta 946*:33-39, 1988.

Dagerlind, Å. and M. Schalling. Localization of dopamine β-hydroxylase (DBH) and neuropeptide tyrosine (NPY) mRNA in the grey monkey locus coeruleus and adrenal medulla. *Acta Physiol Scand 134*:563-564, 1988.

Dahlöf, P., K. Persson, J.M. Lundberg, and C. Dahlöf. Neuropeptide Y (NPY) induced inhibition of preganglionic nerve stimulation evoked release of adrenalin and noradrenaline in the pithed rat. *Acta Physiol Scand 132*:51-57, 1988.

Dahlström, A., S. Bööj, M. Goldstein, and P.-A. Larsson. The synthesis of NPY and DBH is independently regulated in adrenergic nerves after reserpine. *Neurochem Res 12*:221-225, 1987.

Dahmer, M.K. and R.L. Perlman. Insulin and insulin-like growth factors stimulate deoxyribonucleic acid synthesis in PC12 pheochromocytoma cells. *Endocrinology 122*:2109-2113, 1988a.

Dahmer, M.K. and R.L. Perlman. Bovine chromaffin cells have insulin-like growth factor-I (IGF-I) receptors: IGF-I enhances catecholamine secretion. *J Neurochem 51*:321-323, 1988b.

Damon, D.H., P.A. D'Amore, and J.A. Wagner. Sulfated glycosaminoglycans modify growth factor-induced neurite outgrowth in PC12 cells. *J Cell Physiol 135*:293-300, 1988.

Daniels, A.J. and J.F. Reinhard Jr. Energy-driven uptake of the neurotoxin 1-methyl-4-phenylpyridinium into chromaffin granules via the catecholamine transporter. *J Biol Chem 263*:5034-5036, 1988.

Daniels, A.J., J.F. Reinhard Jr., and G.R. Painter. Accumulation of 1-methyl-4 phenylpyridinium (MPP$^+$) into bovine chromaffin granules results in a large restriction of its molecular motion: A ^{13}C and ^{31}P NMR study. *Biochem Biophys Res Commun 156*:1243-1249, 1988.

Darchen, F., D. Scherman, C. Desnos, and J.-P. Henry. Characteristics of the transport of the quaternary ammonium 1-methyl-4-phenylpyridinium by chromaffin granules. *Biochem Pharmacol 37*:4381-4387, 1988.

Darchen, F., D. Scherman, P.M. Laduron, and J.-P. Henry. Ketanserin binds to the monoamine transporter of chromaffin granules and of synaptic vesicles. *Mol Pharmacol 33*:672-677, 1988.

Daugherty, A., K.N. Frayn, W.S. Redfern, and B. Woodward. The role of catecholamines in the production of ischaemia-induced ventricular arrhythmias in the rat *in vivo* and *in vitro*. *Br J Pharmacol 87*:265-277, 1986.

Davidson, D. Circulating vasoactive substances and hemodynamic adjustments at birth in lambs. *J Appl Physiol 63*:676-684, 1987.

Davis, L.H. and F.C. Kauffman. Calcium-dependent activation of glycogen phosphorylase in rat pheochromocytoma PC12 cells by nerve growth factor. *Biochem Biophys Res Commun* *138*:917-924, 1986.

Davis, L.H. and F.C. Kauffman. Metabolism via the pentose phosphate pathway in rat pheochromocytoma PC12 cells: effects of nerve growth factor and 6-aminonicotinamide. *Neurochem Res 12*:521-527, 1987.

Dean, G.E., P.J. Nelson, W.S. Agnew, and G. Rudnick. Hydrodynamic properties of the chromaffin granule hydrogen ion pumping adenosinetriphosphatase. *Biochemistry* *26*:2301-2305, 1987.

Dean, G.E., P.J. Nelson, and G. Rudnick. Characterization of native and reconstituted hydrogen ion pumping adenosine-triphosphatase of chromaffin granules. *Biochemistry 25*:4918-4925, 1986.

Debinski, W., O. Kuchel, N.T. Buu, R. Garcia, M. Cantin, and J. Genest. Involvement of the adrenal glands in the action of the atrial natriuretic factor. *Proc Soc Exp Biol Med 181*:318-324, 1986.

De Block, J. and W.P. De Potter. The cell-free interaction between chromaffin granules and plasma membranes: An in vitro model for exocytosis? *FEBS Lett 222*:358, 1987a.

De Block, J. and W.P. De Potter. The cell-free interaction between chromaffin granules and plasma membranes: An in vitro model for exocytosis? [letter]. *Biochem Biophys Res Commun 148*:896-897, 1987b.

De Boer, S.F., J.L. Slangen, and J. Van Der Gugten. Adaptation of plasma catecholamine and corticosterone responses to short-term repeated noise stress in rats. *Physiol Behav 44*:273-280, 1988.

DeBold, C.R., E.E. Mufson, J.K. Menefee, and D.N. Orth. Proopiomelanocortin gene expression in a pheochromocytoma using upstream transcription initiation sites. Biochem Biophys Res Commun 155:895-900, 1988.

de Champlain, J., M. Bouvier, and G. Drolet. Abnormal regulation of the sympathoadrenal system in deoxycorticosterone acetate-salt hypertensive rats. *Can J Physiol Pharmacol 65*:1605-1614, 1987.

Deftos, L.J., B.T. Björnsson, D.W. Burton, D.T. O'Connor, and D.H. Copp. Chromogranin A is present in and released by fish endocrine tissue. *Life Sci 40*:2133-2136, 1987.

Deftos, L.J., S.S. Murray, D.W. Burton, R.J. Parmer, D.T. O'Connor, A.M. Delegeane, and P.L. Mellon. A cloned chromogranin A (CgA) cDNA detects a 2.3Kb mRNA in diverse neuroendocrine tissues. *Biochem Biophys Res Commun 137*:418-423, 1986.

Delarue, C., F. Leboulenger, M. Morra, F. Héry, A.J. Verhofstad, A. Bérod, L. Denoroy, G. Pelletier, and H. Vaudry. Immunohistochemical and biochemical evidence for the presence of serotonin in amphibian adrenal chromaffin cells.

Brain Res 459:17-26, 1988.

Delgado, J.M. and J.M. Billo. Care of the patient with Parkinson's disease: Surgical and nursing interventions. *J Neurosci Nurs 20*:142-150, 1988.

Delicado, E.G. and M.T. Miras-Portugal. Glucose transporters in isolated chromaffin cells: Effects of insulin and secretagogues. *Biochem J 243*:541-547, 1987.

Delicado, E.G., M. Torres, and M.T. Miras-Portugal. Glucose transporters in chromaffin cells: subcellular distribution and characterization. *FEBS Lett 229*:35-39, 1988.

DeLisle, R.C. and J.A. Williams. Regulation of membrane fusion in secretory exocytosis. *Annu Rev Physiol 48*:225-238, 1986.

DeLorme, E.M. and R. McGee Jr. Effects of prolonged depolarization on the nicotinic acetylcholine receptors of PC12 cells. *J Neurochem 50*:1248-1252, 1988.

DeLorme, E.M., C.S. Rabe, and R. McGee Jr. Regulation of the number of functional voltage-sensitive Ca^{++} channels on PC12 cells by chronic changes in membrane potential. *J Pharmacol Exp Ther 244*:838-843, 1988.

Demeneix, B. and N.J. Grant. α-Melanocyte stimulating hormone promotes neurite outgrowth in chromaffin cells. FEBS Lett 226:337-342, 1988.

Dennerll, T.J., H.C. Joshi, V.L. Steel, R.E. Buxbaum, and S.R. Heidemann. Tension and compression in the cytoskeleton of PC12 neurites II: Quantitative measurements. *J Cell Biol 107*:665-674, 1988.

De Oliveria Almeida, H., V. De Paula Antunes Teixeira, G. De Oliveria, G., M. Da Costa Brandão, and H. Gobbi. [Adrenal medullitis in cases of human rabies.] *Mem Inst Oswaldo Cruz 81*:439-442, 1986.

deQuidt, M.E. and P.C. Emson. Neuropeptide Y in the adrenal gland: Characterisation, distribution and drug effects. *Neuroscience 19*:1011-1022, 1986.

Derer, M., O. Grynszpan-Winograd, and J. Thibault. [Neurite-like process outgrowth of chromaffin cells in adult mouse adrenal medulla in culture. Influence of non-chromaffin cells.] *C R Acad Sci [III] 305*:363-368, 1987.

Deschepper, C.F. and W.F. Ganong. Renin and angiotensin in endocrine glands, pp. 79-98. In: **Frontiers in Neuroendocrinology**, Vol. 10. (L. Martini and W.F. Ganong, eds). Raven Press, New York, 1988.

Deupree, J.D. and J.J. Hitchcock. Effects of the sulfhydryl reagent *N*-ethylmaleimide on reserpine binding to the catecholamine transporter in chromaffin granule membranes. *Cell Mol Neurobiol 8*:217-224, 1988.

Deus Fombellida, J. [Primary hyperplasia of the adrenal medulla: Precursor stages of pheochromocytomas?] *Arch Esp Urol 40*:385-388, 1987.

DeWolf, W.E. Jr., S.A. Carr, A. Varrichio, P.J. Goodhart, M.A.

Mentzer, G.D. Roberts, C. Southan, R.E. Dolle, and L.I. Kruse. Inactivation of dopamine β-hydroxylase by p-cresol: Isolation and characterization of covalently modified active site peptides. *Biochemistry* 27:9093-9101, 1988.

Dhawan, S., L.T. Duong, R.L. Ornberg, and P.J. Fleming. Subunit exchange between membranous and soluble forms of bovine dopamine β–hydroxylase. *J Biol Chem* 262:1869-1875, 1987.

Dhawan, S., P. Hensley, J.C. Osborne Jr., and P.J. Fleming. Adenosine 5'-diphosphate-dependent subunit dissociation of bovine dopamine β–hydroxylase. *J Biol Chem* 261:7680-7684, 1986.

Dhingra, S., B. Bhatia, G.S. Chhina, and B. Singh. Effect of demedullation on freezing injury in hind limbs of rats. *Int J Biometeorol* 31:191-199, 1987.

Dicke, T.E., M.L. Henry, and J.P. Minton. Intracaval extension of pheochromocytoma simulating pulmonary embolism. *J Surg Oncol* 34:160-164, 1987.

Diener, R.M. Case History. Pheochromocytomas and reserpine: Review of carcinogenicity bioassay. *J Am Coll Toxicol* 7:95-105, 1988.

Diliberto Jr., E.J., F.S. Menniti, J. Knoth, A.J. Daniels, J.S. Kizer, and O.H. Viveros. Adrenomedullary chromaffin cells as a model to study the neurobiology of ascorbic acid: From monooxygenation to neuromodulation. *Ann NY Acad Sci* 498:28-53, 1987.

Dillen, L., M. Claeys, and W.P. De Potter. Determination of dopamine-β–hydroxylase in cerebrospinal fluid by high-performance liquid chromatography with electrochemical detection. *J Pharmacol Methods* 15:51-63, 1986.

Di Paolo, T., P. Bédard, M. Daigle, and R. Boucher. Long-term effects of MPTP on central and peripheral catecholamine and indoleamine concentrations in monkeys. *Brain Res* 379:286-293, 1986.

Di Virgilio, F., D. Milani, A. Leon, J. Meldolesi, and T. Pozzan. Voltage-dependent activation and inactivation of calcium channels in PC12 cells: Correlation with neurotransmitter release. *J Biol Chem* 262:9189-9195, 1987.

Di Virgilio, F., T. Pozzan, C.B. Wollheim, L.M. Vicentini, and J. Meldolesi. Tumor promoter phorbol myristate acetate inhibits Ca^{2+} influx through voltage-gated Ca^{2+} channels in two secretory cell lines, PC12 and RINm5F. *J Biol Chem* 261:32-35, 1986.

Dix, T.A., D.M. Kuhn, and S.J. Benkovic. Mechanism of oxygen activation by tyrosine hydroxylase. *Biochemistry* 26:3354-3361, 1987.

D'Mello, S.R., E.P. Weisberg, M.K. Stachowiak, L.M. Turzai, A.E. Gioio, and B.B. Kaplan. Isolation and nucleotide sequence of a cDNA clone encoding bovine adrenal tyrosine hydroxylase: Comparative analysis of tyrosine hydroxylase gene products. *J Neurosci Res 19*:440-449, 1988.

Dobner, P.R., A.S. Tischler, Y.C. Lee, S.R. Bloom, and S.R. Donahue. Lithium dramatically potentiates neurotensin-neuromedin N gene expression. *J Biol Chem 263*:13983-13986, 1988.

Doherty, P., P. Seaton, T.P. Flanigan, and F.S. Walsh. Factors controlling the expression of the NGF receptor in PC12 cells. *Neurosci Lett 92*:222-227, 1988.

Dohi, T., K. Morita, S. Kitayama, and A. Tsujimoto. Forskolin enhancement of acetylcholine-evoked cyclic AMP formation and catecholamine release in perfused dog adrenals. *J Neural Transm 67*:57-66, 1986.

Dominiak, P., F. Kees, and H. Grobecker. Circulating and tissue catecholamines in rats with chronic neurogenic hypertension. *Basic Res Cardiol 81*:20-28, 1986.

Donnerer, J. Reflex activation of the adrenal medulla during hypoglycemia and circulatory dysregulations is regulated by capsaicin-sensitive afferents. *Naunyn Schmiedebergs Arch Pharmacol 338*:282-286, 1988.

Donoso, A.O. and M.B. Barontini. Increase in plasma catecholamines by intraventricular injection of histamine in conscious rats. *Naunyn Schmiedebergs Arch Pharmacol 334*:188-192, 1986.

Doris, P.A. Immunological evidence that the adrenal gland is a source of an endogenous digitalis-like factor. *Endocrinology 123*:2440-2444, 1988.

Drolet, G. and P. Gauthier. Similar poststimulatory pressor responses with different mechanisms in response to excitation of the locus coeruleus before and after acute adrenalectomy. *Can J Physiol Pharmacol 65*:1136-1141, 1987.

Drubin, D., S. Kobayashi, D. Kellogg, and M. Kirschner. Regulation of microtubule protein levels during cellular morphogenesis in nerve growth factor-treated PC12 cells. *J Cell Biol 106*:1583-1591, 1988.

Drubin, D., S. Kobayashi, and M. Kirschner. Association of tau protein with microtubules in living cells. *Ann NY Acad Sci 466*:257-268, 1986.

Drucker-Colín, R. and I. Madrazo-Navarro. Adrenal grafts for Parkinson's disease [letter]. *N Engl J Med 317*:1093, 1987.

Drust, D.S. and C.E. Creutz. Aggregation of chromaffin granules by calpactin at micromolar levels of calcium. *Nature 331*:88-91, 1988.

D'Souza, N.B. and I. Lindberg. Evidence for the phosphorylation of a proenkephalin-derived peptide, peptide B. *J Biol Chem 263*:2548-2552, 1988.

Dwork, A.J., G. Pezzoli, V. Silani, S. Fahn, and R. Hill. Transplantation of fetal substantia nigra and adrenal medulla to the caudate nucleus in two patients with Parkinson's disease [letter]. *N Eng J Med 319*:370-371., 1988.

Dymecki, J., M. Póltorak, D. Markiewicz, M. Hauptman, W. Dziedziak, and W. Kostowski. Effects of intracerebral transplantation of the adrenal medulla in rats with experimental Parkinson's disease. *Neuropatol Pol 24*:29-42, 1986.

Dzhabarov, B.D. [Vegetative and trophic disorders and functional status of the sympatho-adrenal system in Rossolimo-Curschmann-Steinert Batten myotonic dystrophy.] *Zh Nevropatol Psikhiatr 86*:333-335, 1986.

Eberhard, D.A. and R.W. Holz. Cholinergic stimulation of inositol phosphate formation in bovine adrenal chromaffin cells: Distinct nicotinic and muscarinic mechanisms. *J Neurochem 49*:1634-1643, 1987.

Eberhard, D.A. and R.W. Holz. Intracellular Ca^{2+} activates phospholipase C. *Trends Neurosci 11*:517-520, 1988.

Edwards, A.V., D. Hansell, and C.T. Jones. Effects of synthetic adrenocorticotrophin on adrenal medullary responses to splanchnic nerve stimulation in conscious calves. *J Physiol (Paris) 379*:1-16, 1986.

Edwards, A.V. and C.T. Jones. The effect of splanchnic nerve stimulation on adrenocortical activity in conscious calves. *J Physiol (Lond) 382*:385-396, 1987.

Edwards, A.V. and C.T. Jones. Secretion of corticotrophin releasing factor from the adrenal during splanchnic nerve stimulation in conscious calves. *J Physiol (Lond) 400*:89-100, 1988.

Ehrhart, M., D. Grube, M.-F. Bader, D. Aunis, and M. Gratzl. Chromogranin A in the pancreatic islet: Cellular and subcellular distribution. *J Histochem Cytochem 34*:1673-1682, 1986.

Ehrhart, M., A. Jörns, D. Grube, and M. Gratzl. Cellular distribution and amount of chromogranin A in bovine endocrine pancreas. *J Histochem Cytochem 36*:467-472, 1988.

Ehrlich, Y.H. and E. Kornecki. Extracellular protein phosphorylation systems in the regulation of cellular responsiveness. *Prog Clin Biol Res 249*:193-204, 1987.

Eiden, L.E. Is chromogranin a prohormone? *Nature 325*:301, 1987a.

Eiden, L.E. The enkephalin-containing cell: Strategies for polypeptide synthesis and secretion throughout the neuroendocrine system. *Cell Mol Neurobiol 7*:339-352, 1987b.

Eiden, L.E., W.B. Huttner, J. Mallet, D.T. O'Connor, H. Winkler, and A. Zanini. A nomenclature proposal for the chromogranin/secretogranin proteins. *Neuroscience 21*:1019-1021, 1987.

Eiden, L.E., A.L. Iacangelo, C.-M. Hsu, A.J. Hotchkiss, M.-F. Bader, and D. Aunis. Chromogranin A synthesis and secretion in chromaffin cells. *J Neurochem 49*:65-74, 1987.

Eiden, L.E. and N. Zamir. Metorphamide levels in chromaffin cells increase after treatment with reserpine. *J Neurochem*

46:1651-1654, 1986.

Ekker, M. and T.L. Sourkes. Developmental pattern of ornithine decarboxylase activity, *S*-adenosylmethionine decarboxylase, and polyamines of rat adrenal glands. *Biol Neonate 51*:260-267, 1987a.

Ekker, M. and T.L. Sourkes. Decreased activity of adrenal *S*-adenosylmethionine decarboxylase in rats subjected to dopamine agonists, metabolic stress, or bodily immobilization. *Endocrinology 120*:1299-1307, 1987b.

Ekker, M., T.L. Sourkes, and R. Gabor. Reduced amounts of *S*-adenosylmethionine decarboxylase in the adrenal glands of rats following administration of piribedil or 2-deoxyglucose. *Biochem Pharmacol 37*:3613-3618, 1988.

Ekman, R., A. Bjartell, E. Ekblad, and F. Sundler. Immunoreactive delta sleep-inducing peptide in pituitary adrenocorticotropin/α-melanotropin cells and adrenal medullary cells of the pig. *Neuroendocrinology 45*:298-304, 1987.

El-Maghraby, M.Z. Differentiation of adreno-chromaffin cells in the newborn rat, as detected by formaldehyde-induced fluorescence, compared with the chromaffin reaction. *Acta Anat (Basel) 131*:103-107, 1988.

El-Refai, M.F. and T.M. Chan. Effects of adrenalectomy on binding to and actions of adrenergic receptors. *Biochem J 237*:527-531, 1986.

Engeland, W.C., D.F. Bereiter, and D.S. Gann. Sympathetic control of adrenal secretion of enkephalins after hemorrhage in awake dogs. *Am J Physiol 251*:R341-R348, 1986.

Engeland, W.C., M.P. Lilly, T.O. Bruhn, and D.S. Gann. Comparison of corticotropin-releasing factor and acetylcholine on catecholamine secretion in dogs. *Am J Physiol 253*:R209-R215, 1987.

Eremina, S.A. and E.I. Belyakova. [Phases of primary reaction to stress in sympathoadrenal system.] *Biull Eksp Biol Med 104*:155-157, 1987.

Ertl, S., H. Deckart, A. Blottner, and M. Tautz. Radiopharmacokinetics and radiation absorbed dose calculations from ^{131}I-meta-iodobenzylguanidine. *Nucl Med Commun 8*:643-653, 1987.

Esler, M., G. Jennings, P. Korner, I. Willett, F. Dudley, G. Hasking, W. Anderson, and G. Lambert. Assessment of human sympathetic nervous system activity from measurements of norepinephrine turnover. *Hypertension 11*:3-20, 1988.

Evans, C.J., E. Erdelyi, J. Hunter, and J.D. Barchas. Co-localization and characterization of immunoreactive peptides derived from two opioid precursors in guinea pig adrenal glands. *J Neurosci 5*:3423-3427, 1985.

Evéquoz, D., B. Waeber, J.-F. Aubert, J.-P. Flückiger, J. Nussberger, and H.R. Brunner. Neuropeptide Y prevents the blood pressure fall induced by endotoxin in conscious rats with adrenal medullectomy. *Circ Res 62*:25-30, 1988.

Evéquoz, D., B. Waeber, G. Matsueda, J. Nussberger, and H.R. Brunner. Differential blood pressure response to atriopeptin III and sodium nitroprusside in conscious rats with adrenal medullectomy. *Eur J Pharmacol 144*:217-221, 1987.

Evinger, M.J., D.H. Park, E.E. Baetge, D.J. Reis, and T.H. Joh. Strain-specific differences in levels of the mRNA for the epinephrine synthesizing enzyme phenylethanolamine *N*-methyltransferase. *Mol Brain Res 1*:63-73, 1986.

Factor, S.A., J. Sanchez-Ramos, and W.J. Weiner. Adrenal grafts for Parkinson's disease [letter]. *N Engl J Med 317*:1091, 1987.

Falke, T.H.M., L. te Strake, M.I. Shaff, M.P. Sandler, M.V. Kulkarni, C.L. Partain, A.C. Nieuwenhuizen-Kruseman, and A.E. James Jr. MR imaging of the adrenals: Correlation with computed tomography. *J Comput Assist Tomogr 10*:242-253, 1986.

Farber, L.H., F.J. Wilson, and D.J. Wolff. Calmodulin-dependent phosphatases of PC12, GH_3, and C_6 cells: Physical, kinetic, and immunochemical properties. *J Neurochem 49*:404-414, 1987.

Faucon Biguet, N., M. Buda, A. Lamouroux, D. Samolyk, and J. Mallet. Time course of the changes of TH mRNA in rat brain and adrenal medulla after a single injection of reserpine. *EMBO J 5*:287-291, 1986.

Fauquet, M., B. Grima, A. Lamouroux, and J. Mallet. Cloning of quail tyrosine hydroxylase: Amino acid homology with other hydroxylases discloses functional domains. *J Neurochem 50*:142-148, 1988.

Favre, L., A. Forster, M. Fathi, and M.B. Vallotton. Calcium-channel inhibition in pheochromocytoma. *Acta Endocrinol (Copenh) 113*:385-390, 1986.

Favre, R., M. deHaut, Y. Dalmaz, J.M. Pequignot, and L. Peyrin. Peripheral distribution of free dopamine and its metabolites in the rat. *J Neural Transm 66*:135-149, 1986.

Fernández-Ruiz, J.J., A.R. Bukhari, R. Martínez-Arrieta, J.A.F. Tresguerres, and J.A. Ramos. Effects of estrogens and progesterone on the catecholaminergic activity of the adrenal medulla in female rats. *Life Sci 42*:1019-1028, 1988.

Fernández-Ruiz, J.J., M. Cebeira, C. Agrasal, J.A.F. Tresguerres, A.I. Esquifino, and J.A. Ramos. Effect of elevated prolactin levels on the synthesis and release of catecholamines from the adrenal medulla in female rats. *Neuroendocrinology 45*:208-211, 1987.

Fernández-Ruiz, J.J., A.I. Esquifino, R.W. Steger, A. Bartke, and S.L. Hodges. Effects of prolactin on the adrenal medulla in mice and hamsters. *Neuroendocrinol Lett 10*:5-13, 1988.

Fernández-Ruiz, J.J., R. Martínez-Arrieta, M.L. Hernandez, and J.A. Ramos. Possible direct effect of prolactin on catecholamine synthesis and release in rat adrenal medulla: *In vitro* studies. *J Endocrinol Invest 11*:603-608, 1988.

Fernyhough, P. and D.N. Ishii. Nerve growth factor modulates tubulin transcript levels in pheochromocytoma PC12 cells. *Neurochem Res 12*:891-899, 1987.

Ferris, C.F., R.E. Carraway, K. Brandt, and S.E. Leeman. Chromatographic and immunochemical characterization of neurotensin in cat adrenal gland and its release during splanchnic nerve stimulation. *Neuroendocrinology 43*:352-358, 1986.

Féty, R., V. Misére, L. Lambás-Señas, and B. Renaud. Central and peripheral changes in catecholamine-synthesizing enzyme activities after systemic administration of the neurotoxin DSP-4. *Eur J Pharmacol 124*:197-202, 1986.

Fiandaca, M.S., J.H. Kordower, J.T. Hansen, and D.M. Gash. Adrenal medullary tissue grafting in Parkinson's disease. *J Neurosurg 69*:150-152, 1988a.

Fiandaca, M.S., J.H. Kordower, J.T. Hansen, S.-S. Jiao, and D.M. Gash. Adrenal medullary autografts into the basal ganglia of Cebus monkeys: Injury-induced regeneration. *Exp Neurol 102*:76-91, 1988b.

Fischbein, L., A. Rio, J.J. Wenger, J. Tongio, and C. Brechenmacher. [Pheochromocytoma of the organ of Zuckerkandl.] *Arch Mal Coeur 79*:1090-1092, 1986.

Fischer, M. and H. Vetter. [Adrenal gland scintigraphy.] *Radiologe 26*:181-185, 1986.

Fischer-Colbrie, R., J. Diez-Guerra, P.C. Emson, and H. Winkler. Bovine chromaffin granules: Immunological studies with antisera against neuropeptide Y, [Met]enkephalin and bombesin. *Neuroscience 18*:167-174, 1986.

Fischer-Colbrie, R., C. Hagn, L. Kilpatrick, and H. Winkler. Chromogranin C: A third component of the acidic proteins in chromaffin granules. *J Neurochem 47*:318-321, 1986.

Fischer-Colbrie, R., A. Iacangelo, and L.E. Eiden. Neural and humoral factors separately regulate neuropeptide Y, enkephalin, and chromogranin A and B mRNA levels in rat adrenal medulla. *Proc Natl Acad Sci USA 85*:3240-3244, 1988.

Fischer-Colbrie, R. and M. Schober. Isolation and characterization of chromogranins A, B, and C from bovine chromaffin granules and a rat pheochromocytoma. *J Neurochem 48*:262-270, 1987.

Fiscus, R.R., B.T. Robles, S.A. Waldman, and F. Murad. Atrial natriuretic factors stimulate accumulation and efflux of cyclic GMP in C6-2B rat glioma and PC12 rat pheochromocytoma cell cultures. *J Neurochem 48*:522-528, 1987.

Fisher, B.M., G.J.A. MacPhee, D.L. Davies, S.G. McPherson, I.L. Brown, and A. Goldberg. A case of watery diarrhoea syndrome due to an adrenal phaeochromocytoma secreting vasoactive intestinal polypeptide with coincidental autoimmune thyroid disease. *Acta Endocrinol (Copenh)*

114:340-344, 1987.

Fitzpatrick, P.F. The pH dependence of binding of inhibitors to bovine adrenal tyrosine hydroxylase. *J Biol Chem 263*:16058-16062, 1988.

Fitzpatrick, P.F., M.R. Harpel, and J.J. Villafranca. Use of alternate substrates to probe the order of substrate addition to dopamine β-hydroxylase. *Arch Biochem Biophys 249*:70-75, 1986.

Fitzpatrick, P.F. and J.J. Villafranca. The mechanism of inactivation of dopamine β-hydroxylase by hydrazines. *J Biol Chem 261*:4510-4518, 1986.

Fitzpatrick, P.F. and J.J. Villafranca. Mechanism-based inhibitors of dopamine β-hydroxylase. *Arch Biochem Biophys 257*:231-250, 1987.

Flügge, G., T. Inagami, and E. Fuchs. Atrial natriuretic peptide detected by immunocytochemistry in peripheral organs of *Tupaia belangeri. Histochemistry 86*:479-483, 1987.

Foley, P.J., W.A. Tacker, W.D. Voorhees, S.H. Ralston, and M.A. Elchisak. Effects of naloxone on the adrenomedullary response during and after cardiopulmonary resuscitation in dogs. *Am J Emerg Med 5*:357-361, 1987.

Foley, P.J., W.A. Tacker, J. Wortsman, S. Frank, and P.E. Cryer. Plasma catecholamine and serum cortisol responses to experimental cardiac arrest in dogs. *Am J Physiol 253*:E283-E289, 1987.

Fontenot, G.K., W.A. Cass, and P.R. Vulliet. Changes in rat adrenal tyrosine hydroxylase mRNA levels following chronic amphetamine administration. *Proc West Pharmacol Soc 30*:263-265, 1987.

Fonteríz, R.I., L. Gandía, M.G. Lopez, C.R. Artalejo, and A.G. García. Dihydropyridine chirality at the chromaffin cell calcium channel. *Brain Res 408*:359-362, 1987.

Forsberg, E.J., G. Feuerstein, E. Shohami, and H.B. Pollard. Adenosine triphosphate stimulates inositol phospholipid metabolism and prostacyclin formation in adrenal medullary endothelial cells by means of P_2-purinergic receptors. *Proc Natl Acad Sci USA 84*:5630-5634, 1987.

Forsberg, E.J. and H.B. Pollard. Ba^{2+}-induced ATP release form adrenal medullary chromaffin cells is mediated by Ba^{2+} entry through both voltage- and receptor-gated Ca^{2+} channels. *Neuroscience 27*:711-715, 1988.

Forsberg, E.J., E. Rojas, and H.B. Pollard. Muscarinic receptor enhancement of nicotine-induced catecholamine secretion may be mediated by phosphoinositide metabolism in bovine adrenal chromaffin cells. *J Biol Chem 261*:4915-4920, 1986.

Fosha-Dolezal, S.R., T.B. Avery, W.C. Wagner, and M.R. Fedde. Changes in serum potassium concentration with exercise in Hereford calves: Effects of adrenalectomy. *Comp Biochem Physiol [A] 91*:135-139, 1988.

Foucart, S., J. De Champlain, and R. Nadeau. *In vivo* interactions between prejunctional α_2- and β_2-adrenocepters at the level of the adrenal medulla. *Can J Physiol Pharmacol 66*:1340-1343, 1988.

Foucart, S., R. Nadeau, and J. deChamplain. The release of catecholamines from the adrenal medulla and its modulation by α_2-adrenoceptors in the anaesthetized dog. *Can J Pharmacol 65*:550-557, 1987.

Foucart, S., R. Nadeau, and J. deChamplain. Local modulation of adrenal catecholamines release by β_2-adrenoceptors in the anaesthetized dog. *Naunyn Schmiedebergs Arch Pharmacol 337*:29-34, 1988.

Fouchier, F., P. Bastiani, T. Baltz, D. Aunis, and G. Rougon. Glycosylphosphatidylinositol is involved in the membrane attachment of proteins in granules of chromaffin cells. *Biochem J 256*:103-108, 1988.

Fournier, R.D., C.C. Chiueh, I.J. Kopin, J.J. Knapka, D. DiPette, and H.G. Preuss. Refined carbohydrate increases blood pressure and catecholamine excretion in SHR and WKY. *Am J Physiol 250*:E381-E385, 1986.

Fournier, S. and J.-M. Trifaró. Calmodulin-binding proteins in chromaffin cell plasma membranes. *J Neurochem 51*:1599-1609, 1988b.

Fournier, S. and J.-M. Trifaró. A similar calmodulin-binding protein expressed in chromaffin, synaptic, and neurohypo-physeal secretory vesicles. *J Neurochem 50*:27-37, 1988a.

Fraeyman, N.H., E.J. Van de Velde, and F.H. De Smet. Molecular forms of dopamine β-hydroxylase in rat superior cervical ganglion and adrenal gland. *Experientia 44*:746-749, 1988.

Frank, F., C. Sturiale, G. Gaist, and V. Manetto. Adrenal medullary autograft in human brain for Parkinson's disease [letter]. *Acta Neurochir (Wien) 94*:162-163, 1988.

Franke, W.W., C. Grund, and T. Achtstätter. Co-expression of cytokeratins and neurofilament proteins in a permanent cell line: Cultured rat PC12 cells combine neuronal and epithelial features. *J Cell Biol 103*:1933-1943, 1986.

Freed, C.R. Transplantation of fetal substantia nigra and adrenal medulla to the caudate nucleus in two patients with Parkinson's disease [letter]. *N Engl J Med 319*:370, 1988.

Freed, W.J. Adrenal medulla grafts in animals [letter]. *Science 241*:275, 1988.

Freed, W.J., H.E. Cannon-Spoor, and E. Krauthamer. Intrastriatal adrenal medulla grafts in rats: Long-term survival and behavioral effects. *J Neurosurg 65*:664-670, 1986.

Freed, W.J., U. Patel-Vaidya, and H.M. Geller. Properties of PC12 pheochromocytoma cells transplanted to the adult rat brain. *Exp Brain Res 63*:557-566, 1986.

Fricker, L.D., C.J. Evans, F.S. Esch, and E. Herbert. Cloning and sequence analysis of cDNA for bovine carboxypeptidase E [letter]. *Nature* 323:461-464, 1986.

Fried, G., M. Wikström, and H. Lagercrantz. Postnatal development of catecholamines and response to hypoxia in adrenals and paraganglia of rabbits. *J Auton Nerv Syst* 24:65-70, 1988.

Friedlander, D.R., M. Grumet, and G.M. Edelman. Nerve growth factor enhances expression of neuron-glia cell adhesion molecule in PC12 cells. *J Cell Biol* 102:413-419, 1986.

Friedman, J. Recent research advances in Parkinson's disease: Part II. Studies in lower animals indicate a future for brain implants. *RI Med J* 69:361-363, 1986.

Friedman, J.E., P.I. Lelkes, K. Rosenheck, and A. Oplatka. Control of stimulus-secretion coupling in adrenal medullary chromaffin cells by microfilament-specific macromolecules. *J Biol Chem* 261:5745-5750, 1986.

Friedrich, U. and H. Bönisch. The neuronal noradrenaline transport system of PC-12 cells: Kinetic analysis of the interaction between noradrenaline, Na^+ and Cl^- in transport. *Naunyn Schmiedebergs Arch Pharmacol* 333:246-252, 1986.

Frigon, R.P. Dopamine β-monooxygenase from human plasma. *Methods Enzymol* 142:603-607, 1987.

Fröhlich, O. The "tunneling" mode of biological carrier-mediated transport. *J Membr Biol* 101:189-198, 1988.

Fuchs, E., K. Shigematsu, and J.M. Saavedra. Binding sites of atrial natriuretic peptide in tree shrew adrenal gland. *Peptides* 7:873-876, 1986.

Fujii, T., N. Yamamoto, and K. Sato. Adrenal enucleation abolishes the development of hypersensitivity to haloperidol in rats. *Proc Jpn Acad* 63:381-384, 1987.

Fujimoto, M., Y. Kataoka, A. Guidotti, and I. Hanbauer. Effect of γ-aminobutyric acid$_A$ receptor agonists and antagonists on the release of enkephalin-containing peptides from dog adrenal gland. *J Pharmacol Exp Ther* 243:195-199, 1987.

Fujisawa, H. and S. Okuno. Tyrosine 3-monooxygenase from rat adrenals. *Methods Enzymol* 142:63-71, 1987.

Fujita, T., K. Ando, H. Noda, Y. Ito, and Y. Sato. Effects of increased adrenomedullary activity and taurine in young patients with borderline hypertension. *Circulation* 75:525-532, 1987.

Fujita, T., T. Kanno, and S. Kobayashi. **The Paraneuron.** Springer-Verlag, Tokyo, 1988.

Fuller, R.W. Norepinephrine *N*-methyltransferase from rabbit adrenal glands. *Methods Enzymol* 142:655-660, 1987.

Gaardsvoll, H., D. Obendorf, H. Winkler, and E. Bock. Demonstration of immunochemical identity between the synaptic vesicle proteins synaptin and synaptophysin/p38. *FEBS Lett* 242:117-120, 1988.

Gagnon, C., K. Veeraragavan, and R. Coulombe. Protein-carboxyl methylation in adrenal medullary cells. *Cell Mol Neurobiol* 8:95-103, 1988.

Gaillard, G., O. Rascol, and J.L. Montastruc. Adrenal catecholamine concentration after domperidone. *Clin Neuropharmacol* 10:479-481, 1987.

Gaillard, G., M.A. Tran, M. Rostin, R. Salvayre, and J.L. Montastruc. Effect of clonidine on adrenal medulla catecholamine levels in rats. *Fundam Clin Pharmacol 1*:1-5, 1987.

Gallo, V.P., A. Civinini, and L. Mastrolia. Comparative data on acetylcholinesterase cytochemistry in various chromaffin cell types of *Discoglossus pictus* (Amphibia, Anura). *Basic Appl Histochem 31*:135-142, 1987.

Gallo-Payet, N., P. Pothier, and H. Isler. On the presence of chromaffin cells in the adrenal cortex: Their possible role in adrenocortical function. *Biochem Cell Biol 65*:588-592, 1987.

Gandía, L., M.G. López, R.I. Fonteríz, C.R. Artalejo, and A.G. García. Relative sensitivities of chromaffin cell calcium channels to organic and inorganic calcium antagonists. *Neurosci Lett 77*:333-338, 1987.

García, A.G., C.R. Artalejo, R. Borges, J.A. Reig, and F. Sala. Pharmacological properties of the chromaffin cell calcium channel. *Adv Exp Med Biol 211*:139-157, 1987.

García-Arrarás, J.E., M. Fauquet, M. Chanconie, and J. Smith. Coexpression of somatostatin-like immunoreactivity and catecholaminergic properties in neural crest derivatives: Comodulation of peptidergic and adrenergic differentiation in cultured neural crest. *Dev Biol 114*:247-257, 1986.

Gash, D.M. and J.R. Sladek (eds). **Transplantation Into the Mammalian CNS.** Progess in Brain Research, Vol 78. Elsevier Science Publishers, Amsterdam, 1988.

Gasnier, B., J.C. Ellory, and J.-P. Henry. Functional molecular mass of binding sites for [^{3}H]dihydrotetrabenazine and [^{3}H]reserpine and of dopamine β–hydroxylase and cytochrome b_{561} from chromaffin granule membrane as determined by radiation inactivation. *Eur J Biochem 165*:73-78, 1987.

Gasnier, B., M.-P. Roisin, D. Scherman, S. Coornaert, G. Desplanches, and J.-P. Henry. Uptake of *meta*-iodobenzylguanidine by bovine chromaffin granule membranes. *Mol Pharmacol 29*:275-280, 1986.

Gasnier, B., D. Scherman, and J.-P. Henry. Inactivation of the catecholamine transporter during the preparation of chromaffin granule membrane 'ghosts'. *FEBS Lett 222*:215-219, 1987.

Gatti, G., L. Madeddu, A. Pandiella, T. Pozzan, and J. Meldolesi. Second-messenger generation in PC12 cells. Interactions between cyclic AMP and Ca^{2+} signals. *Biochem J 255*:753-760, 1988.

Gaumann, D.M. and T.L. Yaksh. Effects of hemorrhage and

opiate antagonists on adrenal release of neuropeptides in cats. *Peptides 9*:393-405, 1988.

Gaumann, D.M., T.L. Yaksh, M.K. Dousa, G.M. Tyce, D.L. Lucas, and V.S. Hench. Effects of hemorrhage and naloxone on adrenal release of methionine-enkephalin and catecholamines in halothane anesthetized dogs. *J Auton Nerv Syst 21*:29-41, 1987.

Gaumann, D.M., T.L. Yaksh, G.M. Tyce, and D.L. Lucas. Opioids preserve the adrenal medullary response evoked by severe hemorrhage: Studies on adrenal catecholamine and met-enkephalin secretion in halothane anesthetized cats. *Anesthesiology 68*:743-753, 1988a.

Gaumann, D.M., T.L. Yaksh, G.M. Tyce, and S.L. Stoddard. Sympathetic stimulating effects of sufentanil in the cat are mediated centrally. *Neurosci Lett 91*:30-35, 1988b.

Gauthier, P. Adrenomedullary pressor responses to stimulation of the rostral hypothalamus in the rat: Influence of adrenaline-induced vasodilation and reflex cardioinhibition. *Can J Physiol Pharmacol 66*:213-221, 1988.

Geertsen, S., R. Afar, J.-M. Trifaró, and M. Quik. Regulation of α-bungarotoxin sites in chromaffin cells in culture by nicotinic receptor ligands, K^+, and cAMP. *Mol Pharmacol 34*:549-556, 1988.

Geisow, M.J., U. Fritsche, J.M. Hexham, B. Dash, and T. Johnson. A consensus amino-acid sequence repeat in *Torpedo* and mammalian Ca^{2+}-dependent membrane-binding proteins. *Nature 320*:636-638, 1986.

Gemmell, R.T. and J. Nelson. The ultrastructure of the pituitary and the adrenal gland of three newborn marsupials (*Dasyurus hallucatus, Trichosurus vulpecula* and *Isoodon macrourus*). *Anat Embryol (Berl) 177*:395-402, 1988.

George, J.C. and T.M. John. Canada goose: Ultrastructural changes in the adrenal gland in relation to migration, breeding and moult. *Cytobios 54*:105-128, 1988.

Georges, E., J.-M. Trifaró, and W.E. Mushynski. Hypophosphorylated neurofilament subunits in the cytoskeletal and soluble fractions of cultured bovine adrenal chromaffin cells. *Neuro-science 22*:753-763, 1987.

Gharib, C., G. Gauquelin, J.M. Pequignot, G. Geelen, C.-A. Bizollon, and A. Guell. Early hormonal effects of head-down tilt (-10°) in humans. *Aviat Space Environ Med 59*:624-629, 1988.

Gherut, A. The adrenal medulla and sudden cardiac death [editorial]. *Endocrinologie 25*:59-60, 1987.

Ghosh, A. Vertebrate endocrinology. pp. 49-70. In: **Animal Physiology: Status series report 6**. (Nagabhushanam, R., ed). Indian National Science Academy, New Delhi, 1986.

Ghosh, A. and S. Chatterjee. The avian adrenal medulla and the effects of ascorbic acid on adrenomedullary hormones. *Indian*

Rev Life Sci 6:67-73, 1986.

Ghosh, A. and B. Guha. The subcellular morphology of avian adrenal medulla. *Proc Indian Acad Sci (Anim Sci)* 97:329-337, 1988.

Gibbs, D.M. Oxytocin inhibits ACTH and peripheral catecholamine secretion in the urethane-anesthetized rat. *Regul Pept 14*:125-132, 1986.

Giraud, P., H.-U. Affolter, A. Hotchkiss, E. Castanas, C. Oliver, and L.E. Eiden. [Regulation of the biosynthesis of enkephalins in the chromaffin cells.] *Ann Endocrinol (Paris)* 47:40-42, 1986.

Giuliani, S., C.A. Maggi, and A. Meli. Differences in cardiovascular responses to peripherally administered GABA as influenced by basal conditions and type of anaesthesia. *Br J Pharmacol 88*:659-670, 1986.

Glavinovic, M.I. and J.-M. Trifaró. Quinine blockade of currents through Ca^{2+}-activated K^+ channels in bovine chromaffin cells. *J Physiol 399*:139-152, 1988.

Glazer, G.M. MR imaging of the liver, kidneys, and adrenal glands. *Radiology 166*:303-312, 1988.

Glazer, G.M., I.R. Francis, and L.E. Quint. Imaging of the adrenal glands. *Invest Radiol 23*:3-11, 1988.

Glazer, G.M., E.J. Woolsey, J. Borrello, I.R. Francis, A.M. Aisen, F. Bookstein, M.A. Amendola, M.D. Gross, R.L. Bree, and W. Martel. Adrenal tissue characterization using MR imaging. *Radiology 158*:73-79, 1986.

Glazer, H.S. and J.K.T. Lee. Adrenal gland-MR imaging [letter]. *Radiology 166*:285, 1988.

Glazer, W.M., H. Morgenstern, D.V. Jeste, G. Zahner, H.M. Hafez, and C.L. Benarroche. Serum dopamine β-hydroxylase activity and tardive dyskinesia. *Psychoneuroendocrinology 12*:289-294, 1987.

Goedert, M. and S.P. Hunt. The cellular localization of preprotachykinin A messenger RNA in the bovine nervous system. *Neuroscience 22*:983-992, 1987.

Gold, P.E. Glucose modulation of memory storage processing. *Behav Neural Biol 45*:342-349, 1986.

Gold, P.E. Sweet memories. *Am Sci 75*:151-155, 1987.

Golding, D.W. and D.V. Pow. 'Neurosecretion' by a classic cholinergic innervation apparatus: A comparative study of adrenal chromaffin glands in four vertebrate species (teleosts, anurans, mammals). *Cell Tissue Res 249*:421-425, 1987.

Goldstein, M. and L.A. Greene. Activation of tyrosine hydroxylase by phosphorylation. pp. 75-80. In: **Psychopharmacology: The Third Generation of Progress.** (H.Y. Meltzer ed). Raven Press, New York, 1987.

Gomperts, B.D. Calcium shares the limelight in stimulus-secretion coupling. *Trends Biochem Sci 11*:290-292, 1986.

Gonzalez, C.B., E. Caorsi, and O.T. Berrios. Spectrophotometric determination of cytochrome b_{561} in membranes of neurosecretory granules. *IRCS Med Sci-Biochem 14*:607-608, 1986.

González, M.C., A.R. Artalejo, C. Montiel, P.P. Hervás, and A.G. García. Characterization of a dopaminergic receptor that modulates adrenomedullary catecholamine release. *J Neurochem 47*:382-388, 1986.

González, M.P., M.J. Oset-Gasque, A. Gimenez Solves, and S. Cañadas. Succinic semialdehyde dehydrogenase activity in bovine adrenal medulla and blood platelets: A comparative study with the brain enzyme. *Comp Biochem Physiol [B] 86*:489-492, 1987.

Gonzalez-Brito, A., R.J. Reiter, M.G. Tannenbaum, and D.J. Jones. Adrenalectomy but not gonadectomy affects rat pineal β-adrenergic receptor density. *Neurosci Lett 92*:330-334, 1988.

Goodhart, P.J., W.E. DeWolf Jr., and L.I. Kruse. Mechanism-based inactivation of dopamine β–hydroxylase by *p*-cresol and related alkylphenols. *Biochemistry 26*:2576-2583, 1987.

Gordon, D., D. Merrick, V. Auld, R. Dunn, A.L. Goldin, N. Davidson, and W.A. Catterall. Tissue-specific expression of the RI and RII sodium channel subtypes. *Proc Natl Acad Sci USA 84*:8682-8686, 1987.

Gospodarowicz, D., A. Baird, J. Cheng, G.M. Lui, F. Esch, and P. Bohlen. Isolation of fibroblast growth factor from bovine adrenal gland: Physicochemical and biological characterization. *Endocrinology 118*:82-90, 1986.

Graham, A.D.M., L.D. Longo, and C.Y. Cheung. Catecholamine secretion from the adrenal medulla of the fetus, regulation by hormones. *J Dev Physiol 8*:227-235, 1986.

Graif, M., A.W.-L. Leung, R.E. Steiner, and I.R. Young. Magnetic resonance imaging of the retroperitoneum. *Clin Radiol 37*:441-449, 1986.

Grandori, C. and H. Hanafusa. p60[c-src] is complexed with a cellular protein in subcellular compartments involved in exocytosis. *J Cell Biol 107*:2125-2135, 1988.

Grant, N.J., D. Aunis, and M.-F. Bader. Morphology and secretory activity of digitonin- and α-toxin-permeabilized chromaffin cells. *Neuroscience 23*:1143-1155, 1987.

Grant, N.J., B. Demeneix, D. Aunis, and O.K. Langley. Induction of neurofilament phosphorylation in cultured chromaffin cells. *Neuroscience 27*:717-726, 1988.

Grassi Milano, E. and F. Accordi. Evolutionary trends in adrenal gland of anurans and urodeles. *J Morphol 189*:249-259, 1986.

Gray, T.S. and J.E. Morley. Neuropeptide Y: Anatomical distribution and possible function in mammalian nervous system. *Life Sci 38*:389-401, 1986.

Green, D.P.L. Granule swelling and membrane fusion in exocytosis. *J Cell Sci 88*:547-549, 1987.

Green, S.H., R.E. Rydel, J.L. Connolly, and L.A. Greene. PC12 cell mutants that possess low- but not high-affinity nerve growth factor receptors neither respond to nor internalize nerve growth factor. *J Cell Biol 102*:830-843, 1986.

Greenberg, D.A., C.L. Carpenter, and R.O. Messing. Depolarization-dependent binding of the calcium channel antagonist,(+)-[^{3}H]PN200-100, to intact cultured PC12 cells. *J Pharmacol Exp Ther 238*:1021-1027, 1986.

Greenberg, D.A., C.L. Carpenter, and R.O. Messing. Ethanol-induced component of ^{45}Ca^{2+} uptake in PC12 cells is sensitive to Ca^{+2} channel modulating drugs. *Brain Res 410*:143-146, 1987.

Greiner, C.A.M., A.T. Lloyd, and G. Guroff. Ontogeny of the nerve growth factor receptor in primary cultures of neural crest cells. *Dev Brain Res 26*:145-150, 1986.

Grima, B., A. Lamouroux, C. Boni, J.-F. Julien, F. Javoy-Agid, and J. Mallet. A single human gene encoding multiple tyrosine hydroxylases with different predicted functional characteristics. *Nature 326*:707-711, 1987.

Grimaldi, K.A., J.C. Hutton, and K. Siddle. Production and characterization of monoclonal antibodies to insulin secretory granule membranes. *Biochem J 245*:557-566, 1987.

Gripois, D. and M. Valens. Adrenal dopamine β–hydroxylase induction in the young rat: Influence of thyroid hormones. *Mol Cell Endocrinol 45*:77-80, 1986a.

Gripois, D. and M. Valens. Adrenal phenylethanolamine-*N*-methyltransferase activity in the young rat: Influence of thyroid hormones and insulin-induced hypoglycemia. *Biol Neonate 49*:219-223, 1986b.

Gripois, D., M. Valens, and A. Diarra. Adrenal medullary responses to insulin-induced hypoglycaemia in the young rat. Influence of thyroid hormones. *J Auton Nerv Syst 15*:165-178, 1986.

Gripois, D., M. Valens, P. Richter, and J. Genermont. Adrenal dopamine content of the developing rat. Influence of the thyroid status and insulin-hypoglycaemia. *J Auton Nerv Syst 21*:181-184, 1987.

Grof, E., G.M. Brown, P. Grof, and G.R. Van Loon. Effects of lithium administration on plasma catecholamines. *Psychiatry Res 19*:87-92, 1986.

Grønberg, M. and T. Flatmark. Studies on Mg^{2+}-dependent ATPase in bovine adrenal chromaffin granules. With special reference to the effect of inhibitors and energy coupling. *Eur J Biochem 164*:1-8, 1987.

Grønberg, M. and T. Flatmark. Inhibition of the H$^+$-ATPase in bovine adrenal chromaffin granule ghosts by diethylstilbestrol: Evidence for a tight coupling between ATP hydrolysis and proton translocation. *FEBS Lett 229*:40-44, 1988.

Gross, M.D., B. Shapiro, J.C. Sisson, and A. Zweifler. Clonidine-induced suppression of plasma catecholamines in states of adrenal medulla hyperfunction. *J Endocrinol Invest 10*:359-364, 1987.

Gross, R., G. Bertrand, G. Ribes, P. Petit, and M.M. Loubatieres-Mariani. Epinephrine potentiates adenosine-stimulating effect on glucagon secretion. *Am J Physiol* *252*:E426-E430, 1987.

Grube, D., D. Aunis, F. Bader, Y. Cetin, A. Jörns, and S. Yoshie. Chromogranin A (CGA) in the gastro-entero-pancreatic (GEP) endocrine system: I. CGA in the mammalian endocrine pancreas. *Histochemistry 85*:441-452, 1986.

Grynszpan-Winograd, O. and M. Jousselin-Hosaja. [Synapses between chromaffin cells in the mouse adrenal medulla transplanted into the brain.] *C R Acad Sci [III] 306*:17-23, 1988.

Guerrero, I., H. Wong, A. Pellicer, and D.E. Burstein. Activated N-*ras* gene induces neuronal differentiation of PC12 rat pheochromocytoma cells. *J Cell Physiol 129*:71-76, 1986.

Guidotti, A. and I. Hanbauer. Participation of GABA/benzo-diazepine receptor system in the adrenal chromaffin cell function. *Adv Biochem Psychopharmacol 42*:165-172, 1986.

Guillot, M.D., A. Guillot, F. Martínez-Soriano, and V. Smith-Agreda. Morphological changes of the adrenal gland following soft-laser irradiation of the pineal gland. *Anat Anz 167*:289-296, 1988.

Günther, R.W. [Sonography in adrenal gland diseases.] *Radiologe 26*:174-180, 1986.

Gupta, M., M.A. England, and M.F.D. Notter. Effects of antimitotic drugs on the morphological features of PC12 cells in culture--A light and EM study. *Brain Res Bull 18*:555-561, 1987.

Gusovsky, F., J.W. Daly, T. Yasumoto, and E. Rojas. Differential effects of maitotoxin on ATP secretion and on phospho-inositide breakdown in rat pheochromocytoma cells. *FEBS Lett 233*:139-142, 1988.

Gutierrez, L.M., J.J. Ballesta, M.J. Hidalgo, L. Gandia, A.G. García, and J.A. Reig. A two-dimensional electrophoresis study of phosphorylation and dephosphorylation of chromaffin cell proteins in response to a secretory stimulus. *J Neurochem 51*:1023-1030, 1988.

Haavik, J., K.K. Andersson, L. Petersson, and T. Flatmark. Soluble tyrosine hydroxylase (tyrosine 3-monooxygenase) from bovine adrenal medulla: Large-scale purification and physicochemical properties. *Biochim Biophys Acta 953*:142-156, 1988.

Haavik, J. and T. Flatmark. Isolation and characterization of tetrahydropterin oxidation products generated in the tyrosine 3-monooxygenase (tyrosine hydroxylase) reaction. *Eur J Biochem 168*:21-26, 1987.

Hacker, G.W., A.E. Bishop, G. Terenghi, I.M. Varndell, J. Aghahowa, K. Pollard, J. Thurner, and J.M. Polak. Multiple peptide production and presence of general neuroendocrine markers detected in 12 cases of human phaeochromocytoma and in mammalian adrenal glands. *Virchows Arch [A] 412*:399-411, 1988.

Hadjiconstantinou, M., S. Tjioe, H. Alho, C. Miller, and N.H. Neff. 1-methyl-4-phenyl-1,2,3,6-tetrahydropyridine (MPTP) accelerates the accumulation of lipofuscin in mouse adrenal gland. *Neurosci Lett 83*:1-6, 1987.

Hagag, N., S. Halegoua, and M. Viola. Inhibition of growth factor-induced differentiation of PC12 cells by microinjection of antibody to *ras* p21. *Nature 319*:680-682, 1986.

Häggendal, J., L. Jönsson, G. Johansson, S. Bjurström, and J. Carlsten. Disordered catecholamine release in pigs susceptible to malignant hyperthermia. *Pharmacol Toxicol 63*:257-261, 1988.

Hagn, C., R.L. Klein, R. Fischer-Colbrie, B.H. Douglas II, and H. Winkler. An immunological characterization of five common antigens of chromaffin granules and of large dense-cored vesicles of sympathetic nerve. *Neurosci Lett 67*:295-300, 1986.

Hagn, C., K.W. Schmid, R. Fischer-Colbrie, and H. Winkler. Chromogranin A, B, and C in human adrenal medulla and endocrine tissues. *Lab Invest 55*:405-411, 1986.

Håkanson, R., C. Wahlestedt, E. Ekblad, L. Edvinsson, and F. Sundler. Neuropeptide Y: Coexistence with noradrenaline. Functional implications. *Prog Brain Res 68*:279-287, 1986.

Halbrügge, T., T. Gerhardt, J. Ludwig, E. Heidbreder, and K.-H. Graefe. Assay of catecholamines and dihydroxyphenyl-ethylene-glycol in human plasma and its application in orthostasis and mental stress. *Life Sci 43*:19-26, 1988.

Hale, P.J., V. Suarez, A. Williams, R.M. Baddeley, and M. Nattrass. Insulinoma and ganglioneuroma. *Br J Surg 74*:1183, 1987.

Hall, F.L., P. Fernyhough, D.N. Ishii, and P.R. Vulliet. Suppression of nerve growth factor-directed neurite outgrowth in PC12 cells by sphingosine, an inhibitor of protein kinase C. *J Biol Chem 263*:4460-4466, 1988.

Hama, T., K.-P. Huang, and G. Guroff. Protein kinase C as a component of a nerve growth factor-sensitive phosphorylation system in PC12 cells. *Proc Natl Acad Sci USA 83*:2353-2357, 1986.

Hamaji, M., H. Nakaba, M. Nakamura, M. Izukura, M. Miyata, and Y. Kawashima. Impaired adrenal vascular response to hemorrhagic hypotension by an adreno-femoral venous shunt. *J Surg Res 43*:234-238, 1987.

Hamaji, M., M. Nakamura, M. Izukura, H. Nakaba, T. Hashimoto, Y. Tanaka, T. Tumori, M. Miyata, Y. Kawashima, and T.S. Harrison. Autoregulation and regional blood flow of the dog during hemorrhagic shock. *Circ Shock 19*:245-255, 1986.

Hamann, M., M. Desarmenien, E. Desaulles, M.-F. Bader, and P.

Feltz. Quantitative evaluation of the properties of a pyridazinyl GABA derivative (SR 95531) as a GABA$_A$ competitive antagonist: An electrophysiological approach. *Brain Res 442*:287-296, 1988.

Hamilton, J.W., L.L.H. Chu, J.B. Rouse, K. Reddig, and R.R. MacGregor. Structural characterization of adrenal chromogranin A and parathyroid secretory protein-I as homologs. *Arch Biochem Biophys 244*:16-26, 1986.

Hamman, H.C., L.C. Gaffey, K.R. Lynch, and C.E. Creutz. Cloning and characterization of a cDNA encoding bovine endonexin (Chromobindin 4). *Biochem Biophys Res Commun 156*:660-667, 1988.

Hamos, J., P.R. Desai, and J.J. Villafranca. Characterization and kinetic studies of deglycosylated dopamine β-hydroxylase. *FASEB J 1*:143-148, 1987.

Han, V.K.M., J. Snouweart, A.C. Towle, P.K. Lund, and J.M. Lauder. Cellular localization of tyrosine hydroxylase mRNA and its regulation in the rat adrenal medulla and brain by in situ hybridization with an oligodeoxyribonucleotide probe. *J Neurosci Res 17*:11-18, 1987.

Hano, T., M. Kuchii, M. Umemoto, I. Nishio, and Y. Masuyama. Epinephrine release from the perfused adrenal medulla of spontaneously hypertensive rats. *J Hypertens 4*:S209-S211, 1986.

Hansen, J.T., G. Bing, M.F.D. Notter, and D.M. Gash. Paraneuronal grafts in unilateral 6-hydroxydopamine-lesioned rats: Morphological aspects of adrenal chromaffin and carotid body glomus cell implants, Chapter 65, pp. 1-5. In: **Progress in Brain Research**, Vol. 78. (D.M. Gash and J.R. Sladek Jr., eds). Elsevier Science Publishers, Amsterdam, 1988.

Hansen, J.T., J.H. Kordower, M.S. Fiandaca, S.-S. Jiao, M.F.D. Notter, and D.M. Gash. Adrenal medullary autografts into the basal ganglia of Cebus monkeys: Graft viability and fine structure. *Exp Neurol 102*:65-75, 1988.

Hansen, J.T., M.F.D. Notter, S.-H. Okawara, and D.M. Gash. Organization, fine structure, and viability of the human adrenal medulla: Considerations for neural transplantation. *Ann Neurol 24*:599-609, 1988.

Hansson, H.-A., A. Nilsson, J. Isgaard, H. Billig, O. Isaksson, A. Skottner, I.K. Andersson, and B. Rozell. Immuno-histochemical localization of insulin like growth factor I in the adult rat. *Histochemistry 89*:403-410, 1988.

Harish, O.E., L.-S. Kao, R. Raffaniello, A.R. Wakade, and A.S. Schneider. Calcium dependence of muscarinic receptor-mediated catecholamine secretion from the perfused rat adrenal medulla. *J Neurochem 48*:1730-1735, 1987.

Harper, J.C., C. Pagonis, and J.M. Littleton. Altered characteristics of catecholamine release from rat cortical slices and bovine adrenal chromaffin cells in culture after chronic exposure to ethanol and the effect of the dihydropyridine drugs on these systems. *Alcohol Alcohol Suppl 1*:725-729, 1987.

Harrington, C.A., E.J. Lewis, D. Krzemien, and D.M. Chikaraishi. Identification and cell type specificity of the tyrosine hydroxylase gene promoter. *Nucleic Acids Res 15*:2363-2384, 1987.

Harris, B., T.R. Cheek, and R.D. Burgoyne. Effects of metallo-endoproteinase inhibitors on secretion and intracellular free calcium in bovine adrenal chromaffin cells. *Biochim Biophys Acta 889*:1-5, 1986.

Harris, K.M., S. Kongsamut, and R.J. Miller. Protein kinase C mediated regulation of calcium channels in PC-12 pheochromocytoma cells. *Biochem Biophys Res Commun 134*:1298-1305, 1986.

Haselbacher, G.K., J.-C. Irminger, J. Zapf, W.H. Ziegler, and R.E. Humbel. Insulin-like growth factor II in human adrenal pheochromocytomas and Wilms tumors: Expression at the mRNA and protein level. *Proc Natl Acad Sci USA 84*:1104-1106, 1987.

Hashimoto, S. K-252a, a potent protein kinase inhibitor, blocks nerve growth factor-induced neurite outgrowth and changes in the phosphorylation of proteins in PC12h cells. *J Cell Biol 107*:1531-1539, 1988.

Hashimoto, S., C. Iwasaki, H. Kuzuya, and G. Guroff. Regulation of nerve growth factor action on Nsp100 phosphorylation in PC12h cells by calcium. *J Neurochem 46*:1599-1604, 1986.

Hawthorn, J., S.S. Nussey, J.R. Henderson, and J.S. Jenkins. Immunohistochemical localization of oxytocin and vasopressin in the adrenal glands of rat, cow, hamster and guinea pig. *Cell Tissue Res 250*:1-6, 1987.

Hawthorne, J.N. Does receptor-linked phosphoinositide metabolism provide messengers mobilizing calcium in nervous tissue? *Int Rev Neurobiol 28*:241-273, 1986.

Hayashi, Y., S. Miwa, K. Lee, K. Koshimura, A. Kamei, K. Hamahata, and M. Fujiwara. A nonisotopic method for determination of the *in vivo* activities of tyrosine hydroxylase in the rat adrenal gland. *Anal Biochem 168*:176-183, 1988.

Haycock, J.W., M.D. Browning, and P. Greengard. Cholinergic regulation of protein phosphorylation in bovine adrenal chromaffin cells. *Proc Natl Acad Sci USA 85*:1677-1681, 1988.

Haycock, J.W., P. Greengard, and M.D. Browning. Cholinergic regulation of protein III phosphorylation in bovine adrenal chromaffin cells. *J Neurosci 8*:3233-3239, 1988.

Hearn, S.A. Electron microscopic localization of chromogranin A in osmium-fixed neuroendocrine cells with a protein A-gold technique. *J Histochem Cytochem 35*:795-801, 1987.

Hedner, T. and J. Cassuto. Opioids and opioid receptors in peripheral tissues. *Scand J Gastroenterol [Suppl] 130*:27-46, 1987.

Heinemann, S.H., F. Conti, W. Stühmer, and E. Neher. Effects of hydrostatic pressure on membrane processes: Sodium channels, calcium channels, and exocytosis. *J Gen Physiol 90*:765-778, 1987.

Heisler, S. and E. Morrier. Bovine adrenal medullary cells contain functional atrial natriuretic peptide receptors. *Biochem Biophys Res Commun 150*:781-787, 1988.

Helman, L.J., T.G. Ahn, M.A. Levine, A. Allison, P.S. Cohen, M.J. Cooper, D.V. Cohn, and M.A. Israel. Molecular cloning and primary structure of human chromogranin A (secretory protein I) cDNA. *J Biol Chem 263*:11559-11563, 1988.

Helman, L.J., C.J. Thiele, W.M. Linehan, B.D. Nelkin, S.B. Baylin, and M.A. Israel. Molecular markers of neuroendocrine development and evidence of environmental regulation. *Proc Natl Acad Sci USA 84*:2336-2339, 1987.

Hendley, E.D., M.A. Cierpial, and R. McCarty. Sympathetic-adrenal medullary response to stress in hyperactive and hypertensive rats. *Physiol Behav 44*:47-51, 1988.

Henry, D.P. and R.R. Bowsher. An improved radioenzymatic assay for plasma norepinephrine using purified phenylethanolamine *N*-methyltransferase. *Life Sci 38*:1473-1483, 1986.

Henry, J.-P., D. Scherman, M.-P. Roisin, B. Gasnier, and M.-F. Isambert. [Molecular pharmacology of the catecholamine transporter of chromaffin granules from bovine adrenal medulla.] *Biochimie 68*:451-458, 1986.

Herbert, E., M. Comb, A. Seasholtz, M. Martin, and D. Liston. Use of gene transfer approaches to study regulation of expression of opioid peptide genes. *Res Publ Assoc Res Nerv Ment Dis 64*:33-46, 1986.

Herman, H.H., K. Wimalasena, L.C. Fowler, C.A. Beard, and S.W. May. Demonstration of the ascorbate dependence of membrane-bound dopamine β-monooxygenase in adrenal chromaffin granule ghosts. *J Biol Chem 263*:666-672, 1988.

Hexum, T.D., E.A. Majane, L.R. Russett, and H.-Y.T. Yang. Neuropeptide Y release from the adrenal medulla after cholinergic receptor stimulation. *J Pharmacol Exp Ther 243*:927-930, 1987.

Hexum, T.D. and L.R. Russett. Plasma enkephalin-like peptide response to chronic nicotine infusion in guinea pig. *Brain Res 406*:370-372, 1987.

Heym, C., (ed). **Histochemistry and Cell Biology of Autonomic Neurons and Paraganglia.** *Exp Brain Res, Series 16*:1-360, 1987.

Higgins, L.S. and D.K. Berg. Immunological identification of a nicotinic acetylcholine receptor on bovine chromaffin cells. *J Neurosci 7*:1792-1798, 1987.

Higgins, L.S. and D.K. Berg. A desensitized form of neuronal acetylcholine receptor detected by ^{3}H-nicotine binding on bovine adrenal chromaffin cells. *J Neurosci 8*:1436-1446,

1988a.

Higgins, L.S. and D.K. Berg. Metabolic stability and antigenic modulation of nicotinic acetylcholine receptors on bovine adrenal chromaffin cells. *J Cell Biol 107*:1147-1156, 1988b.

Higgins, L.S. and D.K. Berg. Cyclic AMP-dependent mechanism regulates acetylcholine receptor function on bovine adrenal chromaffin cells and discriminates between new and old receptors. *J Cell Biol 107*:1157-1165, 1988c.

Higuchi, H., E. Costa, and H.-Y.T. Yang. Neuropeptide Y inhibits the nicotine-mediated release of catecholamines from bovine adrenal chromaffin cells. *J Pharmacol Exp Ther 244*:468-474, 1988.

Higuchi, H., H.-Y.T. Yang, and E. Costa. Age-related bidirectional changes in neuropeptide Y peptides in rat adrenal glands, brain, and blood. *J Neurochem 50*:1879-1886, 1988.

Higuchi, K., H. Nawata, K.-I. Kato, H. Ibayashi, and H. Matsuo. α-Human atrial natriuretic polypeptide (α-hANP) specific binding sites in bovine adrenal gland. *Biochem Biophys Res Commun 137*:657-663, 1986.

Hilse, H. and P. Oehme. The effect of bilateral adrenal demedullation on blood pressure and heart rate of spontaneously hypertensive (SHR) and normotensive Wistar Kyoto (WKY) rats. *Pharmazie 42*:353, 1987.

Himeno, A., A.J. Nazarali, and J.M. Saavedra. Quantitative *in vitro* autoradiographic characterization of [^{125}I] angiotensin III binding sites in rat adrenal gland. *Regul Pept 23*:127-133, 1988.

Himmelhoch, S. and G. Rossi. Ultracryotomy for the study of unfixed membranes. *Histochemistry 84*:191-195, 1986.

Hinson, J.P., G.P. Vinson, and B.J. Whitehouse. The relationship between perfusion medium flow rate and steroid secretion in the isolated perfused rat adrenal gland *in situ*. *J Endocrinol 111*:391-396, 1986.

Hinson, J.P., G.P. Vinson, B.J. Whitehouse, and G.M. Price. Effects of stimulation on steroid output and perfusion medium flow rate in the isolated perfused rat adrenal gland in situ. *J Endocrinol 109*:279-285, 1986.

Hirano, T. Neural regulation of adrenal chromaffin cell function in the mouse: Fate and distribution of [^{3}H]dopamine in denervated adrenal medulla. *J Auton Nerv Syst 15*:285-295, 1986.

Hirano, T. Distribution of [^{3}H]deoxyglucose and [^{3}H]dopamine in the adrenal medulla and nerve endings of the mouse. *J Auton Nerv Syst 25*:79-82, 1988.

Hirano, T., Y. Kidokoro, and H. Ohmori. Acetylcholine dose-response relation and the effect of cesium ions in the rat adrenal chromaffin cell under voltage clamp. *Pflugers Arch 408*:401-407, 1987.

Hoelzer, D.R., G.P. Dalsky, W.E. Clutter, S.D. Shah, J.O.

Holloszy, and P.E. Cryer. Glucoregulation during exercise: Hypoglycemia is prevented by redundant glucoregulatory systems, sympathochromaffin activation, and changes in islet hormone secretion. *J Clin Invest* 77:212-221, 1986.

Hoffer, B.J., A.-C. Granholm, J.O. Stevens, and L. Olson. Catecholamine-containing grafts in parkinsonism: Past and present. *Clin Res 36*:189-195, 1988.

Hoffman, B.B. Observations in New England Deaconess Hospital rats harboring pheochromocytoma. *Clin Invest Med 10*:555-560, 1987.

Hofmann, H. -D., C. Ebener, and K. Unsicker. Age-dependent differences in ^{125}I-nerve growth factor binding properties of rat adrenal chromaffin cells. *J Neurosci Res 18*:574-577, 1987.

Hökfelt, T., O. Johansson, V. Holets, B. Meister, and T. Melander. Distribution of neuropeptides with special reference to their coexistence with classical transmitters. pp.401-416. In: **Psychopharmacology: The Third Generation of Progress**, (H.Y. Meltzer, ed). Raven Press, New York, 1987.

Hollingsworth, E.B., D. Ukena, and J.W. Daly. The protein kinase C activator phorbol-12-myristate-13-acetate enhances cyclic AMP accumulation in pheochromocytoma cells. *FEBS Lett 196*:131-134, 1986.

Höllt, V. Opioid peptide processing and receptor selectivity. *Annu Rev Pharmacol Toxicol 26*:59-77, 1986.

Holz, R.W. The role of osmotic forces in exocytosis from adrenal chromaffin cells. *Annu Rev Physiol 48*:175-189, 1986.

Holz, R.W. Control of exocytosis from adrenal chromaffin cells. *Cell Mol Neurobiol 8*:259-268, 1988.

Holz, R.W. and R.A. Senter. Effects of osmolality and ionic strength on secretion from adrenal chromaffin cells permeabilized with digitonin. *J Neurochem 46*:1835-1842, 1986.

Holz, R.W. and R.A. Senter. Effects of trypsin on secretion stimulated by micromolar Ca^{2+} and phorbol ester in digitonin-permeabilized adrenal chromaffin cells. *Cell Mol Neurobiol 8*:115-128, 1988.

Hook, V.Y.H. Regulation of carboxypeptidase H by inhibitory and stimulatory mechanisms during neuropeptide precursor processing. *Cell Mol Neurobiol 8*:49-55, 1988.

Hook, V.Y.H. and E.F. LaGamma. Product inhibition of carboxypeptidase H. *J Biol Chem 262*:12583-12588, 1987.

Hook, V.Y.H. and D. Liston. Distribution of enkephalin-containing peptides with bovine chromaffin granules. *Neuropeptides 9*:263-267, 1987.

Horwitz, J. and R.L. Perlman. Measurement of inositol phospholipid metabolism in PC12 pheochromocytoma cells. *Methods Enzymol 141*:169-175, 1987.

Hosaka, Y., L.M. Rainwater, C.S. Grant, G.M. Farrow, J.A. van Heerden, and M.M. Lieber. Pheochromocytoma: Nuclear deoxyribonucleic acid patterns studied by flow cytometry.

Surgery 100:1003-1008, 1986.

Hosang, M. and E.M. Shooter. The internalization of nerve growth factor by high-affinity receptors on pheochromocytoma PC12 cells. *EMBO J 6*:1197-1202, 1987.

Hoshi, T., S.S. Garber, and R.W. Aldrich. Effect of forskolin on voltage-gated K^+ channels is independent of adenylate cyclase activation. *Science 240*:1652-1655, 1988.

Hoshi, T. and S.J. Smith. Large depolarization induces long openings of voltage-dependent calcium channels in adrenal chromaffin cells. *J Neurosci 7*:571-580, 1987.

Houchi, H., J.M. Masserano, and N. Weiner. Bradykinin activates tyrosine hydroxylase in rat pheochromocytoma PC-12 cells. *FEBS Lett 235*:137-140, 1988.

Houchi, H., K. Morita, K. Minakuchi, and M. Oka. Inhibition by hydralazine of catecholamine biosynthesis in cultured bovine adrenal medullary cells. *Biochem Pharmacol 35*:2047-2049, 1986.

Houchi, H., M. Oka, M. Misbahuddin, K. Morita, and A. Nakanishi. Stimulation by vasoactive intestinal polypeptide of catecholamine synthesis in isolated bovine adrenal chromaffin cells. *Biochem Pharmacol 36*:1551-1554, 1987.

Hull, C.J. Phaeochromocytoma: Diagnosis, preoperative preparation and anaesthetic management. *Br J Anaesth 58*:1453-1468, 1986.

Husebye, E.S. and T. Flatmark. Characterization of phospholipase activities in chromaffin granule ghosts isolated from the bovine adrenal medulla. *Biochim Biophys Acta 920*:120-130, 1987.

Husebye, E.S. and T. Flatmark. Phosphatidylinositol kinase of bovine adrenal chromaffin granules: Kinetic properties and inhibition by low concentrations of Ca^{2+}. *Biochim Biophys Acta 968*:261-265, 1988a.

Husebye, E.S. and T. Flatmark. Phosphatidylinositol kinase of bovine adrenal chromaffin granules. Modulation by hydrophilic and amphiphilic cations. *Biochem Pharmacol 37*:4149-4156, 1988b.

Huttner, W.B. and U.M. Benedum. Chromogranin A and pancreastatin [letter]. *Nature 325*:305, 1987.

Hutton, J.C. Calcium-binding proteins and secretion. *Cell Calcium 7*:339-352, 1986.

Hutton, J.C., H.W. Davidson, and M. Peshavaria. The mechanism of chromogranin A processing. *Nature 325*:766, 1987.

Hutton, J.C., E. Nielsen, and W. Kastern. The molecular cloning of the chromogranin A-like precursor of β-granin and pancreastatin from the endocrine pancreas. *FEBS Lett 236*:269-274, 1988.

Hutton, J.C., M. Peshavaria, H.W. Davidson, K. Grimaldi, R.P. VonStrandmann, and K. Siddle. The insulin secretory granule: Features and functions in common with other endocrine granules. *Adv Exp Med Biol 211*:385-396, 1987.

Hwang, B.H., J.-Y. Wu, and W.B. Severs. Effects of chronic dehydration on angiotensin II receptor binding in the subfornical organ, paraventricular hypothalamic nucleus and adrenal medulla of Long-Evans rats. *Neurosci Lett 65*:35-40, 1986.

Hyman, S.A., W.D. Rogers, D.W. Smith, R.J. Maciunas, G.S. Allen, and M.L. Berman. Perioperative management for transplant of autologous adrenal medulla to the brain for parkinsonism. *Anesthesiology 69*:618-622, 1988.

Iacangelo, A.L., H.-U. Affolter, L.E. Eiden, E. Herbert, and M. Grimes. Bovine chromogranin A sequence and distribution of its messenger RNA in endocrine tissues. *Nature 323*:82-86, 1986.

Iacangelo, A.L., R. Fischer-Colbrie, K.J. Koller, M.J. Brownstein, and L.E. Eiden. The sequence of porcine chromogranin A messenger RNA demonstrates chromogranin A can serve as the precursor for the biologically active hormone, pancreastatin. *Endocrinology 122*:2339-2341, 1988.

Iacangelo, A.L., H. Okayama, and L.E. Eiden. Primary structure of rat chromogranin A and distribution of its mRNA. *FEBS Lett 227*:115-121, 1988.

Iacono, R.P. and R. Sandyk. Adrenal medullary tissue transplantation in Parkinson's disease. *J Neurosurg 68*:158-159, 1988.

Iadecola, C., P.M. Lacombe, M.D. Underwood, T. Ishitsuka, and D.J. Reis. Role of adrenal catecholamines in cerebrovasodilation evoked from brain stem. *Am J Physiol 252*:H1183-H1191, 1987.

Iadecola, C., P.M. Lacombe, M.D. Underwood, A.F. Sved, and D.J. Reis. Adrenal catecholamines participate in the cerebrovascular vasodilation elicited by electrical stimulation of the dorsal medullary reticular formation in rat. *Acta Physiol Scand [Suppl] 552*:74-77, 1986.

Ignatius, M.J., E.M. Shooter, R.E. Pitas, and R.W. Mahley. Lipoprotein uptake by neuronal growth cones *in vitro*. *Science 236*:959-962, 1987.

Iguchi, A., M. Gotoh, H. Matsunaga, A. Yatomi, A. Honmura, M. Yanase, and N. Sakamoto. Relative contributions of the nervous system and hormones to CNS-mediated hyperglycemia. *Am J Physiol 255*:E920-E927, 1988.

Iguchi, H., S. Natori, K.-I. Kato, H. Nawata, and M. Chrétien. Different processing of chromogranin B into GAWK-immunoreactive fragments in the bovine adrenal medulla and pituitary gland. *Life Sci 43*:1945-1952, 1988.

Iguchi, H., S. Natori, H. Nawata, K.-I. Kato, H. Ibayashi, J.S.D. Chan, N.G. Seidah, and M. Chrétien. Presence of the novel pituitary protein "7B2" in bovine chromaffin granules: Possible co-release of 7B2 and catecholamine as induced by nicotine. *J Neurochem 49*:1810-1814, 1987.

Ikeda, K., E.C. Weir, M. Mangin, P.S. Dannies, B. Kinder, and L.J. Deftos. Expression of messenger ribonucleic acids encoding a parathyroid hormone-like peptide in normal human and animal tissues with abnormal expression in human parathyroid adenomas. *Mol Endocrinol 2*:1230-1236, 1988.

Improta, T., A.M. Salvatore, A. Di Luzio, G. Romeo, E.M. Coccia, and P. Calissano. IFN-γ facilitates NGF-induced neuronal differentiation in PC12 cells. *Exp Cell Res 179*:1-9, 1988.

Inagaki, S. and S. Kito. Peptides in the peripheral nervous system. *Prog Brain Res 66*:269-316, 1986.

Inagaki, S., Y. Kubota, S. Kito, K. Kangawa, and H. Matsuo. Immunoreactive atrial natriuretic polypeptides in the adrenal medulla and sympathetic ganglia. *Regul Pept 15*:249-260, 1986.

Inglis, G.C., C.J. Kenyon, J.A.M. Hannah, J.M.C. Connell, and S.G. Ball. Does dopamine regulate aldosterone secretion in the rat? *Clin Sci 73*:93-97, 1987.

Inoue, K. and J.G. Kenimer. Muscarinic stimulation of calcium influx and norepinephrine release in PC12 cells. *J Biol Chem 263*:8157-8161, 1988.

Inoue, M., T. Kimura, K. Matsui, K. Ota, M. Shoji, K. Iitake, and K. Yoshinaga. Responses of vasopressin and enkephalins to hemorrhage in adrenalectomized dogs. *Am J Physiol 253*: R467-R474, 1987.

Inturrisi, C.E., A.D. Branch, H.D. Robertson, R.D. Howells, S.O. Franklin, J.R. Shapiro, S.E. Calvano, and B.C. Yoburn. Glucocorticoid regulation of enkephalins in cultured rat adrenal medulla. *Mol Endocrinol 2*:633-640, 1988.

Inturrisi, C.E., S.O. Franklin, J.R. Shapiro, S.E. Calvano, and B.C. Yoburn. Adrenal enkephalin biosynthsis regulated by glucocorticoid. *Natl Inst Drug Abuse Res Monogr Ser 81*:129-135, 1988.

Inturrisi, C.E., E.F. LaGamma, S.O. Franklin, T. Huang, T.J. Nip, and B.C. Yoburn. Characterization of enkephalins in rat adrenal medullary explants. *Brain Res 448*:230-236, 1988.

Ismael, Z., T.J. Millar, D.H. Small, and I.W. Chubb. Acetylcholinesterase generates enkephalin-like immunoreactivity when it degrades the soluble proteins (chromogranins) from adrenal chromaffin granules. *Brain Res 376*:230-238, 1986.

Israel, A., Y. Barbella, and J.M. Saavedra. Compensatory increase in adrenomedullary angiotensin-converting enzyme activity (kininase II) after unilateral adrenalectomy. *Regul Pept 16*:97-105, 1986.

Ito, K., A. Sato, Y. Sato, and H. Suzuki. Increases in adrenal catecholamine secretion and adrenal sympathetic nerve unitary activities with aging in rats. *Neurosci Lett 69*:263-268, 1986.

Ito, Y., T. Fujita, and K. Yamashita. [Increased sympathoadrenomedullary activity in young patients with borderline hypertension: Hyperresponses to isometric stress and glucagon injection.] *Nippon Naibunpi Gakkai Zasshi 63*:1267-1277, 1987.

Izumi, F., Y. Toyohira, N. Yanagihara, A. Wada, and H. Kobayashi. Barium-evoked release of catecholamines from digitonin-permeabilized adrenal medullary cells. *Neurosci Lett 69*:172-175, 1986.

Izumi, F., A. Wada, N. Yanagihara, H. Kobayashi, and Y. Toyohira. Monensin-induced influx of ^{22}Na and the release of catecholamines in cultured bovine adrenal medulla cells and isolated chromaffin granules. *Biochem Pharmacol 35*:2937-2940, 1986.

Izumi, F., N. Yanagihara, A. Wada, Y. Toyohira, and H. Kobayashi. Lysis of chromaffin granules by phospholipase A_2-treated plasma membranes: A cell-free model for exocytosis in adrenal medulla. *FEBS Lett 196*:349-352, 1986.

Izumi, M. [Changes in the potentiation effect of [D-Ala^2, Met^5]-enkephalin on pressor responses to l-adrenaline in adrenal-enucleated rats.] *Folia Pharmacol Japon 91*:301-308, 1988.

Jacobs, J.R. and J.K. Stevens. Experimental modification of PC12 neurite shape with the microtubule-depolymerizing drug Nocodazole: A serial electron microscopic study of neurite shape control. *J Cell Biol 103*:907-915, 1986a.

Jacobs, J.R. and J.K. Stevens. Changes in the organization of the neuritic cytoskeleton during nerve growth factor-activated differentiation of PC12 cells: A serial electron microscopic study of the development and control of neurite shape. *J Cell Biol 103*:895-906, 1986b.

Jacobs, S.C., J.W. Mason, T.R. Kosten, V. Wahby, S.V. Kasl, and A.M. Ostfeld. Bereavement and catecholamines. *J Psychosom Res 30*:489-496, 1986.

Jaeger, C.B. Morphological and immunocytochemical characteristics of PC12 cell grafts in rat brain. *Ann NY Acad Sci 495*:334-349, 1987.

Jaffee, R.B., J.J. Mulchahey, A.M. Di Blasio, and M.C. Martin. Peptide regulation of pituitary and target tissue function and growth in the primate fetus. *Recent Prog Horm Res 44*:431-549, 1988.

Jankovic, J. Parkinson's disease: Recent advances in therapy. *South Med J 81*:1021-1027, 1988.

Janocko, L. and M.J. Mycek. Prolonged hypothermia after phenobarbital-induced anesthesia in adrenalectomized rats. *Pharmacol Biochem Behav 24*:1651-1658, 1986.

Jansson, S., L.-E. Tisell, M. Fjälling, S. Lindberg, L. Jacobsson, and B.F. Zachrisson. Early diagnosis of and surgical strategy for adrenal medullary disease in MEN II gene carriers. *Surgery 103*:11-18, 1988.

Jansson, S., L.-E. Tisell, and G. Hansson. Morphology of the adrenal medulla indicating multiple neuroectodermal abnormalities in pheochromocytoma patients. *Acta Med Austriaca 15*:99-100, 1988.

Jaques, S. Jr., and M.C. Tobes. Morphologic and biochemical variability of tissue and cultured cells from human pheochromocytoma. *J Cell Physiol 128*:261-270, 1986.

Jaques, S. Jr., M.C. Tobes, and J.C. Sisson. Sodium dependency of uptake of norepinephrine and *m*-iodobenzylguanidine into cultured human pheochromocytoma cells: Evidence for uptake-one. *Cancer Res 47*:3920-3928, 1987.

Jefferys, D. and J.W. Funder. Glucocorticoids, adrenal medullary opioids, and the retention of a behavioral response after stress. *Endocrinology 121*:1006-1009, 1987.

Jiao, S., Y. Wang, Q. Cai, and Y. Liu. Intracephalic transplantation of adrenal medulla improve Parkinsonism. I. Basic experimental study. *J Capital Inst Med 8*:1-6, 1987.

Joh, T.H. and O. Hwang. Dopamine β-hydroxylase: Biochemistry and molecular biology. *Ann NY Acad Sci 493*:342-350, 1987.

Joh, T.H., O. Hwang, and C. Abate. Phenylalanine hydroxylase, tyrosine hydroxylase, and tryptophan hydroxylase. pp. 1-32. In: **Neuromethods Series I: Neurotransmitter enzymes**, Vol. 5. (A.A. Boulton, G.B. Baker, and P.H. Yu, eds). Humana Press, Clifton, NJ, 1986.

Johnson, J.D., W.G. Conroy, and G.E. Isom. Alteration of cytosolic calcium levels in PC12 cells by potassium cyanide. *Toxicol Appl Pharmacol 88*:217-224, 1987.

Johnson, R.G. Jr. Accumulation of biological amines into chromaffin granules: A model for hormone and neurotransmitter transport. *Physiol Rev 68*:232-307, 1988.

Johnson, R.G. Jr. (ed). Cellular and molecular biology of hormone- and neurotransmitter-containing secretory vesicles. *Ann NY Acad Sci 493*:1-590, 1987.

Johnson, T.S., P.B. Rock, J.B. Young, C.S. Fulco, and L.A. Trad. Hemodynamic and sympathoadrenal responses to altitude in humans: Effect of dexamethasone. *Aviat Space Environ Med 59*:208-212, 1988.

Jones, C.T., M.M. Roebuck, D.W. Walker, and B.M. Johnston. The role of the adrenal medulla and peripheral sympathetic nerves in the physiological responses of the fetal sheep to hypoxia. *J Dev Physiol 10*:17-36, 1988.

Jones, C.T., M.M. Roebuck, D.W. Walker, H. Lagercrantz, and B.M. Johnston. Cardiovascular, metabolic and endocrine effects of chemical sympathectomy and of adrenal demedullation in fetal sheep. *J Dev Physiol 9*:347-367, 1987.

Jones, S.B., M.V. Westfall, and M.M. Sayeed. Plasma catecholamines during *E. coli* bacteremia in conscious rats. *Am J Physiol 254*:R470-R477, 1988.

Joshi, B.N., M.E. Troiani, J. Milin, F. Nürnburger, and R.J. Reiter. Adrenal-mediated depression of *N*-acetyltransferase activity and melatonin levels in the rat pineal gland. *Life Sci 38*:1573-1580, 1986.

Jousselin-Hosaja, M. Effects of transplantation on mouse adrenal chromaffin cells. *J Endocrinol 116*:149-153, 1988a.

Jousselin-Hosaja, M. Ultrastructural evidence for the development of adrenal medullary grafts in the brain. *Exp Brain Res 73*:637-647, 1988b.

Joynt, R.J. and D.M. Gash. Neural transplants: Are we ready? [editorial]. *Ann Neurol 22*:455-456, 1987.

Jungmann, E. [Endogenous dopaminergic regulation of aldosterone secretion. Studies of pathophysiologic value.] *Fortschr Med 105*:337-339, 1987.

Jungmann, E., G. Germann, I. Austin, B. Mack, D. Storp-Wenke, W. Fassbinder, U. Schwedes, K.-H. Usadel, A. Encke, and K. Schöffling. The role of adrenal medulla in endogenous dopaminergic inhibition of aldosterone secretion. *Res Exp Med (Berl) 186*:427-434, 1986.

Kachi, T. Pineal actions on the autonomic system. *Pineal Res Rev 5*:217-263, 1987.

Kachi, T., T.K. Banerji, and W.B. Quay. Quantitative cytological analysis of functional changes in adrenomedullary chromaffin cells in normal, sham-operated, and pinealectomized rats in relation to time of day: II. Nuclear-cytoplasmic ratio, nuclear size, and pars granulosa of nucleolus. *J Pineal Res 5*:141-159, 1988a.

Kachi, T., T.K. Banerji, and W.B. Quay. Quantitative cytological analysis of functional changes in adrenomedullary chromaffin cells in normal, sham-operated, and pinealectomized rats in relation to time of day: III. Nuclear density. *J Pineal Res 5*:527-534, 1988b.

Kage, R., L. Thim, W. Creutzfeldt, and J.M. Conlon. Post-translational processing of preprotachykinins: Isolation of protachykinin-(I-37)-peptide from human adrenal-medullary phaeochromocytoma tissue. *Biochem J 253*:203-207, 1988.

Kalra, S.P., M.G. Dube, and P.S. Kalra. Continuous intraventricular infusion of neuropeptide Y evokes episodic food intake in satiated female rats: Effects of adrenalectomy and cholecystokinin. *Peptides 9*:723-728, 1988.

Kamikubo, K., H. Murase, M. Murayama, M. Matsuda, and K. Miura. Evidence for disulfide bonds in membrane bound and solubilized opioid receptors. *J Neurochem 50*:503-509, 1988.

Kamikubo, K., H. Murase, M. Murayama, and K. Miura. Solubilization of adrenal medullary opioid receptors. *Endocrinol Jpn 34*:897-901, 1987.

Kamikubo, K., H. Murase, M. Murayama, K. Miura, M. Nozaki, and K. Tsurumi. Opioid and non-opioid binding of β-endorphin to bovine adrenal medullary membranes. *Regul Pept 15*:155-162, 1986.

Kamikubo, K., H. Murase, M. Niwa, K. Miura, M. Nozaki, and K.

Tsurumi. Adrenal medullary opioid receptors are linked to GTP-binding proteins, pertussis toxin substrates. *Natl Inst Drug Abuse Res Monogr Ser 75*:129-132, 1986.

Kamikubo, K., H. Murase, M. Niwa, K. Miura, M. Nozaki, and K. Tsurumi. Coupling of adrenal medullary opioid receptors to islet-activating protein-sensitive GTP-binding proteins. *Life Sci 40*:1791-1797, 1987.

Kamiya, T., T. Yasui, O. Suzuki, H. Nakamoto, T. Maruyama, Y. Kawamura, H. Ookawa, Y. Kiryu, and K. Shimizu. [von Recklinghausen's disease associated with pheochromocytoma -- an autopsy case report and review of 24 cases in Japanese literature.] *Gan No Rinsho 32*:1872-1878, 1986.

Kamo, H., S.U. Kim, P.L. McGeer, H. Tago, and D.H. Shin. Transplantation of cultured human adrenal chromaffin cells into 6-hydroxydopamine-lesioned rat brain. *Synapse 1*:324-328, 1987.

Kanamatsu, T., C.D. Unsworth, E.J. Diliberto Jr., O.H. Viveros, and J.S. Hong. Reflex splanchnic nerve stimulation increases levels of proenkephalin A mRNA and proenkephalin A-related peptides in the rat adrenal medulla. *Proc Natl Acad Sci USA 83*:9245-9249, 1986.

Kaneda, N., H. Ichinose, K. Kobayashi, K. Oka, F. Kishi, A. Nakazawa, Y. Kurosawa, K. Fujita, and T. Nagatsu. Molecular cloning of cDNA and chromosomal assignment of the gene for human phenylethanolamine *N*-methyltransferase, the enzyme for epinephrine biosynthesis. *J Biol Chem 263*:7672-7677, 1988.

Kaneda, N., K. Kobayashi, H. Ichinose, F. Kishi, A. Nakazawa, Y. Kurosawa, K. Fujita, and T. Nagatsu. Isolation of a novel cDNA clone for human tyrosine hydroxylase: Alternative RNA splicing produces four kinds of mRNA from a single gene. *Biochem Biophys Res Commun 146*:971-975, 1987.

Kanner, B.I. and S. Schuldiner. Mechanism of transport and storage of neurotransmitters. *CRC Crit Rev Biochem 22*:1-38, 1987.

Kao, L.-S. Calcium homeostasis in digitonin-permeabilized bovine chromaffin cells. *J Neurochem 51*:221-227, 1988.

Kao, L.-S. and A.S. Schneider. Calcium mobilization and catecholamine secretion in adrenal chromaffin cells. A Quin-2 fluorescence study. *J Biol Chem 261*:4881-4888, 1986.

Karimovo, M. Kh. and M.F. Bondarenko. [Function of the adrenal medullary layer in rats with experimental alloxan diabetes.] *Probl Endokrinol (Mosk) 32*:61-63, 1986.

Karlberg, B.E., L. Hedman, S. Lennquist, and T. Pollare. The value of the clonidine-suppression test in the diagnosis of pheochromocytoma. *World J Surg 10*:753-761, 1986.

Kasaian, M.T. and K.E. Neet. Internalization of nerve growth factor by PC12 cells: A description of cellular pools. *J Biol Chem 263*:5083-5090, 1988.

Katafuchi, T., Y. Oomura, and M. Kurosawa. Effects of chemical stimulation of paraventricular nucleus on adrenal and renal nerve activity in rats. *Neurosci Lett 86*:195-200, 1988.

Katafuchi, T., H. Yoshimatsu, Y. Oomura, and A. Sato. Responses of adrenal catecholamine secretion to lateral hypothalamic stimulation and lesion in rats. *Brain Res 363*:141-144, 1986.

Kataoka, Y., M. Fujimoto, H. Alho, A. Guidotti, M. Geffard, G.D. Kelly, and I. Hanbauer. Intrinsic γ-aminobutyric acid receptors modulate the release of catecholamine from canine adrenal gland *in situ. J Pharmacol Exp Ther 239*:584-590, 1986.

Kataoka, Y., M. Ohara-Imaizumi, S. Ueki, and K. Kumakura. Stimulatory action of γ-aminobutyric acid on catecholamine secretion from bovine adrenal chromaffin cells measured by a real-time monitoring system. *J Neurochem 50*:1765-1768, 1988.

Katoh-Semba, R., S. Kitajima, Y. Yamazaki, and M. Sano. Neuritic growth from a new subline of PC12 pheochromocytoma cells: Cyclic AMP mimics the action of nerve growth factor. *J Neurosci Res 17*:36-44, 1987.

Katoh-Semba, R., S.D. Skaper, and S. Varon. G_{m1} ganglioside treatment of PC12 cells stimulates ganglioside, glycolipid, and lipid, but not glycoprotein synthesis independently from the effects of nerve growth factor. *J Neurochem 46*:574-582, 1986.

Katopodis, A.G. and S.W. May. A new facile trinitrophenylated substrate for peptide α-amidation and its use to characterize PAM activity in chromaffin granules. *Biochem Biophys Res Commun 151*:499-505, 1988.

Katzenstein, G.E., D. Lund, P. Schultz, and R.V. Lewis. Target tissue distribution of the proenkephalin peptides F, E, and B. *Biochem Biophys Res Commun 146*:1184-1190, 1987.

Kawada, T., S.-I. Sakabe, T. Watanabe, M. Yamamoto, and K. Iwai. Some pungent principles of spices cause the adrenal medulla to secrete catecholamine in anesthetized rats. *Proc Soc Exp Biol Med 188*:229-233, 1988.

Kawai, K., M. Senba, and H. Tsuchiyama. Eosionophilic globules in pheochromocytoma of the adrenal medulla. A histochemical, immunohistochemical and ultrastructural study. *APMIS 96*:911-916, 1988.

Kawate, R., M. Kumekawa, K. Meguro, and A. Sato. [Effects of halothane anesthesia on sympatho-adrenal function in rats.] *Masui 36*:1349-1356, 1987.

Kazama, Y.-I., T. Noguchi, T. Kawabe, and T. Onaya. Cushing's syndrome associated with adrenomedullary hyperplasia. *JAMA 55*:2446, 1986.

Kecorius, E., D.H. Small, and B.G. Livett. Characterization of a dipeptidyl aminopeptidase from bovine adrenal medulla. *J Neurochem 50*:38-44, 1988.

Keith, C.H. Slow transport of tubulin in the neurites of differentiated PC12 cells. *Science 235*:337-339, 1987.

Kelley, P.M. and D. Njus. Cytochrome b_{561} spectral changes associated with electron transfer in chromaffin-vesicle ghosts. *J Biol Chem 261*:6429-6432, 1986.

Kelley, P.M. and D. Njus. A kinetic analysis of electron transport across chromaffin vesicle membranes. *J Biol Chem 263*:3799-3804, 1988.

Kelner, K.L., K. Morita, J.S. Rossen, and H.B. Pollard. Restricted diffusion of tyrosine hydroxylase and phenylethanolamine *N*-methyltransferase from digitonin-permeabilized adrenal chromaffin cells. *Proc Natl Acad Sci USA 83*:2998-3002, 1986.

Kemp, G. and M. Edge. Cholinergic function and α-bungarotoxin binding in PC12 cells. *Mol Pharmacol 32*:356-363, 1987.

Kenney, P.J. and R.J. Stanley. Calcified adrenal masses. *Urol Radiol 9*:9-15, 1987.

Kenney, P.J., D.P. Streeten, and G.H. Anderson. Difficulties in the prospective diagnosis of functional adrenal diseases by CT. *Urol Radiol 8*:184-189, 1986.

Kent, U.M. and P.J. Fleming. Purified cytochrome b_{561} catalyzes transmembrane electron transfer for dopamine β–hydroxylase and peptidyl glycine α-amidating monooxygenase activities in reconstituted systems. *J Biol Chem 262*:8174-8178, 1987.

Kesse, W.K., T.L. Parker, and R.E. Coupland. The innervation of the adrenal gland. I. The source of pre- and postganglionic nerve fibres to the rat adrenal gland. *J Anat 157*:33-41, 1988.

Khaisman, E.B., V.A. Arefolov, and L.A. Malikova. [The role of peripheral catecholaminergic systems in antistress action of neuropeptides.] *Biull Eksp Biol Med 105*:302-305, 1988.

Khalid, B.A.K., P. Morat, and Z. Merican. The effects of naloxone, dexamethasone, deoxycorticosterone and 17-hydroxyprogesterone on blood pressure responses of normal and adrenalectomized rats during hypovolaemic shock. *Clin Exp Pharmacol Physiol 14*:111-117, 1987.

Khalil, Z., B.G. Livett, and P.D. Marley. The role of sensory fibres in the rat splanchnic nerve in the regulation of adrenal medullary secretion during stress. *J Physiol (Lond) 370*:201-215, 1986.

Khalil, Z., B.G. Livett, and P.D. Marley. Sensory fibres modulate histamine-induced catecholamine secretion from the rat adrenal medulla and sympathetic nerves. *J Physiol (Lond) 391*:511-526, 1987.

Khalil, Z., P.D. Marley, and B.G. Livett. Elevation in plasma catecholamines in response to insulin stress is under both neuronal and nonneuronal control. *Endocrinology 119*:159-167, 1986.

Khalil, Z., P.D. Marley, and B.G. Livett. Effect of substance P on nicotine-induced desensitization of cultured bovine adrenal chromaffin cells: Possible receptor subtypes. *Brain Res 459*:282-288, 1988a.

Khalil, Z., P.D. Marley, and B.G. Livett. Mammalian tachykinins modulate the nicotinic secretory response of cultured bovine adrenal chromaffin cells. *Brain Res 459*:289-297, 1988b.

Kilpatrick, I.C., M.W. Jones, and O.T. Phillipson. A semi-automated analysis method for catecholamines, indoleamines, and some prominent metabolites in microdissected regions of the nervous system: An isocratic HPLC technique employing coulometric detection and minimal sample preparation. *J Neurochem 46*:1865-1876, 1986.

Kim, S.H., F.M. Sessler, and R.L. Malvin. Multiple renin forms in the adrenal gland. *Am J Physiol 255*:E531-E536, 1988.

Kim, Y.I. and E. Neher. IgG from patients with Lambert-Eaton syndrome blocks voltage-dependent calcium channels. *Science 239*:405-408, 1988.

Kimura, N., N. Sasano, and K. Ishioka. Use of formaldehyde-induced fluorescence for cytological diagnosis of pheochromocytoma. *Acta Pathol Jpn 36*:1049-1054, 1986.

Kimura, T., M. Katoh, and S. Satoh. Inhibition by opioid agonists and enhancement by antagonists of the release of catechol-amines from the dog adrenal gland in response to splanchnic nerve stimulation: Evidence for the functional role of opioid receptors. *J Pharmacol Exp Ther 244*:1098-1102, 1988.

King, B.F., K.K. Kopecky, M.K. Baker, and S.A. Clark. Extramedullary hematopoiesis in the adrenal glands: CT characteristics. *J Comput Assist Tomogr 11*:342-343, 1987.

King, S.C., T.E. Ellenberger, and S.M. Goldin. Biochemical and immunological evidence for a calcium pump in chromaffin granules. *Biochem Biophys Res Commun 155*:656-663, 1988.

Kirby, R.F. and R. McCarty. Ontogeny of functional sympathetic innervation to the heart and adrenal medulla in the preweanling rat. *J Auton Nerv Syst 19*:67-75, 1987.

Kitayama, S. [Role of γ-aminobutyric acid in the process of catecholamine secretion from adrenal medulla.] *Hiroshima Daigaku Shigaku Zasshi 17*:21-34, 1985.

Kitayama, S., Y. Kóyama, K. Morita, T. Dohi, and A. Tsujimoto. Increase in catecholamine release and $^{45}Ca^{2+}$ uptake induced by GABA in cultured bovine adrenal chromaffin cells. *Eur J Pharmacol 131*:145-147, 1986.

Kitayama, S., K. Morita, T. Dohi, and A. Tsujimoto. The nature of the stimulatory action of γ-aminobutyric acid in the isolated perfused dog adrenals. *Naunyn Schmiedebergs Arch Pharmacol 326*:106-110, 1984.

Kitayama, S., K. Morita, T. Dohi, and A. Tsujimoto. Potassium ion is indispensable to the catecholamine releasing response of dog adrenals to γ-aminobutyric acid. *Naunyn Schmiedebergs Arch Pharmacol 332*:66-69, 1986.

Kitayama, S., K. Morita, T. Dohi, and A. Tsujimoto. Benzodiazepine inhibition of the stimulation-evoked catechol-amine release from bovine adrenomedullary cells in culture. *Neurochem Int 13*:265-270, 1988.

Kitayama, S. and A. Tsujimoto. Involvement of GABAergic mechanisms in the catecholamine secretion from adrenal medulla. pp. 249-259. In: **GABAergic Mechanisms in the Mammalian Periphery**, (S.L. Erdó and N.G. Bowery, eds.), Raven Press, New York, 1986.

Kittner, B., M. Bräutigam, and H. Herken. PC12 cells: A model system for studying drug effects on dopamine synthesis and release. *Arch Int Pharmacodyn Ther 286*:181-194, 1987.

Kiuchi, K., M. Hagiwara, H. Hidaka, and T. Nagatsu. Effect of the 1-methyl-4-phenylpyridinium ion on phosphorylation of tyrosine hydroxylase in rat pheochromocytoma PC12h cells. *Neurosci Lett 89*:209-215, 1988.

Kiuchi, K., K. Kiuchi, A. Togari, and T. Nagatsu. Effect of spermine on tyrosine hydroxylase activity before and after phosphorylation by cyclic AMP-dependent protein kinase. *Biochem Biophys Res Commun 148*:1460-1467, 1987.

Kjaer, M., P.A. Farrell, N.J. Christensen, and H. Galbo. Increased epinephrine response and inaccurate glucoregulation in exercising athletes. *J Appl Physiol 61*:1693-1700, 1986.

Kjaer, M. and H. Galbo. Effect of physical training on the capacity to secrete epinephrine. *J Appl Physiol 64*:11-16, 1988.

Kjaer, M., N.H. Secher, and H. Galbo. Physical stress and catecholamine release. *Baillieres Clin Endocrinol Metab 1*:279-298, 1987.

Kjellman, B.F., J. Beck-Friis, J.G. Ljunggren, S.B. Ross, F. Undén, and L. Wetterberg. Serum dopamine-β–hydroxylase activity in patients with major depressive disorders. *Acta Psychiatr Scand 73*:266-270, 1986.

Klein, R.L., N. Yabuno, D.F. Peeler, A. Thureson-Klein, B.H. Douglas II, R.B. Duff, and W.E. Clayton. Adrenal enkephalin and catecholamine contents following subarachnoid hemorrhage in cats. *Neuropeptides 8*:143-158, 1986.

Kley, N. Multiple regulation of proenkephalin gene expression by protein kinase C. *J Biol Chem 263*:2003-2008, 1988.

Kley, N., J.-P. Loeffler, and V. Höllt. Ca^{2+}-dependent histaminergic regulation of proenkephalin mRNA levels in cultured adrenal chromaffin cells. *Neuroendocrinology 46*:89-92, 1987.

Kley, N., J.-P. Loeffler, C.W. Pittius, and V. Höllt. Proenkephalin A gene expression in bovine adrenal chromaffin cells is regulated by changes in electrical activity. *EMBO J 5*:967-970, 1986.

Kley, N., J.-P. Loeffler, C.W. Pittius, and V. Höllt. Involvement of ion channels in the induction of proenkephalin A gene expression by nicotine and cAMP in bovine chromaffin cells. *J Biol Chem 262*:4083-4089, 1987.

Klimovskaya, L.D. and L.V. Kokoreva. [Reactivity of the sympatho-adrenal system and exercise tolerance during repeated exposure to a constant magnetic field.] *Kosm Biol Aviakosm Med 20*:70-72, 1986.

Knardahl, S., B.J. Sanders, and A.K. Johnson. Effects of adrenal demedullation on stress-induced hypertension and cardio-vascular responses to acute stress. *Acta Physiol Scand 133*:477-483, 1988.

Knight, D.E. Calcium and exocytosis. *Ciba Found Symp 122*:250-270, 1986a.

Knight, D.E. Botulinum toxin types A, B and D inhibit catechol-amine secretion from bovine adrenal medullary cells. *FEBS Lett 207*:222-226, 1986b.

Knight, D.E. Calcium and diacylglycerol control of secretion. *Biosci Rep 7*:355-367, 1987.

Knight, D.E. An appreciation of Peter Baker. *J Membr Biol 103*:1-6, 1988.

Knight, D.E. and P.F. Baker. Observations on the muscarinic activation of catecholamine secretion in the chicken adrenal. *Neuroscience 19*:357-366, 1986.

Knight, D.E. and M.C. Scrutton. Gaining access to the cytosol: The technique and some applications of electropermea-bilization. *Biochem J 234*:497-506, 1986.

Knight, D.E., D. Sugden, and P.F. Baker. Evidence implicating protein kinase C in exocytosis from electropermeabilized bovine chromaffin cells. *J Membr Biol 104*:21-34, 1988.

Knoth, J., O.H. Viveros, and E.J. Diliberto Jr. Evidence for the release of newly acquired ascorbate and α-aminoisobutyric acid from the cytosol of adrenomedullary chromaffin cells through specific transporter mechanisms. *J Biol Chem 262*:14036-14041, 1987.

Kobayashi, H., F. Izumi, and J. Meldolesi. Rat adrenal chromaffin cells become sensitive to α-latrotoxin when cultured *in vitro*: The effect of nerve growth factor. *Neurosci Lett 65*:114-118, 1986.

Koide, M., A.K. Cho, and B.D. Howard. Characterization of xylamine binding to proteins of PC12 pheochromocytoma cells. *J Neurochem 47*:1277-1285, 1986.

Koike, T. Potentiation of nerve growth factor (NGF)-mediated neurite outgrowth in high K^+ medium is associated with increased binding of iodinated NGF in PC12 cells. *Cell Biol Int Rep 11*:423-428, 1987a.

Koike, T. Depolarization-induced increase in surface binding and internalization of ^{125}I-nerve growth factor by PC12 pheochromocytoma cells. *J Neurochem 49*:1784-1789, 1987b.

Koike, T. and A. Takashima. Cell cycle-dependent modulation of biosynthesis and stimulus-evoked release of catecholamines in PC12 pheochromocytoma cells. *J Neurochem 46*:1493-1500, 1986.

Koivunen, D.G., J.F. Seaton, S.S. Scaer, D.M. Potter, and T.S. Harrison. Peripherally delivered renin supports reflex adrenal catecholamine secretion in anephric dogs. *Curr Surg 44*:27-31, 1987.

Koizumi, M., K. Endo, H. Sakahara, T. Nakashima, Y. Nakano,

K. Nakao, and K. Torizuka. Computed tomography and ^{131}I-MIBG scintigraphy in the diagnosis of pheochromocytoma. *Acta Radiol (Stockh)* 27:305-309, 1986.

Koizumi, S., M.L. Contreras, Y. Matsuda, T. Hama, P. Lazarovici, and G. Guroff. K-252a: A specific inhibitor of the action of nerve growth factor on PC12 cells. *J Neurosci* 8:715-721, 1988.

Kolchinskaya, L.I., A.D. Chaika, and L.A. Kravets. [Interaction of the chromaffin-cell membrane fragments with artificial phospholipid membranes.] *Biol Membrany* 5:836-842, 1988.

Komatsu, H., N. Shirasu, K. Takei, N. Tanabe, T. Ishihama, Y. Yamada, K. Kobayashi, and A. Ueno. Right adrenal pheochromocytoma with anterolateral displacement of the inferior vena cava: Skin incision and approach. *J Urol* 137:477-479, 1987.

Kondo, H. [Functions of the sympathetic-adrenomedullary system and adrenocortex in patients with vibration syndrome. Investigations of plasma and urine catecholamine and plasma cortisol during cold exposure.] *Sangyo Igaku* 30:256-265, 1988.

Kondo, H., H. Kuramoto, and T. Fujita. An immuno-electron-microscopic study of the localization of VIP-like immuno-reactivity in the adrenal gland of the rat. *Cell Tissue Res* 245:531-538, 1986.

Kondratiev, A.D., V.Yu. Alakhov, V. A. Movsesyan, A.A. Chernyi, L.B. Kaminir, and E.S. Severin. Nerve growth factor inhibits ADP-ribosylation in pheochromocytoma PC12 cell line. *Bioorg Khim* 12:736-740, 1986.

Konecka, A.M. [Adrenal opioid peptides.] *Postepy Biochem* 33:561-579, 1987.

Konecki, D.S., U.M. Benedum, H.-H. Gerdes, and W.B. Huttner. The primary structure of human chromogranin A and pancreastatin. *J Biol Chem* 262:17026-17030, 1987.

Kongsamut, S. and R.J. Miller. Nerve growth factor modulates the drug sensitivity of neurotransmitter release from PC12 cells. *Proc Natl Acad Sci USA* 83:2243-2247, 1986.

Konings, F. Comment on the letter of Drs. De Block and De Potter. *FEBS Lett* 222:359, 1987.

Kordower, J.H., R.L. Dean III, H.C. White, and F.F. Marciano. Transplantation into the mammalian CNS: A meeting report on the sixth Schmitt neurological sciences symposium. *Neurobiol Aging* 9:127-137, 1988.

Korfali, E., M. Doygun, I.H. Ulus, C. Rakunt, and K. Aksoy. Effects of neuronotrophic factors on adrenal medulla grafts implanted into adult rat brains. *Neurosurgery* 22:994-998, 1988.

Kotani, S., H. Murofushi, S. Maekawa, C. Sato, and H. Sakai. Characterization of microtubule-associated proteins isolated from bovine adrenal gland. *Eur J Biochem* 156:23-29, 1986.

Kovarik, M.F., S.B. Jones, and F.D. Romano. Plasma catechol-amines following cecal ligation and puncture in the rat. *Circ Shock* 22:281-290, 1987.

Kóyama, Y., S. Kitayama, T. Dohi, and A. Tsujimoto. Evidence that prostaglandins activate calcium channels to enhance basal and stimulation-evoked catecholamine release from bovine adrenal chromaffin cells in culture. *Biochem Pharmacol* 37:1725-1730, 1988.

Kozlowski, M.R., M.P. Rosser, and E. Hall. Identification of ^{3}H-bradykinin binding sites in PC12 cells and brain. *Neuropeptides* 12:207-211, 1988.

Krahe, Th., A. Steudel, W. Dewes, V. Nicolas, D. Klingmüller, and K. Lackner. [Magnetic resonance tomography (MRT) of the adrenal gland.] *Fortschr Rontgenstr* 146:520-526, 1987.

Krane, N.K. Clinically unsuspected pheochromocytomas: Experience at Henry Ford Hospital and a review of the literature. *Arch Intern Med* 146:54-57, 1986.

Kress, H.G., H. Eckhardt-Wallasch, P.W.L. Tas, and K. Koschel. Volatile anesthetics depress the depolarization-induced cytoplasmic calcium rise in PC12 cells. *FEBS Lett* 221:28-32, 1987.

Krisch, K., G. Horvat, I. Krisch, G. Wengler, H. Alibeik, N. Neuhold, W. Ulrich, O. Braun, and M. Hochmeister. Immunochemical characterization of a novel secretory protein (defined by monoclonal antibody HISL 19) of peptide hormone producing cells which is distinct from chromogranin A, B, and C. *Lab Invest* 58:411-420, 1988.

Kristensen, P., D.M. Hougaard, L.S. Nielsen, and K. Danø. Tissue-type plasminogen activator in rat adrenal medulla. *Histochemistry* 85:431-436, 1986.

Krubsack, A.J., M.A. Arnaout, T.C. Hagen, J.S. Zielonka, H. Choi, R. Akhtar, and W.J. Schulte. Zona fasciculata cortical adenoma and adrenal medullary hyperplasia in MEN II patient: Unique concurrent presentation. *Clin Nucl Med* 13:730-733, 1988.

Krum, J.M. and J.M. Rosenstein. Patterns of angiogenesis in neural transplant models: I. Autonomic tissue transplants. *J Comp Neurol* 258:420-434, 1987.

Kruse, L.I., W.E. DeWolf Jr., P.A. Chambers, and P.J. Goodhart. Design and kinetic characterization of multisubstrate inhibitors of dopamine β-hydroxylase. *Biochemistry* 25:7271-7278, 1986.

Kruse, L.I., C. Kaiser, W.E. DeWolf Jr., J.S. Frazee, R.W. Erickson, M. Ezekiel, E.H. Ohlstein, R.R. Ruffolo Jr., and B.A. Berkowitz. Substituted 1-benzylimidazole-2-thiols as potent and orally active inhibitors of dopamine β-hydroxylase [letter]. *J Med Chem* 29:887-889, 1986a.

Kruse, L.I., C. Kaiser, W.E. DeWolf Jr., J.S. Frazee, E. Garvey, E.L. Hilbert, W.A. Faulkner, K.E. Flaim, J.L. Sawyer, and B.A. Berkowitz. Multisubstrate inhibitors of dopamine β-hydroxylase. 1. Some 1-phenyl and 1-phenyl-bridged deriva-tives of imidazole-2-thione. *J Med Chem* 29:2465-2472, 1986b.

Kruse, L.I., C. Kaiser, W.E. DeWolf Jr., J.S. Frazee, S.T. Ross, J.

Wawro, M. Wise, K.E. Flaim, J.L. Sawyer, R.W. Erickson, M. Ezekiel, E.H. Ohlstein, and B.A. Berkowitz. Multisubstrate inhibitors of dopamine β–hydroxylase. 2. Structure-activity relationships at the phenethylamine binding site. *J Med Chem* 30:486-494, 1987.

Kuchel, O., N.T. Buu, K. Racz, A. DeLéan, O. Serri, and J. Kyncl. Role of sulfate conjugation of catecholamines in blood pressure regulation. *Fed Proc* 45:2254-2259, 1986.

Kudo, T., E.Q. Wei, R. Inoki, M. Terasawa, and T. Nakao. Influence of Y-20003, an analgesic agent, on the endogenous opioid peptide system in rats. *Natl Inst Drug Abuse Res Monogr Ser* 75:236-239, 1986.

Kuhn, D.M. and M.L. Billingsley. Tyrosine hydroxylase: Purification from PC12 cells, characterization and production of antibodies. *Neurochem Int* 11:463-475, 1987.

Kuhn, L.J., M. Hadman, and E.L. Sabban. Effect of monensin on synthesis, post-translational processing, and secretion of dopamine β–hydroxylase from PC12 pheochromocytoma cells. *J Biol Chem* 261:3816-3825, 1986.

Kumakura, K., M. Ohara, and G.P. Satô. Real-time monitoring of the secretory function of cultured adrenal chromaffin cells. *J Neurochem* 46:1851-1858, 1986.

Kumakura, K., A. Sato, and H. Suzuki. Direct recording of total catecholamine secretion from the adrenal gland in response to splanchnic nerve stimulation in rats. *J Neurosci Methods* 24:39-43, 1988.

Kunze, D.L., S.L. Hamilton, M.J. Hawkes, and A.M. Brown. Dihydropyridine binding and calcium channel function in clonal rat adrenal medullary tumor cells. *Mol Pharmacol* 31:401-409, 1987.

Kuramoto, H. An immunohistochemical study of chromaffin cells and nerve fibers in the adrenal gland of the bull-frog, *Rana catesbeiana*. *Arch Histol Jpn* 50:15-38, 1987.

Kuramoto, H., H. Kondo, and T. Fujita. Neuropeptide tyrosine (NPY)-like immunoreactivity in adrenal chromaffin cells and intraadrenal nerve fibers of rats. *Anat Rec* 214:321-328, 1986.

Kuramoto, H., H. Kondo, and T. Fujita. Calcitonin gene-related peptide (CGRP)-like immunoreactivity in scattered chromaffin cells and nerve fibers in the adrenal gland of rats. *Cell Tissue Res* 247:309-315, 1987.

Kurosawa, M., A. Sato, Y. Sato, and H. Suzuki. Undiminished reflex responses of adrenal sympathetic nerve activity to stimulation of baroreceptors and cutaneous mechanoreceptors in aged rats. *Neurosci Lett* 77:193-198, 1987.

Kurosawa, M., A. Sato, Y. Sato, and H. Suzuki. The sympatho-adrenal medullary functions in aged rats under anesthesia. *Ann NY Acad Sci* 515:329-342, 1988.

Kurosawa, M., A. Sato, R.S. Swenson, and Y. Takahashi. Sympathoadrenal medullary functions in response to intracerebroventricularly injected corticotropin-releasing factor in anesthetized rats. *Brain Res 367*:250-257, 1986.

Kvetnansky, R., F.J.H. Tilders, I.D. vanZoest, M. Dobrakovova, F. Berkenbosch, J. Culman, P. Zeman, and P.G. Smelik. Sympathoadrenal activity facilitates β-endorphin and α-MSH secretion but does not potentiate ACTH secretion during immobilization stress. *Neuroendocrinoloy 45*:318-324, 1987.

Laasberg, T., A. Pihlak, T. Neuman, H. Paves, and M. Saarma. Nerve growth factor increases the cyclic GMP level and activates the cyclic GMP phosphodiesterase in PC12 cells. *FEBS Lett 239*:367-370, 1988.

Ladona, M.G., D. Aunis, L. Gandía, and A.G. García. Dihydropyridine modulation of the chromaffin cell secretory response. *J Neurochem 48*:483-490, 1987.

Ladona, M.G., M.-F. Bader, and D. Aunis. Influence of hypertonic solutions on catecholamine release from intact and permeabilized cultured chromaffin cells. *Biochim Biophys Acta 927*:18-25, 1987.

La Gamma, E.F. and J.E. Adler. Glucocorticoids regulate adrenal opiate peptides. *Molec Brain Res 2*:125-130, 1987.

La Gamma, E.F. and J.E. Adler. Development of transsynaptic regulation of adrenal enkephalin. *Dev Brain Res 39*:177-182, 1988.

La Gamma, E.F., J.D. White, J.F. McKelvy, and I.B. Black. Second messenger mechanisms governing opiate peptide transmitter regulation in the rat adrenal medulla. *Brain Res 441*:292-298, 1988.

Lagercrantz, H. and T.A. Slotkin. The "stress" of being born. *Sci Am 254*:100-107, 1986.

Lairmore, M.D., A.P. Knight, and J.C. DeMartini. Three primary neoplasms in a goat: Hepatocellular carcinoma, phaeochromocytoma and leiomyoma. *J Comp Pathol 97*:267-271, 1987.

Laliberte, F., M.-F. Laliberte, F. Alhenc-Gelas, and C. Chevillard. Cellular and subcellular immunohistochemical localization of angiotensin-converting enzyme in the rat adrenal gland. *Lab Invest 56*:364-371, 1987.

Lamouroux, A.A., A. Vigny, N. Faucon Biguet, M.C. Darmon, R. Franck, J.-P. Henry, and J. Mallet. The primary structure of human dopamine β-hydroxylase: Insights into the relationship between the soluble and the membrane-bound forms of the enzyme. *EMBO J 6*:3931-3937.

Langley, O.K., M.C. Aletsee, and M. Gratzl. Endocrine cells share expression of N-CAM with neurones. *FEBS Lett 220*:108-112, 1987.

Langley, O.K. and D. Aunis. Surface expression of neural cell adhesion molecule in cultured bovine paraneurones: Immunogold and immunoperoxidase methods compared. *Neurosci Lett 64*:151-156, 1986.

Langley, O.K., D. Perrin, and D. Aunis. α-Fodrin in the adrenal gland: Localization by immunoelectron microscopy. *J Histochem Cytochem 34*:517-525, 1986.

Lascar, G. and J. Taxi. [Structural characteristics and innervation of the chromaffin tissue in the adrenal gland of the Axolotl.] *C R Acad Sci [III] 303*:211-216, 1986.

Laschinski, G., B. Kittner, and M. Bräutigam. Direct inhibition of tyrosine hydroxylase from PC12 cells by catechol derivatives. *Naunyn Schmiedebergs Arch Pharmacol 332*:346-350, 1986.

Laslop, A., R. Fischer-Colbrie, V.Y.H. Hook, D. Obendorf, and H. Winkler. Identification of two glycoproteins of chromaffin granules as the carboxypeptidase H. *Neurosci Lett 72*:300-304, 1986.

Lassmann, H., C. Hagn, R. Fischer-Colbrie, and H. Winkler. Presence of chromogranin A, B and C in bovine endocrine and nervous tissues: A comparative immunohistochemical study. *Histochem J 18*:380-386, 1986.

Lathers, C.M., R.F. Flax, and L.J. Lipka. The effect of C1 spinal cord transection or bilateral adrenal vein ligation on thioridazine-induced arrhythmia and death in the cat. *J Clin Pharmacol 26*:515-523, 1986.

Lau, C. Activation of the adrenal cortex or the peripheral sympathoadrenomedullary system does not necessarily influence milk ejection in the rat. *J Endocrinol 118*:399-405, 1988.

Lau, C., J.V. Bartolome, M.B. Bartolome, and T.A. Slotkin. Central and sympatho-adrenal responses to insulin in adult and neonatal rats. *Dev Brain Res 36*:277-280, 1987.

Lau, C., A.M. Cameron, O. Irsula, L.L. Antolick, C. Langston, and R.J. Kavlock. Teratogenic effects of nitrofen on cellular and functional maturation of the rat lung. *Toxicol Appl Pharmacol 95*:412-422, 1988.

Lau, C., M. Franklin, L. McCarthy, A. Pylypiw, and L.L. Ross. Thryoid hormone control of preganglionic innervation of the adrenal medulla and chromaffin cell development in the rat. An ultrastructural, morphometric and biochemical evaluation. *Dev Brain Res 44*:109-117, 1988.

Lau, C., L.L. Ross, W.L. Whitmore, and T.A. Slotkin. Regulation of adrenal chromaffin cell development by the central monoaminergic system: Differential control of norepinephrine and epinephrine levels and secretory responses. *Neuroscience 22*:1067-1075, 1987.

Lau, C., F.J. Seidler, A.M. Cameron, H.A. Navarro, J.M. Bell, J.V. Bartolome, and T.A. Slotkin. Nutritional influences on adrenal chromaffin cell development: Comparison with central neurons. *Pediatr Res 24*:583-587, 1988.

Laury, M.C., M. Beauvallet, L. Zizine, P. Delost, and R. Portet. Hypophysectomy and the sympatho-adrenal system in cold acclimation. *Experientia 42*:1252-1253, 1986.

Lauweryns, J.M., L. vanRanst, R.V. Lloyd, and D.T. O'Connor. Chromogranin in bronchopulmonary neuroendocrine cells.

Immunocytochemical detection in human, monkey, and pig respiratory mucosa. *J Histochem Cytochem 35*:113-118, 1987.

Layfield, L.J., B.J. Glasgow, M.H. Du Puis, and S. Bhuta. Aspiration cytology and immunohistochemistry of a pheochromocytoma-ganglioneuroma of the adrenal gland. *Acta Cytol (Baltimore) 31*:33-39, 1987.

Lazarovici, P., G. Dickens, H. Kuzuya, and G. Guroff. Long-term, heterologous down-regulation of the epidermal growth factor receptor in PC12 cells by nerve growth factor. *J Cell Biol 104*:1611-1621, 1987.

Leapman, R.D. and R.L. Ornberg. Quantitative electron energy loss spectroscopy in biology. *Ultramicroscopy 24*:251-268, 1988.

Leboulenger, F., A. Cupo, E. Castanas, M. Benyamina, G. Pelletier, and H. Vaudry. Immunohistochemical and biochemical evidence for the presence of the pentapeptide met-enkephalin and the heptapeptide met-enkephalin-Arg[6]-Phe[7] but not the octapeptide met-enkephalin-Arg[6]-Gly[7]-Leu[8] in amphibian chromaffin cells. *Neurochem Int 8*:303-309, 1986.

Le Douarin, N.M. and J. Smith. Development of the peripheral nervous system from the neural crest. *Annu Rev Cell Biol 4*:375-404, 1988.

Lee, K., S. Miwa, M. Fujiwara, T. Magaribuchi, and M. Fujiwara. Differential effects of hypoxia on the turnover of norepinephrine and epinephrine in the heart, adrenal gland, submaxillary gland and stomach. *J Pharmacol Exp Ther 240*:954-958, 1987.

Lee, M., H. Nohta, Y. Umegae, and Y. Ohkura. Assay for tyrosine hydroxylase by high-performance liquid chromatography with fluorescence detection. *J Chromatogr 415*:289-296, 1987.

Lee, S.A. and R.W. Holz. Protein phosphorylation and secretion in digitonin-permeabilized adrenal chromaffin cells. Effects of micromolar Ca^{2+}, phorbol esters, and diacylglycerol. *J Biol Chem 261*:17089-17098, 1986.

Lee, S.A., R.W. Holz, and D.R. Hathaway. Effects of purified myosin light chain kinase on myosin light chain phosphorylation and catecholamine secretion in digitonin-permeabilized chromaffin cells. *Biosci Rep 7*:323-332, 1987.

Le Floch, J.P., M. Thomsen, J. Joly, and J.P. Luton. [Adrenal medulla secretion and peripheral adrenergic receptors in familial dysautonomia.] *Presse Med 16*:1891-1894, 1987.

Lehmann, M., A. Berg, and J. Keul. Sex-related differences in free plasma catecholamines in individuals of similar performance ability during graded ergometric exercise. *Eur J Appl Physiol 55*:54-58, 1986.

Lelkes, P.I., J.E. Friedman, K. Rosenheck, and A. Oplatka. Destabilization of actin filaments as a requirement for the secretion of catecholamines from permeabilized chromaffin cells. *FEBS Lett 208*:357-363, 1986.

Lelkes, P.I. and H.B. Pollard. Oligopeptide inhibitors of metalloendoprotease activity inhibit catecholamine secretion from bovine adrenal chromaffin cells by modulating intracellular calcium homeostasis. *J Biol Chem* 262:15496-15505, 1987.

Lemaire, S., L. Chouinard, P. Mercier, and R. Day. Bombesin-like immunoreactivity in bovine adrenal medulla. *Regul Pept* 13:133-146, 1986.

Lemaire, S., M. Dumont, P. Mercier, I. Lemaire, and R. Calvert. Biochemical characterization of various populations of isolated bovine adrenal chromaffin cells. *Neurochem Int* 5:193-200, 1983.

Lemaire, S., M. Dumont, and S. Nolet. Sensitive method of detection, quantitation and purification of peptides using pre-column derivatization with phenyl isothiocyanate. *J Chromatogr* 425:77-86, 1988.

Lembeck, F. Substance P: From extract to excitement. *Acta Physiol Scand* 133:435-454, 1988.

Lenz, H.J. CNS regulation of gastric and autonomic functions in dogs by gastrin-releasing peptide. *Am J Physiol* 255:G298-G303, 1988.

Lenzen, S., M. Freisinger-Treichel, and U. Panten. Monoamine oxidase in rat and bovine endocrine tissues. *J Neurochem* 49:1183-1190, 1987.

Leonard, D.G.B., E.B. Ziff, and L.A. Greene. Identification and characterization of mRNAs regulated by nerve growth factor in PC12 cells. *Mol Cell Biol* 7:3156-3167, 1987.

Letiec, A., D. Guilloteau, F. Huguet, J.-L. Baulieu, J.-C. Besnard, and C. Viel. [Iodo carboxamidino-1 phenyl-4 piperazine, a new adrenomedullary imaging agent: Comparison with metaiodobenzylguanidine. *Int J Nucl Med Biol* 12:495-496, 1986.

Levi, A., S. Biocca, A. Cattaneo, and P. Calissano. The mode of action of nerve growth factor in PC12 cells. *Mol Neurobiol* 2:201-226, 1988.

Levin, B.E. and A.C. Sullivan. Glucose, insulin and sympatho-adrenal activation. *J Auton Nerv Syst* 20:233-242, 1987.

Levine, E., P. deVries, and L.H. Wetzel. MR imaging of inferior vena caval recurrence of extraadrenal pheochromocytoma. *J Comput Assist Tomogr* 11:717-718, 1987.

Levine, M., W. Hartzell, and A. Bdolah. Ascorbic acid and Mg-ATP co-regulate dopamine β-monooxygenase activity in intact chromaffin granules. *J Biol Chem* 263:19353-19362, 1988.

Levine, M.A. New concepts in the biology and biochemistry of ascorbic acid. *N Engl J Med* 314:892-902, 1986a.

Levine, M.A. Ascorbic acid specifically enhances dopamine β-monooxygenase activity in resting and stimulated chromaffin cells. *J Biol Chem* 261:7347-7356, 1986b.

Levine, M.A. and W. Hartzell. Ascorbic acid: The concept of

optimum requirements. *Ann NY Acad Sci 498*:424-444, 1987.

Levine, R.A. and M.P. Galloway. The regulation of biogenic amine synthesis and role of tetrahydrobiopterin. *Topics Neurochem Neuropharmacol 1*:119-130, 1987.

Lewin, R. Brain grafts benefit Parkinson's patients. *Science 236*:149, 1987.

Lewin, R. Cloud over Parkinson's therapy. *Science 240*:390-392, 1988a.

Lewin, R. Disappointing brain graft results. *Science 240*:1407, 1988b.

Lewin, R. Caution continues over transplants. *Science 242*:1379, 1988c.

Lewinski, A., A. Bartke, A.I. Esquifino, E. Sewerynek, and R.W. Steger. Adrenal catecholamine content: effects of congenital GH, PRL and TSH deficiency and of hormone replacement therapy in the male mouse. *Exp Clin Endocrinol 87*:176-182, 1986.

Lewis, E.J., C.A. Harrington, and D.M. Chikaraishi. Transcriptional regulation of the tyrosine hydroxylase gene by glucocorticoid and cyclic AMP. *Proc Natl Acad Sci USA 84*:3550-3554, 1987.

Lieberman, A.N. The use of adrenal medullary and fetal grafts as a treatment for Parkinson's disease. *NY State J Med 88*:287-289, 1988.

Lieberman, A.N., M. Koslow, and P. Maestrone. Adrenal grafts for Parkinson's disease [letter]. *N Engl J Med 317*:1092, 1987a.

Lieberman, A.N., M. Koslow, and P. Maestrone. Adrenal medullary grafts transplanted into the brain as a treatment for Parkinson's disease. *NY State J Med 87*:377-379, 1987b.

Liebisch, D., M. Bommer, M. Schimak, and A. Herz. Inhibition of nicotine-induced secretion from bovine chromaffin cells by the amidated C-terminal sequence of the opioid peptide amidorphin. *Biochem Biophys Res Commun 143*:545-551, 1987.

Liebisch, D., E. Weber, B. Kosicka, C. Gramsch, A. Herz, and B.R. Seizinger. Isolation and structure of a C-terminally amidated nonopioid peptide amidorphin-(8-26), from bovine striatum: A major product of proenkephalin in brain but not in adrenal medulla. *Proc Natl Acad Sci USA 83*:1936-1940, 1986.

Liggett, S.B., J.C. Marker, S.D. Shah, C.L. Roper, and P.E. Cryer. Direct relationship between mononuclear leukocyte and lung β-adrenergic receptors and apparent reciprocal regulation of extravascular, but not intravascular, α- and β-adrenergic receptors by the sympathochromaffin system in humans. *J Clin Invest 82*:48-56, 1988.

Lilly, M.P., W.C. Engeland, and D.S. Gann. Adrenal medullary responses to repeated hemorrhage in conscious dogs. *Am J Physiol 251*:R1193-R1199, 1986.

Lima, L. and T.L. Sourkes. Reserpine and the monoaminergic

regulation of adrenal dopamine β–hydroxylase activity. *Neuroscience 17*:235-245, 1986a.

Lima, L. and T.L. Sourkes. Pharmacological analysis of the neurotransmitter mechanisms regulating phenylethanolamine *N*-methyltransferase in the adrenal gland. *Biochem Pharmacol 35*:3965-3969, 1986b.

Lima, L. and T.L. Sourkes. Cholinergic and GABAergic regulation of dopamine β-hydroxylase activity in the adrenal gland of the rat. *J Pharmacol Exp Ther 237*:265-270, 1986c.

Lima, L. and T.L. Sourkes. Effect of corticotropin-releasing factor on adrenal DBH and PNMT activity. *Peptides 8*:437-441, 1987.

Lima, L., T.L. Sourkes, and B. Dubrovsky. Differential roles of raphe nuclei in the regulation of dopamine β–hydroxylase in the adrenal gland of the rat. *J Neurochem 46*:317-320, 1986.

Linares, O.A. and J.B. Halter. Sympathochromaffin system activity in the elderly. *J Am Geriatr Soc 35*:448-453, 1987.

Lindberg, I. Reserpine-induced alterations in the processing of proenkephalin in cultured chromaffin cells: Increased amidation. *J Biol Chem 261*:16317-16322, 1986.

Lindberg, I. The chromaffin granule trypsinlike protease -- a kallikrein? *Proteases Biolog Control Biotechnol 57*:269-275, 1987.

Lindberg, S., M. Fjälling, L. Jacobsson, S. Jansson, and L.-E. Tisell. Methodology and dosimetry in adrenal medullary imaging with iodine[131] MIGB. *J Nucl Med 29*:1638-1643, 1988.

Lindenbaum, M.H., S. Carbonetto, and W.E. Mushynski. Nerve growth factor enhances the synthesis, phosphorylation, and metabolic stability of neurofilament proteins in PC12 cells. *J Biol Chem 262*:605-610, 1987.

Lindvall, O., E.-O. Backlund, L. Farde, G. Sedvall, R. Freedman, B.J. Hoffer, A. Nobin, A. Seiger, and L. Olson. Transplantation in Parkinson's disease: Two cases of adrenal medullary grafts to the putamen. *Ann Neurol 22*:457-468, 1987.

Linstedt, A.D. and R.B. Kelly. Overcoming barriers to exocytosis. *Trends Neurosci 10*:446-448, 1987.

Lishajko, F. Binding of catecholamines to phospholipids isolated from adrenal medullary granules. *Acta Physiol Scand 127*:177-185, 1986.

Livett, B.G. and P.D. Marley. Effects of opioid peptides and morphine on histamine-induced catecholamine secretion from cultured, bovine adrenal chromaffin cells. *Br J Pharmacol 89*:327-334, 1986.

Livett, B.G., P.D. Marley, K.I. Mitchelhill, D.C.-C. Wan, and T.D. White. Assessment of adrenal chromaffin cell secretion: Presentation of four techniques. pp. 177-204. In: **In Vitro Methods for Studying Secretion**, Vol 3. (A.M. Poisner and J.M. Trifaró, eds). Elsevier Science Publishers, Amsterdam, 1987.

Livett, B.G., K.I. Mitchelhill, and D.M. Dean. Adrenal chromaffin cells--their isolation and culture. pp. 171-175. In: **In Vitro Methods for Studying Secretion**, Vol. 3. (A.M. Poisner and J.M. Trifaró, eds). Elsevier Science Publishers, Amsterdam, 1987.

Ljones, T. Dopamine β-monooxygenase from bovine adrenal medulla. *Methods Enzymol 142*:596-602, 1987.

Lloyd, R.V. Immunohistochemical localization of chromogranin in normal and neoplastic endocrine tissues. *Pathol Annu 22*:69-90, 1987.

Lloyd, R.V. Immunohistochemical localization of catecholamines, catecholamine synthesizing enzymes, and chromogranins in neuroendocrine cells and tumors. pp. 317-340. In: **Advances in Immunohistochemistry**. (R.A. DeLellis, ed). Raven Press, New York, 1988.

Lloyd, R.V., M. Cano, P. Rosa, A. Hille, and W.B. Huttner. Distribution of chromogranin A and secretogranin I (chromogranin B) in neuroendocrine cells and tumors. *Am J Pathol 130*:296-304, 1988.

Lloyd, R.V., J.C. Sisson, B. Shapiro, and A.A.J. Verhofstad. Immunohistochemical localization of epinephrine, norepinephrine, catecholamine-synthesizing enzymes, and chromogranin in neuroendocrine cells and tumors. *Am J Pathol 125*:45-54, 1986.

Lopatin, A.N., A.D. Kondratiev, B.V. Loginov, and E.I. Melnik. Potassium channels of clonal line pheochromocytoma. *Biol Membrany 5*:711-717, 1988.

Lorrain, J., E.D. Dipaola, F. Lhoste, and I. Cavero. Effects of pergolide on the cardiovascular responses to sinoaortic deafferentation in dogs with intact or surgically decentralized adrenal glands. *J Pharmacol Exp Ther 240*:288-293, 1987.

Lotshaw, D.P., H.Z. Ye, and C. Edwards. Endocytosis of surface bound dopamine β-hydroxylase and plasma membrane following catecholamine secretion by bovine adrenal chromaffin cells. *Neurochem Int 9*:391-399, 1986.

Lowe, A.W., L. Madeddu, and R.B. Kelly. Endocrine secretory granules and neuronal synaptic vesicles have three integral membrane proteins in common. *J Cell Biol 106*:51-59, 1988.

Lukas, R.J. Characterization of curaremimetic neurotoxin binding sites on cellular membrane fragments derived from the rat pheochromocytoma PC12. *J Neurochem 47*:1768-1773, 1986.

Lundberg, J.M., A. Hemsén, G. Fried, E. Theodorsson-Norheim, and H. Lagercrantz. Co-release of neuropeptide Y (NPY)-like immunoreactivity and catecholamines in newborn infants. *Acta Physiol Scand 126*:471-473, 1986.

Lundberg, J.M., T. Hökfelt, A. Hemsén, E. Theodorsson-Norheim, J. Pernow, B. Hamberger, and M. Goldstein. Neuropeptide Y-like immunoreactivity in adrenaline cells of adrenal medulla and in tumors and plasma of pheochromocytoma patients. *Regul Pept 13*:169-182, 1986.

Lütgemeier, I., F.C. Luft, T. Unger, U. Ganten, R.E. Lang, K.H. Gless, and D. Ganten. Blood pressure, electrolyte and adrenal responses in swim-trained hypertensive rats. *J Hypertens 5*:241-247, 1987.

Luyendijk, W. The future role of neurosurgery in medicine. *Neurosurg Rev 9*:7-12, 1986.

Lydén-Sokolowski, A., B.S. Larsson, and N.G. Lindquist. Disposition of 1-methyl-4-phenyl-1,2,3,6-tetrahydropyridine (MPTP) in mice before and after monoamine oxidase and catecholamine reuptake inhibition. *Pharmacol Toxicol 63*:75-80, 1988.

Lynch, D.R., K.M. Braas, and S.H. Snyder. Atrial natriuretic factor receptors in rat kidney, adrenal gland, and brain: Autoradiographic localization and fluid balance dependent changes. *Proc Natl Acad Sci USA 83*:3557-3561, 1986.

Lyon, R.A., M. Titeler, L. Bigornia, and A.S. Schneider. D_2 dopamine receptors on bovine chromaffin cell membranes: Identification and characterization by [^{3}H]*N*-methylspiperone binding. *J Neurochem 48*:631-635, 1987.

Ma, X.-L., Y.-M. Zang, M.-Z. Zhu, and Y.-M. Wang. [Role of the adrenal medulla in hemorheologic changes during myocardial ischemia in dogs.] *Sheng Li Hsueh Pao 40*:140-144, 1988.

MacLean, M.R. and A. Ungar. Effects of the renin-angiotensin system on the reflex response of the adrenal medulla to hypotension in the dog. *J Physiol (Lond) 373*:343-352, 1986.

Macquin-Mavier, I., C. Clerici, M.L. Franco-Montoya, and A. Harf. Mechanism of histamine-induced epinephrine release in guinea pig. *J Pharmacol Exp Ther 247*:706-709, 1988.

Madias, N.E., W.E. Goorno, and S. Herson. Severe lactic acidosis as a presenting feature of pheochromocytoma. *Am J Kidney Dis 10*:250-253, 1987.

Madrazo, I., R. Drucker-Colín, V. Díaz, J. Martínez-Mata, C. Torres, and J.J. Becerril. Open microsurgical autograft of adrenal medulla to the right caudate nucleus in two patients with intractable Parkinson's disease. *N Engl J Med 316*:831-834, 1987.

Madrazo, I., R. Drucker-Colín, V. León, and C. Torres. Adrenal medulla transplanted to caudate nucleus for treatment of Parkinson's disease: Report of 10 cases. *Surg Forum 38*:510-512, 1987.

Madrazo, I., V. León, C. Torres, M.d.C. Aguilera, G. Varela, F. Alvarez, A. Fraga, R. Drucker-Colín, F. Ostrosky, M. Skurovich, and R. Franco. Transplantation of fetal substantia nigra and adrenal medulla to the caudate nucleus in two patients with Parkinson's disease [letter]. *N Engl J Med 318*:51, 1988.

Magalhães, M.C., A. Bonito-Vitor, and M.M. Magalhães. Autoradiographic study of RNA synthesis in adrenal cells of the young rat in primary culture. *Cell Biol Int Rep 12*:109-114, 1988.

Magalhães, M.C., T.A.M. Serra, and M.M. Magalhães. RNA synthesis inhibitors on young rat adrenal in primary culture. An ultrastructural study. *Tissue Cell 19*:167-175, 1987.

Maggi, C.A. and A. Meli. Suitability of urethane anesthesia for physiopharmacological investigations in various systems. Part 1: General considerations. *Experientia 42*:109-114, 1986.

Mahapatra, M.S., S.K. Mahata, and B.R. Maito. Influence of age on diurnal rhythms of adrenal norepinephrine, epinephrine, and corticosterone levels in soft-shelled turtles (*Lissemys punctata punctata*). *Gen Comp Endocrinol 67*:279-281, 1987.

Mahata, S.K. and A. Ghosh. Influence of age on adrenomedullary catechol hormones content in twenty-five avian species. *Boll Zool 53*:63-67, 1986a.

Mahata, S.K. and A. Ghosh. Influence of splanchnic nerve on the resynthesis of adrenomedullary catecholamines post-reserpine-induced depletion in the pigeon, *Columba livia*. *Proc Indian natn Sci Acad B52*:346-350, 1986b.

Mahata, S.K. and A. Ghosh. Influence of splanchnic nerve and age on the action of insulin in the adrenomedullary catecholamine content and blood glucose level in pigeon. *Arch Biol (Bruxelles) 97*:443-454, 1986c.

Mahata, S.K., A. Mandal, and A. Ghosh. Influence of age and splanchnic nerve on the action of melatonin in the adrenomedullary catecholamine content and blood glucose level in the avian group. *J Comp Physiol [B] 158*:601-608, 1988.

Maitra, S.K. Seasonal changes in the adrenal medulla of male blossomheaded parakeet in relation to environmental and reproductive cycles. *J Interdiscipl Cycle Res 18*:43-48, 1987.

Majewski, H., P.I. Alade, and M.J. Rand. Adrenaline and stress-induced increases in blood pressure in rats. *Clin Exp Pharmacol Physiol 13*:283-288, 1986.

Makinen, K.K. and M.M. Hamalainen. Biochemical and morphological changes of the rat adrenal medulla induced by peroral xylitol. *Arch Toxicol 59*:192-193, 1986.

Malhotra, R.K. and A.R. Wakade. Vasoactive intestinal polypeptide stimulates the secretion of catecholamines from the rat adrenal gland. *J Physiol (Lond) 388*:285-294, 1987a.

Malhotra, R.K. and A.R. Wakade. Non-cholinergic component of rat splanchnic nerves predominates at low neuronal activity and is eliminated by naloxone. *J Physiol (Lond) 383*:639-652, 1987b.

Malhotra, R.K., T.D. Wakade, and A.R. Wakade. Comparison of secretion of catecholamines from the rat adrenal medulla during continuous exposure to nicotine, muscarine or excess K. *Neuroscience 26*:313-320, 1988a.

Malhotra, R.K., T.D. Wakade, and A.R. Wakade. Vasoactive intestinal polypeptide and muscarine mobilize intracellular

Ca^{2+} through breakdown of phosphoinositides to induce catecholamine secretion: Role of IP$_3$ in exocytosis. *J Biol Chem 263*:2123-2126, 1988b.

Malviya, A.N., M.M. Gabellec, and G. Rebel. Plasma membrane lipids of bovine adrenal chromaffin cells. *Lipids 21*:417-419, 1986.

Mandel, M., Y. Moriyama, J.D. Hulmes, Y.-C.E. Pan, H. Nelson, and N. Nelson. cDNA sequence encoding the 16-kDa proteolipid of chromaffin granules implies gene duplication in the evolution of H$^+$-ATPases. *Proc Natl Acad Sci USA 85*:5521-5524, 1988.

Manger, W.M. Pheochromocytoma [editorial]. *West J Med 145*:382-384, 1986.

Mangner, T.J., M.C. Tobes, D.W. Wieland, J.C. Sisson, and B. Shapiro. Metabolism of iodine[131] metaiodobenzylguanidine in patients with metastatic pheochromocytoma. *J Nucl Med 27*:37-44, 1986.

Man in 't Veld, A.J., F. Boomsma, P. Moleman, and M.A.D.H. Schalekamp. Congenital dopamine-β-hydroxylase deficiency: A novel orthostatic syndrome. *Lancet 1*:183-188, 1987.

Mannelli, M., M.L. DeFeo, M. Maggi, P. Geppetti, E. Baldi, C. Pupilli, and M. Serio. Effect of verapamil on catecholamine secretion by human pheochromocytoma [letter]. *Hypertension 8*:813-814, 1986.

Mannelli, M., M. Maggi, M.L. DeFeo, M. Boscaro, G. Opocher, F. Mantero, E. Baldi, and G. Giusti. Opioid modulation of normal and pathological human chromaffin tissue. *J Clin Endocrinol Metab 62*:577-582, 1986.

Marcinkiewicz, M., R. Fischer-Colbrie, J.P. Falgueyret, S. Benjannet, N.G. Seidah, C. Lazure, H. Winkler, and M. Chrétien. Two-dimensional immunoblotting analysis and immunocytochemical localization of the secretory polypeptide 7B2 in adrenal medulla. *Neurosci Lett 95*:81-87, 1988.

Maret, G.E., D. Blanot, M. Welti, and J.-L. Fauchére. Synthesis and use of peptide substrates for the isolation and characterization of proenkephalin processing enzymes. *Int J Pept Protein Res 29*:734-738, 1987.

Maret, G.E. and J.-L. Fauchére. The prohormone processing activity is enriched in a low-density subpopulation of chromaffin granules. *FEBS Lett 195*:258-260, 1986.

Maret, G.E. and J.-L. Fauchére. Purification of an endopeptidase from bovine adrenal medulla granules which cleaves *in vitro* at paired but not at single basic residues. *Anal Biochem 172*:248-258, 1988.

Margolis, R.K., B. Goossen, and R.U. Margolis. Phosphatidyl-inositol-anchored glycoproteins of PC12 pheochromocytoma cells and brain. *Biochemistry 27*:3454-3458, 1988.

Margolis, R.K., L.A. Greene, and R.U. Margolis. Poly(*N*-acetyllactosaminyl) oligosaccharides in glycoproteins of PC12 pheochromocytoma cells and sympathetic neurons. *Biochemistry* 25:3463-3468, 1986.

Margolis, R.K., J.A. Ripellino, B. Goossen, R. Steinbrich, and R.U. Margolis. Occurrence of the HNK-1 epitope (3-sulfo-glucuronic acid) in PC12 pheochromocytoma cells, chromaffin granule membranes, and chondroitin sulfate proteoglycans. *Biochem Biophys Res Commun 145*:1142-1148, 1987.

Margolis, R.U., R. Fischer-Colbrie, and R.K. Margolis. Poly(*N*-Acetyllactosaminyl) oligosaccharides of chromaffin granule membrane glycoproteins. *J Neurochem 51*:1819-1824, 1988.

Margulies, K.B. and S.G. Sheps. Carney's triad: Guidelines for management. *Mayo Clin Proc 63*:496-502, 1988.

Marker, J.C., D.A. Arnall, R.K. Conlee, and W.W. Winder. Effect of adrenodemedullation on metabolic responses to high-intensity exercise. *Am J Physiol 251*:R552-R559, 1986.

Markossian, K.A., V.Z. Melkonyan, N.A. Paitian, and R.M. Nalbandyan. On the copper transfer between dopamine β-monooxygenase and Cu-thionein. *Biochem Biophys Res Commun 153*:558-563, 1988.

Markossian, K.A., N.A. Paitian, M.V. Mikaelyan, and R.M. Nalbandyan. On the physiological role of neurocuprein: Aponeurocuprein is an inhibitor of dopamine β-monooxygenase. *Biochem Biophys Res Commun 138*:1-8, 1986.

Markossian, K.A., N.A. Paitian, and R.M. Nalbandyan. The reactivation of apodopamine β-monooxygenase by vanadyl ions. *FEBS Lett 238*:401-404, 1988.

Marley, P.D. New insights into the non-nicotinic regulation of adrenal medullary function. *Trends Pharmacol Sci 8*:411-413, 1987.

Marley, P.D. Desensitization of the nicotinic secretory response of adrenal chromaffin cells. *Trends Pharmacol Sci 9*:102-107, 1988.

Marley, P.D. and S.J. Bunn. Lack of effect of opioid compounds on angiotensin II responses of bovine adrenal medullary cells. *Neurosci Lett 90*:343-348, 1988.

Marley, P.D., S.J. Bunn, and B.G. Livett. Prostanoid responses of bovine adrenal medullary cells: Lack of effect of opioids. *Eur J Pharmacol 145*:173-181, 1988.

Marley, P.D. and B.G. Livett. Neuropeptides in the autonomic nervous system. *CRC Crit Rev Clin Neurobiol 1*:201-283, 1985.

Marley, P.D. and B.G. Livett. Differences between the mechanisms of adrenaline and noradrenaline secretion from isolated, bovine, adrenal chromaffin cells. *Neurosci Lett 77*:81-86, 1987a.

Marley, P.D. and B.G. Livett. Effects of opioid compounds on desensitization of the nicotinic response of isolated bovine adrenal chromaffin cells. *Biochem Pharmacol 36*:2937-2944, 1987b.

Marley, P.D., K.I. Mitchelhill, and B.G. Livett. Metorphamide, a novel endogenous adrenal opioid peptide, inhibits nicotine-induced secretion from bovine adrenal chromaffin cells. *Brain Res 363*:10-17, 1986a.

Marley, P.D., K.I. Mitchelhill, and B.G. Livett. Effects of opioid peptides containing the sequence of Met[5]-enkephalin or Leu[5]-enkephalin on nicotine-induced secretion from bovine adrenal chromaffin cells. *J Neurochem 46*:1-11, 1986b.

Marriott, D., M. Adams, and M.R. Boarder. Effect of forskolin and prostaglandin E_1 on stimulus secretion coupling in cultured bovine adrenal chromaffin cells. *J Neurochem 50*:616-623, 1988.

Martin, S.M., T.J. Malkinson, L.G. Bauce, W.L. Veale, and Q.J. Pittman. Plasma catecholamines in conscious rabbits after central administration of vasopressin. *Brain Res 457*:192-195, 1988.

Martin, W.H. and C.E. Creutz. Chromobindin A: A Ca^{2+}- and ATP-regulated chromaffin granule binding protein. *J Biol Chem 262*:2803-2810, 1987.

Martinez J.L. Jr., G. Schulteis, P.H. Janak, and S.B. Weinberger. Behavioral assessment of forgetting in aged rodents and its relationship to peripheral sympathetic function. *Neurobiol Aging 9*:697-708, 1988.

Martinez, P., A. Gimenez, E. Castro, M.J. Oset-Gasque, S. Cañadas, and M.P. González. GABA binding in bovine adrenal medulla membranes is sensitive to baclofen. *Comp Biochem Physiol [C] 88*:155-157, 1987.

Maruoka, Y., H. Saito, H. Tanaka, T. Nakajima, and K. Yazawa. Determination of catecholamines and their metabolites in urine of *Suncus murinus*. *Biogenic Amines 5*:483-488, 1988.

Marzotko, D. and D. Pankow. [Effect of chronic alcohol administration on the adrenals thyroid, pancreatic islands and liver of male albino rats.] *Gegenbaurs Morphol Jahrb 133*:185-202, 1987a.

Marzotko, D. and D. Pankow. Effect of single dichloromethane administration on the adrenal medulla of male albino rats. *Acta Histochem (Jena) 82*:177-183, 1987b.

Masiakowski, P. and E.M. Shooter. Nerve growth factor induces the genes for two proteins related to a family of calcium-binding proteins in PC12 cells. *Proc Natl Acad Sci USA 85*:1277-1281, 1988.

Mason, D.F.J., S. Medbak, and L.H. Rees. Circulating [Met]enkephalin and catecholamine responses to acute hypotension and hypertension in anaesthetized greyhounds. *Br J Pharmacol 91*:103-111, 1987.

Mastrolia, L., R. Bichi, M. Arizzi, and H. Manelli. Acetylcholinesterase and pseudocholinesterases in developing chick adrenal. *Basic Appl Histochem 30*:97-108, 1986.

Mastrolia, L., H. Manelli, V.P. Gallo, and M. Arizzi. Acetylcholinesterase activity in the adrenal chromaffin cells of *Triturus cristatus, Siren lacertina* and *Desmognathus quadramaculatus (Amphibia Urodela). Cell Mol Biol 32*:511-517, 1986.

Mathieson, P.W., L.M. Sandler, G.F. Joplin, J. Lynn, and J.M.T. Willoughby. Coexistent adrenal and extra-adrenal phaeochromocytomas accurately localized with non-invasive investigations. *Br J Surg 73*:41-42, 1986.

Mathieu, E., E. Despres, N. Delepine, and A. Taieb. MR imaging of the adrenal gland in Sipple disease. *J Comput Assist Tomogr 11*:790-794, 1987.

Matsuda, M. Autotransplantation of the adrenal gland using microvascular anastomosis. *Transplant Proc 19*:2581-2587, 1987.

Matsuda, Y. and G. Guroff. Purification and mechanism of activation of a nerve growth factor-sensitive S6 kinase from PC12 cells. *J Biol Chem 262*:2832-2844, 1987.

Matsuda, Y., N. Nakanishi, G. Dickens, and G. Guroff. A nerve growth factor-sensitive S6 kinase in cell-free extracts from PC12 cells. *J Neurochem 47*:1728-1734, 1986.

Matsui, H. Effect of subthalamic stimulation on adrenal epinephrine and norepinephrine secretion in the rat. *Brain Res 417*:158-160, 1987.

Matsumoto, S., K. Tanaka, A. Yamamoto, H. Nakada, M. Uchida, and Y. Tashiro. Immunoelectron microscopic localization of dopamine β–hydroxylase and chromogranin A in adrenomedullary chromaffin cells. *Cell Struct Funct 12*:483-496, 1987.

Matsumoto, T., K. Sumikawa, and T. Kashimoto. Mechanism of the inhibitory effect of thiopentone on stimulus-secretion coupling in chromaffin cells. *J Pharm Pharmacol 38*:595-599, 1986.

Matsuo, H. Isolation and identification of opioid peptides. *Natl Inst Drug Abuse Res Monogr Ser 70*:92-108, 1986.

Matsuoka, I., R. Satake, and K. Kurihara. Cholinergic differentiation of clonal rat pheochromocytoma cells (PC12) induced by factors contained in glioma-conditioned medium: Enhancement of high-affinity choline uptake system and reduction of norepinephrine uptake system. *Dev Brain Res 24*:145-152, 1986a.

Matsuoka, I., R. Satake, and K. Kurihara. Promotion of cell-substratum adhesion of clonal rat pheochromocytoma cells (PC12) by factors contained in glioma-conditioned medium (GCM): Separation of two active factors contained in GCM. *Dev Brain Res 24*:133-143, 1986b.

Matsuoka, I., B. Syuto, K. Kurihara, and S. Kubo. ADP-ribosylation of specific membrane proteins in pheochromocytoma and primary-cultured brain cells by botulinum neurotoxins type C and D. *FEBS Lett 216*:295-299, 1987.

Matsuuchi, L., K.M. Buckley, A.W. Lowe, and R.B. Kelly. Targeting of secretory vesicles to cytoplasmic domains in AtT-20 and PC12 cells. *J Cell Biol 106*:239-251, 1988.

Matthies, H.J.G., H.C. Palfrey, and R.J. Miller. Calmodulin- and protein phosphorylation-independent release of catecholamines from PC12 cells. *FEBS Lett 229*:238-242, 1988.

Maurer, R. and J.C. Reubi. Distribution and coregulation of three peptide receptors in adrenals. *Eur J Pharmacol 125*:241-247, 1986a.

Maurer, R. and J.C. Reubi. Somatostatin receptors in the adrenal. *Mol Cell Endocrinol 45*:81-90, 1986b.

May, S.W., H.H. Herman, S.F. Roberts, and M.C. Ciccarello. Ascorbate depletion as a consequence of product recycling during dopamine β-monooxygenase catalyzed selenoxidation. *Biochemistry 26*:1626-1633, 1987.

May, S.W., K. Wimalasena, H.H. Herman, L.C. Fowler, M.C. Ciccarello, and S.H. Pollock. Novel antihypertensives targeted at dopamine β-monooxygenase: Turnover-dependent cofactor depletion by phenyl aminoethyl selenide. *J Med Chem 31*:1066-1068, 1988.

Mazzanti, L., V. Scalori, M. Mian, M.G. Alessandrí, and L. Giovannini. Aclacinomycin tissue distribution in the rat. *Chemioterapia 7*:49-52, 1988.

McCafferty, B. and R.H. Angeletti. Microsequencing of dopamine β-monooxygenase. *J Neurosci Res 18*:289-292, 1987.

McCarty, R. Age-related alterations in sympathetic-adrenal medullary responses to stress. *Gerontology 32*:172-183, 1986.

McCarty, R. Sympathetic nervous system development in hypertensive rats. *J Hypertens 4 [Suppl 3]*:S155-S157, 1986.

McCarty, R., M.A. Cierpial, R.F. Kirby, and T.J. Jenal. Development of cardiac sympathetic and adrenal-medullary responses in borderline hypertensive rats. *J Auton Nerv Syst 21*:43-49, 1987.

McCarty, R., K. Horwatt, and M. Konarska. Chronic stress and sympathetic-adrenal medullary responsiveness. *Soc Sci Med 26*:333-341, 1988.

McCarty, R., R.F. Kirby, M.A. Cierpial, and T.J. Jenal. Accelerated development of cardiac sympathetic responses in spontaneously hypertensive (SHR) rats. *Behav Neural Biol 48*:321-333, 1987.

McCracken, J., P.R. Desai, N.J. Papadopoulos, J.J. Villafranca, and J. Peisach. Electron spin-echo studies of the copper(II) binding sites in dopamine β-hydroxylase. *Biochemistry 27*:4133-4137, 1988.

McGahan, J.P. Adrenal gland: MR imaging. *Radiology 166*:284-285, 1988.

McHugh, E.M. and R. McGee Jr. Direct anesthetic-like effects of forskolin on the nicotinic acetylcholine receptors of PC12 cells. *J Biol Chem 261*:3103-3106, 1986.

McIlhinney, R.A.J., S.J. Bacon, and A.D. Smith. A simple and rapid method for the production of cholera B chain coupled to horseradish peroxidase for neuronal tracing. *J Neurosci*

Methods 22:189-194, 1988.

McKay, D.B., M.J. Cobianchi, and A.S. Schneider. Comparison of the effects of colchicine and β-lumicolchicine on cultured adrenal chromaffin cells: Lack of evidence for an action of colchicine on receptor-associated microtubules. *Pharmacology* 35:155-162, 1987.

McMillen, I.C., H.M. Mulvogue, C.L. Coulter, C.A. Browne, and P.R.C. Howe. Ontogeny of catecholamine-synthesizing enzymes and enkephalins in the sheep adrenal medulla: An immunocytochemical study. *J Endocrinol* 118:221-226, 1988.

McRae-Degueurce, A., H.L. Klawans, R.D. Penn, A. Dahlström, C.M. Tanner, C.G. Goetz, and P.M. Carvey. An antibody in the CSF of Parkinson's disease patients disappears following adrenal medulla transplantation. *Neurosci Lett* 94:192-197, 1988.

Meers, P., K. Hong, and D. Papahadjopoulos. Free fatty acid enhancement of cation-induced fusion of liposomes: Synergism with synexin and other promoters of vesicle aggregaiton. *Biochemistry* 27:6784-6794, 1988.

Mekhedova, A.Y. and A.M. Ghadirian. The influence of experimental neurosis on the conditional reflexes and the content of blood catecholamines and acetylcholine in dogs. *Pavlov J Biol Sci* 14:79-85, 1979.

Meldolesi, J., G. Gatti, A. Ambrosini, T. Pozzan, and E.W. Westhead. Second-messenger control of catecholamine release from PC12 cells. Role of muscarinic recepters and nerve-growth-factor-induced cell differentiation. *Biochem J* 255:761-768, 1988.

Meldolesi, J., A. Malgaroli, C.B. Wollheim, and T. Pozzan. Ca^{2+} transients and secretion: Studies with quin2 and other Ca^{2+} indicators. pp. 289-313. In: *In Vitro* **Methods for Studying Secretion**, Vol. 3. (A.M. Poisner and J.-M. Trifaró, eds). Elsevier Science Publishers, Amsterdam, 1987.

Meldolesi, J. and T. Pozzan. Pathways of Ca^{2+} influx at the plasma membrane: Voltage-, receptor-, and second messenger-operated channels. *Exp Cell Res* 171:271-283, 1987.

Meldolesi, J., P. Volpe, and T. Pozzan. The intracellular distribution of calcium. *Trends Neurosci* 11:449-452, 1988.

Melnikova, T.S., A.N. Strizhakov, V.M. Strugatskîi, Podzolkova, N.M., and O.S. Brusov. [Transcutaneous electro-stimulation for sympathoadrenal crises.] *Sov Med* 5:71-75, 1987.

Mendoza, E.A., M.C. Lueg, F.A. Puyau, and C.M. Sutherland. Pheochromocytoma: Localization with CT scan and selective angiogram. *J La State Med Soc* 138:45-47, 1986.

Menniti, F.S., J. Knoth, and E.J. Diliberto Jr. Role of ascorbic acid in dopamine β-hydroxylation. The endogenous enzyme cofactor and putative electron donor for cofactor regeneration. *J Biol Chem* 261:16901-16908, 1986.

Menniti, F.S., J. Knoth, D.S. Peterson, and E.J. Diliberto Jr. The *in situ* kinetics of dopamine β–hydroxylase in bovine adrenomedullary chromaffin cells. Intravesicular compartmentation reduces apparent affinity for the cofactor ascorbate. *J Biol Chem* 262:7651-7657, 1987.

Mercuro, G., G. Gessa, C.A. Rivano, L. Lai, and A. Cherchi. Evidence for a dopaminergic control of sympathoadrenal catecholamine release. *Am J Cardiol* 62:827-828, 1988.

Merz, B. Adrenal-to-brain transplants improve the prognosis for Parkinson's disease. *JAMA* 257:2691-2692, 1987a.

Merz, B. Adrenal-to-brain transplants continue. *JAMA* 258:881-882, 1987b.

Merz, B. 'Slow down and proceed with caution' is new rule for brain-graft surgeons. *JAMA* 260:449-450, 1988.

Messing, R.O., C.L. Carpenter, and D.A. Greenberg. Inhibition of calcium flux and calcium channel antagonist binding in the PC12 neural cell line by phorbol esters and protein kinase C. *Biochem Biophys Res Commun* 136:1049-1056, 1986.

Metters, K.M. and J. Rossier. Affinity purification of synenkephalin-containing peptides, including a novel 23.3-kilodalton species. *J Neurochem* 49:721-728, 1987.

Metters, K.M., J. Rossier, J. Paquin, M. Chrétien, and N.G. Seidah. Selective cleavage of proenkephalin-derived peptides (<23,300 daltons) by plasma kallikrein. *J Biol Chem* 263:12543-12553, 1988.

Mezey, E., N.G. Seidah, M. Chrétien, and M.J. Brownstein. Demonstration of the vasopressin associated glycopeptide in the brain and peripheral tissues of the Brattleboro rat. *Neuropeptides* 7:79-85, 1986.

Mezrich, R., M.P. Banner, and H.M. Pollack. Magnetic resonance imaging of the adrenal glands. *Urol Radiol* 8:127-138, 1986.

Michener, M.L., W.B. Dawson, and C.E. Creutz. Phosphorylation of a chromaffin granule-binding protein in stimulated chromaffin cells. *J Biol Chem* 261:6548-6555, 1986.

Miettinen, M. and A. Saari. Pheochromocytoma combined with malignant schwannoma: Unusual neoplasm of the adrenal medulla. *Ultrastruct Pathol* 12:513-527, 1988.

Mikaelyan, M.V., N.A. Grigoryan, and R.M. Nalbandyan. Similarities of physico-chemical and antigenic properties of neurocupreins and extremely acidic copper proteins from chromaffin granules. *Biochim Biophys Acta* 924:548-555, 1987.

Mikaelyan, M.V., K.A. Markossian, N.A. Paitian, S.G. Sharoyan, and R.M. Nalbandyan. Secretory granules from different glands contain neurocuprein-like protein. *Biochem Biophys Res Commun* 155:1430-1436, 1988.

Milbrandt, J. Nerve growth factor rapidly induces c-*fos* mRNA in PC12 rat pheochromocytoma cells. *Proc Natl Acad Sci USA* 83:4789-4793, 1986.

Millaruelo, A.I., M.R. Sagarra, E.G. Delicado, M. Torres, and M.T. Miras-Portugal. Enzymes and pathways of glucose utilization in bovine adrenal medulla. *Mol Cell Biochem* 70:67-76, 1986.

Miller, L.G., A.S. Tischler, J.E. Jumblatt, and D.J. Greenblatt. Benzodiazepine binding sites on PC12 cells: Modulation by nerve growth factor and forskolin. *Neurosci Lett* 89:342-348, 1988.

Milner, J.D. and R.J. Wurtman. Catecholamine synthesis: Physiological coupling to precursor supply. *Biochem Pharmacol* 35:875-881, 1986.

Min, K.-W., A. Clemens, J. Bell, and H. Dick. Malignant peripheral nerve sheath tumor and pheochromocytoma. A composite tumor of the adrenal. *Arch Pathol Lab Med* 112:266-270, 1988.

Minamino, N., A. Uehara, and A. Arimura. Biological and immunological characterization of corticotropin-releasing activity in the bovine adrenal medulla. *Peptides* 9:37-45, 1988.

Minenko, A., B. Gabrysiak, and P. Oehme. Decreased SP-stimulated diesteratic hydrolysis of inositol phospholipids in adrenal medulla slices from spontaneously hypertensive rats. *Biomed Biochim Acta* 47:31-37, 1988.

Minenko, A., G. Kiselev, E. Tulkova, and P. Oehme. Nicotinic stimulation of polyphosphoinositide turnover in rat adrenal medulla slices. *Pharmazie* 42:341-344, 1987.

Minenko, A. and P. Oehme. Substance P action on inositol phospholipids in rat adrenal medulla slices. *Biomed Biochim Acta* 46:461-467, 1987.

Minenko, A., E. Tjulkova, and P. Oehme. Effects of substance P (SP) and the N-terminal SP-analogue SP (1-4) on inositol phospholipids in rat adrenal medulla slices. *Biomed Biochim Acta* 47:871-879, 1988.

Ming, L. and B. Renaud. Effects of α-methyldopa on systolic pressure and activities of tyrosine hydroxylase, dopamine-β-hydroxylase and phenylethanolamine-*N*-methyltransferase in nervous system of spontaneously hypertensive rats. *Chung Kuo Yao Li Hsueh Pao* 6:244-248, 1985.

Minth, C.D., P.C. Andrews, and J.E. Dixon. Characterization, sequence, and expression of the cloned human neuropeptide Y gene. *J Biol Chem* 261:11974-11979, 1986.

Miras-Portugal, M.T., M. Torres, P. Rotlan, and D. Aunis. Adenosine transport in bovine chromaffin cells in culture. *J Biol Chem* 261:1712-1719, 1986.

Misbahuddin, M. Muscarinic receptor-mediated catecholamine secretion from isolated guinea pig adrenal chromaffin cells. *Tokushima J Exp Med* 34:101-105, 1987.

Misbahuddin, M., H. Houchi, A. Nakanishi, K. Morita, and M. Oka. Stimulation by vasoactive intestinal polypeptide of muscarinic receptor-mediated catecholamine secretion from isolated guinea pig adrenal medullary cells. *Neurosci Lett* 72:315-319, 1986.

Misbahuddin, M. and M. Oka. Muscarinic stimulation of guinea pig adrenal chromaffin cells stimulates catecholamine secretion without significant increase in Ca^{2+} uptake. *Neurosci Lett 87*:266-270, 1988.

Misbahuddin, M., M. Oka, A. Nakanishi, and K. Morita. Stimulatory effect of vasoactive intestinal polypeptide on catecholamine secretion from isolated guinea pig adrenal chromaffin cells. *Neurosci Lett 92*:202-206, 1988.

Mitchell, G., J. Hattingh, and M. Ganhao. Stress in cattle assessed after handling, after transport and after slaughter. *Vet Rec 123*:201-205, 1988.

Mitchell, J.P. and P.R. Vulliet. Altered reactivity of the rat adrenal medulla following periods of chronic stress. *Adv Exp Med Biol 221*:367-374, 1987.

Mitsuma, T., D.H. Sun, T. Nogimori, M. Chaya, K. Ohtake, and Y. Hirooka. Dopamine inhibits thyrotropin-releasing hormone release from rat adrenal gland *in vitro*. *Horm Res 25*:223-227, 1987.

Miyamoto, S. and S. Fujime. Regulation by Ca^{2+} of membrane elasticity of bovine chromaffin granules. *FEBS Lett 238*:67-70, 1988.

Miyamoto, Y., M. Ozaki, and H. Yamamoto. Effects of adrenalectomy on pharmacokinetics and antinociceptive activity of morphine in rats. *Jpn J Pharmacol 46*:379-386, 1988.

Miyauchi, A., F. Matsuzaka, K. Kuma, K. Endo, T. Ogihara, and M. Maeda. [Diagnosis of adrenal medullary diseases in patients with sporadic or hereditary medullary thyroid carcinoma. A report of 37 cases with a maximum 8-year follow-up study.] *Nippon Geka Gakkai Zasshi 88*:1423-1429, 1987.

Mód, L., Z. Rihmer, I. Magyar, M. Arató, A. Alfoldi, and G. Bagdy. Serum DBH activity in psychotic vs. nonpsychotic unipolar and bipolar depression [letter]. *Psychiatry Res 19*:331-333, 1986.

Mohamed, A.A., T.L. Parker, and R.E. Coupland. The innervation of the adrenal gland. II. The source of spinal afferent nerve fibres to the guinea-pig adrenal gland. *J Anat 160*:51-58, 1988.

Monkhouse, W.S. The effect of *in vivo* hydrocortisone administration on the labelling index and size of chromaffin tissue in the postnatal and adult mouse. *J Anat 144*:133-144, 1986.

Monkhouse, W.S. and I. Fussell. A fraction of labelled mitoses study on adrenal chromaffin tissue in the newborn mouse and the effect of hydrocortisone. *J Anat 157*:105-109, 1988.

Monkhouse, W.S. and A. Khalique. The adrenal and renal veins of man and their connections with azygos and lumbar veins. *J Anat 146*:105-115, 1986.

Monstein, H.-J. and T. Geijer. A highly sensitive northern blot assay detects multiple proenkephalin A-like mRNAs in human caudate nucleus and pheochromocytoma. *Biosci Rep 8*:255-261, 1988.

Montiel, C., A.R. Artalejo, P.M. Bermejo, and P. Sánchez-García. A dopaminergic receptor in adrenal medulla as a possible site of action for the droperidol-evoked hypertensive response. *Anesthesiology 65*:474-479, 1986.

Moore, R.Y. Parkinson's disease - A new therapy? [editorial]. *N Engl J Med 316*:872-873, 1987.

Morel, G., J.-G. Chabot, T. Garcia-Caballero, F. Gossard, F. Dihl, M. Belles-Isles, and S. Heisler. Synthesis, internalization, and localization of atrial natriuretic peptide in rat adrenal medulla. *Endocrinology 123*:149-158, 1988.

Moreland, S., M.P. Ushay, S.D. Kimball, J.R. Powell, and R.S. Moreland. Pressor responses induced by BAY K 8644 involve both release of adrenal catecholamines and calcium channel activation. *Br J Pharmacol 93*:994-1004, 1988.

Morelli, A., M. Grasso, and P. Calissano. Effect of nerve growth factor on glucose utilization and nucleotide content of pheochromocytoma cells (clone PC12). *J Neurochem 47*:375-381, 1986.

Morihisa, J.M., R.K. Nakamura, W.J. Freed, M. Mishkin, and R.J. Wyatt. Transplantation techniques and the survival of adrenal medulla autografts in the primate brain. *Ann NY Acad Sci 495*:599-605, 1987.

Morita, K., T. Dohi, S. Kitayama, Y. Koyama, and A. Tsujimoto. Enhancement of stimulation-evoked catecholamine release from cultured bovine adrenal chromaffin cells by forskolin. *J Neurochem 48*:243-247, 1987a.

Morita, K., T. Dohi, S. Kitayama, Y. Koyama, and A. Tsujimoto. Stimulation-evoked Ca^{2+} fluxes in cultured bovine adrenal chromaffin cells are enhanced by forskolin. *J Neurochem 48*:248-252, 1987b.

Morita, K., H. Houchi, A. Nakanishi, K. Minakuchi, and M. Oka. Inhibitory action of hydralazine on catecholamine-synthesizing enzymes prepared from bovine adrenal medulla. *Jpn J Pharmacol 40*:445-453, 1986.

Morita, K., H. Houchi, A. Nakanishi, K. Minakuchi, K. Teraoka, and M. Oka. Inositol phospholipids cause the activation of adrenal tyrosine hydroxylase through their electrostatic action on the enzyme. *Int J Biochem 18*:857-860, 1986.

Morita, K., S. Ishii, H. Uda, and M. Oka. Requirement of ATP for exocytotic release of catecholamines from digitonin-permeabilized adrenal chromaffin cells. *J Neurochem 50*:644-648, 1988.

Morita, K., M.A. Levine, and H.B. Pollard. Stimulatory effect of ascorbic acid on norepinephrine biosynthesis in digitonin-permeabilized adrenal medullary chromaffin cells. *J Neurochem 46*:939-945, 1986.

Morita, K., A. Nakanishi, H. Houchi, M. Oka, K. Teraoka, K.

Minakuchi, S. Hamano, and Y. Murakumo. Modulation by basic polypeptides of ATP-induced activation of tyrosine hydroxylase prepared from bovine adrenal medulla. *Arch Biochem Biophys* 247:84-90, 1986.

Morita, K., A. Nakanishi, Y. Murakumo, M. Oka, and K. Teraoka. Effects of hypoglycemic sulfonylureas on catecholamine secretion and calcium uptake in cultured bovine adrenal chromaffin cells. *Biochem Pharmacol* 37:983-985, 1988.

Morita, K., M. Oka, and S. Hamano. Effects of cytoskeleton-disrupting drugs on ouabain-stimulated catecholamine secretion from cultured adrenal chromaffin cells. *Biochem Pharmacol* 37:3357-3359, 1988.

Morita, K., K. Teraoka, and M. Oka. Interaction of cytoplasmic tyrosine hydroxylase with chromaffin granule: *In vitro* studies on association of soluble enzyme with granule membranes and alteration in enzyme activity. *J Biol Chem* 262:5654-5658, 1987.

Moriyama, Y. and N. Nelson. The purified ATPase from chromaffin granule membranes is an anion-dependent proton pump. *J Biol Chem* 262:9175-9180, 1987a.

Moriyama, Y. and N. Nelson. Nucleotide binding sites and chemical modification of the chromaffin granule proton ATPase. *J Biol Chem* 262:14723-14729, 1987b.

Moriyama, Y. and N. Nelson. Internal anion binding site and membrane potential dominate the regulation of proton pumping by the chromaffin granule ATPase. *Biochem Biophys Res Commun* 149:140-144, 1987c.

Moriyama, Y. and N. Nelson. The vacuolar H^+-ATPase, a proton pump controlled by a slip. *Prog Clin Biol Res* 273:387-394, 1988a.

Moriyama, Y. and N. Nelson. Purification and properties of a vanadate- and *N*-ethylmaleimide-sensitive ATPase from chromaffin granule membranes. *J Biol Chem* 263:8521-8527, 1988b.

Moriyama, Y. and N. Nelson. Inhibition of vacuolar H^+-ATPases by fusidic acid and suramin. *FEBS Lett* 234:383-386, 1988c.

Morris, M., V. Kapoor, and J. Chalmers. Plasma neuropeptide Y concentration is increased after hemorrhage in conscious rats: Relative contributions of sympathetic nerves and the adrenal medulla. *J Cardiovas Pharmacol* 9:541-545, 1987.

Morton, A.J. An investigation of the role of adrenergic innervation in the regulation of the extraneuronal uptake of [^{3}H]-isoprenaline into rat vasa deferentia and atria. *Br J Pharmacol* 91:333-346, 1987.

Mosher, A.H. and C.H. Kircher. Proliferative lesions of the adrenal medulla in rats treated with zomepirac sodium. *J Am Coll Toxicol* 7:83-91, 1988.

Motti, E.D.F., G. Pezzoli, V. Silani, and G. Scarlato. Surgical lesions, Parkinsonism, and brain graft operations. *Lancet* 2:346, 1988.

Moulonguet, M. [Adrenal gland tumors.] *Rev Infirm* 37:49-52, 1987.

Mpoy, M. and J. Kolanowski. Urinary catecholamine excretion in patients with secondary adrenocortical insufficiency. *J Endocrinol Invest* 9:253-255, 1986.

Mukherjee, M. The effect of catecholamines on the testis of weaver bird (*Ploceus philippinus*). *Mikroskopie* 42:39-44, 1985.

Mukoyama, M., K. Nakao, N. Morii, S. Shiono, H. Itoh, A. Sugawara, T. Yamada, Y. Saito, H. Arai, and H. Imura. Atrial natriuretic polypeptide in bovine adrenal medulla. *Hypertension* 11:692-696, 1988.

Muldoon, S.M., J.E. McKenzie, and F.J. Collins. Pressor effect of nalbuphine in hemorrhagic shock is dependent on the sympathoadrenal system. *Circ Shock* 26:89-98, 1988.

Mulinari, R.A., M.T. Zanella, E.M.M. Guerra, O. Kohlmann Jr., C.I. Saad, A. Andriollo, J.G.R. Carvalho, and A.B. Ribeiro. The clonidine test for the diagnosis of pheochromocytoma: The usefulness of urinary metanephrine measurements. *Braz J Med Biol Res* 20:43-46, 1987.

Müller, T.H. and K. Unsicker. Nerve growth factor and dexamethasone modulate synthesis and storage of catecholamines in cultured rat adrenal medullary cells: Dependence on postnatal age. *J Neurochem* 46:516-524, 1986.

Mundy, D., T. Hermann, and W.J. Strittmatter. Specific inhibitors implicate a soluble metalloendoproteinase in exocytosis. *Cell Mol Neurobiol* 7:425-437, 1987.

Muñoz, J.A., J. García-Estañ, M. Canteras, T. Quesada, and M.T. Miras-Portugal. Dopamine-β-hydroxylase activity in plasma, spleen and adrenal gland of streptozotocin-diabetic rats: Correlation with cataracts. *Rev Esp Fisiol* 42:63-70, 1986.

Murase, H., K. Kamikubo, M. Murayama, K. Yasuda, K. Tsurumi, and K. Miura. [Characterization of adrenal medullary opioid receptors. II. Coupling of adrenal medullary opioid receptors to GTP binding proteins.] *Nippon Naibunpi Gakkai Zasshi* 63:741-751, 1987a.

Murase, H., K. Kamikubo, M. Murayama, K. Yasuda, K. Tsurumi, and K. Miura. [Characterization of adrenal medullary opioid receptors. I. Binding of opioids to adrenal medullary opioid receptors.] *Nippon Naibunpi Gakkai Zasshi* 63:727-740, 1987b.

Murofushi, H., A. Ikai, K. Okuhara, S. Kotani, H. Aizawa, K. Kumakura, and H. Sakai. Purification and characterization of kinesin from bovine adrenal medulla. *J Biol Chem* 263:12744-12750, 1988.

Murphy, C.A., M.A. Cierpial, A.H. Borom, and R. McCarty. Sympathetic responses of the heart and adrenal medulla in developing Dahl hypertensive rats. *Physiol Behav* 39:733-737, 1987.

Murray, S.S., L.L. Deaven, D.W. Burton, D.T. O'Connor, P.L.

Mellon, and L.J. Deftos. The gene for human chromogranin A (CgA) is located on chromosome 14. *Biochem Biophys Res Commun 142*:141-146, 1987.

Mutoh, T., B.B. Rudkin, S. Koizumi, and G. Guroff. Nerve growth factor, a differentiating agent, and epidermal growth factor, a mitogen, increase the activities of different S6 kinases in PC 12 cells. *J Biol Chem 263*:15853-15856, 1988.

Nagahama, S., Y.-F. Chen, M.D. Lindheimer, and S. Oparil. Mechanism of the pressor action of LY171555, a specific dopamine D_2 receptor agonist, in the conscious rat. *J Pharmacol Exp Ther 236*:735-742, 1986.

Nagatsu, T. Dopamine β-hydroxylase. pp. 79-115. In: **Neuromethods, Series I: Neurotransmitter Enzymes**, Vol. 5. (A.A. Boulton, G.B. Baker, and P.H. Yu, eds). Humana Press, Clifton, NJ, 1986.

Nagatsu, T. and Y. Hirata. Inhibition of the tyrosine hydroxylase system by MPTP, 1-methyl-4-phenylpyridinium ion (MPP^+) and the structurally related compounds *in vitro* and *in vivo*. *Eur Neurol 26 [Suppl 1]*:11-15, 1987.

Nagatsu, T. and K. Oka. Tyrosine 3-monooxygenase from bovine adrenal medulla. *Methods Enzymol 142*:56-62, 1987.

Nagatsu, T., H. Suzuki, K. Kiuchi, M. Saitoh, and H. Hidaka. Effects of myosin light-chain kinase inhibitor on catecholamine secretion from rat pheochromocytoma PC12h cells. *Biochem Biophys Res Commun 143*:1045-1048, 1987.

Nakada, T., H. Furuta, and T. Katayama. Catecholamine metabolism in pheochromocytoma and normal adrenal medullae. *J Urol 140*:1348-1351, 1988.

Nakajo, M. [Role of the scintigram in the diagnosis of adrenal diseases.] *Rinsho Hoshasen 31*:495-498, 1986.

Nakaki, T., N. Sasakawa, S. Yamamoto, and R. Kato. Functional shift from muscarinic to nicotinic cholinergic receptors involved in inositol trisphosphate and cyclic GMP accumulation during the primary culture of adrenal chromaffin cells. *Biochem J 251*:397-403, 1988.

Nakamura, M., K. Kamata, H. Inoue, and M. Inaba. Effect of intracerebroventricular administration of opioid peptides on basal serum adrenaline levels in conscious rats. *Jpn J Pharmacol 47*:151-157, 1988a.

Nakamura, M., K. Kamata, H. Inoue, K. Matsuyama, and M. Inaba. The effects of intracerebroventricular administration of adrenergic agonists and antagonists on adrenaline secretion from the adrenal medulla in stressed conscious rats. J *Pharmacobiodyn 11*:600-606, 1988b.

Nakanishi, A., K. Morita, Y. Murakumo, Y. Ishimura, M. Oka, and S. Hamano. Inhibitory action of hydralazine on catecholamine secretion from cultured bovine adrenal chromaffin cells. *Pharmacol Res Commun 18*:895-908, 1986.

Nakanishi, A., K. Morita, and M. Oka. Influence of reduction of cytoplasmic ATP on catecholamine secretion from intact and digitonin-permeabilized adrenal chromaffin cells. *Jpn J Pharmacol 46*:109-115, 1988.

Nakanishi, A., K. Morita, M. Oka, and N. Katsunuma. Evidence against a possible involvement of the serine, and thiol proteases in the exocytotic mechanism of catecholamine secretion in cultured bovine adrenal medullary cells. *Biochem Int 13*:799-807, 1986.

Nakata, H. and H. Fujisawa. Adenosine binding sites of rat pheochromocytoma PC12 cell membranes: Partial characterization and solubilization. *J Biochem (Tokyo) 104*:457-460, 1988.

Nakazato, Y., A. Ohga, M. Oleshansky, U. Tomita, and Y. Yamada. Voltage-independent catecholamine release mediated by the activation of muscarinic receptors in guinea-pig adrenal glands. *Br J Pharmacol 93*:101-109, 1988.

Nakazato, Y., A. Ohga, and Y. Yamada. Facilitation of transmitter action on catecholamine output by cardiac glycoside in perfused adrenal gland of guinea-pig. *J Physiol (Lond) 374*:475-491, 1986.

Naoi, M., H. Ichinose, T. Takahashi, I. Nagatsu, and T. Nagatsu. Reduction of aromatic L-amino acid decarboxylase activity in clonal pheochromocytoma PC12h cells by culture in the presence of N-methyl-4-phenylpyridinium ion (MPP^+). *Neurosci Lett 95*:229-235, 1988.

Naoi, M., K. Shibahara, H. Suzuki, and T. Nagatsu. Specific binding of glycosylated β-galactosidase to rat clonal pheochromo-cytoma PC12h cells: Effects of nerve growth factor and gangliosides. *J Neurochem 50*:1297-1301, 1988.

Naoi, M., H. Suzuki, K. Kiuchi, T. Takahashi, and T. Nagatsu. Effect of N-methyl-4-phenylpyridinium ion in monoamine oxidase in a clonal rat pheochromocytoma cell line, PC12h. *J Neurochem 48*:1912-1916, 1987.

Naoi, M., H. Suzuki, T. Takahashi, K. Shibahara, and T. Nagatsu. Ganglioside GM1 causes expression of type B monoamine oxidase in a rat clonal pheochromocytoma cell line, PC12h. *J Neurochem 49*:1602-1605, 1987.

Naoi, M., T. Takahashi, H. Ichinose, and T. Nagatsu. Inhibition of aromatic L-amino acid decarboxylase in clonal pheochromo-cytoma PC12h cells by N-methyl-4-phenylpyridinium ion (MPP^+). *Biochem Biophys Res Commun 152*:15-21, 1988a.

Naoi, M., T. Takahashi, H. Ichinose, K. Wakabayashi, T. Sugimura, and T. Nagatsu. Reduction of enzyme activity of tyrosine hydroxylase and aromatic L-amino acid decarboxylase in clonal pheochromocytoma PC12h cells by carcinogenic heterocyclic amines. *Biochem Biophys Res Commun 157*:494-499, 1988b.

Naoi, M., T. Takahashi, and T. Nagatsu. Metabolism of N-methyl-4-phenyl-1,2,3,6-tetrahydropyridine in a rat pheochromocytoma cell line, PC12h. *Life Sci 41*:2655-2661,

1987.

Naoi, M., T. Takahashi, and T. Nagatsu. Effect of 1-methyl-4-phenylpyridinium ion (MPP$^+$) on catecholamine levels and activity of related enzymes in clonal rat pheochromocytoma PC12h cells. *Life Sci 43*:1485-1491, 1988.

Naranjo, J.R., I. Mocchetti, J.P. Schwartz, and E. Costa. Permissive effect of dexamethasone on the increase of proenkephalin mRNA induced by depolarization of chromaffin cells. *Proc Natl Acad Sci USA 83*:1513-1517, 1986.

Naranjo, J.R., B.C. Wise, B. Mellstrom, and E. Costa. Negative feedback regulation of the content of proenkephalin mRNA in chromaffin cell cultures. *Neuropharmacology 27*:337-343, 1988.

Nassar-Gentina, V., H.B. Pollard, and E. Rojas. Electrical activity in chromaffin cells of intact mouse adrenal gland. *Am J Physiol 254*:C675-C683, 1988.

Natelson, B.H., D. Creighton, R. McCarty, W.N. Tapp, D. Pitman, and J.E. Ottenweller. Adrenal hormonal indices of stress in laboratory rats. *Physiol Behav 39*:117-125, 1987.

Negishi, M., S. Ito, T. Tanaka, H. Yokohama, H. Hayashi, T. Katada, M. Ui, and O. Hayaishi. Covalent cross-linking of prostaglandin E receptor from bovine adrenal medulla with a pertussis toxin-insensitive guanine nucleotide-binding protein. *J Biol Chem 262*:12077-12084, 1987.

Negishi, M., S. Ito, H. Yokohama, H. Hayashi, T. Katada, M. Ui, and O. Hayaishi. Functional reconstitution of prostaglandin E receptor from bovine adrenal medulla with guanine nucleotide binding proteins. *J Biol Chem 263*:6893-6900, 1988.

Nelson, N. The vacuolar proton-ATPase of eukaryotic cells. *Bioessays 7*:251-254, 1987.

Nelson, N., S. Cidon, and Y. Moriyama. Chromaffin granule proton pump. *Methods Enzymol 157*:619-633, 1988.

Nelson, T.J. and S. Kaufman. Interaction of tyrosine hydroxylase with ribonucleic acid and purification with DNA-cellulose or poly(A)-Sepharose affinity chromatography. *Arch Biochem Biophys 257*:69-84, 1987.

Neufeld, G., D. Gospodarowicz, L. Dodge, and D.K. Fujii. Heparin modulation of the neurotropic effects of acidic and basic fibroblast growth factors and nerve growth factor on PC12 cells. *J Cell Physiol 131*:131-140, 1987.

Nguyen, G.-K. Percutaneous fine-needle aspiration biopsy cytology of the kidney and adrenal. *Pathol Annu 22*:163-191, 1987.

Nguyen, M.H., D. Harbour, and C. Gagnon. Secretory proteins from adrenal medullary cells are carboxyl-methylated *in vivo* and released under their methylated form by acetylcholine. *J Neurochem 49*:38-44, 1987.

Nguyen, T.-T. and A. De Léan. Nonadrenergic modulation by clonidine of the cosecretion of catecholamines and enkephalins in adrenal chromaffin cells. *Can J Physiol*

Pharmacol 65:823-827, 1987.

Nguyen, T.-T., H. Ong, and A. De Léan. Secretion and biosynthesis of atrial natriuretic factor by cultured adrenal chromaffin cells. *FEBS Lett 231*:393-396, 1988.

Nichol, C.A., G.K. Smith, J.F. Reinhard Jr., E.C. Bigham, M.M. Abou-Donia, O.H. Viveros, and D.S. Duch. Regulation of tetrahydrobiopterin biosynthesis and cofactor replacement by tetrahydropterins. pp. 81-106. In: **Unconjugated Pterins in Neurobiology: Clinical Implications.** (W. Lovenberg and R. Levine, eds). Taylor & Francis, London, 1988.

Nicholson, W.E., G.S. DeCherney, R.V. Jackson, and D.N. Orth. Pituitary and hypothalamic hormones in normal and neoplastic adrenal medullae: Biologically active corticotropin-releasing hormone and corticotropin. *Regul Pept 18*:173-188, 1987.

Nieber, K. and P. Oehme. Effect of naloxone on the basal ^{3}H-noradrenaline outflow from adrenal gland slices of normotensive and spontaneously hypertensive rats. *Biomed Biochim Acta 45*:655-659, 1986.

Nieber, K. and P. Oehme. Effect of substance P (SP) and the N-terminal SP-analogue SP(1-4) on the pre- and postsynaptic transmitter release in rat adrenal gland slices. *Biomed Biochim Acta 46*:103-109, 1987.

Nieber, K., P. Oehme, V.A. Arefolov, and A.V. Valdman. Effect of the N-terminal tetrapeptide of substance P SP (1-4) and tuftsin on the pre- and postsynaptic transmitter outflow in rat adrenal gland slices. *Biomed Biochim Acta 47*:663-666, 1988.

Nielsen, F.C. and S. Gammeltoft. Insulin-like growth factors are mitogens for rat pheochromocytoma PC12 cells. *Biochem Biophys Res Commun 154*:1018-1023, 1988.

Niijima, A. [Nervous regulation of the metabolism.] *Nippon Seirigaku Zasshi 47*:735-745, 1985.

Niijima, A. Neural control of blood glucose level. *Jpn J Physiol 36*:827-841, 1986.

Nilaver, G., M.C. Beinfeld, C.T. Bond, D. Daikh, B. Godfrey, and J.P. Adelman. Heterogeneity of motilin immunoreactivity in mammalian tissue. *Synapse 2*:266-275, 1988.

Nilsson, O., A. Dahlström, L.-E. Tisell, J.M. Lundberg, E. Theodorsson-Norheim, M. Goldstein, and H. Ahlman. Growth of human pheochromocytomas in the anterior eye chamber of the rat. A histochemical study on amine and peptide content of pheochromocytoma tumour cells. *Regul Pept 15*:129-141, 1986.

Nir, S., A. Stutzin, and H.B. Pollard. Effect of synexin on aggregation and fusion of chromaffin granule ghosts at pH 6. *Biochim Biophys Acta 903*:309-318, 1987.

Nishigaki, I., H. Ichinose, K. Tamai, and T. Nagatsu. Purification of aromatic L-amino acid decarboxylase from bovine brain with a monoclonal antibody. *Biochem J 252*:331-335, 1988.

Nishino, H., T. Ono, R. Shibata, S. Kawamata, H. Watanabe, S. Shiosaka, M. Tohyama, and Z. Karadi. Adrenal medullary cells transmute into dopaminergic neurons in dopamine-depleted rat caudate and ameliorate motor disturbances. *Brain Res 445*:325-337, 1988.

Nissinen, E. Determination of phenylethanolamine-*N*-methyl-transferase activity by high-performance liquid chromatography with on-line radiochemical detection. *J Chromatogr 371*:37-42, 1986.

Njus, D., P.M. Kelley, and G.J. Harnadek. Bioenergetics of secretory vesicles. *Biochim Biophys Acta 853*:237-265, 1986.

Njus, D., N.J. Kusnetz, Y.V. Pacquing, and P.M. Kelley. Cytochrome b_{561} and the maintenance of redox poise in secretory vesicles. pp. 303-312. In: **Plasma Membrane Oxidoreductases in Control of Animal and Plant Growth.** (F.L. Crane, ed). Plenum, New York, 1988.

Noble, E.P., M. Bommer, D. Liebisch, and A. Herz. H_1-histaminergic activation of catecholamine release by chromaffin cells. *Biochem Pharmacol 37*:221-228, 1988.

Noble, E.P., M. Bommer, E. Sincini, T. Costa, and A. Herz. H_1-Histaminergic activation stimulates inositol-1-phosphate accumulation in chromaffin cells. *Biochem Biophys Res Commun 135*:566-573, 1986.

Noguchi, S., T. Muramatsu, and T. Koike. Temporal increase of two proteins in PC12 pheochromocytoma cells induced by nerve growth factor. *Cell Struct Funct 12*:127-130, 1987.

Noronha-Blob, L., R.P. Marshall, W.J. Kinnier, and D.C. U'Prichard. Pharmacological profile of adenosine A_2 receptor in PC12 cells. *Life Sci 39*:1059-1067, 1986.

Notter, M.F.D., M. Gupta, and D.M. Gash. Neuronal properties of monkey adrenal medulla in vitro. *Cell Tissue Res 244*:69-76, 1986.

Notter, M.F.D., I. Irwin, J.W. Langston, and D.M. Gash. Neurotoxicity of MPTP and MPP$^+$ in vitro: Characterization using specific cell lines. *Brain Res 456*:254-262, 1988.

Novíc, M., L.J. Mijatov-Ukropina, R. Marinkovíc, B. Lazetic, and I. Popov. [Selective influence of antiepileptics on the cortex and medulla of adrenal gland.] *Srp Arh Celok Lek 113*:983-990, 1985.

Novíc, M., L.J. Mijatov-Ukropina, N. Mihíc, and R. Markinovíc. [Adrenal catecholamines in hypoxic stress.] *Srp Arh Celok Lek 114*:917-926, 1986.

Nussey, S.S., R.A. Prysor-Jones, A.H. Taylor, V.T.Y. Ang, and J.S. Jenkins. Arginine vasopressin and oxytocin in the bovine adrenal gland. *J Endocrinol 115*:141-149, 1987.

Nutr. Rev. An ascorbate shuttle drives catecholamine formation by adrenal chromaffin granules. *Nutr Rev 44*:248-250, 1986.

Obata, A., H. Tanaka, and H. Kawazura. Magnetic resonance studies on the copper site of dopamine β-monooxygenase in the presence of cyanide and azide anions. *Biochemistry* 26:4962-4968, 1987.

Obata, K., N. Kojima, H. Nishiye, H. Inoue, T. Shirao, S.C. Fujita, and K. Uchizono. Four synaptic vesicle-specific proteins: Identification by monoclonal antibodies and distribution in the nervous tissue and the adrenal medulla. *Brain Res 404*:169-179, 1987.

Obendorf, D., U. Schwarzenbrunner, R. Fischer-Colbrie, A. Laslop, and H. Winkler. Immunological characterization of a membrane glycoprotein of chromaffin granules: Its presence in endocrine and exocrine tissues. *Neuroscience 25*:343-351, 1988a.

Obendorf, D., U. Schwarzenbrunner, R. Fischer-Colbrie, A. Laslop, and H. Winkler. In adrenal medulla synaptophysin (protein p38) is present in chromaffin granules and in a special vesicle population. *J Neurochem 51*:1573-1580, 1988b.

O'Connor, D.T. and L.J. Deftos. Secretion of chromogranin A by peptide-producing endocrine neoplasms. *N Engl J Med 314*:1145-1151, 1986.

O'Connor, E.F. The adrenal medulla. pp. 182-193. In: **Lithium and the Endocrine System. Lithium Therapy Monographs**, Vol. 2. (F.N. Johnson, ed). 1988.

O'Connor, E.F., S.K. Naylor, R.H. Cox, and J.E. Lawler. Lithium chloride stabilizes systolic blood pressure and increases adrenal catecholamines in the spontaneously hypertensive rat. *Physiol Behav 44*:69-74, 1988.

Odink, J., J.T.N.M. Thissen, J.H. Zeegers, and W.H.P. Schreurs. Diurnal variations in, and age and sex dependency on, the excretion of free catecholamines, homovanillic acid, vanilmandelic acid, and 5-hydroxyindole-3-acetic acid in urine of children and adolescents. *Biogenic Amines 3*:257-265, 1986.

Ogawa, M., T. Ishikawa, and H. Ohta. Transdifferentiation of endocrine chromaffin cells into neuronal cells. *Curr Top Dev Biol 20*:99-110, 1986.

O'Grady, T.J. Bile, magnetism, and dopamine: Simple answers to difficult problems. *Semin Oncol 13*:201-214, 1986.

Ögüs, A. Possible role of glutathione in cofactor regeneration of dopamine β-hydroxylase. *Biochem Med Metab Biol 37*:1-4, 1987.

Ohara-Imaizumi, M. and K. Kumakura. Receptor desensitization and the transient nature of the secretory response of adrenal chromaffin cells. *Neurosci Lett 70*:250-254, 1986.

Ohara-Imaizumi, M., Y. Miyakawa, and K. Kumakura. Pertussis toxin attenuates clonidine inhibition of catecholamine release in adrenal chromaffin cells. *Neurosci Lett 93*:294-299, 1988.

Oka, M., M. Misbahuddin, Y. Ishimura, and K. Morita. Effects of DJ-7141, a new α_2-adrenoceptor agonist, on catecholamine secretion from isolated bovine adrenal medullary cells. *Biochem Pharmacol 36*:609-612, 1987.

Okuma, Y., K. Yokotani, and Y. Osumi. Sympathoadreno-medullary system mediation of the bombesin-induced central inhibition of gastric acid secretion. *Eur J Pharmacol 139*:73-78, 1987.

Okumura-Noji, K., T. Kato, and R. Tanaka. Ca^{2+}-Dependent phosphorylation by endogenous kinases of Mr 95K and 50K-55K proteins in PC12 pheochromocytoma cells. *Neurochem Res 11*:1583-1595, 1986.

Olson, L., E.-O. Backlund, W.J. Freed, M. Herrera-Marschitz, B.J. Hoffer, A. Seiger, and I. Strömberg. Transplantation of monoamine-producing cell systems *in oculo* and intracranially: Experiments in search of a treatment for Parkinson's disease. *Ann NY Acad Sci 457*:105-126, 1985.

Olson, L., E.-O. Backlund, G. Gerhardt, B.J. Hoffer, O. Lindvall, G. Rose, A. Seiger, and I. Strömberg. Nigral and adrenal grafts in Parkinsonism: Recent basic and clinical studies. *Adv Neurol 45*:85-94, 1987.

Olson, L., I. Strömberg, M. Bygdeman, A.-Ch. Granholm, B.J. Hoffer, R. Freedman, and A. Seiger. Human fetal tissues grafted to rodent hosts: Structural and functional observations of brain, adrenal and heart tissues *in oculo*. *Exp Brain Res 67*:163-178, 1987.

O'Malley, K.L. Sequence homology between tyrosine hydroxylase mRNA and a rat adrenal medullary cDNA. *J Neurosci Res 16*:3-12, 1986.

O'Malley, K.L., M.J. Anhalt, B.M. Martin, J.R. Kelsoe, S.L. Winfield, and E.I. Ginns. Isolation and characterization of the human tyrosine hydroxylase gene: Identification of 5' alternative splice sites responsible for multiple mRNAs. *Biochemistry 26*:6910-6914, 1987.

Ong, H., C. Lazure, T.-T. Nguyen, N. McNicoll, N.G. Seidah, M. Chrétien, and A. De Léan. Bovine adrenal chromaffin granules are a site of synthesis of atrial natriuretic factor. *Biochem Biophys Res Commun 147*:957-963, 1987.

Oppedal, B.R., P. Brandtzaeg, and J.T. Kemshead. Immuno-histochemical performance testing of monoclonal antibodies to neuroblastoma cells on normal adrenals, spinal and sympathetic ganglia, and neural crest tumours. *Histopathology 11*:351-362, 1987.

Ornberg, R.L. Cryoultramicrotomy and electron microanalysis of isolated bovine adrenal chromaffin cells. pp. 135-139. In: **The Science of Biological Specimen Preparation for Microscopy and Microanalysis.** (M. Mueller, R.P. Becker, A. Boyde, and J.J. Wolosewick, eds). SEM Inc., AMF O'Hare (Chicago), IL, 1985.

Ornberg, R.L., L.T. Duong, and H.B. Pollard. Intragranular vesicles: New organelles in the secretory granules of adrenal chromaffin cells. *Cell Tissue Res 245*:547-553, 1986.

Ornberg, R.L., G.A.J. Kuijpers, and R.D. Leapman. Electron

probe microanalysis of the subcellular compartments of bovine adrenal chromaffin cells: Comparison of chromaffin granules *in situ* and *in vitro*. *J Biol Chem 263*:1488-1493, 1988.

Orts, A., C. Orellana, T. Cantó, V. Ceña, C. González-García, and A.G. García. Inhibition of adrenomedullary catecholamine release by propranolol isomers and clonidine involving mechanisms unrelated to adrenoceptors. *Br J Pharmacol 92*:795-801, 1987.

Osamura, R.Y., Y. Tsutsumi, N. Yanaihara, H. Imura, and K. Watanabe. Immunohistochemical studies for multiple peptide-immunoreactivities and co-localization of Met-enkephalin-Arg^6-Gly^7-Leu^8, neuropeptide Y and somatostatin in human adrenal medulla and pheochromocytomas. *Peptides 8*:77-87, 1987.

Osserman, E.F. Adrenal grafts for Parkinson's disease [letter]. *N Engl J Med 317*:1093, 1987.

Ostrosky-Solís, F., L. Quintanar, I. Madrazo, R. Drucker-Colín, R. Franco-Bourland, and V. Leon-Meza. Neuropsychological effects of brain autograft of adrenal medullary tissue for the treatment of Parkinson's disease. *Neurology 38*:1442-1450, 1988.

O'Sullivan, A.J. and R.D. Burgoyne. The role of cytoplasmic pH in the inhibitory action of high osmolarity on secretion from bovine adrenal chromaffin cells. *Biochim Biophys Acta 969*:211-216, 1988.

Oyama, M. [The roles of endogenous amines in anaphylactic shock in rats.] *Kanagawa Shigaku 20*:307-319, 1985.

Ozawa, S. and O. Sand. Electrophysiology of excitable endocrine cells. *Physiol Rev 66*:887-952, 1986.

Ozutsumi, K., N. Sugimoto, and M. Matsuda. *Clostridium perfringens* type A enterotoxin induces release of noradrenaline from the neurosecretory PC12 cell line. *Biochem Biophys Res Commun 144*:217-223, 1987.

Pácha, J., J. Mejsnar, and R. Kvetnansky. Lymphatic transport rate of noradrenaline during adrenergic thermogenesis. *Comp Biochem Physiol [C] 83*:161-164, 1986.

Padbury, J.F., Y. Agata, J.K. Ludlow, M. Ikegami, B. Baylen, and J.A. Humme. Effect of fetal adrenalectomy on catecholamine release and physiologic adaptation at birth in sheep. *J Clin Invest 80*:1096-1103, 1987.

Palatini, P. Anion activation of dopamine β-hydroxylase: A kinetic model. *Int J Biochem 20*:427-433, 1988.

Paling, M.R., J.R. Brookeman, and J.P. Mugler. Tumor detection with phase-contrast imaging: An evaluation of clinical potential. *Radiology 162*:199-203, 1987.

Pandey, S.C., M.P. Dubey, and B.N. Dhawan. Ontogenic studies on biogenic amines in heart, lung and adrenal glands of *Mastomys natalensis*. Indian J Biochem Biophys 23:348-350, 1986.

Pandiella-Alonso, A., A. Malgaroli, L.M. Vicentini, and J.

Meldolesi. Early rise of cytosolic Ca^{2+} induced by NGF in PC12 and chromaffin cells. *FEBS Lett 208*:48-51, 1986.

Panerai, A.E. Plasma [Met]enkephalin concentrations after endocrine and pharmacological modifications. *Pharmacol Res Commun 20*:195-200, 1988.

Panula, P., L. Kivipelto, O. Nieminen, E.A. Majane, and H.-Y.T. Yang. Neuroanatomy of morphine-modulating peptides. *Med Biol 65*:127-135, 1987.

Pappas, G.D. and J. Sagen. Fine structure of PC12 cell implants in the rat spinal cord. *Neurosci Lett 70*:59-64, 1986.

Pappas, G.D. and J. Sagen. Fine structural correlates of vascular permeability of chromaffin cell transplants in CNS pain modulatory regions. *Exp Neurol 102*:280-289, 1988a.

Pappas. G.D. and J. Sagen. Morphological correlates of chromaffin cell transplants in CNS pain modulatory regions. pp. 283-288. In: **Tissue Engineering**, (R. Skalak and C.F. Fox, eds). Alan R. Liss, Inc., New York, 1988b.

Pappas, G.D. and J. Sagen. The fine structure of chromaffin cell implants in the pain modulatory regions of the rat periaqueductal gray and spinal cord. pp. 513-520, In: **Progress in Brain Research**, Vol. 78. (D.M. Gash and J.R. Sladek, eds). Elsevier Science Publishers, Amsterdam, 1988c.

Park, D.H. Monoamine methyltransferases: Phenylethanolamine *N*-methyltransferase and catechol *O*-methyltransferase. pp. 117-145. In: **Neuromethods, Series I: Neurotransmitter Enzymes**, Vol. 5. (A.A. Boulton, G.B. Baker and P.H. Yu, eds). Humana Press, Clifton, NJ, 1986.

Park, D.H., M. Ehrlich, M.J. Evinger, and T.H. Joh. Strain differences in distribution of phenylethanolamine *N*-methyltransferase activity from rat brain and adrenal gland. *Brain Res 372*:185-188, 1986.

Parmer, R.J. and D.T. O'Connor. Enkephalins in human phaeochromocytomas: localization in immunoreactive, high molecular weight form to the soluble core of chromaffin granules. *J Hypertens 6*:187-198, 1988.

Parsons, S.J. and C.E. Creutz. p60c-*src* activity detected in the chromaffin granule membrane. *Biochem Biophys Res Commun 134*:736-742, 1986.

Parti, R., E.D. Özkan, G.J. Harnadek, and D. Njus. Inhibition of norepinephrine transport and reserpine binding by reserpine derivatives. *J Neurochem 48*:949-953, 1987.

Parvez, H., K. Ichihara, S. Parvez, K. Sakai, Y. Abiko, and T. Nagatsu. Myocardial ischemia stimulates catecholamine synthesis and catabolism in the dog adrenal medulla. *Jpn Heart J 27*:345-354, 1986.

Parysek, L.M. and R.D. Goldman. Characterization of intermediate filaments in PC12 cells. *J Neurosci 7*:781-791, 1987.

Parysek, L.M. and R.D. Goldman. Distribution of a novel 57 kDa intermediate filament (IF) protein in the nervous system. *J*

Neurosci 8:555-563, 1988.

Patil, A., K. Fillmore, J. Valentine, and D. Hill. The study of the effect of human chorionic gonadotrophic (HCG) hormone on the survival of adrenal medulla transplant in brain. Preliminary study. *Acta Neurochir (Wien) 87*:76-78, 1987.

Patzak, A., D. Aunis, and O.K. Langley. Membrane recycling after exocytosis: An ultrastructural study of cultured chromaffin cells. *Exp Cell Res 171*:346-356, 1987.

Patzak, A. and H. Winkler. Exocytotic exposure and recycling of membrane antigens of chromaffin granules: Ultrastructural evaluation after immunolabeling. *J Cell Biol 102*:510-515, 1986.

Paves, H., T. Neuman, M. Metsis, and M. Saarma. Nerve growth factor induces rapid redistribution of F-actin in PC12 cells. *FEBS Lett 235*:141-143, 1988.

Pearce, J.M.S. Adrenal and nigral transplants for Parkinson's disease. *Br Med J 296*:1211-1212, 1988.

Pearse, A.G.E. The diffuse neuroendocrine system: Peptides, amines, placodes and the APUD theory. *Prog Brain Res 68*:25-31, 1986.

Pearse, B.M.F. Clathrin and coated vesicles. *EMBO J 6*:2507-2512, 1987.

Pelto-Huikko, M. Correlative microscopy of catecholamines and neuropeptides in adrenal medulla. pp. 143-155. In: **Correlative Microscopy in Biology: Instrumentation and Methods**. (M.A. Hayat, ed). Academic Press Inc, San Diego, 1987.

Pelto-Huikko, M., T. Salminen, and A. Hervonen. Localization of enkephalin- and neurotensin-like immunoreactivities in cat adrenal medulla. *Histochemistry 88*:31-36, 1987.

Penn, R.D., C.G. Goetz, C.M. Tanner, H.L. Klawans, K.M. Shannon, C.L. Comella, and T.R. Witt. The adrenal medullary transplant operation for Parkinson's disease: Clinical observations in five patients. *Neurosurgery 22*:999-1004, 1988.

Penner, R., E. Neher, and F. Dreyer. Intracellularly injected tetanus toxin inhibits exocytosis in bovine adrenal chromaffin cells. *Nature 324*:76-78, 1986.

Penner, R. and E. Neher. The role of calcium in stimulus-secretion coupling in excitable and non-excitable cells. *J Exp Biol 139*:329-345, 1988.

Peppers, S.C. and R.W. Holz. Catecholamine secretion from digitonin-treated PC12 cells: Effects of Ca^{2+}, ATP, and protein kinase C activators. *J Biol Chem 261*:14665-14669, 1986.

Percy, J.M. and D.K. Apps. Proton-translocating adenosine triphosphatase of chromaffin-granule membranes: The active site is in the largest (70 kDa) subunit. *Biochem J 239*:77-81, 1986.

Perin, M.S., V.A. Fried, C.A. Slaughter, and T.C. Südhof. The structure of cytochrome b_{561}, a secretory vesicle-specific electron transport protein. *EMBO J* 7:2697-2703, 1988.

Perlow, M.J. Brain grafting as a treatment for Parkinson's disease. *Neurosurgery* 20:335-342, 1987.

Perrin, D., O.K. Langley, and D. Aunis. Anti-α-fodrin inhibits secretion from permeabilized chromaffin cells. *Nature* 326:498-501, 1987.

Peskind, E.R., M.A. Raskind, C.W. Wilkinson, D.E. Flatness, and J.B. Halter. Peripheral sympathectomy and adrenal medullectomy do not alter cerebrospinal fluid norepinephrine. *Brain Res* 367:258-264, 1986.

Peters, J.A., E.F. Kirkness, H. Callachan, J.J. Lambert, and A.J. Turner. Modulation of the GABA$_A$ receptor by depressant barbituates and pregnane steroids. *Br J Pharmacol* 94:1257-1269, 1988.

Pezzoli, G., S. Fahn, A. Dwork, D.D. Truong, J.G. de Yebenes, V. Jackson-Lewis, J. Herbert, and J.L. Cadet. Non-chromaffin tissue plus nerve growth factor reduces experimental parkinsonism in aged rats. *Brain Res* 459:398-403, 1988.

Pfeiffer, G.L., W.S. Brimijoin, D.R. Studelska, and S.W. Carmichael. Electron microscopy of isolated mitochondrion-chromaffin vesicle complexes. *Z Mikrosk Anat Forsch* 101:567-576, 1987.

Pigeon, D., P. Ferrara, F. Gros, and J. Thibault. Rat pheochromo-cytoma tyrosine hydroxylase is phosphorylated on serine 40 by an associated protein kinase. *J Biol Chem* 262:6155-6158, 1987.

Pinto, J.E.B., P.M. Politi, and B. Fernandez. Effects of doxorubicin on the release of catecholamines from the bovine adrenal medulla. *Arch Int Pharmacodyn Ther* 289:140-148, 1987.

Plevin, R. and M.R. Boarder. Stimulation of formation of inositol phosphates in primary cultures of bovine adrenal chromaffin cells by angiotensin II, histamine, bradykinin, and carbachol. *J Neurochem* 51:634-641, 1988.

Pocock, G. and C.D. Richards. The action of pentobarbitone on stimulus-secretion coupling in adrenal chromaffin cells. *Br J Pharmacol* 90:71-80, 1987.

Pocock, G. and C.D. Richards. The action of volatile anaesthetics on stimulus-secretion coupling in bovine adrenal chromaffin cells. *Br J Pharmacol* 95:209-217, 1988.

Pocock, G. and T.J.B. Simons. Effects of lead ions on events associated with exocytosis in isolated bovine adrenal medullary cells. *J Neurochem* 48:376-382, 1987.

Pocotte, S.L. and R.W. Holz. Effects of phorbol ester on tyrosine hydroxylase phosphorylation and activation in cultured bovine adrenal chromaffin cells. *J Biol Chem* 261:1873-1877, 1986.

Pocotte, S.L., R.W. Holz, and T. Ueda. Cholinergic receptor-mediated phosphorylation and activation of tyrosine hydroxylase in cultured bovine adrenal chromaffin cells. *J Neurochem* 46:610-622, 1986.

Pollard, H.B., K.W. Brocklehurst, E.J. Forsberg, A. Stutzin, G. Lee, and A.L. Burns. Calcium regulation of membrane fusion during hormone secretion. *Adv Exp Med Biol 211*:369-383, 1987.

Pollard, H.B., A.L. Burns, and E. Rojas. A molecular basis for synexin-driven, calcium-dependent membrane fusion. *J Exp Biol 139*:267-286, 1988.

Porter, I.D., B.J. Whitehouse, A.H. Taylor, and S.S. Nussey. Effect of arginine vasopressin and oxytocin on acetylcholine-stimulation of corticosteroid and catecholamine secretion from the rat adrenal gland perfused *in situ*. *Neuropeptides 12*:265-271, 1988.

Powis, D.A. and P.F. Baker. α_2-Adrenoceptors do not regulate catecholamine secretion by bovine adrenal medullary cells: A study with clonidine. *Mol Pharmacol 29*:134-141, 1986.

Pozzan, T., F. DiVirgilio, L.M. Vicentini, and J. Meldolesi. Activation of muscarinic receptors in PC12 cells: Stimulation of Ca^{2+} influx and redistribution. *Biochem J 234*:547-553, 1986.

Prentice, H.M., S.E. Moore, J.G. Dickson, P. Doherty, and F.S. Walsh. Nerve growth factor-induced changes in neural cell adhesion molecule (N-CAM) in PC12 cells. *EMBO J 6*:1859-1863, 1987.

Printz, M.P. and V.L. Boyd. Studies of the angiotensin II receptor on cultured nervous system cells: Chromaffin cells and C6 glioma. *J Hypertens 4 [Suppl 6]*:S416-S418, 1986.

Prokocimer, P.G., M. Maze, and B.B. Hoffman. Role of the sympathetic nervous system in the maintenance of hypertension in rats harboring pheochromocytoma. *J Pharmacol Exp Ther 241*:870-874, 1987.

Proye, C., P. Fossati, P. Fontaine, J. Lefebvre, M. Decoulx, J.L. Wemeau, D. Dewailly, E. Rwamasirabo, and P. Cecat. Dopamine-secreting pheochromocytoma: An unrecognized entity? Classification of pheochromocytomas according to their type of secretion. *Surgery 100*:1154-1162, 1986.

Pruss, R.M. Monoclonal antibodies to chromaffin cells can distinguish proteins specific to or specifically excluded from chromaffin granules. *Neuroscience 22*:141-147, 1987.

Pruss, R.M., E. Mezey, D.S. Forman, L.E. Eiden, A.J. Hotchkiss, D.A. DiMaggio, and T.L. O'Donohue. Enkephalin and neuropeptide Y: Two colocalized neuropeptides are independently regulated in primary cultures of bovine chromaffin cells. *Neuropeptides 7*:315-327, 1986.

Pruss, R.M. and E.A. Shepard. Cytochrome b_{561} can be detected in many neuroendocrine tissues using a specific monoclonal antibody. *Neuroscience 22*:149-157, 1987.

Pruss, R.M. and K.A. Stauderman. Voltage-regulated calcium channels involved in the regulation of enkephalin synthesis are blocked by phorbol ester treatment. *J Biol Chem 263*:13173-13178, 1988.

Pruss, R.M. and N. Zamir. Regulated expression of atrial natriuretic peptide-like immunoreactivity in cultured bovine adrenomedullary chromaffin cells. *Neurochem Int 11*:299-304, 1987.

Pryde, J.G. and J.H. Phillips. Fractionation of membrane proteins by temperature-induced phase separation in Triton X-114: Application to subcellular fractions of the adrenal medulla. *Biochem J 233*:525-533, 1986.

Quik, M., L. Bergeron, H. Mount, and J. Philie. Dopamine D2 receptor binding in adrenal medulla: Characterization using [^{3}H]spiperone. *Biochem Pharmacol 36*:3707-3713, 1987.

Quik, M., S. Fournier, and J.-M. Trifaró. Modulation of the nicotinic α-bungarotoxin site in chromaffin cells in culture by a factor(s) endogenous to neuronal tissue. *Brain Res 372*:11-20, 1986.

Quik, M., S. Geertsen, and J.-M. Trifaró. Marked up-regulation of the α-bungarotoxin site in adrenal chromaffin cells by specific nicotinic antagonists. *Mol Pharmacol 31*:385-391, 1987.

Rabe, C.S., E. Delorme, and F.F. Weight. Muscarine-stimulated neurotransmitter release from PC12 cells. *J Pharmacol Exp Ther 243*:534-541, 1987.

Rabe, C.S. and F.F. Weight. Effects of ethanol on neurotransmitter release and intracellular free calcium in PC12 cells. *J Pharmacol Exp Ther 244*:417-422, 1988.

Rabin, R.A. Differential response of adenylate cyclase and ATPase activities after chronic ethanol exposure of PC12 cells. *J Neurochem 51*:1148-1155, 1988.

Racz, K., O. Kuchel, and N.T. Buu. Bromocriptine decreases blood pressure of spontaneously hypertensive rats without affecting the adrenomedullary synthesis of catecholamines. *J Cardiovasc Pharmacol 8*:676-680, 1986.

Racz, K., O. Kuchel, W. Debinski, and N.T. Buu. Improved liquid chromatographic determination of dopamine-β-hydroxylase activity in tissues and plasma. *J Chromatogr 382*:117-125, 1986.

Radin, D.R., P.W. Ralls, W.D. Boswell Jr., P.M. Colletti, S.A. Lapin, and J.M. Halls. Pheochromocytoma: Detection by unenhanced CT. *Am J Radiol 146*:741-744, 1986.

Rainey, W.E., J.I. Mason, C. Cochet, and B.R. Carr. Protein kinase-C in the human fetal adrenal gland. *J Clin Endocrinol Metab 67*:908-914, 1988.

Ramachandran, A.V. and M.M. Patel. Seasonal histomorphological alterations of adrenals and thyroid in normal and pinealectomized domestic pigeons *Columba livia*. *Indian J Exp Biol 24*:755-759, 1986.

Rasilo, M.-L. and T. Yamagata. Occurrence of glucose polymer in undifferentiated PC12 cells. *FEBS Lett 227*:191-194, 1988a.

Rasilo, M.-L. and T. Yamagata. Characterization of a glucose polymer from PC12 cells and neuronal cells of rat embryo. *J Biochem (Tokyo) 104*:742-754, 1988b.

Rasmussen, B.L. and O. Thorlacius-Ussing. Ultrastructural localization of mercury in adrenals from rats exposed to methyl mercury. *Virchows Arch [B]* 52:529-538, 1987.

Ratge, D., E. Knoll, and H. Wisser. Plasma free and conjugated catecholamines in clinical disorders. *Life Sci* 39:557-564, 1986.

Rathinavelu, A. and G.E. Isom. High affinity receptors for atrial natriuretic factor in PC12 cells. *Biochem Biophys Res Commun* 156:78-85, 1988.

Rauch, A.L. and W.G. Campbell Jr. Adrenal and vascular tyrosine hydroxylase activity in Goldblatt hypertension. *Hypertension* 12:434-442, 1988.

Rausch, D.M., A.L. Iacangelo, and L.E. Eiden. Glucocorticoid- and nerve growth factor-induced changes in chromogranin A expression define two different neuronal phenotypes in PC12 cells. *Mol Endocrinol* 2:921-927, 1988.

Raza-Bukhari, A., J.J. Fernández-Ruiz, J.R. Pais, C. Alvarez Sanz, and J.A. Ramos. Changes in catecholamine metabolism in adrenal medulla of adult male rats with experimentally induced hyperprolactinemia. *Biogenic Amines* 5:103-109, 1988.

Reed, J.K. and D. England. The effect of nerve growth factor on the development of sodium channels in PC12 cells. *Biochem Cell Biol* 64:1153-1159, 1986.

Regunathan, S. and T.L. Sourkes. Effect of vasopressin on the induction of adrenal tyrosine hydroxylase and phenylethanol-amine N-methyltransferase. *Regul Pept* 23:217-226, 1988.

Reiffen, F.U. and M. Gratzl. Chromogranins, widespread in endocrine and nervous tissue, bind Ca^{2+}. *FEBS Lett* 195:327-330, 1986a.

Reiffen, F.U. and M. Gratzl. Ca^{2+} binding to chromaffin vesicle matrix proteins: Effect of pH, Mg^{2+}, and ionic strength. *Biochemistry* 25:4402-4406, 1986b.

Reinhard, J.F. Jr., E.J. Diliberto Jr., O.H. Viveros, and A.J. Daniels. Subcellular compartmentalization of 1-methyl-4-phenylpyridinium with catecholamines in adrenal medullary chromaffin vesicles may explain the lack of toxicity to adrenal chromaffin cells. *Proc Natl Acad Sci USA* 84:8160-8164, 1987.

Reinig, J.W. and J.L. Doppman. Magnetic resonance imaging of the adrenal. *Radiologe* 26:186-190, 1986.

Reinig, J.W., J.L. Doppman, A.J. Dwyer, and J. Frank. MRI of indeterminate adrenal masses. *Am J Radiol* 147:493-496, 1986a.

Reinig, J.W., J.L. Doppman, A.J. Dwyer, A.R. Johnson, and R.H. Knop. Adrenal masses differentiated by MR. *Radiology* 158:81-84, 1986b.

Reubi, J.C., R. Maurer, K. von Werder, J. Torhorst, J.G.M. Klijn, and S.W.J. Lamberts. Somatostatin receptors in human endocrine tumors. *Cancer Res* 47:551-558, 1987.

Richard, F., M. Buda, and F. Legay. Purifications of rat adrenal dopamine-β–hydroxylase: Immunological analysis of its soluble and membrane-bound forms with the use of an antibody raised against the soluble form. *J Neurochem 51*:1140-1147, 1988.

Richter, R. and Ch. Grunwald. Effect of substance P on the membrane potential of rat adrenal chromaffin cells. *Biomed Biochim Acta 46*:837-840, 1987.

Richter-Landsberg, C. and B. Jastorff. The role of cAMP in nerve growth factor-promoted neurite outgrowth in PC12 cells. *J Cell Biol 102*:821-829, 1986.

Ricordi, C., S.D. Shah, P.E. Lacy, W.E. Clutter, and P.E. Cryer. Delayed extra-adrenal epinephrine secretion after bilateral adrenalectomy in rats. *Am J Physiol 254*:E52-E53, 1988.

Rieker, S., R. Fischer-Colbrie, L.E. Eiden, and H. Winkler. Phylogenetic distribution of peptides related to chromogranins A and B. *J Neurochem 50*:1066-1073, 1988.

Rindi, G., R. Buffa, F. Sessa, O. Tortora, and E. Solcia. Chromogranin A, B and C immunoreactivities of mammalian endocrine cells: Distribution, distinction from costored hormones/prohormones and relationship with the argyrophil component of secretory granules. *Histochemistry 85*:19-28, 1986.

Rindlisbacher, B., M. Reist, and P. Zahler. Diacylglycerol breakdown in plasma membranes of bovine chromaffin cells is a two-step mechanism mediated by a diacylglycerol lipase and a monoacylglycerol lipase. *Biochim Biophys Acta 905*:349-357, 1987.

Rink, T.J. and D.E. Knight. Stimulus-secretion coupling: A perspective highlighting the contributions of Peter Baker. *J Exp Biol 139*:1-30, 1988.

Riopelle, R.J. Hypothesis - Adrenal medulla autografts in Parkinson's disease: A proposed mechanism of action. *Can J Neurol Sci 15*:366-370, 1988.

Riphagen, C.L., L.G. Bauce, W.L. Veale, and Q.J. Pittman. The effects of intrathecal administration of arginine-vasopressin and substance P on blood pressure and adrenal secretion of epinephrine in rats. *J Auton Nerv Syst 16*:91-99, 1986.

Rochford, J. and J.L. Henry. Lack of effect of adrenal denervation on analgesia elicited by continuous and intermittent cold water swim in the rat. *Brain Res 445*:404-406, 1988.

Rodriguez del Castillo, A., M. Torres, E.G. Delicado, and M.T. Miras-Portugal. Subcellular distribution studies of diadenosine polyphosphates--Ap_4A and Ap_5A--in bovine adrenal medulla: Presence in chromaffin granules. *J Neurochem 51*:1696-1703, 1988.

Rogawski, M.A., M. Pieniek, S. Suzuki, and J.M.H. ffrench-Mullen. Phencyclidine selectively blocks the sustained voltage-dependent potassium conductance in PC12 cells. *Brain Res 456*:38-48, 1988.

Rohrer, H., A. Acheson, J. Thibault, and H. Thoenen. Developmental potential of quail dorsal root ganglion cells analyzed *in vitro* and *in vivo*. *J Neurosci 6*:2616-2624, 1986.

Rojas, E., E.J. Forsberg, and H.B. Pollard. Optical detection of calcium dependent ATP release from stimulated medullary chromaffin cells. *Adv Exp Med Biol 211*:7-29, 1987.

Rojas, E. and H.B. Pollard. Membrane capacity measurements suggest a calcium-dependent insertion of synexin into phosphatidylserine bilayers. *FEBS Letters 217*:25-31, 1987.

Rökaeus, A. and M.J. Brownstein. Construction of a porcine adrenal medullary cDNA library and nucleotide sequence analysis of two clones encoding a galanin precursor. *Proc Natl Acad Sci USA 83*:6287-6291, 1986.

Rökaeus, A. and M. Carlquist. Nucleotide sequence analysis of cDNAs encoding a bovine galanin precursor protein in the adrenal medulla and chemical isolation of bovine gut galanin. *FEBS Lett 234*:400-406, 1988.

Romano, C., R.A. Nichols, and P. Greengard. Synapsin I in PC12 cells. II. Evidence for regulation by NGF of phosphorylation at a novel site. *J Neurosci 7*:1300-1306, 1987.

Romano, C., R.A. Nichols, P. Greengard, and L.A. Greene. Synapsin I in PC12 cells. I. Characterization of the phosphoprotein and effect of chronic NGF treatment. *J Neurosci 7*:1294-1299, 1987.

Rose, F.C. and L. Nashef. Parkinson's disease: Further steps forward. *Gerontology 33*:369-373, 1987.

Rosenberg, M.B., E. Hawrot, and X.O. Breakefield. Biotinylated β-nerve growth factor binds to PC12 cells. *Ann NY Acad Sci 463*:214-216, 1986.

Rosenberg, P.H., E. Nissinen, P.T. Mannisto, L. Tuomisto, and J.E. Heavner. Differential effect of tetracaine and bupivacaine on catecholamine and 5-hydroxytryptamine turnover. *Pain 29*:239-245, 1987.

Rosenheck, K. and P.I. Lelkes, (eds). **Stimulus-Secretion Coupling in Chromaffin Cells.** Volumes I and II. CRC Press, Boca Raton, FL. 1987.

Rosenheck, K. and H. Plattner. Ultrastructural and cytochemical characterization of adrenal medullary plasma membrane vesicles and their interaction with chromaffin granules. *Biochim Biophys Acta 856*:373-382, 1986.

Rosenstein, J.M. Vascular and glial alterations after autonomic tissue grants into the brain. *Ann NY Acad Sci 495*:86-100, 1987a.

Rosenstein, J.M. Adrenal medulla grafts produce blood-brain barrier dysfunction. *Brain Res 414*:192-196, 1987b.

Roske, I., R. Rathsack, K. Nieber, H. Hilse, and P. Oehme. Influence of adrenal demedullation on stress-related behaviour in Wistar rats. *Pharmazie 42*:253-255, 1987.

Roske, I., E.A. Ûmatov, E.I. Pevcova, R. Rathsack, and P. Oehme. Influence of adrenal demedullation on stress induced alterations within the general adaptation behaviour of August-rats. *Pharmazie* 43:339-343, 1988.

Roskoski, R. Jr., and L.M. Roskoski. Activation of tyrosine hydroxylase in PC12 cells by the cyclic GMP and cyclic AMP second messenger systems. *J Neurochem* 48:236-242, 1987.

Roskoski, R. Jr., P.R. Vulliet, and D.B. Glass. Phosphorylation of tyrosine hydroxylase by cyclic GMP-dependent protein kinase. *J Neurochem* 48:840-845, 1987.

Ross, S.T., L.I. Kruse, E.H. Ohlstein, R.W. Erickson, M. Ezekiel, K.E. Flaim, J.L. Sawyer, and B.A. Berkowitz. Inhibitors of dopamine β-hydroxylase. 3. Some 1-(pyridylmethyl)imidazole-2-thiones. *J Med Chem* 30:1309-1313, 1987.

Roth, K.A., D.M. Wilson, J.H. Eberwine, R.I. Dorin, K. Kovacs, K.G. Bensch, and A.R. Hoffman. Acromegaly and pheochromocytoma: A multiple endocrine syndrome caused by a plurihormonal adrenal medullary tumor. *J Clin Endocrinol Metab* 63:1421-1426, 1986.

Rothschild, A.J., A.F. Schatzberg, P.J. Langlais, J.E. Lerbinger, M.M. Miller, and J.O. Cole. Psychotic and nonpsychotic depressions: I. Comparison of plasma catecholamines and cortisol measures. *Psychiatry Res* 20:143-153, 1987.

Rowell, L.B., G.L. Brengelmann, and P.R. Freund. Unaltered norepinephrine-heart rate relationship in exercise with exogenous heat. *J Appl Physiol* 62:646-650, 1987.

Rowland, E.A., T.H. Müller, M. Goldstein, and L.A. Greene. Cell-free detection and characterization of a novel nerve growth factor-activated protein kinase in PC12 cells. *J Biol Chem* 262:7504-7513, 1987.

Rubio, M.C. and M.B. Barontini. Abnormal catecholamine metabolism in pheochromocytoma. *Medicina (B Aires)* 45:631-636, 1985.

Rudy, B., B. Kirschenbaum, A. Rukenstein, and L.A. Greene. Nerve growth factor increases the number of functional Na channels and induces TTX-resistant Na channels in PC12 pheochromocytoma cells. *J Neurosci* 7:1613-1625, 1987.

Ruff, F. and M.C. Santais. [The H1 and H2 histamine receptors.] *Allerg Immunol (Paris)* 20:317-325, 1988.

Rundle, S., P. Somogyi, R. Fischer-Colbrie, C. Hagn, H. Winkler, and I.W. Chubb. Chromogranin A, B, and C: Immunohisto-chemical localization in ovine pituitary and the relationship with hormone-containing cells. *Regul Pept* 16:217-233, 1986.

Rungby, J. Exogenous silver in dorsal root ganglia, peripheral nerve, enteric ganglia, and adrenal medulla. *Acta Neuropathol (Berl)* 69:45-53, 1986.

Russell, J.T., M.A. Levine, and D. Njus. Electron transfer across posterior pituitary neurosecretory vesicle membranes. *J Biol Chem* 260:226-231, 1985.

Rutter, P.C., S.J. Potocnik, and J. Ludbrook. Sympathoadrenal mechanisms in cardiovascular responses to naloxone after hemorrhage. *Am J Physiol* 252:H40-H46, 1987.

Ruzzier, F., S. Lee, W.F. Dryden, and R. Miledi. *In vitro* reinnervation of adult rat muscle fibres by foreign neurons and transformed chromaffin PC12 cells. *Proc R Soc Lond [Biol] 234*:1-9, 1988.

Rydel, R.E. and L.A. Greene. Acidic and basic fibroblast growth factors promote stable neurite outgrowth and neuronal differentiation in cultures of PC12 cells. *J Neurosci 7*:3639-3653, 1987.

Saadat, S., A.D. Stehle, A. Lamouroux, J. Mallet, and H. Thoenen. Influence of cell-cell contact on levels of tyrosine hydroxylase in cultured bovine adrenal chromaffin cells. *J Biol Chem 262*:13007-13014, 1987.

Saadat, S., A.D. Stehle, A. Lamouroux, J. Mallet, and H. Thoenen. Predicted amino acid sequence of bovine tyrosine hydroxylase and its similarity to tyrosine hydroxylases from other species. *J Neurochem 51*:572-578, 1988.

Saadat, S. and H. Thoenen. Selective induction of tyrosine hydroxylase by cell-cell contact in bovine adrenal chromaffin cells is mimicked by plasma membranes. *J Cell Biol 103*:1991-1997, 1986.

Saadat, S. and H. Thoenen. Regulation of tyrosine hydroxylase mRNA levels in rat pheochromocytoma PC12 cells by cell-cell contact. *Exp Cell Res 176*:187-193, 1988.

Saarma, M., U. Toots, E. Raukas, A. Zhelkovsky, A. Pivazian, and T. Neuman. Nerve growth factor induces changes in (2'-5')oligo(A) synthetase and 2'-phosphodiesterase activities during differentiation of PC12 pheochromocytoma cells. *Exp Cell Res 166*:229-236, 1986.

Saavedra, J.M. and N. Alexander. Angiotensin II receptors in adrenal gland, pituitary gland and brain of sino-aortic denervated rats. *J Hypertens 4*:S154-S157, 1986.

Saavedra, J.M. and E.M. Krieger. Early increase in adrenomedullary catecholamine synthesis in sinoaortic denervated rats. *J Auton Nerv Syst 18*:181-183, 1987.

Sabatine, J.M. and C.J. Coffee. The identification and characterization of two cyclic nucleotide phosphodiesterases from bovine adrenal medulla. *Arch Biochem Biophys 249*:95-105, 1986.

Sagen, J. and G.D. Pappas. Morphological and functional correlates of chromaffin cell transplants in CNS pain modulatory regions. *Ann NY Acad Sci 495*:306-333, 1987.

Sagen, J. and G.D. Pappas. Pharmacologic consequences of the vascular permeability of chromaffin cell transplants in CNS pain modulatory regions. *Exp Neurol 102*:290-297, 1988a.

Sagen, J. and G.D. Pappas. Pain reduction by chromaffin cell grafts in the CNS. pp. 289-294. In: **Tissue Engineering**, (R. Skalak and C.F. Fox, eds). Alan R. Liss, Inc., New York, 1988b.

Sagen, J., G.D. Pappas, and M.J. Perlow. Adrenal medullary

tissue transplants in the rat spinal cord reduce pain sensitivity. *Brain Res 384*:189-194, 1986.

Sagen, J., G.D. Pappas, and M.J. Perlow. Alterations in nociception following adrenal medullary transplants into the rat periaqueductal gray. *Exp Brain Res 67*:373-379, 1987a.

Sagen, J., G.D. Pappas, and M.J. Perlow. Fine structure of adrenal medullary grafts in the pain modulatory regions of the rat periaqueductal gray. *Exp Brain Res 67*:380-390, 1987b.

Sagen, J., G.D. Pappas, and H.B. Pollard. Analgesia induced by isolated bovine chromaffin cells implanted in rat spinal cord. *Proc Natl Acad Sci USA 83*:7522-7526, 1986.

Saito, Y. and S. Kawashima. Enhancement of neurite outgrowth in PC12h cells by a protease inhibitor. *Neurosci Lett 89*:102-107, 1988.

Sajovic, P., E. Moraru, L.A. Greene, and M.L. Shelanski. Selective staining of large projection neurons by monoclonal antibody to a glycoprotein of PC12 cells. *J Neurosci 6*:82-93, 1986.

Sakai, M., M. Tadokoro, N. Makino, T. Ishigaki, T. Abe, and S. Sakuma. [Diagnostic efficacy of m-[^{131}I] iodobenzylguanidine (I-131 MIBG) scintigraphy in locating pheochromocytoma: comparison with computed tomography and ultrasonography.] *Kaku Igaku 24*:7-14, 1987.

Sakata, S. and J. Iriuchijima. Adrenomedullary origin of the hindquarter vasodilation during the transposition response of the rat. *Can J Physiol Pharmacol 66*:18-21, 1988.

Sakellariou, G., M. Markianos, N. Tsichlakis, and D. Kartakis. A family study of plasma dopamine-β-hydroxylase in schizophrenia. *Psychiatry Res 20*:221-227, 1987.

Sakharov, D.A. and L.M. Zhuravleva. [The origin of chromaffin tissue in vertebrates: The chromaffin cells of ascidians.] *Zh Obshch Biol 49*:218-226, 1988.

Sala, F., R.I. Fonteríz, R. Borges, and A.G. García. Inactivation of potassium-evoked adrenomedullary catecholamine release in the presence of calcium, strontium or BAY K 8644. *FEBS Lett 196*:34-38, 1986.

Samaan, N.A. and R.C. Hickey. Pheochromocytoma. *Semin Oncol 14*:297-305, 1987.

Samsoondar, J. and J.E. Kudlow. Partial purification of an adrenal growth factor produced by normal bovine anterior pituitary cells in culture. *Endocrinology 120*:929-935, 1987.

Sanberg, P.R. and A.B. Norman. Adrenal transplants for Huntington's disease? *Nature 335*:122, 1988.

Sano, M., K.-I. Kato, T. Totsuka, and R. Katoh-Semba. A convenient bioassay for NGF using a new subline of PC12 pheochromocytoma cells (PC12D). *Brain Res 459*:404-406, 1988.

Sarafian, T., D. Aunis, and M.-F. Bader. Loss of proteins from digitonin-permeabilized adrenal chromaffin cells essential for exocytosis. *J Biol Chem 262*:16671-16676, 1987.

Sargent, S.M., P.E. Aydelott, and H.J. Doll. Autologous adrenal medulla transplant. Investigational treatment for Parkinson's disease. *AORN J 47*:682-694, 1988.

Sasakawa, N., K. Ishii, and R. Kato. Nicotinic receptor-mediated intracellular calcium release in cultured bovine adrenal chromaffin cells. *Neurosci Lett 63*:275-279, 1986.

Sasakawa, N., K. Ishii, S. Yamamoto, and R. Kato. Differential effects of protein kinase C activators on carbamylcholine- and high K^+-induced rises in intracellular free calcium concentration in cultured adrenal chromaffin cells. *Biochem Biophys Res Commun 139*:903-909, 1986.

Sasakawa, N., T. Nakaki, S. Yamamoto, and R. Kato. Inositol trisphosphate accumulation by high K^+ stimulation in cultured adrenal chromaffin cells. *FEBS Lett 223*:413-416, 1987.

Sasakawa, N., S. Yamamoto, T. Nakaki, and R. Kato. Effects of islet-activating protein on the catecholamine release, Ca^{2+} mobilization and inositol trisphosphate formation in cultured adrenal chromaffin cells. *Biochem Pharmacol 37*:2485-2487, 1988.

Sastry, B.R., S.S. Chirwa, P.B.Y. May, and H. Maretic. Substances released during tetanic stimulation of rabbit neocortex induce neurite growth in PC12 cells and long-term potentiation in guinea pig hippocampus. *Neurosci Lett 91*:101-105, 1988.

Sato, A. Neural mechanisms of somatic sensory regulation of catecholamine secretion from the adrenal gland. *Adv Biophys 23*:39-80, 1987.

Sato, A., Y. Sato, and R.F. Schmidt. Catecholamine secretion and adrenal nerve activity in response to movements of normal and inflamed knee joints in cats. *J Physiol (Lond) 375*:611-624, 1986.

Sato, A., Y. Sato, K. Shimamura, and H. Suzuki. An increase in the sympathoadrenal medullary function in stroke-prone spontaneously hypertensive rats under anesthetized and resting conditions. *Neurosci Lett 72*:309-314, 1986.

Sato, A., Y. Sato, and H. Suzuki. Changes in sympatho-adrenal medullary functions during aging. *Neurol Neurobiol 31*:27-36, 1987.

Satoh, T., M. Endo, S. Nakamura, and Y. Kaziro. Analysis of guanine nucleotide bound to *ras* protein in PC12 cells. *FEBS Lett 236*:185-189, 1988.

Sautel, M., J. Sacquet, M. Vincent, and J. Sassard. NE turnover in genetically hypertensive rats of Lyon strain. II. Peripheral organs. *Am J Physiol 255*:H736-H741, 1988.

Schadt, J.C. and R.R. Gaddis. Role of adrenal medulla in hemodynamic response to hemorrhage and naloxone. *Am J Physiol 254*:R559-R565, 1988.

Schäfer, T., U.O. Karli, E.K.-M. Gratwohl, F.E. Schweizer, and M.M. Burger. Digitonin-permeabilized cells are exocytosis competent. *J Neurochem 49*:1697-1707, 1987.

Schäfer, T., U.O. Karli, F.E. Schweizer, and M.M. Burger. Docking of chromaffin granules--A necessary step in exocytosis? *Biosci Rep 7*:269-279, 1987.

Schalling, M., A. Dagerlind, S. Brené, H. Hallman, M. Djurfeldt, H. Persson, L. Terenius, M. Goldstein, D. Schlesinger, and T. Hökfelt. Coexistence and gene expression of phenylethanolamine *N*-methyltransferase, tyrosine hydroxylase, and neuropeptide tyrosine in the rat and bovine adrenal gland: Effects of reserpine. *Proc Nat Acad Sci USA 85*:8306-8310, 1988.

Schalling, M., A. Dagerlind, S. Brené, R. Petterson, S. Kvist, M. Brownstein, S.E. Hyman, L. Mucke, H.M. Goodman, T.H. Joh, M. Goldstein, and T. Hökfelt. Localization of mRNA for phenylethanolamine *N*-methyltransferase (PNMT) using *in situ* hybridization. *Acta Physiol Scand 131*:631-632, 1987.

Schalling, M., T. Hökfelt, B. Wallace, M. Goldstein, D. Filer, C. Yamin, and D.H. Schlesinger. Tyrosine 3-hydroxylase in rat brain and adrenal medulla: Hybridization histochemistry and immunohistochemistry combined with retrograde tracing. *Proc Natl Acad Sci USA 83*:6208-6212, 1986.

Schalling, M., K. Seroogy, T. Hökfelt, S.Y. Chai, H. Hallman, H. Persson, D. Larhammar, A. Ericsson, L. Terenius, J. Graffi, J. Massoulié, and M. Goldstein. Neuropeptide tyrosine in the rat adrenal gland--immunohistochemical and *in situ* hybridization studies. *Neuroscience 24*:337-349, 1988.

Scharrer, B. Neurosecretion: Beginnings and new directions in neuropeptide research. *Annu Rev Neurosci 10*:1-17, 1987.

Scherman, D., F. Darchen, C. Desnos, and J.-P. Henry. 1-Methyl-4-phenylpyridinium is a substrate of the vesicular monoamine uptake system of chromaffin granules. *Eur J Pharmacol 146*:359-360, 1988.

Scherman, D., B. Gasnier, P. Jaudon, and J.-P. Henry. Hydrophobicity of the tetrabenazine-binding site of the chromaffin granule monoamine transporter. *Mol Pharmacol 33*:72-77, 1988.

Scherman, D., A. Soumarmon, and J.-P. Henry. The acylphosphate present in chromaffin granule membrane preparations is not associated with the proton-pump. *Biochimie 68*:1303-1309, 1986.

Scherman, D. and M.J. Weber. Characterization of the vesicular monoamine transporter in cultured rat sympathetic neurons: Persistence upon induction of cholinergic phenotypic traits. *Dev Biol 119*:68-74, 1987.

Schilling, K. and M. Gratzl. Quantification of p38/synaptophysin in highly purified adrenal medullary chromaffin vesicles. *FEBS Lett 233*:22-24, 1988.

Schmid, K.W., R. Fischer-Colbrie, C. Hagn, B. Jasani, E.D. Williams, and H. Winkler. Chromogranin A and B and secretogranin II in medullary carcinomas of the thyroid. *Am J*

Surg Pathol 11:551-556, 1987.

Schmid-Antomarchi, H., M. Hugues, and M. Lazdunski. Properties of the apamin-sensitive Ca^{2+}-activated K^+ channel in PC12 pheochromocytoma cells which hyper-produce the apamin receptor. *J Biol Chem 261*:8633-8637, 1986.

Schmidt, W.E., E.G. Siegel, H. Kratzin, and W. Creutzfeldt. Isolation and primary structure of tumor-derived peptides related to human pancreastatin and chromogranin A. *Proc Natl Acad Sci USA 85*:8231-8235, 1988.

Schmidt, W.E., E.G. Siegel, R. Lamberts, B. Gallwitz, and W. Creutzfeldt. Pancreastatin: Molecular and immunocyto-chemical characterization of a novel peptide in porcine and human tissues. *Endocrinology 123*:1395-1404, 1988.

Schober, M., R. Fischer-Colbrie, K.W. Schmid, G. Bussolati, D.T. O'Connor, and H. Winkler. Comparison of chromogranins A, B, and secretogranin II in human adrenal medulla and pheochromocytoma. *Lab Invest 57*:385-391, 1987.

Schömig, E. and H. Bönisch. Solubilization and characterization of the ^{3}H-desipramine binding site of rat phaeochromocytoma cells (PC12-cells). *Naunyn Schmiedebergs Arch Pharmacol 334*:412-417, 1986.

Schroeder, B.A., R.G. Wells, and J.R. Sty. Comparative imaging: Pheochromocytoma. *Clin Nucl Med 12*:148-149, 1987.

Schuijers, J.A., D.W. Walker, C.A. Browne, and G.D. Thorburn. Effect of hypoxemia on plasma catecholamines in intact and immunosympathectomized fetal lambs. *Am J Physiol 251*:R893-R900, 1986.

Schurter, H.R. and C. Hedinger. [Cryptococcoma of the adrenal gland.] *Schweiz Med Wochenschr 117*:1186-1190, 1987.

Schwarting, G.A., A. Gajewski, L. Barbero, A.S. Tischler, and D. Costopoulos. Complex glycosphingolipids of the pheochro-mocytoma cell line PC12: Enhanced fucosylglycolipid synthesis following nerve growth factor treatment. *Neuroscience 19*:647-656, 1986.

Schwartz, J.P. and E. Costa. Hybridization approaches to the study of neuropeptides. *Annu Rev Neurosci 9*:277-304, 1986.

Schwartz, T.W., S.P. Sheikh, and M.M. O'Hare. Receptors on phaeochromocytoma cells for two members of the PP-fold family--NPY and PP. *FEBS Lett 225*:209-214, 1987.

Scott, R.A., R.J. Sullivan, W.E. DeWolf Jr., R.E. Dolle, and L.I. Kruse. The copper sites of dopamine β-hydroxylase: An X-ray absorption spectroscopic study. *Biochemistry 27*:5411-5417, 1988.

Sedl'ak, J. Effect of denervation on glutathione and oxidized glutathione in rat adrenal cortex and medulla after repeated stress. *Endocrinol Exp (Bratisl) 21*:263-268, 1987.

Seidah, N.G., G.N. Hendy, J. Hamelin, J. Paquin, C. Lazure, K.M. Metters, J. Rossier, and M. Chrétien. Chromogranin A can act as a reversible processing enzyme inhibitor: Evidence from the inhibition of the IRCM-serine protease 1 cleavage of pro-enkephalin and ACTH at pairs of basic amino acids. *FEBS Lett 211*:144-150, 1987.

Seidl, K., M. Manthorpe, S. Varon, and K. Unsicker. Differential effects of nerve growth factor and ciliary neuronotrophic factor on catecholamine storage and catecholamine synthesizing enzymes of cultured rat chromaffin cells. *J Neurochem 49*:169-174, 1987.

Seidler, F.J. and T.A. Slotkin. Non-neurogenic adrenal catecholamine release in the neonatal rat: Exocytosis or diffusion? *Brain Res 28*:274-277, 1986a.

Seidler, F.J. and T.A. Slotkin. Ontogeny of adrenomedullary responses to hypoxia and hypoglycemia: Role of splanchnic innervation. *Brain Res Bull 16*:11-14, 1986b.

Selvaggi, F.P., G. Pace, A. Zambonin Zallone, A. Teti, and A. Furino. Ultrastructural study of human adrenal pheochromo-cytoma by open needle biopsy. *Appl Pathol 2*:255-263, 1984.

Semenenko, F.M., A.C. Cuello, M. Goldstein, K.Y. Lee, and E. Sidebottom. A monoclonal antibody against tyrosine hydroxylase: Application in light and electron microscopy. *J Histochem Cytochem 34*:817-821, 1986.

Serck-Hanssen, G., and O. Søvik. Specific insulin binding in bovine chromaffin cells; demonstration of preferential binding to adrenalin-storing cells. *Life Sci 41*:2799-2806, 1987.

Serck-Hanssen, G., O. Søvik, and R. Lie. Characterization of specific insulin binding sites on chromaffin cells from bovine adrenal medulla. *Int J Biochem 20*:1435-1441, 1988.

Serov, R.A. and N.N. Sokolova. [The adrenal medulla and adrenergic innervation of the cardiac ventricles of the white rat postnatal ontogenesis (histofluorescent investigation).] *Arkh Anat Gistol Embriol 92*:36-39, 1987.

Serventi, I.M. and C.J. Coffee. Characterization of myosin light-chain kinase from bovine adrenal medulla. *Arch Biochem Biophys 245*:379-388, 1986.

Shabunina, N.D. Combined effect of industrial and residential noise on the status of the human sympatho-adrenal system. *Gig Sanit 2*:21-23, 1987.

Shaposhnikov, V.M. [Morphofunctional peculiarities of the adenohypophyseal secretory cells and those of the adrenal glands in rats at ageing.] *Arkh Anat Gistol Embriol 90*:71-77, 1986.

Sharma, T.R., T.D. Wakade, R.K. Malhotra, and A.R. Wakade. Secretion of catecholamines from the perfused adrenal gland of the rat is not regulated by α-adrenoceptors. *Eur J Pharmacol 122*:167-172, 1986.

Shaw, T.J. and P.C. Letourneau. Chromaffin cell heterogeneity of process formation and neuropeptide content under control and nerve growth factor-altered conditions in cultures of chick embryonic adrenal gland. *J Neurosci Res 16*:337-355, 1986.

Shen, R.-s., E.L. Hamilton-Byrd, P.R. Vulliet, S.-W. Kwan, and C.W. Abell. A simplified $^{14}CO_2$-trapping microassay for tyrosine hydroxylase activity. *J Neurosci Methods 16*:163-173, 1986.

Shigetomi, S., S. Ueno, H. Tosaki, H. Kohno, S. Hashimoto, and S. Fukuchi. [Increased activity of sympathoadrenomedullary system and decreased renal dopamine receptor contents after short-term and long-term sodium loading in rats.] *Nippon Naibunpi Gakkai Zasshi 62*:776-783, 1986.

Shigetomi, S., S. Ueno, H. Tosaki, H. Kohno, K. Mori, K. Satoh, K. Katoh, K. Tanaka, Y. Haga, and S. Fukuchi. [Effects of dopamine and synthetic atrial natriuretic polypeptide on the release of norepinephrine and epinephrine from the adrenal medulla of rats.] *Nippon Naibunpi Gakkai Zasshi 63*:836-845, 1987.

Shimamura, M., T. Hayase, M. Ito, M.-L. Rasilo, and T. Yamagata. Characterization of a major neutral glycolipid in PC12 cells as III^{3}Gal α-globotriaosylceramide by the method for determining glycosphingolipid saccharide sequence with endoglycoceramidase. *J Biol Chem 263*:12124-12128, 1988.

Shorr, R.G.L., M.D. Minnich, A. Varrichio, M.W. Strohsacker, L. Gotlib, L.I. Kruse, W.E. DeWolf Jr., and S.T. Crooke. Immuno-cross-reactivity suggests that catecholamine biosynthesis enzymes and β-adrenergic receptors may be related. *Mol Pharmacol 32*:195-200, 1987.

Shulkin, B.L., B. Shapiro, M.C. Tobes, S.-W. Shen, D.M. Wieland, L.J. Meyers, H.T. Lee, N.A. Petry, J.C. Sisson, and W.H. Beierwaltes. Iodine[123]-4-amino-3-iodobenzylguanidine, a new sympathoadrenal imaging agent: Comparison with iodine[123] metaiodobenzylguanidine. *J Nucl Med 27*:1138-1142, 1986.

Shulkin, B.L., S.-W. Shen, J.C. Sisson, and B. Shapiro. Iodine[31] MIBG scintigraphy of the extremities in metastatic pheochromocytoma and neuroblastoma. *J Nucl Med 28*:315-318, 1987.

Shvalev, V.N., A.M. Vikhert, R.A. Stropus, A.A. Sosunov, E.R. Pavlovich, R.A. Kargina-Terentyeva, N.I. Zhuchkova, A.Y. Anikin, and K.L. Maryan. Changes in neural and humoral mechanisms of the heart in sudden death due to myocardial abnormalities. *J Am Coll Cardiol 8*:55A-64A, 1986.

Sidey, F.M., H.G. Dean, and B.L. Furman. Role of the adrenal medulla in stress-induced hyperinsulinaemia in normal mice and in mice infected with *Bordetella pertussis* or treated with pertussis toxin. *J Endocrinol 118*:135-140, 1988.

Siegel, D.L. Chromaffin cell synapsin? *Nature 327*:467-468, 1987.

Siegel, R.E., A.L. Iacangelo, J. Park, and L.E. Eiden. Chromogranin A biosynthetic cell populations in bovine endocrine and neuronal tissues: Detection by *in situ* hybridization histochemistry. *Mol Endocrinol 2*:368-374, 1988.

Sietzen, M., M. Schober, R. Fischer-Colbrie, D. Scherman, G.

Sperk, and H. Winkler. Rat adrenal medulla: Levels of chromogranins, enkephalins, dopamine β-hydroxylase and of the amine transporter are changed by nervous activity and hypophysectomy. *Neuroscience 22*:131-139, 1987.

Silani, V., A. Pizzuti, O. Strada, A. Falini, G. Pezzoli, and G. Scarlato. Cryopreservation of human fetal adrenal medullary cells. *Brain Res 454*:383-386, 1988.

Simasko, S.M., J.A. Durkin, and G.A. Weiland. Effects of substance P on nicotinic acetylcholine receptor function in PC12 cells. *J Neurochem 49*:253-260, 1987.

Simasko, S.M., J.R. Soares, and G.A. Weiland. Two components of carbamylcholine-induced loss of nicotinic acetylcholine receptor function in the neuronal cell line PC12. *Mol Pharmacol 30*:6-12, 1986.

Simon, J.-P., M.-F. Bader, and D. Aunis. Secretion from chromaffin cells is controlled by chromogranin A-derived peptides. *Proc Natl Acad Sci USA 85*:1712-1716, 1988.

Simons, T.J.B. and G. Pocock. Lead enters bovine adrenal medullary cells through calcium channels. *J Neurochem 48*:383-389, 1987.

Simonyi, A., B. Kanyicska, T. Szentendrei, and M.I.K. Fekete. Effect of chronic morphine treatment on the adrenaline biosynthesis in adrenals and brain regions of the rat. *Biochem Pharmacol 37*:749-752, 1988.

Singh, D.N.P. and T.C. Mathew. Adrenomedullary chromaffin cells of the rat: An ultrastructural study. *Acta Anat (Basel) 129*:329-332, 1987.

Sirimanne, S.R., H.H. Herman, and S.W. May. Interaction of dopamine β-mono-oxygenase with substituted imidazoles and pyrazoles: Catalysis and inhibition. *Biochem J 242*:227-233, 1987.

Skattebøl, A., and R.A. Rabin. Effects of ethanol on $^{45}Ca^{2+}$ uptake in synaptosomes and in PC12 cells. *Biochem Pharmacol 36*:2227-2229, 1987.

Sladek J.R. Jr., D.E. Redmond Jr., and R.H. Roth. Transplantation of fetal neurons in primates. *Clin Res 35*:200-204, 1988.

Sladek J.R. Jr., and I. Shoulson. Neural transplantation: A call for patience rather than patients. *Science 240*:1386-1388, 1988.

Slotkin, T.A. Development of the sympathoadrenal axis. pp.69-96. In: **Developmental Neurobiology of the Autonomic Nervous System**, (P.M. Gootman, ed). Humana Press, Clifton, NJ, 1986.

Slotkin, T.A. and J.V. Bartolome. Role of ornithine decarboxylase and the polyamines in nervous system development: A review. *Brain Res Bull 17*:307-320, 1986.

Slotkin, T.A., L. Orband-Miller, and K.L. Queen. Do catecholamines contribute to the effects of neonatal hypoxia on development of brain and heart? Influence of concurrent α–adrenergic blockade on ornithine decarboxylase activity. *Int J Dev Neurosci* 5:135-143, 1987.

Slotkin, T.A. and F.J. Seidler. Adrenomedullary catecholamine release in the fetus and newborn: Secretory mechanisms and their role in stress and survival. *J Dev Physiol 10*:1-16, 1988.

Small, D.H., Z. Ismael, and I.W. Chubb. Acetylcholinesterase hydrolyses chromogranin A to yield low molecular weight peptides. *Neuroscience 19*:289-295, 1986.

Smith, E.J., G.A.D. McPherson, and J. Lynn. Inferior vena cava involvement by a phaeochromocytoma. *Br J Surg 74*:597, 1987.

Smith, G.K. and C.A. Nichol. Synthesis, utilization, and structure of the tetrahydropterin intermediates in the bovine adrenal medullary *de novo* biosynthesis of tetrahydrobiopterin. *J Biol Chem 261*:2725-2737, 1986.

Smoliakova, G.P. and M.I. Rodivoz. [Assessment of curative action of DOPA and phosphaden in experimental dystrophic changes of the retina induced by disturbances in the functional state of the sympathoadrenal system.] *Oftalmol Zh 1*:42-45, 1988.

Snyder, S.H. Parkinson's disease. A cure using brain transplants? *Nature 326*:824-825, 1987.

Soltanov, V.V. and N.V. Karpovitch. [Existence of peripheral neural mechanism of interoceptive influences on the adrenal medullary function.] *Fiziol Zh SSSR 73*:1253-1261, 1987.

Song, S.-L., W.R. Crowley, and C.E. Grosvenor. Evidence for involvement of an adrenal catecholamine in the β-adrenergic inhibition of oxytocin release in lactating rats. *Brain Res 457*:303-309, 1988.

Sontag, J.-M., D. Aunis, and M.-F. Bader. Peripheral actin filaments control calcium-mediated catecholamine release from streptolysin-*O*-permeabilized chromaffin cells. *Eur J Cell Biol 46*:316-326, 1988.

Sorimachi, M., S. Nishimura, and K. Yano. Dihydropyridine BAY K 8644 enables reduction of Ca concentration to induce catecholamine secretion from the perfused cat adrenal. *Jpn J Physiol 35*:871-874, 1985.

Sorimachi, M., S. Nishimura, and K. Yano. Possible role of surface potential in the gating mechanism of Ca channels concerned with catecholamine secretion in the adrenal. *Jpn J Physiol 36*:321-337, 1986.

Sourkes, T.L. Aromatic-L-amino acid decarboxylase. *Methods Enzymol 142*:170-187, 1987.

Souto, M., R.S. Piezzi, and R. Bianchi. Catecholaminergic responses of neonatal adrenal gland to insulin. *J Neural Transm 73*:115-120, 1988.

Spagnoli, D.B., R.G. Frederickson, R.L. Robinson, and S.W. Carmichael. Opossum adrenal medulla: II. Differentiation of the chromaffin cell. *Am J Anat 179*:220-231, 1987.

Sparrow, R.A. and R.E. Coupland. Blood flow to the adrenal

gland of the rat: Its distribution between the cortex and the medulla before and after haemorrhage. *J Anat 155*:51-61, 1987.

Srinivasan, M., Y. Yamamoto, and H. Lagercrantz. Ventilatory effects of naloxone via the sympathoadrenal system in the neonate? *Neurosci Lett 90*:159-164, 1988.

Stachowiak, M.K., S.J. Fluharty, E.M. Stricker, M.J. Zigmond, and B.B. Kaplan. Molecular adaptations in catecholamine biosynthesis induced by cold stress and sympathectomy. *J Neurosci Res 16*:13-24, 1986.

Stachowiak, M.K., P.H.K. Lee, R.J. Rigual, O.H. Viveros, and J.S. Hong. Roles of the pituitary-adrenocortical axis in control of the native and cryptic enkephalin levels and proenkephalin mRNA in the sympathoadrenal system of the rat. *Mol Brain Res 3*:263-273, 1988.

Stachowiak, M.K., R.J. Rigual, P.H.K. Lee, O.H. Viveros, and J.S. Hong. Regulation of tyrosine hydroxylase and phenylethanolamine *N*-methyltransferase mRNA levels in the sympathoadrenal system by the pituitary-adrenocortical axis. *Mol Brain Res 3*:275-286, 1988.

Stachowiak, M.K., E.M. Stricker, M.J. Zigmond, and B.B. Kaplan. A cholinergic antagonist blocks cold stress-induced alterations in rat adrenal tyrosine hydroxylase mRNA. *Mol Brain Res 3*:193-196, 1988.

Stanley, E.F., G. Ehrenstein, and J.T. Russell. Evidence for anion channels in secretory vesicles. *Neuroscience 25*:1035-1039, 1988.

Stein, R., S. Orit, and D.J. Anderson. The induction of a neural-specific gene, SCG10, by nerve growth factor in PC12 cells is transcriptional, protein synthesis dependent, and glucocorticoid inhibitable. *Dev Biol 127*:316-325, 1988.

Stemple, D.L., N.K. Mahanthappa, and D.J. Anderson. Basic FGF induces neuronal differentiation, cell division, and NGF dependence in chromaffin cells: A sequence of events in sympathetic development. *Neuron 1*:517-525, 1988.

Stenström, G. and K. Svärdsudd. Pheochromocytoma in Sweden 1958-1981. An analysis of the National Cancer Registry Data. *Acta Med Scand 220*:225-232, 1986.

Sternberg, H., J. Baudier, K. Akisuki, G.M. Cole, W.H. Martin, C.E. Creutz, P.S. Timiras, and R.D. Cole. Similarities and differences between τ protein and chromobindin A. *Neurochem Intl 13*:149-152, 1988.

Stevens, J.K. and J. Trogadis. Reconstructive three-dimensional electron microscopy: A routine biologic tool. *Anal Quant Cytol Histol 8*:102-107, 1986.

Stevens, J.R., I. Phillips, W.J. Freed, and M. Poltorak. Cerebral transplants for seizures: Preliminary results. *Epilepsia 29*:731-737, 1988.

Stewart, L.C. and J.P. Klinman. Characterization of alternate reductant binding and electron transfer in the dopamine β-monooxygenase. *Biochemistry 26*:5302-5309, 1987.

Stewart, L.C. and J.P. Klinman. Dopamine β-hydroxylase of adrenal chromaffin granules: Structure and function. *Annu Rev Biochem 57*:551-592, 1988a.

Stewart, L.C. and J.P. Klinman. Bovine membranous dopamine β-hydroxylase is not anchored via covalently attached phosphatidylinositol. *J Biol Chem 263*:12183-12186, 1988b.

Stewart, R.E., S.E. Swithers, and R. McCarty. Alterations in binding sites for atrial natriuretic factor in kidneys and adrenal glands of Dahl hypertension-sensitive rats. *J Hypertens 5*:481-487, 1987.

Stewart, R.E., S.E. Swithers, L.M. Plunkett, and R. McCarty. ANF receptors: Distribution and regulation in central and peripheral tissues. *Neurosci Biobehav Rev 12*:151-168, 1988.

Stoddard, S.L., V.K. Bergdall, P.S. Conn, and B.E. Levin. Increases in plasma catecholamines during naturally elicited defensive behavior in the cat. *J Auton Nerv Syst 19*:189-197, 1987.

Stoddard, S.L., V.K. Bergdall, D.W. Townsend, and B.E. Levin. Plasma catecholamines associated with hypothalamically-elicited defense behavior. *Physiol Behav 36*:867-873, 1986a.

Stoddard, S.L., V.K. Bergdall, D.W. Townsend, and B.E. Levin. Plasma catecholamines associated with hypothalamically-elicited flight behavior. *Physiol Behav 37*:709-715, 1986b.

Stoddard, S.L., G.M. Tyce, S.W. Carmichael, D.M. Gaumann, and T.L. Yaksh. Effect of acute and chronic spinal transection on evoked secretion of adrenal medullary catecholamines in the cat. *J Auton Nerv Syst 23*:175-179, 1988.

Stoehr, S.J., J.E. Smolen, R.W. Holz, and B.W. Agranoff. Inositol trisphosphate mobilizes intracellular calcium in permeabilized adrenal chromaffin cells. *J Neurochem 46*:637-640, 1986.

Streit, J. and H.D. Lux. Voltage dependent calcium currents in PC12 growth cones and cells during NGF-induced cell growth. *Pflugers Arch 408*:634-641, 1987.

Strittmatter, S.M., E.B. DeSouza, D.R. Lynch, and S.H. Snyder. Angiotensin-converting enzyme localized in the rat pituitary and adrenal glands by [3H]captopril autoradiography. *Endocrinology 118*:1690-1699, 1986.

Strittmatter, W.J. Molecular mechanisms of exocytosis: The adrenal chromaffin cell as a model system. *Cell Mol Neurobiol 8*:19-25, 1988.

Strömberg, I., A. Hultgårdh-Nilsson, U. Hedin, and T. Ebendal. Fate of intraocular chromaffin cell suspensions: Role of initial nerve growth factor support. *Cell Tissue Res 254*:487-497, 1988.

Stutzin, A. A fluorescence assay for monitoring and analyzing fusion of biological membrane vesicles in vitro. *FEBS Lett 197*:274-280, 1986.

Stutzin, A., Z.I. Cabantchik, P.I. Lelkes, and H.B. Pollard.

Synexin-mediated fusion of bovine chromaffin granule ghosts. Effect of pH. *Biochim Biophys Acta* 905:205-212, 1987.

Sudo, A. Effects of adrenalectomy and chronic guanethidine treatment on tissue adrenaline concentrations in swimming-exposed rats. *Jpn J Pharmacol* 45:197-201, 1987.

Sugawara, I., M. Nakahama, H. Hamada, T. Tsuruo, and S. Mori. Apparent stronger expression in the human adrenal cortex than in the human adrenal medulla of M 170,000-180,000 P-glycoprotein. *Cancer Res* 48:4611-4614, 1988.

Sugimoto, Y., M. Noda, H. Kitayama, and Y. Ikawa. Possible involvement of two signaling pathways in induction of neuron-associated properties by v-Ha-*ras* gene in PC12 cells. *J Biol Chem* 263:12102-12108, 1988.

Sulzer, D., I. Piscopo, F. Ungar, and E. Holtzman. Lead-dependent deposits in diverse synaptic vesicles: Suggestive evidence for the presence of anionic binding sites. *J Neurobiol* 18:467-483, 1987.

Sumikawa, K., Y. Hayashi, K. Fukumitsu, and I. Yoshiya. Selective inhibition by dantrolene of caffeine-induced catecholamine release from perfused dog adrenals. *Res Commun Chem Pathol Pharmacol* 57:45-53, 1987.

Sundaresan, P.R., M.M. Guarnaccia, and J.L. Izzo Jr. Adrenal medullary regulation of rat renal cortical adrenergic receptors. *Am J Physiol* 253:F1063-F1067, 1987.

Supattapone, S., S.M. Strittmatter, L.D. Fricker, and S.H. Snyder. Characterization of a neutral, divalent cation-sensitive endopeptidase: A possible role in neuropeptide processing. *Mol Brain Res* 3:173-182, 1988.

Superti-Furga, A., P.M. Royce, and B. Steinmann. Congenital dopamine β-hydroxylase deficiency. *Lancet* 1:693, 1987.

Suzuki, H., N. Nakanishi, and S. Yamada. Nerve growth factor transiently increases tetrahydrobiopterin and total biopterin contents of pheochromocytoma PC12h cells. *Biochem Biophys Res Commun* 153:382-387, 1988.

Suzuki, H., A.S. Tischler, N.D. Christofides, M. Chrétien, N.G. Seidah, J.M. Polak, and S.R. Bloom. A novel pituitary protein (7B2)-like immunoreactivity is secreted by a rat phaeochro-mocytoma cell line (PC12). *J Endocrinol* 108:151-155, 1986.

Sved, A.F. Peripheral pressor systems in hypertension caused by nucleus tractus solitarius lesions. *Hypertension* 8:742-747, 1986.

Svensson, E., M.A. Thanderz, L. Sjöberg, and M. Gillberg. Military flight experience and sympatho-adrenal activity. *Aviat Space Environ Med* 59:411-416, 1988.

Swilem, A.-M.F., H. Yagisawa, and J.N. Hawthorne. Muscarinic release of inositol trisphosphate without mobilization of calcium in bovine adrenal chromaffin cells. *J Physiol (Paris)* 81:246-251, 1986.

Swithers, S.E., R.E. Stewart, and R. McCarty. Binding sites for atrial natriuretic factor (ANF) in kidneys and adrenal glands of spontaneously hypertensive (SHR) rats. *Life Sci 40*:1673-1681, 1987.

Syvertsen, C., R. Gaustad, K. Schrøder, and T. Ljones. Studies on the binding of copper to dopamine β-monooxygenase and other proteins using the Cu^{2+} ion-selective electrode. *J Inorg Biochem 26*:63-76, 1986.

Syvertsen, C., T.B. Melø, and T. Ljones. Fluorescence studies on dopamine β-monooxygenase: Effects of salts, pH changes, metal-chelating agents and Cu^{2+}. *Biochim Biophys Acta 914*:6-18, 1987.

Szemeredi, K., G. Bagdy, R. Stull, A.E. Calogero, I.J. Kopin, and D.S. Goldstein. Sympathoadrenomedullary inhibition by chronic glucocorticoid treatment in conscious rats. *Endocrinology 123*:2585-2590, 1988.

Szemeredi, K., G. Bagdy, R. Stull, H.R. Keiser, I.J. Kopin, and D.S. Goldstein. Sympathoadrenomedullary hyper-responsive-ness to yohimbine in juvenile spontaneously hypertensive rats. *Life Sci 43*:1063-1068, 1988.

Tabsh, K., R. Rudelstorfer, B. Nuwayhid, and N.S. Assali. Circulatory responses to hypovolemia in the pregnant and nonpregnant sheep after pharmacologic sympathectomy. *Am J Obstet Gynecol 154*:411-419, 1986.

Tachikawa, E., S. Takahashi, C. Shimizu, H. Ban, N. Ohstubo, K. Sato, and T. Kashimoto. Inhibitory effect of polymyxin B on catecholamine secretion from cultured bovine adrenal medullary cells. *Neurosci Lett 82*:95-100, 1987.

Tachikawa, E., A.W. Tank, D.H. Weiner, W.F. Mosimann, N. Yanagihara, and N. Weiner. Tyrosine hydroxylase is activated and phosphorylated on different sites in rat pheochromocytoma PC12 cells treated with phorbol ester and forskolin. *J Neurochem 48*:1366-1376, 1987.

Tachikawa, E., A.W. Tank, N. Yanagihara, W.F. Mosimann, and N. Weiner. Phosphorylation of tyrosine hydroxylase on at least three sites in rat pheochromocytoma PC12 cells treated with 56 mM K^+: Determination of the sites on tyrosine hydroxylase phosphorylated by cyclic AMP-dependent and calcium/calmodulin-dependent protein kinases. *Mol Pharmacol 30*:476-485, 1986.

Takahashi, M., H. Sugino, and Y. Kudo. The mechanism of calcium-independent catecholamine depleting action of monensin from clonal rat pheochromocytoma cells. *Brain Res 382*:332-338, 1986.

Takahashi, T., M. Naoi, and T. Nagatsu. Uptake of *N*-methyl-4-phenylpyridinium ion (MPP$^+$) into PC12h pheochromocytoma cells. *Neurochem Int 11*:89-93, 1987.

Takara, H., A. Wada, M. Arita, K. Sumikawa, and F. Izumi. Ketamine inhibits ^{45}Ca influx and catecholamine secretion by inhibiting ^{22}Na influx in cultured bovine adrenal medullary cells. *Eur J Pharmacol 125*:217-224, 1986.

Tamura, M., T.-T. Lam, and T. Inagami. Specific endogenous Na, K-ATPase inhibitor purified from bovine adrenal. *Biochem Biophys Res Commun* 149:468-474, 1987.

Tan, D.-P. and K. Tsou. New evidence for neuronal function of vasopressin: Sympathetic mediation of intrathecal vasopressin-induced hypertension. *Peptides* 7:569-572, 1986.

Tan, L. -B., C.-H. Huang, Y.-H. Lai, C.-H. Chien, L.-H. Liu, C.-C. Wang, Y.-H. Chou, and C.-P. Chiang. [Experience of surgical treatment of pheochromocytoma - report of 3 cases.] *KaoHsiung J Med Sci* 2:536-542, 1986.

Tanaka, T. Calmodulin-dependent calcium signal transduction. *Jpn J Pharmacol* 46:101-107, 1988.

Tanaka, T., H. Yokohama, M. Negishi, H. Hayashi, S. Ito, and O. Hayaishi. Pertussis toxin facilitates secretagogue-induced catecholamine release from cultured bovine adrenal chromaffin cells. *Biochem Biophys Res Commun* 144:907-914, 1987.

Taniguchi, T., K. Morisawa, M. Ogawa, H. Yamamoto, and S. Fujimoto. Decrease in the level of poly(ADP-ribose) synthetase during nerve growth factor-promoted neurite outgrowth in rat pheochromocytoma PC12 cells. *Biochem Biophys Res Commun* 154:1034-1040, 1988.

Taniuchi, M., E.M. Johnson Jr., P.J. Roach, and J.C. Lawrence Jr. Phosphorylation of nerve growth factor receptor proteins in sympathetic neurons and PC12 cells. *In vitro* phosphorylation by the cAMP-independent protein kinase F_A/GSK-3. *J Biol Chem* 261:13342-13349, 1986.

Tank, A.W., P. Curella, and L. Ham. Induction of mRNA for tyrosine hydroxylase by cyclic AMP and glucocorticoids in a rat pheochromocytoma cell line: Evidence for the regulation of tyrosine hydroxylase synthesis by multiple mechanisms in cells exposed to elevated levels of both inducing agents. *Mol Pharmacol* 30:497-503, 1986.

Tank, A.W., L. Ham, and P. Curella. Induction of tyrosine hydroxylase by cyclic AMP and glucocorticoids in a rat pheochromocytoma cell line: Effect of the inducing agents alone or in combination on the enzyme levels and rate of synthesis of tyrosine hydroxylase. *Mol Pharmacol* 30:486-496, 1986.

Tank, A.W. and N. Weiner. Tyrosine 3-monooxygenase from rat pheochromocytoma. *Methods Enzymol* 142:71-82, 1987.

Tapparelli C., M. Grob, and M.M. Burger. Detergents inhibit exocytosis in PC12 cells: Evidence for an effect on ion fluxes. *J Cell Biochem* 33:289-303, 1987.

Tashiro, N., K. Hirata, S. Maki, and H. Nakao. Cardiac arrhythmias induced in cats by stimulation of the anteromedial hypothalamus. *Int J Psychophysiol* 6:231-240, 1988.

Taugner, G., Ch. Heym, W. Kummer, and I. Wunderlich. Nucleosidetriphosphate-ADP-phosphotransferase: A new enzyme in chromaffin vesicles. *Biogenic Amines* 5:409-426, 1988.

Taylor, H.C., D. Mayes, and A.H. Anton. Clonidine suppression test for pheochromocytoma: Examples of misleading results. *J Clin Endocrinol Metab* 63:238-242, 1986.

Terashima, T., T. Katada, C. Takayama, M. Ui, and Y. Inoue. Immunohistochemical detection of GTP-binding regulatory protein (G_o) in the autonomic nervous system including the enteric nervous system, superior cervical ganglion and adrenal medulla. *Brain Res* 455:353-359, 1988.

TerBush, D.R., M.A. Bittner, and R.W. Holz. Ca^{2+} influx causes rapid translocation of protein kinase C to membranes. Studies of the effects of secretagogues in adrenal chromaffin cells. *J Biol Chem* 263:18873-18879, 1988.

TerBush, D.R. and R.W. Holz. Effects of phorbol esters, diglyceride, and cholinergic agonists on the subcellular distribution of protein kinase C in intact or digitonin-permeabilized adrenal chromaffin cells. *J Biol Chem* 261:17099-17106, 1986.

Terman, G.W., J.W. Lewis, and J.C. Liebeskind. Two opioid forms of stress analgesia: Studies of tolerance and cross-tolerance. *Brain Res* 368:101-106, 1986.

Terman, G.W. and J.C. Liebeskind. Relation of stress-induced analgesia to stimulation-produced analgesia. *Ann NY Acad Sci* 467:300-308, 1986.

Tessari, F., R.A. Travagli, and M. Prosdocimi. Spontaneous diabetes in mice: Catecholamine analysis in various organs. *Neurosci Res Commun* 3:63-68, 1988.

Thieffry, M., J.-F. Chich, D. Goldschmidt, and J.-P. Henry. [Demonstration of a large-conduction ion channel in subcellular fractions of the adrenal medulla.] *C R Acad Sci [III]* 305:193-197, 1987a.

Thieffry, M., J.-F. Chich, D. Goldschmidt, and J.-P. Henry. [Existence of a large ionic channel in sub-cellular fractions from adrenal medulla.] *Cell Biochim* 305:193-197, 1987b.

Thieffry, M., J.-F. Chich, D. Goldschmidt, and J.-P. Henry. Incorporation in lipid bilayers of a large conductance cationic channel from mitochondrial membranes. *EMBO J* 7:1449-1454, 1988.

Thompson, M.L., K.A. Miczek, K. Noda, L. Shuster, and M.S.A. Kumar. Analgesia in defeated mice: Evidence for mediation via central rather than pituitary or adrenal endogenous opioid peptides. *Pharmacol Biochem Behav* 29:451-456, 1988.

Thorlacius-Ussing, O. and B.L. Rasmussen. Light and electron microscopical visualization of selenium in adrenal glands from rats exposed to sodium selenite. *Exp Mol Pathol* 45:59-67, 1986.

Tiercy, J.-M. and E.M. Shooter. Early changes in the synthesis of nuclear and cytoplasmic proteins are induced by nerve growth factor in differentiating rat PC12 cells. *J Cell Biol* 103:2367-

2378, 1986.

Tilders, F.J.H. and F. Berkenbosch. CRF and catecholamines: Their place in the central and peripheral regulation of the stress response. *Acta Endocrinol [Suppl] (Copenh) 276*:63-75, 1986.

Tippett, P.A., A.J. McEwan, and D.M. Ackery. A re-evaluation of dopamine excretion in phaeochromocytoma. *Clin Endocrinol (Oxf) 25*:401-410, 1986.

Tippett, P.A., R.S. West, A.J. McEwan, J.E. Middleton, and D.M. Ackery. A comparison of dopamine and homovanillic acid excretion, as prognostic indicators in malignant phaeochromocytoma. *Clin Chim Acta 166*:123-133, 1987.

Tirrell, J.G. and C.J. Coffee. Identification of 2',3'-cyclic nucleotide 3'-phosphodiesterase in bovine adrenal medulla. *Comp Biochem Physiol [B] 83*:867-873, 1986.

Tischler, A.S., Y. Dayal, K. Balogh, R.B. Cohen, J.L. Connolly, and K. Tallberg. The distribution of immunoreactive chromogranins, S-100 protein, and vasoactive intestinal peptide in compound tumors of the adrenal medulla. *Hum Pathol 18*:909-917, 1987.

Tischler, A.S. and R.A. DeLellis. The rat adrenal medulla. I. The normal adrenal. *J Am Coll Toxicol 7*:1-19, 1988a.

Tischler, A.S. and R.A. DeLellis. The rat adrenal medulla. II. Proliferative lesions. *J Am Coll Toxicol 7*:23-41, 1988b.

Tischler, A.S., R.A. DeLellis, G. Nunnemacher, and H.J. Wolfe. Acute stimulation of chromaffin cell proliferation in the adult rat adrenal medulla. *Lab Invest 58*:733-735, 1988.

Tischler, A.S., Y.C. Lee, D. Costopoulos, V.W. Slayton, W.J. Jason, and S.R. Bloom. Cooperative regulation of neurotensin content in PC12 pheochromocytoma cell cultures: Effects of nerve growth factor, dexamethasone, and activators of adenylate cyclase. *J Neurosci 6*:1719-1725, 1986.

Tischler, A.S., H. Mobtaker, P.W.L. Kwan, W.J. Jason, R.A. DeLellis, and H.J. Wolfe. Hypertrophy of pheochromocytoma cells treated with nerve growth factor and activators of adenylate cyclase. *Cell Tissue Res 249*:161-169, 1987.

Tischler, A.S., H. Mobtaker, K. Mann, G. Nunnemacher, W.J. Jason, Y. Dayal, R.A. DeLellis, L. Adelman, and H.J. Wolfe. Anti-lymphocyte antibody Leu-7 (HNK-1) recognizes a constituent of neuroendocrine granule matrix. *J Histochem Cytochem 34*:1213-1216, 1986.

Tocco, M.D., M.L. Contreras, S. Koizumi, G. Dickens, and G. Guroff. Decreased levels of nerve growth factor receptor on dexamethasone-treated PC12 cells. *J Neurosci Res 20*:411-419, 1988.

Togari, A., S. Ichikawa, and T. Nagatsu. Activation of tyrosine hydroxylase by Ca^{2+}-dependent neutral protease, calpain. *Biochem Biophys Res Commun 134*:749-754, 1986.

Togari, A., T. Murakami, T. Oshima, K. Fujita, and T. Nagatsu.

Effects of polyamines on tyrosine hydroxylase activity in adrenals. *Neurochem Int 9*:281-286, 1986.

Tomaru, A., K. Oh, Y. Miura, H. Nakamura, M. Yoshikawa, and M. Horiguchi. Pheochromocytoma and Ca^{++} channel blocker. *Jpn Heart J 27*:429-436, 1986.

Tomaselli, K.J., C.H. Damsky, and L.F. Reichardt. Interactions of a neuronal cell line (PC12) with laminin, collagen IV, and fibronectin: Identification of integrin-related glycoproteins involved in attachment and process outgrowth. *J Cell Biol 105*:2347-2358, 1987.

Tomlinson, A., J. Durbin, and R.E. Coupland. A quantitative analysis of rat adrenal chromaffin tissue: Morphometric analysis at tissue and cellular level correlated with catecholamine content. *Neuroscience 20*:895-904, 1987.

Tomori, M. [Pharmacological studies on the pressor response in adrenal-enucleated rats.] *Nippon Yakurigaku Zasshi 87*:67-76, 1986.

Toni, R., S. Mosca, L. Favero, S. Ricci, R. Roversi, G. Toni, and P. Vezzadini. Clinical anatomy of the suprarenal arteries: A quantitative approach by aortography. *Surg Radiol Anat 10*:297-302, 1988.

Torres, M., M.F. Bader, D. Aunis, and M.T. Miras-Portugal. Nerve growth factor effect on adenosine transport in cultured chromaffin cells. *J Neurochem 48*:233-235, 1987.

Torres, M., E.G. Delicado, and M.T. Miras-Portugal. Adenosine transporters in chromaffin cells: Subcellular distribution and characterization. *Biochim Biophys Acta 969*:111-120, 1988.

Torres, M., P. Molina, and M.T. Miras-Portugal. Adenosine transporters in chromaffin cells: Quantification by dipyridamol monoacetate. *FEBS Lett 201*:124-128, 1986.

Toutant, M., D. Aunis, J. Bockaert, V. Homburger, and B. Rouot. Presence of three pertussis toxin substrates and G$_o$aimmunoreactivity in both plasma and granule membranes of chromaffin cells. *FEBS Lett 215*:339-344, 1987.

Toutant, M., J. Bockaert, V. Homburger, and B. Rouot. G-proteins in *Torpedo marmorata* electric organ: Differential distribution in pre- and post-synaptic membranes and synaptic vesicles. *FEBS Lett 222*:51-55, 1987.

Tran, M.-A., G. De Saint-Blanquat, P. Valet, L.D. Tran, F. Anglade, G. Gaillard, O. Rascol, A.-M. Brisac, J.-L. Montastruc, and P. Montastruc. Is adrenal medulla involved in the antihypertensive effect of nicardipine? *J Pharmacol Exp Ther 244*:1116-1120, 1988.

Trifaró, J.-M., M.-F. Bader, A. Côté, R.L. Kenigsberg, T. Hikita, and R.W.H. Lee. Cytoskeleton organization and adrenal chromaffin cell function. pp. 459-472. In: **Contractile proteins in muscle and non-muscle cell systems: Biochemistry, physiology and pathology.** (E.E. Alia, N. Arena and M.A. Russo, eds). Praeger Scientific, Philadelphia, 1985.

Trifaró, J.-M. and A.M. Poisner. Electrophysiological properties of secretory cells: An overview. pp. 269-302. In: **The Electrophysiology of the Secretory Cell.** (A.M. Poisner and J.-M. Trifaró, eds). Elsevier Science Publishers, Amsterdam, 1985.

Trifaró, J.-M. and E.P. Seward. Microinjection techniques in the study of the secretory process. pp. 273-288. In: *In Vitro* **Methods for Studying Secretion,** Vol. 3. (A.M. Poisner and J.-M. Trifaró, eds). Elsevier Science Publishing, Amsterdam, 1987.

Trofimov, V.M. and V.V. Gubin. [Chromaffin tumors of the adrenal glands in a 14-year-old boy.] *Vestn Khir 136*:107-108, 1986.

Troncone, L., V. Rufini, P. Campioni, G. Valle, C. Giordano, G. Ala, G. Deleide, and P. Orlando. Evaluation of a new adrenomedullary imaging agent: ^{131}I-meta-iodobenzylguanidine (^{131}I-MIBG). I: Experimental studies on animals. *Rays 10*:117-124, 1985.

Tsujimoto, A., K. Morita, S. Kitayama, and T. Dohi. Facilitation of acetylcholine-evoked catecholamine release by cyclic AMP on isolated perfused dog adrenal glands. *Arch Int Pharmacodyn Ther 279*:304-313, 1986.

Tsujimoto, G., A. Minegishi, T. Ishizaki, B.B. Hoffman, and K. Hashimoto. Characterization of vasopressor response to sympathetic stimulation in the pithed rat: Differential regulation of neuronally stimulated and blood-borne catecholamine-stimulated pressor responses in rats harboring pheochro-mocytoma. *J Pharmacol Exp Ther 242*:637-645, 1987.

Tulipan, N. Brain transplants: A new approach to the therapy of neurodegenerative disease. *Neurol Clin 6*:405-420, 1988.

Tümmers, U., T.H. Müller, R. Schmidt, K. Seidl, K. Lichtwald, P. Vescei, H.-J. Wagner, and K. Unsicker. Destruction of the preganglionic nerves by β-bungarotoxin does not interfere with normal embryonic development of the rat adrenal medulla. *Dev Biol 117*:619-627, 1986.

Tung, C.-S. and Yin T.-H. Clonidine suppression and its adrenoreceptor mediation in schedule-induced polydipsia. *Physiol Behav 40*:317-322, 1987.

Turner, D.C., L.A. Flier, and S. Carbonetto. Magnesium-dependent attachment and neurite outgrowth by PC12 cells on collagen and laminin substrata. *Dev Biol 121*:510-525, 1987.

Turner, M.C., V. DeQuattro, R. Falk, A.N. Ansari, and E. Lieberman. Childhood familial pheochromocytoma: Conflicting results of localization techniques. *Hypertension 8*: 851-858, 1986.

Uchida, Y. and T. Nomoto. Centrally administered thyrotropin-releasing hormone in the regulation of regional blood flow to brown adipose tissue in rats. *Europ J Pharmacol 149*:257-265, 1988.

Unsicker, K. Differentiation and phenotypical conversion of adrenal medullary cells: The effects of neuronotrophic, neurite-promoting, hormonal, and neuronal signals. *Neurol Neurobiol 16*:183-206, 1986.

Unsicker, K., H. Reichert-Preibsch, R. Schmidt, B. Pettmann, G. Labourdette, and M. Sensenbrenner. Astroglial and fibroblast growth factors have neurotrophic functions for cultured peripheral and central nervous system neurons. *Proc Natl Acad Sci USA 84*:5459-5463, 1987.

Unsicker, K., G. Stahnke, and T.H. Müller. Survival, morphology, and catecholamine storage of chromaffin cells in serum-free culture: Evidence for a survival and differentiation promoting activity in medium conditioned by purified chromaffin cells. *Neurochem Res 12*:995-1003, 1987.

Uvnäs, B. and C.-H. Åborg. Concomitant release by ion exchange of catecholamines (CA) and adenosine triphosphate (ATP) from bovine chromaffin granules superfused with isotonic sodium or potassium salt solutions. *Acta Physiol Scand 129*:585-586, 1987a.

Uvnäs, B. and C.-H. Åborg. A mixed cation (IRC-50)-anion (IR-4B) exchanger shows storage properties reminiscent of the storage of catecholamines (CA) and adenosine triphosphate (ATP) in chromaffin granules. *Acta Physiol Scand 129*:587-588, 1987b.

Uvnäs, B. and C.-H. Åborg. Catecholamines (CA) and adenosine triphosphate (ATP) are separately stored in bovine adrenal medulla, both in ionic linkage to granule sites, and not as a non-diffusible CA-ATP-protein complex. *Acta Physiol Scand 132*:297-311, 1988.

Valenta, L.J., A.N. Elias, and H. Eisenberg. ACTH stimulation of adrenal epinephrine and norepinephrine release. *Horm Res 23*:16-20, 1986.

Valet, P., M.A. Tran, C. Damase-Michel, G. de Saint-Blanquat, L. Dang Tran, F. Anglade, E. Koening-Berard, and J.L. Montastruc. Rilmenidine (S 3341) and the sympathoadrenal system: Adrenoreceptors, plasma and adrenal catecholamines in dogs. *J Auton Pharmacol 8*:319-326, 1988.

van Calker, D., K. Assmann, and W. Greil. Stimulation by bradykinin, angiotensin II, and carbachol of the accumulation of inositol phosphates in PC12 pheochromocytoma cells: Differential effects of lithium ions on inositol mono- and polyphosphates. *J Neurochem 49*:1379-1385, 1987.

van Calker, D. and R. Heumann. Nerve growth factor potentiates the agonist-stimulated accumulation of inositol phosphates in PC-12 pheochromocytoma cells. *Eur J Pharmacol 135*:259-260, 1987.

van der Meer, R.A., J.A. Jongejan, and J.A. Duine. Dopamine β-hydroxylase from bovine adrenal medulla contains covalently-bound pyrroloquinoline quinone. *FEBS Lett 231*:303-307,

1988.

Vander Tuig, J.G., K.A. Crist, and D.R. Romsos. Temporal adjustments in sympathoadrenal activity in rats with obesity-producing hypothalamic knife cuts. *Proc Soc Exp Biol Med* *185*:134-140, 1987.

VanHooff, C.O.M., P.N.E. DeGraan, J. Boonstra, A.B. Oestreicher, M.H. Schmidt-Michels, and W.H. Gispen. Nerve growth factor enhances the level of the protein kinase C substrate B-50 in pheochromocytoma PC12 cells. *Biochem Biophys Res Commun* *139*:644-651, 1986.

Vaquero, J., R. Martinez, S. Oya, S. Coca, F.G. Salazar, and M.I. Colado. Transplantation of adrenal medulla into spinal cord for pain relief: Disappointing outcome [letter]. *Lancet* *2*:1315, 1988.

Vaupel, R., H. Jarry, H.-T. Schlömer, and W. Wuttke. Differential response of substance P-containing subtypes of adrenomedullary cells to different stressors. *Endocrinol* *123*:2140-2145, 1988.

Veeraragavan, K., R. Coulombe, and C. Gagnon. Stoichiometric carboxyl methylation of chromogranins from bovine adrenal medullary cells. *Biochem Biophys Res Commun* *152*:732-738, 1988.

Venna, N. and T. Sabin. Adrenal grafts for Parkinson's disease [letter]. *N Engl J Med* *317*:1092-1093, 1987.

Vicentini, L.M., A. Ambrosini, F. DiVirgilio, J. Meldolesi, and T. Pozzan. Activation of muscarinic receptors in PC12 cells. Correlation between cytosolic Ca^{2+} rise and phosphoinositide hydrolysis. *Biochem J* *234*:555-562, 1986.

Vidrio, H. and F. García-Márquez. Role of the sympatho-adrenal system in the reflex tachycardia produced by hydralazine in the anesthetized rat. *Arch Int Pharmacodyn* *283*:94-104, 1986.

Vincent, S.R., C.H.S. McIntosh, P.B. Reiner, and J.C. Brown. Somatostatin immunoreactivity in the cat adrenal medulla: Localization and characterization. *Histochemistry* *87*:483-486, 1987.

Vindrola, O., R. Aloyz, A. Ase, S. Finkielman, and V.E. Nahmod. Adrenal proenkephalin-derived peptides during postnatal development in spontaneously hypertensive rats. *Endocrinology* *123*:810-815, 1988.

Vindrola, O., A. Ase, S. Finkielman, and V.E. Nahmod. Differential release of enkephalin and enkephalin-containing peptides from perfused cat adrenal glands. *J Neurochem* *50*:424-430, 1988.

Vogel, W.H. and R. Jensh. Chronic stress and plasma catecholamine and corticosterone levels in male rats. *Neurosci Lett* *87*:183-188, 1988.

Volknandt, W., M. Schober, R. Fischer-Colbrie, H. Zimmermann, and H. Winkler. Cholinergic nerve terminals in the rat diaphragm are chromogranin A immunoreactive. *Neurosci*

Lett *81*:241-244, 1987.

Volonté, C. Lithium stimulates the binding of GTP to the membranes of PC12 cells cultured with nerve growth factor. *Neurosci Lett 87*:127-132, 1988.

Volonté, C., G.S. Parries, and E. Racker. Stimulation of inositol incorporation into lipids of PC12 cells. *J Neurochem 51*:1156-1162, 1988.

Volonté, C. and E. Racker. Lithium stimulation of membrane-bound phospholipase C from PC12 cells exposed to nerve growth factor. *J Neurochem 51*:1163-1168, 1988.

von Dalnok, G.K. and H.D. Menssen. A quantitative electron microscopic study of the effect of glucocorticoids *in vivo* on the early postnatal differentiation of paraneuronal cells in the carotid body and the adrenal medulla of the rat. *Anat Embryol (Berl) 174*:307-319, 1986.

vonGrafenstein, H., C.S. Roberts, and P.F. Baker. Kinetic analysis of the triggered exocytosis/endocytosis secretory cycle in cultured bovine adrenal medullary cells. *J Cell Biol 103*:2343-2352, 1986.

Wada, A., M. Arita, H. Kobayashi, and F. Izumi. Binding of [^{3}H]saxitoxin to the voltage-dependent Na channels and inhibition of ^{22}Na influx in bovine adrenal medullary cells. *Neuroscience 23*:327-331, 1987.

Wada, A., M. Arita, N. Yanagihara, and F. Izumi. Binding of [^{3}H]phencyclidine to adrenal medullary cells: Inhibition of ^{22}Na influx, ^{45}Ca influx, ^{86}Rb efflux and catecholamine secretion caused by carbachol and veratridine. *Neuroscience 25*:687-696, 1988.

Wada, A., H. Kobayashi, M. Arita, N. Yanagihara, and F. Izumi. Potassium channels in cultured bovine adrenal medullary cells: Effects of high K, veratridine and carbachol on ^{86}rubidium efflux. *Neuroscience 22*:1085-1092, 1987.

Wada, A., H. Takara, N. Yanagihara, H. Kobayashi, and F. Izumi. Inhibition of Na+-pump enhances carbachol-induced influx of ^{45}Ca^{2+} and secretion of catecholamines by elevation of cellular accumulation of ^{22}Na$^+$ in cultured bovine adrenal medullary cells. *Naunyn Schmiedebergs Arch Pharmacol 332*:351-356, 1986.

Waeber, B., J.-F. Aubert, R. Corder, D. Evéquoz, J. Nussberger, R. Gaillard, and H.R. Brunner. Cardiovascular effects of neuropeptide Y. *Am J Hypertens 1*:193-199, 1988.

Wagner, J.A. NIF (Neurite-Inducing Factor): A novel peptide inducing neurite formation in PC12 cells. *J Neurosci 6*:61-67, 1986.

Wakade, A.R. Noncholinergic transmitter(s) maintains secretion of catecholamines from rat adrenal medulla for several hours of continuous stimulation of splanchnic neurons. *J Neurochem 50*:1302-1308, 1988.

Wakade, A.R., R. Kahn, R.K. Malhotra, C.G. Wakade, and T.D.

Wakade. McN-A-343, a specific agonist of M_1-muscarinic receptors, exerts antinicotinic and antimuscarinic effects in the rat adrenal medulla. *Life Sci 39*:2073-2080, 1986.

Wakade, A.R., R.K. Malhotra, T.R. Sharma, and T.D. Wakade. Changes in tonicity of perfusion medium cause prolonged opening of calcium channels of the rat chromaffin cells to evoke explosive secretion of catecholamines. *J Neurosci 6*:2625-2634, 1986.

Wakade, A.R., R.K. Malhotra, and T.D. Wakade. Phorbol ester facilitates ^{45}Ca accumulation and catecholamine secretion by nicotine and excess K^+ but not by muscarine in rat adrenal medulla. *Nature 321*:698-700, 1986.

Wakade, A.R., R.K. Malhotra, T.D. Wakade, and W.R. Dixon. Simultaneous secretion of catecholamines from the adrenal medulla and of [^{3}H]norepinephrine from sympathetic nerves from a single test preparation: Different effects of agents on the secretion. *Neuroscience 18*:877-888, 1986.

Wakade, A.R., T.D. Wakade, and R.K. Malhotra. Restoration of catecholamine content of previously depleted adrenal medulla *in vitro*: Importance of synthesis in maintaining the catecholamine stores. *J Neurochem 51*:820-829, 1988.

Wakefield, L.M., A.E.G. Cass, and G.K. Radda. Functional coupling between enzymes of the chromaffin granule membrane. *J Biol Chem 261*:9739-9745, 1986a.

Wakefield, L.M., A.E.G. Cass, and G.K. Radda. Electron transfer across the chromaffin granule membrane: Use of EPR to demonstrate reduction of intravesicular ascorbate radical by the extravesicular mitochondrial NADH:ascorbate radical oxido-reductase. *J Biol Chem 261*:9746-9752, 1986b.

Wallin, G., J. Cassuto, S. Högström, and T. Hedner. Influence of intraperitoneal anesthesia on pain and the sympathoadrenal response to abdominal surgery. *Acta Anaesthesiol Scand 32*:553-558, 1988.

Walton, K.M., K. Sandberg, T.B. Rogers, and R.L. Schnaar. Complex ganglioside expression and tetanus toxin binding by PC12 pheochromocytoma cells. *J Biol Chem 263*:2055-2063, 1988.

Wan, D.C.-C., D.A. Powis, P.D. Marley, and B.G. Livett. Effects of α- and β-adrenoceptor agonists and antagonists on ATP and catecholamine release and desensitization of the nicotinic response in bovine adrenal chromaffin cells. *Biochem Pharmacol 37*:725-736, 1988.

Wan, D.C.-C., D. Scanlon, B. Power, C.-l. Choi, P. Hudson, and B.G. Livett. Synthesis of a specific oligodeoxyribonucleotide probe and its use for studying proenkephalin A gene expression. *Neurosci Lett 76*:74-80, 1987.

Wang, S.-Y., Y. Moriyama, M. Mandel, J.D. Hulmes, Y.-C.E. Pan, W. Danho, H. Nelson, and N. Nelson. Cloning of cDNA encoding a 32-kDa protein. An accessory polypeptide of the

H[+]-ATPase from chromaffin granules. *J Biol Chem* *263*:17638-17642, 1988.

Wanke, E., A. Ferroni, P. Gattanini, and J. Meldolesi. α-Latrotoxin of the black widow spider venom opens a small, non-closing cation channel. *Biochem Biophys Res Commun* *134*:320-325, 1986.

Warashina, A., N. Fujiwara, T. Hirano, and K. Shimoji. Characteristics of bradykinin-evoked secretory response in the perfused rat adrenal medulla. *Biomed Res* *9*:139-145, 1988.

Warren, J. The adrenal medulla and the airway. *Br J Dis Chest* *80*:1-6, 1986.

Waschek, J.A., J.R. Dave, R.L. Eskay, and L.E. Eiden. Barium distinguishes separate calcium targets for synthesis and secretion of peptides in neuroendocrine cells. *Biochem Biophys Res Commun 146*:495-501, 1987.

Waschek, J.A. and L.E. Eiden. Calcium requirements for barium stimulation of enkephalin and vasoactive intestinal peptide biosynthesis in adrenomedullary chromaffin cells. *Neuropeptides 11*:39-45, 1988.

Watanabe, T., T. Kawada, and K. Iwai. Enhancement by capsaicin of energy metabolism in rats through secretion of catecholamine from adrenal medulla. *Agric Biol Chem (Tokyo) 51*:75-79, 1987.

Watanabe, T., T. Kawada, and K. Iwai. Effect of capsaicin pretreatment on capsaicin-induced catecholamine secretion from the adrenal medulla in rats. *Proc Soc Exp Biol Med 187*:370-374, 1988.

Watanabe, T., T. Kawada, M. Kurosawa, A. Sato, and K. Iwai. Adrenal sympathetic efferent nerve and catecholamine secretion excitation caused by capsaicin in rats. *Am J Physiol 255*:E23-E27, 1988.

Watanabe, T., T. Kawada, M. Yamamoto, and K. Iwai. Capsaicin, a pungent principle of hot red pepper, evokes catecholamine secretion from the adrenal medulla of anesthetized rats. *Biochem Biophys Res Commun 142*:259-264, 1987.

Watkinson, A., G.J. Dockray, and J. Young. N-linked glycosylation of a proenkephalin A-derived peptide: Evidence for the glycosylation of an NH$_2$-terminally extended Met-enkephalin Arg[6]-Gly[7]-Leu[8] variant. *J Biol Chem 263*:7147-7152, 1988.

Waymire, J.C., J.P. Johnston, K. Hummer-Lickteig, A. Lloyd, A. Vigny, and G.L. Craviso. Phosphorylation of bovine adrenal chromaffin cell tyrosine hydroxylase. Temporal correlation of acetylcholine's effect on site phosphorylation, enzyme activation, and catecholamine synthesis. *J Biol Chem 263*:12439-12447, 1988.

Weiler, R., H. Feichtinger, K.W. Schmid, R. Fischer-Colbrie, L. Grimelius, B. Cedermark, M. Papotti, G. Bussolati, and H. Winkler. Chromogranin A and B and secretogranin II in bronchial and intestinal carcinoids. *Virchows Arch [A]* *412*:103-109, 1987.

Weiler, R., R. Fischer-Colbrie, K.W. Schmid, H. Feichtinger, G. Bussolati, L. Grimelius, K. Krisch, H. Kerl, D.T. O'Connor, and H. Winkler. Immunological studies on the occurrence and properties of chromogranin A and B and secretogranin II in endocrine tumors. *Am J Surg Pathol 12*:877-884, 1988.

Weinbach, E.C., J.L. Costa, B.D. Nelson, C.E. Claggett, T. Hundal, D. Bradley, and S.J. Morris. Effects of tricyclic antidepressant drugs on energy-linked reactions in mitochondria. *Biochem Pharmacol 35*:1445-1451, 1986.

Weisberg, E.P., D.K. Batter, W.E. Brown, and B.B. Kaplan. Purification and partial amino acid sequence of bovine adrenal phenylethanolamine *N*-methyltransferase: A comparison of nucleic acid and protein sequence data. *J Neurosci Res 19*:377-382, 1988.

Weiss, C.D. The human immunodeficiency virus and the adrenal medulla. *Ann Intern Med 105*:300, 1986.

Welbourn, R.B. Early surgical history of phaeochromocytoma. *Br J Surg 74*:594-596, 1987.

Weppelman, R.M. Avian dopamine β-monooxygenase. *Methods Enzymol 142*:608-617, 1987.

Wernecke, K. and M. Galanski. [Percutaneous fine-needle biopsy of the adrenal glands.] *Radiologe 26*:191-197, 1986.

Wernert, N., A. Antalffy, and G. Dhom. Effects of estradiol on adrenal cortex and medulla of the rat: Morphometric studies. *Pathol Res Pract 181*:551-557, 1986.

Westerink, B.H.C. and W. Koolstra. Circadian variation of catecholamine excretion in rats: Correlation with locomotor activity and effects of drugs. *Neuropharmacology 25*:1255-1262, 1986.

Westermann, R., F. Stögbauer, K. Unsicker, and R. Lietzke. Calcium-dependence of chromogranin A-catecholamine interaction. *FEBS Lett 239*:203-206, 1988.

White, N.M. and C. Messier. Effects of adrenal demedullation on the conditioned emotional response and on the memory improving action of glucose. *Behav Neurosci 102*:499-503, 1988.

White, P.D., W.A. Lishman, and M.A. Wyke. Phaeochromo-cytoma as a cause of reversible dementia. *J Neurol Neurosurg Psychiatry 49*:1449-1451, 1986.

White, R.J. Adrenal grafts for Parkinson's disease [letter]. *N Engl J Med 317*:1092, 1987.

White, T.D. Role of adenine compounds in autonomic neurotransmission. *Pharmacol Ther 38*:129-168, 1988.

White, T.D., J.E. Bourke, and B.G. Livett. Direct and continuous detection of ATP secretion from primary monolayer cultures of bovine adrenal chromaffin cells. *J Neurochem 49*:1266-1273, 1987.

Whiting, P.J., R. Schoepfer, L.W. Swanson, D.M. Simmons, and

J.M. Lindstrom. Functional acetylcholine receptor in PC12 cells reacts with a monoclonal antibody to brain nicotinic receptors. *Nature* 327:515-518, 1987.

Wiedenmann, B., W.W. Franke, C. Kuhn, R. Moll, and V.E. Gould. Synaptophysin: A marker protein for neuroendocrine cells and neoplasms. *Proc Natl Acad Sci USA* 83:3500-3504, 1986.

Wiedenmann, B., H. Rehm, M. Knierim, and C.-M. Becker. Fractionation of synaptophysin-containing vesicles from rat brain and cultured PC12 pheochromocytoma cells. *FEBS Lett* 240:71-77, 1988.

Wilburn, L.A., P.C. Goldsmith, K.-J. Chang, and R.B. Jaffe. Ontogeny of enkephalin and catecholamine-synthesizing enzymes in the primate fetal adrenal medulla. *J Clin Endocrinol Metab* 63:974-980, 1986.

Wilburn, L.A. and R.B. Jaffe. Quantitative assessment of the ontogeny of met-enkephalin, norepinephrine and epinephrine in the human fetal adrenal medulla. *Acta Endocrinol (Copenh)* 118:453-459, 1988.

Wildmann, J., H. Möhler, W. Vetter, U. Ranalder, K. Schmidt, and R. Maurer. Diazepam and *N*-desmethyldiazepam are found in rat brain and adrenal and may be of plant origin. *J Neural Transm* 70:383-398, 1987.

Wilgus, H. and R. Roskoski Jr. Inactivation of tyrosine hydroxylase activity by ascorbate *in vitro* and in rat PC12 cells. *J Neurochem* 51:1232-1239, 1988.

Williams, M., M. Abreu, M.F. Jarvis, and L. Noronha-Blob. Characterization of adenosine receptors in the PC12 pheochromocytoma cell line using radioligand binding: Evidence for A-2 selectivity. *J Neurochem* 48:498-502, 1987.

Williams, V. Parkinson's disease: Autotransplantation of adrenal medulla to caudate nucleus of the brain. *J Neurosci Nurs* 19:174, 1987.

Wilson, B.S., S.H. Phan, and R.V. Lloyd. Chromogranin from normal human adrenal glands: Purification by monoclonal antibody affinity chromatography and partial *N*-terminal amino acid sequence. *Regul Pept* 13:207-223, 1986.

Wilson, R.B., M.A. Holscher, A.G. Kasselberg, and M. Jones. Leu-enkephalin and somatostatin immunoreactivities in canine and equine pheochromocytomas. *Vet Pathol* 23:96-98, 1986.

Wilson, S.K., D.R. Lynch, and S.H. Snyder. Angiotensin-converting enzyme labeled with [^{3}H]captopril. Tissue localizations and changes in different models of hypertension in the rat. *J Clin Invest* 80:841-851, 1987.

Wilson, S.P. Reserpine increases chromaffin cell enkephalin stores without a concomitant decrease in other proenkephalin-derived peptides. *J Neurochem* 49:1550-1556, 1987a.

Wilson, S.P. Vasoactive intestinal peptide and substance P increase levels of enkephalin-containing peptides in adrenal chromaffin cells. *Life Sci 40*:623-628, 1987b.

Wilson, S.P. Purification of adrenal chromaffin cells on Renografin gradients. *J Neurosci Methods 19*:163-171, 1987c.

Wilson, S.P. Vasoactive intestinal peptide elevates cyclic AMP levels and potentiates secretion in bovine adrenal chromaffin cells. *Neuropeptides 11*:17-21, 1988.

Wilson, S.P. and J.F. Beeler. Catecholamine depletion and accumulation of 1-methyl-4-phenyl-1,2,3,6-tetrahydropyridine (MPTP) and 1-methyl-4phenylpyridinium (MPP$^+$) in adrenal medullary chromaffin cells. *Neurochem Int 13*:333-343, 1988.

Winder, W.W., S.F. Loy, D.S. Burke, and S.J. Hawkes. Liver glycogenolysis during exercise in adrenodemedullated male and female rats. *Am J Physiol 251*:R1151-R1155, 1986.

Winder, W.W. and H.T. Yang. Blood collection and processing for measurement of catecholamines in exercising rats. *J Appl Physiol 63*:418-420, 1987.

Winder, W.W., H.T. Yang, A.W. Jaussi, and C.R. Hopkins. Epinephrine, glucose, and lactate infusion in exercising adrenodemedullated rats. *J Appl Physiol 62*:1442-1447, 1987.

Winkler, H. Occurrence and mechanism of exocytosis in adrenal medulla and sympathetic nerve. pp. 43-118. In: **Handbook of Experimental Pharmacology**, Vol. 90/I. (U. Trendelenburg and N. Weiner, eds). Springer-Verlag, Berlin, 1988.

Winkler, H., D.K. Apps, and R. Fischer-Colbrie. The molecular function of adrenal chromaffin granules: Established facts and unresolved topics. *Neuroscience 18*:261-290, 1986.

Winkler, H., R. Fischer-Colbrie, D. Obendorf, and U. Schwarzenbrunner. Adrenergic and cholinergic vesicles: Are there common antigens and common properties? pp. 305-314. In: **Cellular and Molecular Basis of Synaptic Transmission**. (H. Zimmermann, ed). Springer-Verlag, Berlin, 1988.

Winkler, H., R. Fischer-Colbrie, M. Schober, S. Rieker, and R. Weiler. Catecholamine-storing cells secrete a blended mixture of catecholamines, neuropeptides and chromogranins. pp. 347-359. In: **Molecular Mechanisms in Secretion, Alfred Benzon Symposium 25**. (N.A. Thorn, M. Treiman, and O.H. Petersen, eds). Munksgaard, Copenhagen, 1988.

Winkler, V.P., T. Abel, and K. Helmke. [New signs of echographic identification of the adrenal glands in children and adolescents.] *Fortschr Rontgenstr 148*:150-154, 1988.

Wohlfarter, T., R. Fischer-Colbrie, R. Hogue-Angeletti, L.E. Eiden, and H. Winkler. Processing of chromogranin A within chromaffin granules starts at C- and N-terminal cleavage sites. *FEBS Lett 231*:67-70, 1988.

Wong, D.L., L. Yamasaki, and R.D. Ciaranello. Characterization of the isozymes of bovine adrenal medullary phenylethanolamine *N*-methyltransferase. *Brain Res 410*:32-44, 1987.

Woodruff, N.R. and K.E. Neet. β Nerve growth factor binding to PC12 cells. Association kinetics and cooperative interactions.

Biochemistry 25:7956-7966, 1986a.

Woodruff, N.R. and K.E. Neet. Inhibition of β nerve growth factor binding to PC12 cells by α nerve growth factor and γ nerve growth factor. *Biochemistry* 25:7967-7974, 1986b.

Woolf, P.D. Endocrinology of shock. *Ann Emerg Med 15*:1401-1405, 1986.

Wyatt, R.J., J.M. Morihisa, R.K. Nakamura, and W.J. Freed. Transplanting tissue into the brain for function: Use in a model of Parkinson's disease. pp.199-208. In: **Neuropeptides in Neurologic and Psychiatric Disease**, Vol. 64. (J.B. Martin and J.D. Barchas, eds). Raven Press, New York, 1986.

Yamada, S., K. Morita, T. Dohi, and A. Tsujimoto. A modulating role of prostaglandins in catecholamine release by perfused dog adrenal glands. *Eur J Pharmacol 146*:27-34, 1988.

Yamada, Y., H. Teraoka, Y. Nakazato, and A. Ohga. Intracellular Ca^{2+} antagonist TMB-8 blocks catecholamine secretion evoked by caffeine and acetylcholine from perfused cat adrenal glands in the absence of extracellular Ca^{2+}. *Neurosci Lett 90*:338-342, 1988.

Yamagami, K. and M. Sorimachi. The secretory mechanism in adrenal chromaffin cells by nitrophenol compounds: Possible involvement of the change in the surface potentials. *Jpn J Physiol 37*:643-656, 1987.

Yamaguchi, N. and M. Brassard. A differential effect of yohimbine on adrenal and catecholamine release during bilateral carotid occlusion in the dog. *J Autonom Nerv Syst 25*:141-154, 1988.

Yamakita, N., K. Yasuda, E. Goshima, M. Murayama, H. Murase, Y. Minamori, T. Ishizuka, and K. Miura. Comparative assessment of ultrasonography and computed tomography in adrenal disorders. *Ultrasound Med Biol 12*:23-29, 1986.

Yamamoto, D. Dynamics of strychnine block of single sodium channels in bovine chromaffin cells. *J Physiol (Lond) 370*:395-407, 1986.

Yamanaka, K., S. Kigoshi, and I. Muramatsu. Muscarinic receptor subtypes in bovine adrenal medulla. *Biochem Pharmacol 35*:3151-3157, 1986a.

Yamanaka, K., S. Kigoshi, and I. Muramatsu. Copper-induced alteration of muscarinic binding in bovine adrenal medulla. *Jpn J Pharmacol 41*:415-418, 1986b.

Yamanaka, K., I. Muramatsu, and S. Kigoshi. Tetranitromethane modification of the muscarinic receptors in bovine adrenal medulla. *Jpn J Pharmacol 48*:67-76, 1988.

Yamazaki, H., M. Yoshimura, S. Kanbara, R. Takashina, K. Takeda, H. Takahashi, and H. Ijichi. [Role of adrenal catecholamines on the synthesis of renal catecholamines.] *Nippon Jinzo Gakkai Shi 28*:815-820, 1986a.

Yamazaki, H., M. Yoshimura, S. Kanbara, R. Takashina, H. Takahashi, K. Takeda, and H. Ijichi. [Role of adrenal medulla in elevation of blood pressure induced by sodium loading.] *Nippon Jinzo Gakkai Shi 28*:1371-1376, 1986b.

Yanagihara, N., A.W. Tank, T.A. Langan, and N. Weiner. Enhanced phosphorylation of tyrosine hydroxylase at more than one site is induced by 56mM K^+ in rat pheochromocytoma PC12 cells in culture. *J Neurochem 46*:562-568, 1986.

Yanagihara, N., Y. Uezono, Y. Koda, A. Wada, and F. Izumi. Activation of tyrosine hydroxylase by micromolar concentrations of calcium in digitonin-permeabilized adrenal medullary cells. *Biochem Biophys Res Commun 146*:530-536, 1987.

Yanagihara, N., A. Wada, and F. Izumi. Effects of α_2-adrenergic agonists on carbachol-stimulated catecholamine synthesis in cultured bovine adrenal medullary cells. *Biochem Pharmacol 36*:3823-3828, 1987.

Yanagihara, N., K. Yokota, A. Wada, and F. Izumi. Intracellular pH and catecholamine synthesis in cultured bovine adrenal medullary cells: Effect of extracellular Na^+ removal. *J Neurochem 49*:1740-1746, 1987.

Yanase, T., H. Nawata, M. Haji, K.-I. Kato, and H. Ibayashi. Preproenkephalin A-derived opioid peptides and mRNA of preproenkephalin A in human pheochromocytomas. *Mol Cell Endocrinol 51*:237-241, 1987.

Yanase, T., H. Nawata, K.-I. Kato, and H. Ibayashi. Catecholamines and opioid peptides in human phaeochromocytomas. *Acta Endocrinol (Copenh) 113*:378-384, 1986a.

Yanase, T., H. Nawata, K.-I. Kato, and H. Ibayashi. *In vivo* evidence of regulation by pituitary-adrenal axis of urinary epinephrine excretion in men. *Endocrinol Jpn 33*:449-455, 1986b.

Yanase, T., H. Nawata, K.-I. Kato, and H. Ibayashi. Preproenkephalin B-derived opioid peptides in human phaeochromocytomas. *Acta Endocrinol (Copenh) 114*:446-451, 1987a.

Yanase, T., H. Nawata, K.-I. Kato, H. Ibayashi, and H. Matsuo. Studies on adrenorphin in pheochromocytoma. *J Clin Endocrinol Metab 64*:692-697, 1987b.

Yang, H.T., K.I. Carlson, and W.W. Winder. Insulin does not influence muscle glycogenolysis in adrenodemedullated exercising rats. *Am J Physiol 253*:R535-R540, 1987.

Yang, H.T., R.L. Hammer, T.L. Sellers, J. Arogyasami, D.T. Carrell, and W.W. Winder. Adrenodemedullation affects endurance but not hepatic fructose 2,6-bisphosphate. *Am J Physiol 254*:R572-R577, 1988.

Yashima, N., A. Wada, and F. Izumi. Halothane inhibits the cholinergic-receptor-mediated influx of calcium in primary culture of bovine adrenal medulla cells. *J Anesthesiol 64*:466-472, 1986.

Yavin, E., T. Hama, S. Gil, and G. Guroff. Nerve growth factor and gangliosides stimulate the release of glycoproteins from PC12 pheochromocytoma cells. *J Neurochem 46*:794-803, 1986.

Yi, S.Y. [Care of patients with adrenal medulla tissue implanted into the caudate nucleus.] *Chung Hua Hu Li Tsa Chih 22*:105-106, 1987.

Yntema, O.P. and J. Korf. Transient suppression by stress of haloperidol induced catalepsy by the activation of the adrenal medulla. *Psychopharmacology (Berlin) 91*:131-134, 1987.

Yoburn, B.C., S.O. Franklin, S.E. Calvano, and C.E. Inturrisi. Regulation of rat adrenal medullary enkephalins by glucocorticoids. *Life Sci 40*:2495-2503, 1987.

Yokohama, H., M. Negishi, K. Sugama, H. Hayashi, S. Ito, and O. Hayaishi. Inhibition of prostaglandin E_2-induced phosphoinositide metabolisn by phorbol ester in bovine adrenal chromaffin cells. *Biochem J 255*:957-962, 1988.

Yokohama, H., T. Tanaka, S. Ito, M. Negishi, H. Hayashi, and O. Hayaishi. Prostaglandin E receptor enhancement of catecholamine release may be mediated by phosphoinositide metabolism in bovine adrenal chromaffin cells. *J Biol Chem 263*:1119-1122, 1988.

Yokota, K., N. Yanagihara, F. Izumi, and A. Wada. Effects of protonophores on the synthesis of catecholamines and the intracellular pH in cultured bovine adrenal medullary cells. *J Neurochem 51*:246-251, 1988.

Yokotani, K., Y. Okuma, and Y. Osumi. Sympatho-adrenal system involved in the inhibitory effects of nicotine on the vagally stimulated gastric acid output and mucosal blood flow in rats. *Eur J Pharmacol 129*:253-260, 1986.

Yokotani, K., Y. Okuma, and Y. Osumi. Intracerebroventricularly administered nicotine inhibits vagally stimulated gastric acid output in rats by activating the central sympatho-adrenal outflow. *Jpn J Pharmacol 45*:288-291, 1987.

Yokotani, K., K. Yokotani, Y. Okuma, and Y. Osumi. Sympatho-adrenomedullary system mediation of the prostaglandin E_2-induced central inhibition of gastric acid output in rats. *J Pharmacol Exp Ther 244*:335-340, 1988.

Yoneda, Y. and K. Ogita. Localization of [3H]glutamate binding sites in rat adrenal medulla. *Brain Res 383*:387-391, 1986.

Yoneda, Y. and K. Ogita. Enhancement of [3H]glutamate binding by N-methyl-D-aspartic acid in rat adrenal. *Brain Res 406*:24-31, 1987a.

Yoneda, Y. and K. Ogita. Solubilization of novel binding sites for [3H] glutamate in rat adrenal. *Biochem Biophys Res Commun 142*:609-616, 1987b.

Yoneda, Y., K. Ogita, H. Nakamuta, M. Koida, T. Ohgaki, and H. Meguri. Possible interaction of [3H]glutamate binding sites with anion channels in rat neural tissues. *Neurochem Int 9*:521-531, 1986.

Yoshie, S., C. Hagn, M. Ehrhart, R. Fischer-Colbrie, D. Grube, H.

Winkler, and M. Gratzl. Immunological characterization of chromogranins A and B and secretogranin II in the bovine pancreatic islet. *Histochemistry 87*:99-106, 1987.

Yoshikawa, M. and T. Nakada. High level of adrenal catecholamines in hypertensive subjects with impaired renal function. *Int Urol Nephrol 18*:185-192, 1986.

Yoshikawa, M. and H. Saito. [The localization and release of immunoreactive vasoactive intestinal polypeptide (VIP) in bovine adrenal medulla.] *Nippon Naibunpi Gakkai Zasshi 62*:1151-1162, 1986.

Yoshimatsu, H., Y. Oomura, T. Katafuchi, and A. Niijima. Effects of hypothalamic stimulation and lesion on adrenal nerve activity. *Am J Physiol 253*:R418-R424, 1987.

Yoshioka, M., H. Togashi, M. Minami, and H. Saito. Effects of hydralazine on adrenal and cardiac sympathetic nerve activity in anesthetized rats. *Res Commun Chem Pathol Pharmacol 54*:313-320, 1986.

Youdim, M.B.H., E. Heldman, H.B. Pollard, P.J. Fleming, and E.M. McHugh. Contrasting monoamine oxidase activity and tyramine induced catecholamine release in PC12 and chromaffin cells. *Neuroscience 19*:1311-1318, 1986.

Yu, P.H. Monoamine oxidase. pp. 235-272. In: **Neuromethods, Series I: Neurotransmitter enzymes**, Vol. 5. (A.A. Boulton, G.B. Baker, and P.H. Yu, eds). Humana Press, Clifton, NJ, 1986.

Zabransky, S. [Surgery of the adrenals in childhood-pediatric aspects.] *Z Kinderchir 43*:243-251, 1988.

Zagon, I.S., R.E. Rhodes, and P.J. McLaughlin. Localization of enkephalin immunoreactivity in diverse tissues and cells of the developing and adult rat. *Cell Tissue Res 246*:561-565, 1986.

Zahler, P., M. Reist, M. Pilarska, and K. Rosenheck. Phospholipase C and diacylglycerol lipase activities associated with plasma membranes of chromaffin cells isolated from bovine adrenal medulla. *Biochim Biophys Acta 877*:372-379, 1986.

Zgombick, J.M., V.G. Erwin, and K. Cornell. Ethanol-induced adrenomedullary catecholamine secretion in LS/lbg and SS/lbg mice. *J Pharmacol Exp Ther 236*:634-640, 1986.

Zhang, J.M. Influence of electro-stimulation of central ends of cervical vagi and the right tibial nerve on catecholamine release from the adrenal medulla. *Chung Kuo I Hsueh Ko Hsueh Yuan Hsueh Pao 8*:121-124, 1986.

Zhang, J.M. and Y.Q. Cheng. Effect of electro-stimulation of central ends of cervical vagi on the release of catecholamine from the adrenal medulla in rats. *Chung Kuo I Hsueh Ko Hsueh Yuan Hsueh Pao 8*:125-127, 1986.

Zhang, J.M. and J.X. Yang. Influence of third ventricle injection of ACh and physostigmine on adrenal medulla catecholamine release in cats. *Chung Kuo I Hsueh Ko Hsueh Yuan Hsueh Pao*

8:117-120, 1986.

Zhang, S.-Y. and J.L. McGaugh. [Enkephalin and the modulation of memory consolidation. I. Naloxone and met-enkephalin influence on memory consolidation and its alteration by adrenal denervation.] *Chung Kuo I Hsueh Ko Hsueh Yuan Hsueh Pao* 9:60-64, 1987.

Zhang, S.-Y., J.L. McGaugh, R.G. Juler, and I.B. Introini-Collison. Naloxone and [Met^5]enkephalin effects on retention: Attenuation by adrenal denervation. *Eur J Pharmacol 138*:37-44, 1987.

Zhang, W.C., M.C. Ding, and S.-S. Jiao. [Intracaudate grafting of the adrenal medulla in the treatment of Parkinson's disease.] *Chung Hua Wai Ko Tsa Chih 26*:1-5, 1988.

Zhang, W.C., Y.J. Ding, and J.K. Cao. [Intracerebral transplantation of the adrenal medulla in the treatment of parkinsonism.] *Chung Hua Wai Ko Tsa Chih 25*:650-652, 1987.

Ziegler, M.G., B. Kennedy, and H. Elayan. A sensitive radioenzymatic assay for epinephrine forming enzymes. *Life Sci 43*:2117-2122, 1988.

Zimlichman, R., D.S. Goldstein, S. Zimlichman, and H.R. Keiser. Effects of ouabain on cytosolic calcium in lymphocytes, platelets and adrenomedullary cells. *J Hypertens 5*:605-609, 1987.

Zimlichman, R., D.S. Goldstein, S. Zimlichman, R. Stull, and H.R. Keiser. Angiotensin II increases cytosolic calcium and stimulates catecholamine release in cultured bovine adrenomedullary cells. *Cell Calcium 8*:315-325, 1987.

Zimlichman, R., H.B. Pollard, H.R. Keiser, and E. Heldman. Calcium influx but not cytosolic calcium level triggers exocytotic secretion in cultured chromaffin cells. *J Hypertens [Suppl] 4*:S203-S205, 1986.

Zukowska-Grojec, Z., M. Konarska, and R. McCarty. Differential plasma catecholamine and neuropeptide Y responses to acute stress in rats. *Life Sci 42*:1615-1624, 1988.

Zurawski, G., M. Benedik, B.J. Kamb, J.S. Abrams, S.M. Zurawski, and F.D. Lee. Activation of mouse T-helper cells induces abundant preproenkephalin mRNA synthesis. *Science 232*:772-775, 1986.

AUTHOR INDEX

Ong, H., 201
Ono, T., 290
Ookawa, H., 373
Oomura, Y., 251-252
Oparil, S., 338
Oplatka, A., 89-90
Opocher, G., 48
Oppedal, B.R., 377
Orband-Miller, L., 315
Orellana, C., 38
Orit, S., 271
Orlando, P., 361
Ornberg, R.L., 6, 18, 149, 194-195
Ornstein, D.L., 179
Orth, D.N., 220, 333
Orts, A., 38
Osamura, R.Y., 214
Osborne Jr., J.C., 149
Oset-Gasque, M.J., 43, 172
Oshima, T., 140
Osserman, E.F., 296
Ostfeld, A.M., 306
Ostrosky, F., 1, 300
Ostrosky-Solís, F., 296
O'Sullivan, A.J., 5, 68, 88, 92
Osumi, Y., 257, 325
Ota, K., 350
Otten, U., 264
Ottenweller, J.E., 310
Oya, S., 3, 295
Oyama, M., 355
Ozaki, M., 127
Ozawa, S., 63
Özkan, E.D., 236
Ozutsumi, K., 127

Pace, G., 369
Pácha, J., 324
Pacquing, Y.V., 155
Padbury, J.F., 314-315
Pagonis, C., 120
Painter, G.R., 2, 117
Pais, J.R., 334
Paitian, N.A., 158, 160, 175, 201
Pal, D., 123, 336
Palacios-Araus, L., 118
Palatini, P., 161
Palestini, N., 365
Palfrey, H.C., 95
Paling, M.R., 360
Palmiter, R.D., 6, 162
Pan, Y.-C.E., 165, 167

Pandey, S.C., 17
Pandiella, A., 98
Pandiella-Alonso, A., 266
Panerai, A.E., 210
Pankow, D., 120, 125
Panten, U., 169
Panula, P., 220
Papadopoulos, N.J., 150, 158
Papahadjopoulos, D., 111
Papotti, M., 189
Pappas, G.D., 2, 293-295
Paquin, J., 181, 233
Parada, I.M., 9, 279
Park, D.H., 61, 161, 163
Park, J., 7, 186
Parker, T.L., 8, 243
Parmer, R.J., 179, 208
Parries, G.S., 278
Parsons, S.J., 4, 82, 173
Partain, C.L., 360
Parti, R., 236
Parvez, H., 341
Parvez, S., 341
Parysek, L.M., 91, 200
Patel, M.M., 23
Patel-Vaidya, U., 293
Patil, A., 295
Patzak, A., 113-114
Paves, H., 272-273
Pavlovich, E.R., 342
Pearce, J.M.S., 296
Pearse, A.G.E., 10
Pearse, B.M.F., 199
Peeler, D.F., 352
Peisach, J., 158
Pelletier, G., 13, 209, 280
Pellicer, A., 286
Pelto-Huikko, M., 19, 217
Penn, R.D., 1, 298
Penner, R., 66, 115
Peppers, S.C., 74
Pequignot, J.M., 177, 303
Percy, J.M., 6, 165
Peréz Maestu, J.C.R., 362
Perin, M.S., 6, 156
Perlman, R.L., 10, 101, 122, 281
Perlow, M.J., 2, 294, 301

Pernow, J., 213
Perrett, D., 246
Perrin, D., 106
Perronne, C., 375
Persson, H., 212
Persson, K., 51
Peshavaria, M., 179, 202
Peskind, E.R., 256
Peters, J.A., 43-44
Peterson, D.S., 153
Petersson, L., 145
Petit, P., 358
Petry, N.A., 363
Petterson, R., 131, 163
Pettmann, B., 278
Pevcova, E.I., 310
Peyrin, L., 177
Pezzoli, G., 2, 19, 290, 297, 300
Pfeiffer, G.L., 19, 169
Phan, S.H., 181
Philie, J., 4, 41
Phillips, I., 301
Phillips, J.H., 2, 194
Phillipson, O.T., 177
Piasecki, G.J., 315
Pieniek, M., 79
Piezzi, R.S., 249
Pigeon, D., 143
Pihlak, A., 272
Pilarska, M., 173
Pinto, J.E.B., 128
Piscopo, I., 20
Pitas, R.E., 282
Pitman, D., 310
Pittius, C.W., 8, 223-224
Pittman, Q.J., 258
Pivazian, A., 270
Pizzuti, A., 19
Plattner, H., 5, 17
Plevin, R., 55, 101
Ploska, A., 1, 299
Plouin, P.F., 374
Plunkett, L.M., 50
Pocock, G., 73, 80, 121
Pocotte, S.L., 141
Polak, J.M., 197, 200
Politi, P.M., 128
Polk, D.H., 314
Pollack, H.M., 360
Pollard, H.B., 5, 18, 33, 65, 68, 73-74, 79, 83-85, 93, 100, 105, 110-111, 136, 154, 169, 238, 294

103, 107-108, 110-111, 117-118, 122, 137, 140, 154-155, 165-168, 172, 189, 194-195, 202-203, 234-237, 374

ATPase, 6, 58, 76, 154, 157, 165-168, 175, 199, 235-237

Atrial natriuretic factor, 41, 49-50, 201, 216, 329, 338

Atriopeptin, 338

Atropine, 25-26, 32-33, 35-36, 67, 100, 248, 257

Autograft, 296-299

Autopsy, 10, 182, 299, 327, 366, 371, 373, 375-376

Autoradiography, 44, 46, 49-50, 55-56, 104, 119, 171-172, 244, 280, 284, 326

Avian, 12, 22, 314, 330

Axolotl, 12

Axon, 16, 21, 63, 87, 160, 187, 284, 291

Azide, 158, 168

Baclofen, 43-44, 247

Bacteremia, 358

Bacteria, 66, 173

Barbitone, 339

Barbiturates, 44

Barium, 69, 72, 80-82, 110, 112, 141, 215, 222-223

Baroreceptor, 24, 245, 348, 357

Basement lamina, 15

BAY K 8644, 68, 70-72, 80, 112

Behavior, 9, 169, 185, 203, 290, 292, 303, 305-306, 308, 327, 354

Benzilate, 34-35

Benzodiazepine, 57-58, 202, 267

Benzole, 151

Bereaved, 306-307

Bicuculline, 43, 247

Bioenergetics, 234

Biopsy, 297, 364, 369-370

Biopterin, 144-145, 269

Biotinylated, 262

Birds, 12, 22, 123, 148, 185, 357

Bisoxonol, 65

Bladder, 255, 331

Blood-brain barrier, 2, 247, 255, 291

Bombesin, 211-212, 218, 324-325

Botulinum, 105, 107, 115

Bradycardia, 254, 344

Bradykinin, 54-55, 100, 141, 172, 278

Brainstem, 9, 242, 251, 253, 323

Brattleboro, 217

Bromide, 150, 165

Bromoacetamide, 77

Bromocriptine, 124, 343

Bullfrog, 13

Bungarotoxin, 26-27, 242

Bupivacaine, 121, 310

Cadmium, 72, 112

Caffeine, 58, 69-70, 96, 126

Calcitonin, 13, 63, 219

Calcium-activated, 61, 73, 80, 92, 100, 106

Calcium-binding, 4, 82-83, 105, 181, 189, 265

Calcium-channel, 373

Calcium-dependent, 45, 53, 57, 59, 64, 70, 74-76, 79-87, 92-95, 98, 100, 104-105, 110, 112, 122, 127, 140-141, 222-223, 225, 240, 270

Calcium-independent, 98, 115

Calcium-translocating, 236

Caldesmon, 88

Calelectrin, 82-83

Calf, 138, 193, 244-245, 266, 322, 335

Calmodulin, 83, 85, 88-89, 95-96, 103-104, 107-108, 140, 142, 285

Calpactin, 85-86

Calpain, 140

Calpastatin, 140

cAMP, 28, 60, 95-98, 101, 136, 138, 140-144, 173-174, 224-225, 227, 264-265, 267-268, 272, 275-277, 280, 282, 355, 377

Cancer, 302, 327

Canine, 45, 96, 217, 326

Capacitance, 64, 115

Capsaicin, 62, 250, 326

Captopril, 45, 171-172, 317, 328

Carbachol, 28, 32, 35-36, 37, 39-40, 55, 67, 73, 76-78, 80, 93, 95, 97, 100, 108, 115, 121, 124-125, 130, 132, 140, 215

Carbobenzoxy-Gly-Phe-NH, 73

Carbohydrate, 151, 163, 343

Carboxymethylation, 108

Carboxypeptidase, 191, 208, 231-232

Carcinoid, 188-189, 373

Carcinoma, 181, 184, 188, 214, 362, 373, 377-379

Cardiac, 49, 78, 245, 255, 323, 339-342, 345-346, 356-357

Cardioexcitatory, 220

Cardiotoxic, 128, 342

Cardiovascular, 44, 123, 214, 252, 258, 315-317, 337-341, 349, 374

Carotid, 16, 177, 247-248, 292, 315, 328, 347

Carrier, 156, 166, 199, 237, 241, 246

Castrated, 123, 336

Cat, 8-9, 16, 24, 35, 38, 69-71, 78, 81, 111, 118, 185, 197, 210-212, 214, 217, 244, 251-253, 255, 257, 260, 306, 338, 342, 352, 354, 257, 260, 352, 354

Catalase, 150

Catalepsy, 313
Cataracts, 331-332
Cattle, 312
Caudate, 1, 118, 288, 290, 293-294, 296-297, 299, 301
cDNA, 83-84, 130, 134-136, 140, 143, 147-148, 156, 162, 165, 178-181, 197-198, 207, 221, 232, 265, 271, 378-379
Cerebellum, 198
Cerebral, 159, 255, 260, 292, 297, 325, 327, 352
Cerebrospinal, 159, 161, 177, 256-257, 289, 291, 298
Cervical, 132, 161, 177, 228, 242, 251, 258, 291
Cesium, 81
cGMP, 33, 49-50, 96, 143-144, 272
Chelator, 53, 66-67, 69, 229, 267
Chemodectomas, 368
Chemoreceptor, 24
Chicken, 3, 15, 26, 32, 170, 234, 274
Chimeras, 15
Chloralose, 252-253, 255
Chloride, 33, 41-42, 81, 107, 165-166, 195, 222, 236, 239, 344
Chlorimipramine, 237
Chlorisondamine, 140, 254, 260, 356
Chloroadenosine, 95
Chlorophenethylamine, 152
Cholecystokinin, 211
Cholera, 99, 231, 244, 263, 276
Cholesterol, 190
Choline, 32, 81, 190, 240, 264, 286, 329
Cholinergic, 6, 27, 31-32, 44, 57, 62, 69, 71, 78, 85-86, 99-100, 103-104, 120, 140-141, 176, 186, 213, 215, 223, 240, 246-247, 249, 256-257,

325-326
Choroid, 300
Chorophenylpiperazine, 339
Chromobindin, 4, 82-83
Chromogranin, 7, 10, 19, 51, 56, 108, 176, 178-190, 195, 198, 208, 212, 221, 246
Chromograninomas, 188
Ciliary, 234, 277
Cimetidine, 57
Circadian, 329
Clathrin, 199
Clomipramine, 47
Clonidine, 27-29, 37-40, 123-124, 247, 256, 327, 369
Cobalt, 75
Cocaine, 238
Cocktail, 176
Codons, 149
Colchicine, 86, 127
Cold, 140, 250, 307, 309, 313, 322-324, 358
Collagen, 281, 284-285, 336
Compromised, 299
Computed tomography (CT), 360-365, 370, 377-378
Conductance, 28, 41, 64-66, 69, 112
Congeners, 146
Conscious, 71, 193, 244-245, 256, 258-260, 326, 335, 337-339, 344-345, 349-352
Convertase, 232
Convulsant, 79
Copper, 35, 148, 151, 157-161, 175, 194, 201
Cordycepin, 263
Corticosteroid, 123, 126, 245-246, 304-305, 332
Corticosterone, 159, 227-228, 304, 311, 332, 345, 358
Corticotrophs, 184
Corticotropin (ACTH), 122, 228, 312
Corticotropin-releasing factor (CRF), 6, 60,

159, 193, 247, 252, 259, 318, 333, 335, 352
Cortisol, 305, 316-317, 334, 336, 341
Cosecretion, 37, 117, 129, 201, 216, 350
Countertransport, 81
Cow, 14, 41, 50, 55, 105, 163, 180, 212
Co-localization, 8, 13, 212, 229
Co-release, 213, 227
Creatinine, 329
Cresol, 152
Culture, 49, 59, 108, 111, 119, 153, 169, 223, 227-229, 240-241, 261, 270-271, 274-275, 277, 281-284, 308, 336
Cushing's syndrome, 368, 375
Cyanide, 72, 158
Cyanogen, 149
Cycloheximide, 161
Cyclohydrolase, 144
Cyclosporine-treated, 293
Cytochalasin, 21, 127, 240
Cytochrome b_{561}, 6, 155-157, 176, 199, 235
Cytokeratin, 91
Cytoskeleton, 4-5, 21, 24, 86-88, 91, 106, 127, 272
Cytotoxin, 88, 119, 128

DAGO, 259
Dahl, 51, 346
DAMME, 246
Dantrolene, 125-126, 323
Deamination, 169, 304
Decarboxylase, 119, 122, 129, 132-133, 147-148, 165, 170-171, 262, 315, 345-346, 367
Defensive behavior, 9, 252, 306
Deglycosylated, 151
Dehydroascorbate, 146, 155-156

Denervation, 42, 53, 56, 123, 125, 131, 171, 228-229, 244-245, 248, 250, 290, 304, 308-309, 317, 323, 328, 338, 340, 348, 350-351, 354

Deoxycorticosterone, 347, 351

Deoxyglucose, 122, 171, 203

Dephosphorylation, 91, 102

Depolarization, 27-28, 32, 42-43, 52, 61, 67-68, 70-75, 83, 90, 93-94, 120, 142-143, 213, 221, 223-224, 227, 229, 262, 308, 323

Desensitization, 3, 27-32, 51-54, 57, 111, 375

Desipramine, 117, 239, 304, 340

Desmethyldiazepam, 202

Detergent, 56, 63, 66, 85, 115, 136, 167, 194-195, 239, 272

Dexamethasone, 16, 132, 138, 164, 186, 213, 216, 221, 227-228, 231, 241, 262, 265, 279, 282, 305, 311, 332, 335, 351, 355, 379

Diabetes, 320, 331-332, 372, 376

Diacylglycerol, 58, 74, 92, 95, 100, 102, 142, 173, 192, 282

Diadenosine, 194

Diazepam, 202

Dibutyryl, 142, 265, 275, 280

Dichloromethane, 125

Dideoxyforskolin, 98

Diet, 62, 130, 337, 343, 347-348

Diethylstilbestrol, 168

Digitalis-like, 193

Digitonin, 18, 48, 67, 73-74, 80, 89-92, 94-95, 98, 101, 104-105, 107, 109-110, 112-113, 136, 140, 154,

202, 239

Diglyceride, 64, 92

Digoxin, 193

Dihydropteridine, 145

Dihydropyridine, 67, 69-70, 74, 78, 120, 122, 223

Dihydrotetrabenazine, 234-235, 237

Dihydroxyphenylacetic, 177-178

Dihydroxyphenylglycol, 345

Dihydroxytryptamine, 257

Diltiazem, 69, 373

Dinitrophenol, 107-108

Dioctanoylglycerol, 92

Dipeptides, 174

Diprenorphine, 45-49, 59

Dipyridamole, 239

Disaccharide, 191

Dithiothreitol, 49

Diuresis, 328

Divalent, 71-72, 232

DNA, 134-135, 161-162, 178, 180-181, 211, 370

DNase, 106

DOCA, 347

Dog, 24, 38-39, 41-42, 45, 59-60, 96, 123-124, 126, 185, 193, 217, 247, 305-306, 311, 326, 328, 338, 340-341, 343, 348-350, 352-354, 361

Domperidone, 124, 202, 319, 338

Dopamine, 2, 6, 17, 40-41, 62, 64, 90, 107, 111, 117, 119, 123-124, 130, 143, 146-148, 152, 154, 156, 159, 161, 164, 169, 171, 177-178, 194, 202-203, 239, 244-245, 253, 260, 284-286, 289-293, 295, 297, 301, 304-305, 311, 319, 324, 327-331, 333-334, 336-339, 341, 343, 345, 348, 366-367, 373, 377

Dopamine β-hydroxy-lase (DβH), 5-6, 17, 77, 81, 108, 114, 121, 129, 130-132, 134, 138, 148-161, 164, 176, 187, 191, 213, 225, 235, 246-247, 289, 305-306, 329-332, 335, 348, 367, 378

Dopaminergic, 4, 40, 124, 148, 202, 253, 257, 288-291, 293, 296, 300, 336-338

Dopamine-releasing, 6, 193-194

Dopamine-secreting, 293, 366

DOPEG, 304

Dot-blot, 140

Doxorubicin, 128

Droperidol, 338

Dynorphin, 29, 45, 48, 176, 209, 219

Dyskinesia, 297, 378

Dystrophy, 128, 378

Elastic, 113

Electric, 3, 23, 32, 38, 42, 52, 79, 84, 123, 193, 200, 217, 234, 246, 251, 254-255, 257, 266, 306, 311, 325-326, 343, 378

Electrochemical, 25, 111, 159, 236

Electropermeabili-zation, 63, 93

Electrophoresis, 102-104, 165, 185

ELISA, 263

Embryo, 15, 204, 234, 242-243, 264, 274, 276, 376-377, 379

Enalapril, 45

Enantiomer, 70

Endocytosis, 114

Endoglycosidases, 204

Endonexin, 83

Endopeptidase, 174, 232

Endoplasmic reticulum, 16, 18, 23, 67, 108, 170, 224

Endoproteinases, 108

Endorphin, 48, 256,

259, 309, 312-313
Endothelial, 15, 100, 201, 279, 292
Endotoxemia, 21, 337, 358
Enkephalin, 2, 7-9, 13, 17, 34, 37, 42, 45-47, 53, 121, 129-130, 176, 181-182, 186, 198, 207-214, 217-219, 221-233, 244-246, 259-260, 264, 298, 307-309, 313, 335, 337, 340-341, 350-352
Enkephalinergic, 221
Enolase, 172, 198, 286, 370
Enterochromaffin, 184, 187
Enterotoxin, 127
Epilepsy, 2, 301
Epitopes, 89, 150, 186
EPR, 145, 158
Equine, 217
Escape behavior, 252
Ester, 58-59, 74, 88, 91-95, 99, 101, 104-105, 108, 130, 141-142, 201, 216, 223, 236, 267
Estradiol, 126, 333
Ethanol, 73, 119-120, 175, 356
Ethanol-induced, 73, 175, 356
Ether, 249, 335
Ethrane, 121
Ethylcarboxamide, 61
Ethylene, 304, 323
Ethylmaleimide, 165-166, 235
Etomidate, 374
Etorphine, 29, 45-46, 59, 210
Eubacterial, 167
Exercise, 29, 318-322, 345
Exocrine, 195
Exocytosis, 4-5, 14, 18, 22, 24, 29, 47, 53, 63-64, 66-67, 73-75, 81-82, 85-89, 92-95, 98-99, 103-115, 172, 175-176, 183, 192, 239
Exoglycosidases, 163
Exon, 162, 220

Exotoxin, 94
Explants, 146, 227-229
Extra-adrenal, 57, 247, 337, 361-365, 368, 371, 374, 376
Eye, 288-289, 293
Eyes, 162, 288

Fat, 324, 330
Feeding, 259, 324
Feline, 217, 246
Fenestrated, 15
Ferric, 146
Ferricyanide, 156
Ferrocyanide, 150, 155
Ferrous, 141
Fetal, 1, 17, 19, 129, 174, 204, 215, 230, 264, 288-289, 291, 297, 300, 314-317
Fibrillation, 341
Fibrinogen, 343
Fibroblast, 9, 136, 263, 275, 278-280, 292
Fibronectin, 281-282, 284-285
Filament, 5, 86-89, 91, 105-106, 200
Fish, 3, 185
FITC-dextran, 84
Flow cytometry, 370
Flunarizine, 69
Fluorescence, 12-13, 19-20, 33-34, 65, 68, 75, 83-85, 87, 98-99, 112, 114, 116, 122, 133, 136, 158, 204, 266, 291, 313, 370
Fodrin, 106
Foot-shock, 307, 309-311, 313, 345-346
Formalin, 316
Formate, 166
Forskolin, 60, 80, 89, 95-98, 142, 182, 201, 226, 231, 267, 275-276
Fos, 264-265
Freeze-fracture, 18
Frog, 12-13, 209
Fructose, 321
Fucose, 270
Fumarate, 149
Fura 2, 65
Fusidic, 167
F-actin, 89-90, 273

Galactose, 191, 276
Galactosidase, 274
Galanin, 197-198
Gamma, 227, 229, 232
γ-aminobutyric acid (GABA), 4, 41-44, 202-203, 243, 246-247, 339
Ganglion, 8, 17, 78, 122, 125, 161, 163, 177, 184, 188, 200, 202-203, 212, 215, 219, 228, 234, 242-243, 245, 262, 274, 288-289, 291, 299, 377
Ganglioneuroblastomas, 214
Ganglioneuroma, 188, 368-369, 376
Ganglionic, 140, 254, 260, 292, 326, 332, 349, 355-356
Ganglion-blocking, 57
Ganglioside, 169, 190-192, 270, 274, 276
Gastrin, 187
Gastrin-releasing, 13, 218, 245, 326
GDP, 200
Gelsolin, 105-106
Genome, 17, 134-135, 162-163, 181, 211
Glial, 119, 293
Glioma-conditioned, 240, 286
Globoside, 191
Glucagon, 175, 185, 249, 260, 319-320, 347, 358, 378
Glucocorticoid, 10, 15, 132, 138, 159, 162-164, 221, 227-228, 265, 271, 274, 279, 308, 323, 332, 334-336, 379
Glucopenia, 250
Glucoprivation, 307
Glucoregulation, 319-320
Glucosamine, 102, 192
Glucosidase, 358
Glutamate, 55-56, 252-253
Glutamic, 182
Glutamine, 230, 231
Glutathione, 155, 304

Glycogen, 204, 277, 320-322
Glycogenolysis, 320-322, 324, 326
Glycol, 304, 323
Glycolipid, 190-191, 270, 276
Glycolytic, 172
Glycopeptides, 190
Glycoprotein, 58, 113-114, 151, 190-191, 194-196, 204, 231, 270, 274, 276, 282, 285, 332
Glycosaminoglycans, 277, 278, 284
Glycoside, 151, 193, 274
Glycosphingolipid, 191, 270
Glycosylation, 163, 196, 224, 274
Glycosylphosphatidyl-inositol, 102, 149, 192
Glypiation, 6, 148-149
Goat, 375
Goldblatt, 348
Goldfish, 13
Golgi, 16, 23, 56, 115, 149
Gonadotrophs, 184
Goose, 22
Growth, 9-10, 16, 21, 35, 54, 81-82, 90-91, 94, 98, 119, 122, 125, 175, 186, 190-191, 196, 203, 205, 213, 216, 230-231, 261-282, 285-286, 289-290, 292-293, 295, 316, 333, 372
GTP, 59, 61, 98, 107, 144, 199-200, 278
GTPase, 58
Guanethidine, 256, 312, 316, 322, 351
Guanidine, 49, 238
Guanine, 65, 98, 200
Guanosine, 46-48, 58, 96, 98-99, 143
G-actin, 105
G-protein, 58, 105

Haloperidol, 40, 124-125, 313, 336
Halothane, 120-121, 244, 259-260, 344,

347, 350
Hamster, 13-14, 50, 55, 126
Heart, 118, 121, 160, 199, 245, 247, 253, 255, 258, 314-316, 319-320, 322, 338-347, 349, 351, 355-357, 363, 372
Hematogenous, 376
Hematopoiesis, 364
Hemorrhage, 8, 21, 210-211, 349-354
Heparin, 278-279
Hepatic, 9, 175, 250, 260, 307, 321, 324, 326, 358
Hexamethonium, 25-26, 28, 32-33, 39, 44, 57, 67, 69, 110, 248, 337, 355
Histamine, 17, 26, 45-47, 55-57, 100, 224, 250, 258-259, 355, 377
Histaminergic, 57, 259
Histidine, 146, 198
Homeostasis, 35-36, 62, 74, 125, 307, 315, 337
Homovanillic acid, 366-367
Horseradish peroxidase, 114, 244
Huntington's disease, 2, 301
Hybridization, 5, 7, 130-131, 135-136, 140, 163, 186, 197-198, 206-207, 212, 216, 223-225, 265, 378
Hydralazine, 123, 125, 130-131, 245
Hydrazines, 150, 158
Hydrazone, 158
Hydrocortisone, 16, 126
6-Hydroxydopamine, 57, 123, 256-257, 260, 289, 291-292, 315, 322, 325, 352-353
Hydroxylase, 2, 5-6, 13, 17, 39-40, 77, 81, 103, 108, 110-111, 114, 119, 121, 126, 129, 132-134, 137, 139, 141, 144-145, 148-149, 157-158, 160-

161, 176, 187, 213, 216, 225, 227, 235, 247, 265-266, 277, 289-291, 294, 305, 308, 318, 322, 329, 331, 335, 348, 367, 373, 378
Hypercapnia, 319
Hyperglycemia, 249, 260, 320, 356
Hyperinsulinemia, 249
Hyperplasia, 125, 128, 188, 334, 362, 365, 367-368, 372-373, 375-376
Hyperplastic, 14, 375
Hyperprolactinemia, 334
Hyperreflexia, 255
Hypertension, 9, 40, 47-48, 50, 56, 102, 122, 130, 160, 172, 214, 232, 254, 306, 327-328, 338, 343-349, 362, 364, 366-367, 369, 373-375, 378
Hyperthermia, 125-126, 323
Hyperthyroidism, 164, 329
Hypnotic, 323, 356
Hypoglycemia, 22, 160, 164, 222, 246, 248-250, 257, 311, 317, 320-321, 327, 329-330, 377
Hypophysectomy, 126, 130, 132, 210, 221, 228, 246, 308, 324
Hypotension, 123-124, 211, 250, 336-338, 343, 345, 350-351, 353-354, 374, 377
Hypothalamus, 23, 193, 251-252, 254-255, 259, 306, 310, 312, 324-325, 333, 349, 354-355
Hypothermia, 323, 356
Hypothyroidism, 160, 164, 329-330
Hypotonic, 116
Hypovolemia, 250, 349-351, 353
Hypoxanthine-guanine, 285-286

339-340, 343
Prostacyclin, 100
Prostaglandin, 58-59, 101-102, 325
Protease, 82, 115, 140, 181, 183, 193-194, 231, 286
Proteinase, 74, 108
Proteoglycans, 191
Proteolipid, 167
Proteolysis, 106, 143, 145, 179
Proteolytic, 90, 94, 140, 180, 183, 208, 261
Proteophospholipids, 190
Proton, 6, 67, 90, 155, 157, 165-168, 236, 360
Protonation, 152
Protonophores, 131
Proton-pumping, 157, 165-166
Proton-translocating, 165-167
Pseudocholinesterase, 170
Psychotic, 305-306
Pterin, 138, 141
Purine, 269
Putamen, 1, 288, 299
Putrescine, 140, 202-203
Pyrazole, 152
Pyridine, 269
Pyridinium, 146
Pyridoxal, 119
Pyroglutamic, 230
Pyrroloquinoline, 158

Quail, 15, 17, 135
Quin 2, 34, 65, 67, 99
Quinine, 80
Quinone, 158
Quinonoid, 145

Rabbit, 72, 133, 185, 193, 209, 225, 258, 282, 318, 348, 350-351, 356
Rabies, 376-377
Radioautography, 56
Radioenzymatic, 131, 177, 285
Radioimmunoassay, 188, 193, 197, 200, 208, 212-213, 215-

216, 219-220, 230
Ras, 107, 200, 264, 286
Rauwolscine, 327
Real-time, 4, 25, 29, 42, 111
Recklinghausen's syndrome, 373
Recombinant, 134, 197, 276
Redox, 152, 155-157
Reductase, 145, 155
Reinforcement, 305, 327
Reinnervation, 287, 290, 292
Renin, 204, 303, 317, 319, 352
Renin-angiotensin, 254, 328, 352
Renografin, 283
Reserpine, 12, 132-133, 136-137, 155, 159-160, 164, 210, 212-213, 218-219, 225-227, 234-237, 245-246, 254, 325-326, 340, 349, 363, 367-368
Reticulocyte, 225
Retina, 118, 128, 199
Retrovirus, 10, 286, 375
Rheumatoid, 332
Ribosylation, 105, 107, 271
Rilmenidine, 338
RNA, 21, 132, 134-136, 140, 197, 207, 228, 264, 284, 323

Salamanders, 12
Salt, 20, 168, 346-348
Saponin-permeabilized, 103, 110
Saralasin, 317
Sarcoma, 282, 372
Saxitoxin, 61, 77, 266
SCG, 271
Schizophrenia, 306, 378
Schwab-England, 298
Schwannoma, 369
Sciatic, 78, 255, 291
Scintigraphy, 119, 238, 362-364, 370, 377
Scorpion, 77
SDS, 91, 133
Second-messenger, 28,

36, 66, 223
Secretogranin, 179, 182, 184
Selenium, 20
Semidehydroascorbate, 155-156
Sendai, 116
Septicemia, 376
Serine, 84, 115, 142-143, 181, 231, 358
Serotonergic, 164, 243, 254, 257, 339
Serotonin, 13, 159, 164, 167, 237, 242, 355
Serum-free, 223-224, 271, 275, 281, 283
Sheep, 17, 215, 315-316, 353
Shock, 21, 234, 309, 349, 351, 353, 355
SHR, 343-346
Shrew, 49-50, 185, 216
Sino-aortic, 56, 338, 348
Sinusoids, 15, 126
Sipple's syndrome, 361
Sleep-inducing, 220
Snipe, 22
Sodium, 20, 30-31, 61, 64, 70, 72, 75-81, 90, 125, 128, 131, 143, 173, 175, 195, 223-224, 229, 238-240, 258, 266, 286, 328, 336-338, 347-348
Somatostatin, 50, 55, 185, 214-217, 242, 333
Somatotrophs, 184, 372
Somnolence, 298
Southern blot, 83, 135
Soybean, 286
Sparrow, 353
Spectrofluorometry, 157, 203, 313
Spermidine, 140
Spermine, 111, 138, 140
Sphingomyelinase, 165, 174
Sphingosine, 92, 277
Spider, 64
Spinal, 2, 198, 208, 219, 243-244, 247, 255, 258, 260, 293-295, 318, 342-343, 349, 355
Spinner culture, 270

Tuftsin, 52
Tunneling, 237-238
Turtle, 358
Tyramine, 150-151, 154, 340
Tyrosine, 36, 82, 111, 129-130, 132-133, 137, 143, 145, 154, 233, 284, 332, 358
Tyrosine hydroxylase (TH), 2, 5, 13, 39-40, 103, 110-111, 119, 121, 126, 129-148, 154, 160, 162, 164, 169, 216, 225, 227, 265-266, 277,290-291, 293-295, 308, 318, 322-323, 329, 331, 335, 348, 367, 373
T-helper, 198, 230

Ultimobranchial, 185
Ultrasonography, 365, 372
Umbilical, 314-315
Unanesthetized, 9, 254, 256, 258
Uncouplers, 72, 94, 167
Unitary, 68, 80, 245
Unmyelinated, 247, 318
Uranyl, 192
Urethan, 120-123, 245, 252-253, 257, 325, 330, 339
Uridine, 284
Urine, 125, 161, 178, 256, 300, 303-304, 306-307, 325, 328-329, 332, 334, 343, 347-348, 355-356, 362-363, 366-367, 369, 372, 378
Urokinase, 201

Vagus, 247, 251, 257-258, 325-326
Vanadate, 166, 168, 175
Vanillylmandelic acid, 178, 372-373
Vasoactive intestinal peptide (VIP), 4, 53-54, 97, 130, 188-189, 198, 211, 214-216, 222, 226-227, 245, 370
Vasopressin, 14, 130, 217-218, 232, 254, 258-259, 316, 338-339, 349-350
Vasopressor, 251, 257
Venography, 372
Venom, 77
Ventromedial nucleus, 252
Verapamil, 69, 127
Veratridine, 32, 57, 59, 67-68, 77-78, 80, 97, 223, 262
Videomicroscopy, 353
Vimentin, 91, 200
Vinblastine, 127
Virions, 116
Vitamin, 123, 156, 336
Voltage, 30, 32-33, 35-36, 42-43, 61, 64-66, 69-71, 73-80, 92, 94-95, 110, 120-121, 224, 262, 266, 375
Voltage-clamp, 43, 81

Western blot, 134, 147, 168
Whole-animal, 24
Wistar, 13, 14, 353
Wistar-Kyoto, 50, 335, 343-346

Xylamine, 239
Xylitol, 125
X-ray, 159, 194

Yohimbine, 37, 247, 327, 339, 345, 347

Zinc, 145
Zomepirac, 128